MICROCOMPUTER STRUCTURES

MICROCOMPUTER STRUCTURES

Zvonko G. Vranesic
Safwat G. Zaky

University of Toronto
Ontario, Canada

Saunders College Publishing
A Division of Holt, Rinehart and Winston, Inc.

New York • Chicago • San Francisco • Philadelphia
Montreal • Toronto • London • Sydney • Tokyo

TO ANNE AND SHIRLEY

Publisher   *Ted Buchholz*
Acquisitions Editor   *Deborah Moore*
Senior Project Manager   *Marc Sherman*
Production Manager   *Roger Kasunic*
Design Supervisor   *Judy Allan*

Library of Congress Cataloging-in-Publication Data

Vranesic, Zvonko G.
  Microcomputer Structures / Zvonko G. Vranesic,
Safwat G. Zaky.

    Includes bibliographies and index.
    ISBN 0-03-009739-8
    1. Microcomputers. 2. Microprocessors. I. Zaky,
    Safwat G. II. Title.
    QA76.5.V73 1989
    004.165—dc19   88-10289

Printed in the United States of America

9 0 1 2   016   9 8 7 6 5 4 3 2 1

PREFACE

Microcomputer Structures deals with microprocessors and microprocessor systems, emphasizing understanding at the machine level. Assembly language programming is explained and programming examples are given to illustrate its use and its interaction with the hardware. Various system components and the interfacing circuitry needed for their interconnection are discussed together with current trends in system design and packaging. Included in the presentation are several commercial standards for buses and for connection of peripheral devices. The level of detail should permit the reader to deal with most aspects of design of microcomputer systems and their applications at the hardware level.

Use in Teaching

A wide range of material is covered. We have endeavored to write the first half of the text in a style that is easy to read and requires little background knowledge. Only familiarity with a high-level programming language and the basic concepts of logic circuits is assumed. This part of the book is suitable for a one-semester introductory course, dealing with the functional organization of a microprocessor system, assembly language programming, and input/output techniques. A thorough coverage of these topics is essential for understanding microprocessor systems.

v

The second half of the book covers more specialized hardware topics, including the design of the main memory, backplane buses, and board-level subassemblies. Although most of the discussion is presented at the level of logic gates and interchip signals, an understanding of electronic circuit concepts is needed for a full appreciation of hardware design problems. This material is suitable for an advanced one-semester course, where students have had prior exposure to computer organization. The book may also be used in a two-semester course that covers all of the material.

The choice of topics and the style of presentation are based on our experience in teaching microcomputers to students having a variety of backgrounds. The manuscript has been class-tested in several engineering and computer science courses. Parts of it have also been used in courses taught to engineering students other than electrical, and in professional development courses.

Style of Presentation

The basic concepts needed for understanding microprocessor systems are presented in a machine-independent fashion. Then, for each topic, the general discussion is followed by a detailed presentation of selected commercial products that embody these concepts. The commercial products serve to point out the constraints encountered in practical implementations. Eight-bit machines are illustrated using the Motorola 6809 microprocessor, 16-bit machines by Mototola 68000, and 32-bit machines by Motorola 68020 and 68030. Besides representing microprocessors of different word lengths, the 6809 and 68000 provide good examples of synchronous and asynchronous bus structures, respectively. We believe that it is useful to cover both of these microprocessors. However, because of time limitations in a given course, it may be necessary to restrict the discussion to one microprocessor. We have organized the material such that the treatment of each microprocessor is self-contained, so the material pertaining to either microprocessor may be skipped without loss of continuity.

A major consideration in designing a microprocessor system involves the choice of appropriate interconnection schemes and the design of interface circuitry. Two examples of commercial backplane standards, the VMEbus and the Multibus, are presented in detail. Also, the RS-232-C serial interconnection standard, the IEEE 488 instrumentation bus, and the SCSI bus are introduced as examples of peripheral interconnection schemes. Again, selections may be made from these examples without loss of continuity.

We, like many other instructors, have tried different approaches in teaching our students. There are three main choices of style that one can adopt. The first is to present the ideas in terms of a hypothetical machine, for which a simulator may exist. Unfortunately, while providing a convenient vehicle for the presentation of concepts, such hypothetical models tend to be oversimplified and bear little resemblance to real machines. The second approach is to base the discussion totally on a commercially available microprocessor. Ideally, this is the microprocessor used in the laboratory trainers associated with the course. This approach permits a highly relevant treatment of one popular machine. The drawback is that most modern microprocessors are complex devices, characterized by details that are not necessarily representative of other machines. Also, the need to worry about details may obscure the clarity of the general ideas that the student should learn. At the end of the course, the student is likely to feel comfortable with the example microprocessor, but have little inclination to consider other alternatives that may be more suitable in a particular application.

The third approach, which is the one we have adopted in this text, is a compromise between the two. By initially presenting the basic concepts in a generic style, unnecessary details that result from the peculiarities of a given machine are avoided. Because only the basic concepts are discussed in this manner, the discussion is short and should be easy to understand. Having understood the general ideas, the student will have no difficulty learning the intricacies of specific implementations and any other features that a particular microprocessor described later might have.

Organization of Contents

The book begins with an overview of a microcomputer system, its components, and how they are interconnected. Chapter 1 also explains how a microprocessor executes instructions and performs input/output tasks. In Chapter 2, the principles of machine-level programming are examined. The discussion deals with the general concepts of addressing modes, machine instructions, subroutines, assembly languages, and so forth, without relying on the details of a specific machine. Chapters 3 and 4 show how these concepts are implemented in the 6809 and 68000 microprocessors, respectively. The instruction set of each microprocessor is explained, and simple short programs are given to illustrate its use. Then, more extensive examples show how complex tasks may be handled.

A thorough understanding of input/output and interfacing techniques, from both the software and hardware points of view, is essential to users and designers of microprocessor systems. An extensive explanation of these techniques is provided, spanning several chapters. Basic parallel and serial I/O devices are described in Chapter 5 from the programmer's point of view. Hardware details of these devices and the interface circuits needed to connect them to a microprocessor bus are presented in Chapter 6. Chapter 7 deals with interrupt-controlled I/O and Direct Memory Access (DMA). Priority arbitration schemes and the interaction between hardware and software interrupts are discussed. The material on I/O techniques includes commercial examples of serial and parallel interface chips, as well as DMA controller chips.

The organization of the main memory is discussed in Chapter 8, including the concepts of cache memories and virtual memories. The design of static and dynamic memories is examined, with typical examples of components and systems. Chapter 9 gives a brief presentation of typical secondary storage and peripheral devices used in microprocessor systems, with the aim of illustrating their salient characteristics as seen by the user.

Some practical aspects of the construction of microprocessor systems are discussed in Chapter 10, and an example of a single-board microcomputer is described. Electrical considerations, such as crosstalk and transmission line effects, and their influence on the design and packaging of microprocessor hardware, are examined. Chapter 11 examines the related issue of interconnection standards. The VMEbus and Multibus are presented as examples of system buses, and peripheral interconnection schemes are illustrated by the RS-232-C interface, the IEEE 488 bus, and the SCSI bus.

Chapter 12 gives a brief discussion of system software. It is intended to provide an appreciation of the role of system software and the way in which it facilitates the development of application programs.

Associated Laboratory

We believe strongly that students should perform laboratory exercises as soon as possible. For instance, a generic presentation of addressing modes and simple programming examples can be followed immediately by laboratory exercises illustrating their realizations, using the microprocessor available in the laboratory. *Microcomputer Structures* is organized to facilitate this approach. The generic examples in Chapter 2 are redone using the 6809 and 68000 in Chapters 3 and 4, respectively. The treatment of general I/O techniques is followed by the specific schemes found in Motorola products, etc. Since much of the laboratory work requires an understanding of I/O schemes, this subject is introduced early. A number of the problems included at the end of each chapter can serve as laboratory exercises.

Acknowledgments

The authors wish to express their gratitude to all the people who have helped in preparing and improving the quality of this book. Our students, who used the text in its preliminary form, have made many useful comments. Several of our colleagues at the University of Toronto, including Professors Carl Hamacher, Michael Stumm, John Van de Vegte, Keith Balmain and Ian Dalton, have provided us with constructive feedback. The external reviewers also contributed numerous helpful suggestions: K. Hsu, Rochester Institute of Technology; K. Watson, Texas A&M University; T. Miller, III, North Carolina State University; R. Jordan, University of New Mexico; D. Gustafson, Texas Technological University; F. Looft, Worcester Polytechnic Institute; D. Schroder, University of Texas–El Paso; W. Getty, University of Michigan–Ann Arbor; T. Moore, Drexel University; T. Smay, Iowa State University; L. Marino, San Diego State University; B. Brey, DeVry Institute of Technology; M. Schiffman, Auburn University; C. Weber, Lake Superior State College; M. Kalisky, California Polytechnic State University; and M. Lightner, University of Colorado–Boulder. Their role in the development of the manuscript has been invaluable. Our editor, Deborah Moore, her assistant, Karen Mosman, senior project manager, Marc Sherman, and the staff of Holt, Rinehart and Winston, Inc. handled the editing process efficiently and enthusiastically. Finally, Catherine Cheung and Rosanna Reid helped us with the word processing tasks.

Z. G. Vranesic
S. G. Zaky

CONTENTS

MICROCOMPUTER STRUCTURES

Chapter 1

Computer Concepts

The proliferation of microprocessors has profoundly influenced our way of life. Automation, through computer control, has become necessary for economic survival. It would now be difficult to imagine a world without automated assembly lines, on-line banking, word processing systems, computer-aided design workstations, computerized point-of-sale terminals, music synthesizers, electronic games, and a host of other machinery made possible by exploiting the power of microprocessors. Being surrounded by this technology, it is important to understand the principles on which it is based. This book deals with microprocessors in general, emphasizing the hardware aspects that must be learned in order to gain a proper appreciation of diverse microprocessor systems.

Our intention is to expose the reader to systems considerations as quickly as possible. This chapter will give an overview of the concepts on which microprocessor systems are based. Giving a broad picture first will provide the framework for later discussion of the details. The present chapter will also introduce much of the jargon used in the computer literature.

Microprocessors are general-purpose computing devices. They have essentially all of the fundamental characteristics of general computers regardless of their size.

While it may be tempting to think of microprocessors as being somehow unique in the world of computers, it is more prudent to adopt the viewpoint that all computers are based on the same principles. Computers differ in size, performance, cost, ease of use, and even structure. They exhibit a seemingly large variety of concepts, operating characteristics, and user-observable features. Yet, when subjected to close scrutiny at the most basic functional level, it becomes apparent that computers are an embodiment of just a few key ideas that can be readily identified in the majority of commercially available machines. Microprocessors, while subject to some constraints due to their small size, depict these basic concepts just as larger computers do. For this reason we will concentrate on the principles and structure of computing devices in general, using microprocessors as nothing more than one possible approach to the implementation of the concepts discussed.

1.1 BASIC FUNCTIONAL UNITS IN A COMPUTER

Computers are machines that are capable of accepting data and instructions, performing some operations on the data, and making the results available to the external world. Computing machines can be implemented in a variety of technologies. Our discussion will focus on the implementations that use microelectronic circuits, where the entire computer is realized with just a few chips.

Figure 1.1 shows the most elementary functional diagram of a computer. It includes four main units. The input and output units provide the interface with the outside environment. Information, consisting of data and instructions, enters the computer through the *input* unit. There are many possible sources for such information. For example, it may come from a human operator typing on a keyboard, or it may be provided by another machine connected by means of a communications channel, perhaps an ordinary telephone line. The function of the input unit is to provide the mechanism necessary to connect the computer to external input devices, and to provide the ability to collect the input information and put it in a form suitable for use by the main processing part of the computer.

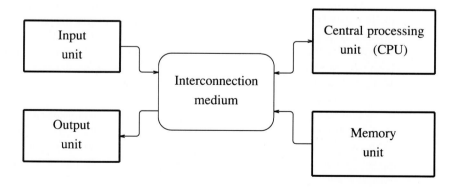

Figure 1.1   Main functional elements in a computer.

Information is processed in the computer by means of electronic circuits that perform relatively simple operations. These operations include the ability to manipulate numbers in some predetermined fashion, typically involving simple tasks such as addition and subtraction. In order to ensure that such operations are carried out in a meaningful manner, it is necessary to be able to control the sequence of events that take place. This is accomplished by means of instructions that govern the behavior of the processing electronic circuits. It is a common practice to concentrate most of the processing circuits in one physically identifiable functional unit, usually called the *central processing unit* (CPU).

Practical computational processes tend to be rather involved (if they weren't we would not need computers!). The electronic circuits in the CPU can handle only simple tasks and can perform operations on only a small number of operands at a time (usually two). This means that a complex computational process must be broken down into a series of steps, each step being simple enough to be executed by the CPU circuits. A list of the required steps, including the data to be operated on, must be available within the computer at the time of computation. Thus, it is necessary to provide a storage medium within the machine. This is the *memory* unit in Figure 1.1. The memory unit may store the data entered through the input unit, the instructions that specify what is to be done with the data, as well as the final processed results.

The results of a computation would be of little value if they could not be made available in some useful form to the user of the machine. It is the function of the *output* unit to meet this requirement. A number of mechanisms may be employed. For example, the results can be printed to provide a hard copy record; they can be displayed on a video terminal for immediate use by the operator; or they can be sent to another machine by transmitting them over a communications channel.

The computer's four basic units—CPU, memory, input, and output—interact closely in the process of computation. Information is frequently passed from one unit to another. This means that a suitable interconnection medium, a set of wires that provides an appropriate interconnection pattern, must be provided. There are several practical ways of organizing the interconnection medium, as will be discussed in Section 1.2.

From the user's point of view it is sometimes difficult to see the physical distinction between the input and the output devices, because they are often packaged within a single physical unit. Consider the case of a video terminal. In its simplest form it consists of a keyboard that serves as an input device and a cathode ray tube (CRT) for display of output information. The two parts are usually housed in the same physical package and, thus, the terminal is in fact both an input and an output device. Indeed, it is a common practice to associate the term "I/O device" with any single unit that provides either an input or an output function, or both.

The previous example of the video terminal suggests that there is a considerable distinction between I/O devices and the processing and storage units in a computer. In fact, the video terminal is best thought of as being a part of a computer system, but not a part of the basic computer, as depicted in Figure 1.2. The computer consists of the CPU, the memory, and an I/O unit that enables it to be connected to I/O devices. Thus, the I/O circuits that provide proper connection between the

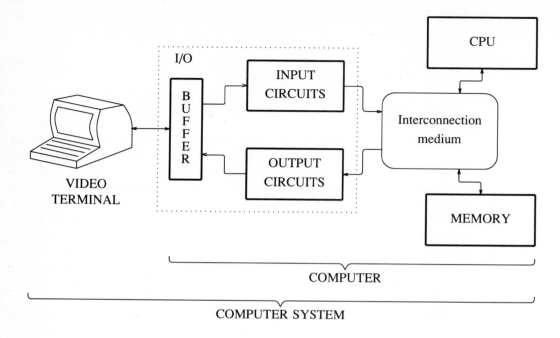

Figure 1.2 A typical I/O connection.

processing and storage units and the I/O devices are an integral part of the computer. These circuits are commonly referred to as the *I/O interface*. However, the actual I/O devices should be considered as functional parts of a computer system.

So far we have not looked at the speed characteristics of computing machines. It was suggested above that processing electronic circuits can perform only simple operations. The power of computers stems from their ability to perform these operations at very high speed. Typically, CPU circuits work at electronic speeds, enabling them to perform several million operations per second. Because a complex computational task can be divided into many simple operations, large problems can be solved in a short time.

It is apparent that fast operation hinges upon the CPU and memory circuits working at electronic speeds. The memory must be fast enough to allow continuous access to the program instructions and data operands at the rate commensurate with the speed of the CPU. But, what about the I/O functions? The situation is much different here. For example, a printer may print upward of 1000 lines per minute. While this is fast in terms of the mechanical constraints, it is not even close to electronic speeds. Other I/O devices can be much slower. Speed performance of a video terminal as an input device is determined by the quickness and typing dexterity of its human operator, which is very slow in comparative terms. An obvious question then arises: how do fast processing circuits interact with slow I/O devices?

The example shown in Figure 1.2 indicates a workable scheme. A *buffer* is inserted between the I/O device and the computer. From the computer side the buffer appears to be just another electronic circuit that can be dealt with at high speed. Input and output circuits are used to gate the information to and from the buffer. The buffer is used as a temporary storage location, where information can be held for relatively long periods as required by the operating speed of the attached I/O device. It is important to note that the buffer and the input and output circuits form a part of the I/O unit; hence they are an integral part of the computer.

In the discussion above we used the word "computer" to describe a collection of functional elements that form a computing machine. The popular jargon is to refer to a small computer as a *minicomputer*. A very small computer should then be called a *microcomputer*. Of course, the basic structure of computers, minicomputers, and microcomputers is essentially the same and is consistent with the diagrams in Figures 1.1 and 1.2. The word microcomputer is often replaced wrongly with the term microprocessor, particularly in the industrial environment and popular literature. A correct interpretation should associate the word *microprocessor* only with the CPU block in Figure 1.1. It is, in fact, a processing unit made up of integrated circuits that is typically fabricated on a single silicon chip and packaged in a single tiny physical package. If one wants to design a full microcomputer, it will normally be necessary to add memory and I/O units to the microprocessor. A complete microcomputer is likely to consist of several integrated circuit packages, only one of which is a microprocessor. However, present technology allows us to place so many integrated circuits on one chip that single chip devices are fabricated which contain not only the microprocessor functions but also a limited amount of memory and some I/O interfaces. A device of this kind is indeed a microcomputer, albeit of somewhat limited capability.

In this text we will use the term "microprocessor" to denote a single-chip CPU functional unit, not to be confused with single-chip microcomputers.

1.2 BUS STRUCTURE

The previous section dealt with the basic functional units in a computer, suggesting that a suitable interconnection medium for these units must be available. This section will provide some insight into a possible structure for interconnections, concentrating on a scheme that is widely used in microcomputer systems.

The most obvious and straightforward way of interconnecting a number of electronic units is to provide a separate set of wires to link each pair. This is also the least attractive approach when several units are involved, as the number of wires needed becomes too large for practical purposes. A more economical alternative is to share some (or most) of the wires as a common link among all units.

Figure 1.3 illustrates a scheme where all functional units of a computer are interconnected through a single common set of wires. The wires that form the common communication path are called a *bus*. Since only one set of such wires is used, the scheme is known as a *single-bus* structure.

The number of wires in a bus depends upon the nature of the equipment that must be interconnected. In the case of microcomputers this number is typically in the range from 30 to 70 wires, but it can exceed 100.

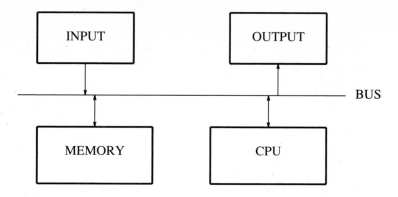

Figure 1.3 Single-bus structure.

Each functional unit connected to a bus requires only a single set of connections linking it to the bus. This is a significantly more attractive scheme from the point of view of cost and wiring complexity than the alternatives that require dedicated connections between units.

While the single-bus structure offers economical advantages and simplicity of connection, it is not without drawbacks. It has a stifling effect on the overall speed of operation of the computer. Since a common set of wires is used for transfer of information between all units, it is apparent that only two units can be meaningfully involved in such a transfer at one time. In a busy system it is likely that several functional units may be ready to transfer information at a given time. Then, there may be contention for the use of the bus, leading to the need for an arbitration mechanism that will allocate the use of the bus to various units. More seriously, some units will be forced to delay their transfers until the bus becomes available, thereby wasting potentially valuable time.

This difficulty can be overcome through introduction of additional interconnection paths. It is quite common to employ *multi-bus* structures, particularly in larger computers where the speed of operation is often the most critical factor. However, the simplicity and cost effectiveness of the single-bus structure make it an attractive choice in systems where the speed of operation is an important but not necessarily critical operating characteristic.

1.3 MEMORY

Typical computation processes involve manipulation of large amounts of information. Much of the information must be held within the computer during computation and be available to the processing circuits in the CPU. As pointed out above, this necessitates the inclusion of adequate memory in the machine.

Performance of a computer is dependent on the speed with which the information can be accessed. Fast CPU circuits are used most advantageously if they can access the required information at a comparably high speed. This means that at least some of the memory elements should be of the type where any particular piece of information can be accessed in less than one microsecond. Moreover, this information should be accessible in random fashion, where the speed of access is not dependent upon either the location in the memory or the order in which the information is accessed. Memory elements which provide this kind of access are called the *main memory*.

In order to achieve operation at electronic speeds, the main memory is constructed with semiconductor integrated circuits. The cost of such circuits is not of great importance when a relatively small amount of memory is needed. However, the cost becomes a very significant factor when large amounts of memory are required, as is often the case. Then, one must look for a cheaper alternative, typically splitting the memory into two parts: a fast main memory and a much larger (and less expensive) secondary storage. The *secondary storage* may be implemented with electromagnetic devices, such as magnetic disks and tapes. The speed of these devices is limited by the mechanical constraints of the driving mechanisms, primarily their rotational speed. But, they can provide very large amounts of relatively inexpensive information storage. They also provide a "nonvolatile" storage medium, because the stored information is not lost when the power supply to the computer is turned off.

The low speed, as measured in terms of the information access times of the secondary storage devices, makes them unsuitable for direct interaction with the processing circuits in the CPU. The usual way of organizing a computer is to have the CPU interact directly with the main memory only. When processing is to be done on information that resides in secondary storage, this information must be transferred into the main memory. Note that this transfer is done via the common bus in our single-bus structure.

The main memory is an indispensable part of the computer. The secondary storage is needed only when memory requirements exceed the size of storage that can be economically provided with the main memory. Note that the secondary storage can also provide inexpensive means for long-term storage of information, because magnetic tapes and some types of disks are removable entities that are easily stored on shelves. When such information is to be used, it is made accessible to the computer simply by mounting the corresponding tape or disk on the driving unit that is a part of the computer system.

The remainder of this section deals with the main memory. Its function is vital and must be thoroughly understood. The main memory consists of a large number of storage cells, each of which can store one bit of information. A *bit* has the value 0 or 1. This small amount of information is seldom processed on an individual basis; it is more useful to handle bits in larger groups, the size of which is usually fixed. The group size of 8 bits is so commonly used that it has been given a special name—a *byte*.

If the information is handled in fixed groups of bits, it is sensible to organize the main memory in a way that allows a desired group to be accessed easily. A convenient approach is to implement the main memory as a set of n-bit groups, called *words*, as shown in Figure 1.4. Having n bits in each word, such memory is said to

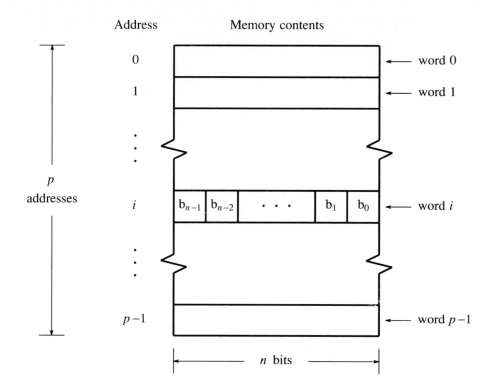

Address   Memory contents

Figure 1.4 Functional structure of the memory.

have a *word length* of n. All n bits in a given word can be accessed simultaneously. Each word is identified by an *address,* which is just a specific number associated with the location of the word. The figure depicts a memory consisting of p words of n bits each. The word addresses range from 0 to p-1. Accessing word i involves reading or writing all n bits of the word with the address i.

We have associated the term "word length" with the structure of the main memory. The term is often associated with the computer as a whole, typically meaning that a machine with a word length of n has a CPU which deals with the information in groups of n bits. However, it is not essential to have the same word length for both the memory and the CPU, although it is always useful that one be a multiple of the other. For example, it is quite feasible to have a memory where 8 bits are accessed simultaneously, working with a CPU that can process 16 bits at a time. In such a case the CPU merely has to access the memory twice for each 16-bit operand that it needs. Typical word lengths in microcomputers range from 8 to 32 bits, almost always being some multiple of 8.

The task of retrieving the information stored in a word of the main memory entails sending the corresponding address (via the bus) and a command that a *read* operation is to be done to the control circuits in the memory unit. The memory responds by making the contents of the selected word available as its output. The *write* operation is similar. The memory is provided with the address of the word, the data to be written into this word, and a command that a write operation is to be performed.

Information stored in the main memory may be of several different types. An *n*-bit word may represent a binary number, or it can denote alphanumeric characters. Alternatively, the contents of a memory word can be interpreted as a command that specifies an operation which is to be executed within the computer. It is customary to refer to the information as *instructions* if it consists of commands and as *data* if it comprises the operands (numbers, characters, logical quantities) used in the operations that are specified by the commands.

Programs consist of instructions and data, both of which are stored in the main memory. From the hardware point of view the memory is just an organized collection of storage cells, filled with bit patterns that may be instructions or data. It is the task of the CPU to interpret the bit patterns that it fetches from the memory in a meaningful way.

1.4 CENTRAL PROCESSING UNIT (CPU)

The CPU's function is to perform the operations needed to execute a program. It does this by fetching instructions from the main memory and executing them one at a time. Since the instructions involve manipulation of operands, the CPU must also be able to get hold of the data to be used as the operands. Such data usually come from the main memory, but can also be provided directly by the I/O units or perhaps reside in temporary storage within the CPU. Storage locations within the CPU are referred to as CPU *registers*.

Figure 1.5 gives a block diagram of the CPU. It indicates the main functional parts, consisting essentially of a collection of registers, some processing circuits, and a control mechanism that orchestrates the overall operation.

The CPU communicates with other units in the computer through a bus. We are assuming that the computer has a single-bus structure. Recall that a bus is just a set of wires, and that it must be possible to transfer information between any two units connected to it without unwanted interference from the rest of the units. This can be achieved if each unit is connected to the bus by means of gating circuits that can be turned on or off as desired. Transmitting and receiving units may operate at substantially different speeds. Since several units share the use of the bus, it should be used for as short a time as possible for any given transfer. A simple scheme for achieving this is to incorporate a buffer in the interface of the units connected to the bus. The buffer consists of a register(s) that can be accessed (i.e., written into or read from) by both the unit and the bus through some gating circuits. Of course, the unit and the bus should not both attempt to load a given register at the same time. The desired speed of transfer is attained by activating a selected pair of registers—a source register transmitting its contents to a destination register via the bus. As soon as the transfer is completed the bus can be used for the next transfer between two different

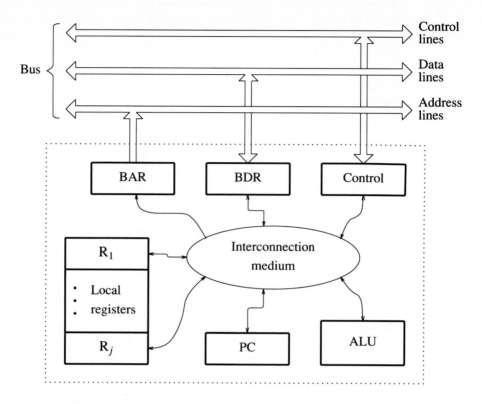

Figure 1.5   General structure of a central processing unit (CPU).

registers. Note that the concept of using the bus buffers is essentially the same as the idea of using I/O buffers described in Section 1.1.

Two buffer registers connect the CPU to the bus. A *bus address register* (BAR) is used to hold the address of a location that the CPU wants to access. A *bus data register* (BDR) handles the data involved in the transfer between the CPU and the addressed location. If the CPU wishes to read the contents of location X in the memory, it places the number X into the BAR and produces a control signal which indicates that a read function is required. The address X and the control signal are transmitted to the memory through the bus. A short time later (dependent upon the speed of the memory circuitry) the memory places the required data on the bus. The data are then loaded into the BDR. Once loaded in the BDR, they become available to the processing circuits in the CPU.

Registers BAR and BDR are also needed when the CPU has to store data in the main memory. Assume that a number Y is to be stored (written) in location X. The process is initiated by the CPU placing the value X in the BAR and the value Y in the BDR. The contents of registers BAR and BDR, along with the "write" control signal, are transmitted via the bus to the memory which then completes the desired storing of information.

We have used the term "bus" to denote the entire set of interconnection wires that links all units in the computer. Since the information transmitted on the bus consists of addresses, data, and control signals, it is customary to refer to the corresponding wires as the address, data, and control lines, respectively. The actual number of lines in each of these groups depends upon the size and structure of the computer. Figure 1.6 shows a typical connection between the CPU and the main memory. If the memory contains p words of n bits, then the bus normally includes n data lines and k address lines, where $p \leq 2^k$. The number and the nature of the control lines depend heavily upon the characteristics of the CPU, as will be detailed

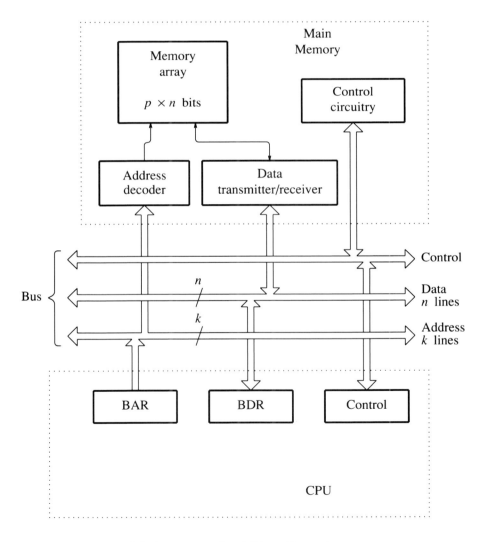

Figure 1.6   Interconnecting CPU and the main memory.

in later chapters. An address decoder circuit in the main memory determines which word is to be accessed, based on the information on the address lines.

The core of the CPU consists of circuits that perform the necessary arithmetic and logic functions. We will refer to them as the *arithmetic and logic unit* (ALU), which typically performs a varied set of functions. Some functions involve two operands (e.g., addition) and others a single operand (e.g., negation).

The operands for the ALU can come from sources external to the CPU, primarily the main memory, or from registers within the CPU. Consider the task of adding a list of numbers stored in the main memory. An obvious way of doing this would be to bring the first two numbers in the list into the ALU, compute their sum, and send the partial sum back to the main memory. Next, the partial sum and the third operand can be brought to the ALU and the process repeated. The scheme has a flaw in having the partial sum unnecessarily shuttled to and from the main memory, requiring the use of the bus, when it would be much more expedient to store the partial sum within the CPU while the next operand to be added to it is fetched from the memory. The flaw can be corrected simply by providing a register in the CPU for this purpose. The name "accumulator" is often associated with such a register.

The previous example suggests the need for one register in the CPU to serve as a temporary storage location. In practical programming applications it is useful to have several such registers available in the CPU to hold frequently used operands. The objective is to save time, since operands held in the CPU registers are immediately available to the ALU while those that must be fetched from the main memory become available only after a delay determined by the time needed to access the memory and carry out the bus transfers. Figure 1.5 includes j such registers. The number of these registers is not large, being less than 20 in many microprocessors and typically less than 50 in most computers.

Computer programs consist of ordered sequences of instructions and data to be manipulated by these instructions. Execution of a program entails taking action specified by the instructions, which are fetched from the main memory and brought into the CPU to be interpreted and executed one at a time. In order to achieve orderly execution of the program, the CPU keeps track of the location in the main memory where the presently executed instruction was fetched from, enabling it to know where the next instruction will be found. A special register is included in the CPU for this purpose, called the *program counter* (PC). Its contents are always updated to reflect the memory address where the next instruction is to be found.

We have discussed several functional parts of the CPU. A fair amount of information transfer takes place among them, even in the process of executing a single instruction. This implies the existence of an interconnection medium. The actual structure of the medium is of little interest to the programmer, who only sees the CPU parts at the functional level.

Finally, we should say a few words about the control of operations in a computer. It is tempting to think of a well-defined single control unit that governs the operation of the entire computer, but such simplified interpretation is valid only from the conceptual point of view. In real implementations the control mechanism is distributed over a number of different units, each unit having sufficient control circuitry to enable it to function effectively within the total system. However, it is common practice to place the bulk of the control functions in the CPU. Control information is routed through the computer by means of the control lines of the bus. Most signals on these lines are originated by the CPU, requiring a sizeable amount of control circuitry in the CPU. Other functional units in the computer have smaller, but readily identifiable, control sections.

As far as the programmer is concerned, much of the control structure is nothing more than particular details of the hardware implementation. Familiarity with these details helps in understanding the finer points of computer operation, but is not essential to the user who merely wants to write and run some programs. On the other hand, a user who wishes to design, alter, or expand a computer system is not likely to be successful without intimate knowledge of the control details.

1.5 INPUT/OUTPUT UNIT

It was stated earlier that the I/O unit in a computer consists of the circuitry necessary to connect I/O devices to a computer system. This circuitry includes the specific interfaces needed for the I/O devices, as well as the control functions that implement the I/O transfers within the computer.

I/O devices usually appear to the computer as passive devices which take action only when instructed to do so. Their behavior and that of other units in the computer must be coordinated in order to perform the desired processing tasks. It is the job of the CPU to provide this coordination.

In a typical operation, the CPU monitors the status of the I/O devices and selects them according to availability and need. Consider an example of an alphanumeric keyboard serving as an input device. The sole function of the keyboard is to recognize when one of its keys is pressed and to make an encoded set of bits available as data that denotes the pressed key. The code normally uses 7 or 8 bits to represent one alphanumeric character.

A computer reads characters from the keyboard by executing a simple input program stored in the main memory. This program implements the following procedure:

1. Monitor the status of the keyboard to detect changes when a key is pressed.

2. When a key is pressed, read the encoded byte of data that corresponds to the activated key.

3. Check the validity of the character received.

4. Repeat the first three steps for the next character.

Note that it is highly desirable that the input program be able to detect certain kinds of manual errors made by the typist. For example, pressing two keys simultaneously may generate an illegal character code that should be recognized as such.

The response time of the system must be short enough to accommodate the fastest typists. The requirement is not difficult to achieve with CPUs that operate at electronic speeds. However, time delays can become a problem if many keyboards and other I/O devices are connected to the same computer, as a single CPU may not be able to service a multitude of devices that are active simultaneously.

Output devices are handled in an analogous manner. If the CPU is to send a line of output text to a printer, it must first check the status of the printer, perhaps repeatedly, until it finds it to be idle. Then it can transfer the data to be printed to the printer and send a command that printing may start.

The underlying theme of the above discussion is that the CPU controls the entire I/O process. It finds out the state of the I/O devices by examining their status indicators. It uses the idle devices as required, or waits for busy devices to become available. If the CPU is to be able to control the devices it must have access to all information associated with them. This information consists of the input and output data, status bits indicating the state of the device, and the control signals that govern the behavior of the device. A convenient way of keeping track of this information is to incorporate a register for each type of information in the interface of the I/O device. Thus, there is a data register to serve as a buffer for the transferred data, a status register whose contents reflect the device status, and a control register whose contents determine the operating mode of the device. Having access to these registers, the CPU may read the status register to determine whether the device is idle or busy (or some other indication of the state). It can write into the control register specific bit patterns that define the desired mode of operation. Finally, it can transfer data to or from the data register, where the transfer typically takes place between the data register in the I/O interface and either a register in the CPU or some location in the main memory.

Since the CPU fully controls the above-described I/O process and performs the necessary steps by executing a program called an I/O routine, the method is known as the *program-controlled I/O* scheme. The scheme is simple to implement; its only drawback is its low speed due to the CPU's intervention at each step by executing one or more instructions in the I/O routine. Of course, the speed factor should be considered only in terms relative to the speed of the I/O device in question. For most devices (e.g., keyboard, printer) the few microseconds that it takes to execute an I/O routine are insignificant when compared with the operating speed of the device. However, if a fast electronic apparatus is used as an I/O device the speed attainable through program-controlled I/O may be inadequate. In such cases one must resort to more complicated techniques that transfer data directly between the I/O device and the main memory, without requiring the CPU intervention for each word of data transferred. Such techniques, called *direct memory access* (DMA), will be discussed in Chapter 7.

Having decided that the CPU must have access to the information in the I/O interface, we should consider the possible means of providing this access. It is quite feasible to set aside a few specialized I/O instructions for the CPU to execute as a part of an I/O routine. An instruction may state

Input   IN23,CPUREG1

indicating that the data in input device number 23 are to be transferred into register 1 of the CPU. Or there may be an instruction

Read_status   IN23,CPUREG2

which transfers the contents of the status register in input device 23 into CPU register 2. Similarly, if a bit-pattern in CPU register 4 is to be transferred into the control register of output device 15, a suitable instruction may be

Set_control   CPUREG4,OUT15

Each such instruction specifies exactly one type of operation. Therein lies a potential problem. Computer instructions are encoded as bit-patterns of predetermined length for storage and execution in the machine. The complexity of these codes and the resultant storage and handling requirements are adversely affected by the increased variety of instruction types. The problem is less troublesome if the instruction repertoire is highly structured into a limited number of instruction types. Thus, it may be useful to devise a scheme where I/O operations do not require specialized instructions, but are handled instead with a set of more general purpose instructions.

Consider the possibility of dealing with the I/O devices as if they were accessible in the same way as the main memory. This means that a unique address must be assigned to each data, status, and control register in all I/O interfaces. Moreover, these addresses cannot coincide with the addresses that identify locations in the main memory. Figure 1.7 shows an example of this scheme. The main memory contains p words, thus needing p addresses. Two I/O interfaces are included, each having three addressable registers. Then, a total of $p+6$ distinct addresses must be recognizable in the system. Since the addresses are encoded and transmitted as binary numbers on the bus, it follows that the bus must have at least $k \geq \log_2(p+6)$ address lines and the CPU must be able to generate the required signals on these lines.

If the I/O interface locations are accessible in the same way as the main memory locations, then all transfers of information on the bus can be accomplished with a single general move instruction

Move   LOC1,LOC2

which transfers the contents of location LOC1 to location LOC2. Consider again the three previously mentioned specialized I/O instructions. Assume that the addressing arrangement of Figure 1.7 is used, where the input device corresponds to IN23 and the output device is OUT15. The addresses of the data, status, and control registers are as shown in the figure. Then, the actions that were previously specified using three different types of instructions can be specified using a general move instruction as follows

Move   p,CPUREG1
Move   p+1,CPUREG2

and

Move   CPUREG3,p+5

The I/O arrangement described above is called *memory-mapped I/O*. If a given CPU can activate k address lines, it can generate 2^k distinct addresses. This represents the addressable space of the computer. The designer of a computer system decides how the addressable space will be used. In a system that uses

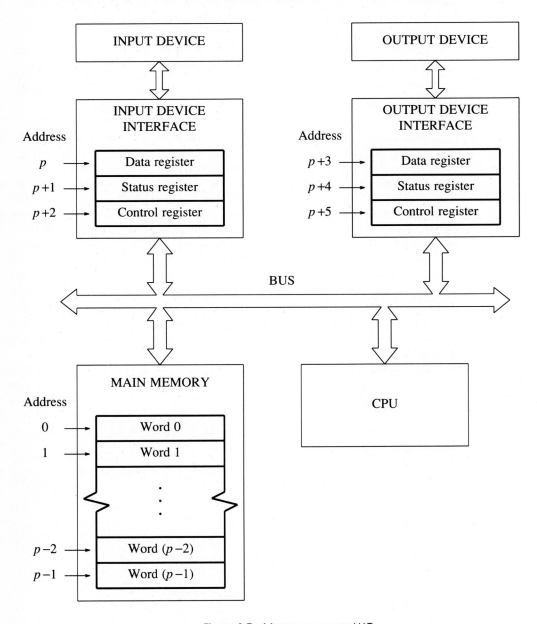

Figure 1.7   Memory-mapped I/O.

memory-mapped I/O, most of the addressable space is likely to be allocated to the main memory, but a sufficient number of addresses must be reserved for dealing with I/O devices. Note that the assignment of some addresses to I/O interfaces reduces the maximum size of the memory that can be addressed directly. This is of little concern, because the I/O locations normally occupy a very small fraction of the addressable space. A microprocessor that uses 16-bit addresses provides an addressable space of $65,536 (= 2^{16})$ locations. Yet, a microcomputer system is most unlikely to need more than 1000 addresses for I/O interface purposes.

Memory-mapped I/O is a simple and efficient concept, which is easy to implement and is highly suitable for use in microcomputer systems. The assignment of addresses to various parts of the system is subject to a few constraints, the main one being that the same address cannot elicit response in two different locations.

1.6 EXAMPLE OF A MICROCOMPUTER SYSTEM

Having considered a number of basic aspects of computers, let us attempt to place them in the perspective of a representative microcomputer system, shown in Figure 1.8. It uses a single-bus structure, and the driving unit is a CPU that is a microprocessor chip. Other units connected to the bus are the main memory and the I/O interfaces.

The main memory is divided into two parts: RAM and ROM. RAM, which stands for *random access memory,* is a memory of the type described in Section 1.3. Information may be written into it as well as read from it during normal execution of a program. The words "random access" indicate that memory locations can be accessed in any order, regardless of their addresses. This is to be contrasted with memory devices where information can be accessed only in sequential order, as in the case of magnetic tapes.

The second part of the main memory in Figure 1.8 is labeled ROM, for *read only memory.* This type of memory is different from RAM only in that its contents can be read during the normal operation of the computer, but cannot be changed. The information stored in ROM is placed there prior to installation in the system, often at the time of manufacture. It should be noted that words in a ROM may be accessed at random, just as in a RAM. The name RAM, however, has been traditionally used to denote a random access memory that can be both read from and written into. This is a somewhat misleading convention, but it is widely used. The reason for including a ROM in a computer is to provide unalterable storage in a part of the main memory. It is used for storage of information that must not be destroyed either inadvertently or through malfunctioning of the system. For example, if the electrical power supply to the computer is interrupted, either through circuit failure or by someone switching it off, the contents of the RAM are destroyed. When the power is turned on, a start-up program must be executed to initialize the various units in the computer into the required starting mode of operation. This program must be loaded into the main memory before it can be executed. If a part of the main memory is implemented as ROM, it can be used to hold the start-up program permanently. Otherwise, in a

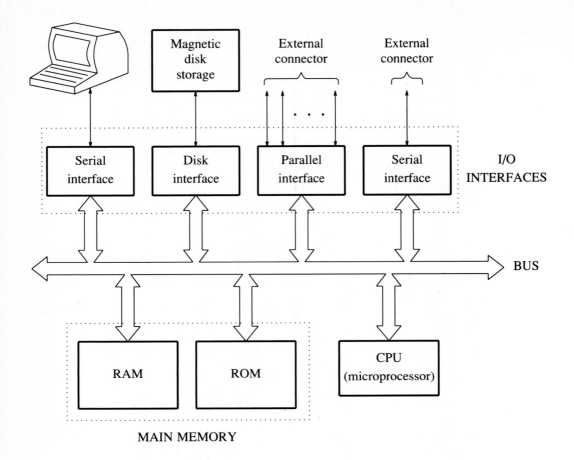

Figure 1.8   An example of a microcomputer system.

computer where the entire memory is of the RAM type, the start-up program will require considerable manual intervention to have it loaded in the memory.

In small systems with low memory requirements the RAM and ROM may consist of single integrated circuit (IC) chips, or they may even share a single IC chip with the CPU, as suggested in Section 1.1. However, when larger amounts of memory are needed, the RAM and ROM parts are likely to involve a number of chips.

Input/output devices are connected to the microcomputer through I/O interfaces. Three different types of interfaces are shown in Figure 1.8. A *serial interface* is used to connect a video terminal. This interface exchanges data with the terminal in serial mode, where data are transmitted one bit at a time along a single communication link. Serial transmission is slow, but inexpensive to implement as far as the number of wires is concerned. While serial transmission is fully acceptable between the terminal and the interface, it is not suitable for the fast operation needed on the computer bus. The function of the serial I/O interface is to deal with the data on the bus in the parallel mode and to communicate with the connected device in the serial mode. If the bus has n data lines, then the interface accepts n bits of data simultaneously from the bus. These n bits are sent to the terminal one bit at a time, requiring n time slots for transmission. The reverse process takes place during reception of data from the terminal.

Some I/O devices can handle data at speeds that cannot be supported with serial interfaces. In such cases a *parallel interface* must be used, where n bits of data are handled simultaneously both on the bus and on the links to the device. This achieves a faster interchange of data at the expense of extra wiring cost. Figure 1.8 includes one parallel interface. The I/O device that may be connected to it is not shown explicitly. Instead, a connector is indicated to which a cable could be attached to connect the desired device. Note that the figure also shows an additional serial interface terminated in an external connector. The two unassigned I/O interfaces in the figure represent a common practice in designing general purpose microcomputer systems. They provide additional I/O capability for the user of the system, who can decide what devices are to be connected to these interfaces.

The last block in the figure, the disk interface, consists of the circuits necessary to interact with a magnetic disk storage unit. In many computer applications it is essential to have some means for long-term storage of information; magnetic disks and tapes are an inexpensive way of achieving this goal. Since the tapes and some disks are manually removable, they provide a convenient medium for on-the-shelf storage of programs and data that are used only occasionally and need not be immediately available in the computer system. Such storage is often referred to as being "off-line," as contrasted to "on-line" storage which comprises the disks and tapes actually connected to the computer at any given time.

Magnetic disk and tape units serve as secondary storage devices, where the primary store is the main memory of the computer. They can also be thought of as I/O devices. A disk or a tape provides a convenient means for distributing software. A disk containing a copy of a particular program generated on one machine can be used to transfer the program to a different machine.

The configuration of Figure 1.8 is representative of typical microcomputer systems. The computer parts, consisting of a number of IC chips, may be placed on a single board with connectors for the terminal, magnetic disk unit and other external devices. The entire system is often encased in a box, giving a portable stand-alone unit, as is done in the case of popular home computers.

1.7 TERMINOLOGY

Like most technical disciplines, the computer field has plenty of jargon. Much of the terminology needed was used in this chapter. Following is a summary of the most important terms and their meanings.

- *Address*: a number generated by the CPU that identifies a location in the memory.

- *Addressable space*: the number of locations that can be identified uniquely by the address bits generated in the CPU.
- *Buffer*: an electronic circuit that facilitates communication between functional units.
- *Bus*: a set of wires that provides a common communications medium for the functional units of a computer.
- *Central Processing Unit (CPU)*: a hardware unit that performs arithmetic and logic operations on the data and controls the flow of execution of programs. In microcomputers, the CPU is usually a microprocessor chip.
- *Data*: operands used in the operations specified by instructions.
- *Direct Memory Access (DMA)*: a scheme for transferring data directly between the main memory and an I/O device, with minimal intervention by the CPU.
- *Instruction*: a command that specifies an operation to be performed by the CPU.
- *I/O device*: a physical device used to input and/or output the information to/from the computer.
- *I/O interface*: an electronic circuit that matches the requirements of an I/O device to the requirements of the computer bus.
- *Main memory*: semiconductor circuits used for storage of a sizeable amount of information. Random access to individual storage locations is provided, with access times corresponding to the speeds at which microprocessors can operate.
- *Memory-mapped I/O*: an arrangement where the registers in the I/O interfaces are given addresses and handled as if they were locations in the main memory.
- *Program-controlled I/O*: a scheme where the CPU transfers individual input or output data units, under control of program instructions.
- *Program Counter*: a register in the CPU that keeps track of the location in the main memory where the instruction to be executed next can be found.
- *Random Access Memory (RAM)*: circuits that constitute a memory where individual storage locations can be read from and written into in random order.
- *Read Only Memory (ROM)*: circuits that constitute a memory whose contents can be read, but cannot be changed.
- *Secondary storage*: electromagnetic devices that allow inexpensive storage of large amounts of information, with access times limited by mechanical constraints. Magnetic disks and tapes are typical examples of this type of storage.
- *Word length*: the number of bits in a basic unit of information manipulated in a computer. The word length of a main memory is the number of bits that can be accessed in one memory access cycle. The word length of a CPU is determined by the size of its internal registers.

1.8 CONCLUDING REMARKS

This chapter introduced a number of basic concepts needed to understand the structure of a computer system in order to provide a quick overview of computer hardware. The detailed discussion of the relevant issues will be given in the chapters that follow.

Chapter 2 deals with the basic ideas of machine-level programming, presented in a generic style because the concepts are applicable to microprocessors in general. Chapters 3 and 4 present Motorola's 6809 and 68000 microprocessors, and show how the concepts of Chapter 2 are implemented in these commercially successful products. A full discussion of I/O techniques is given in Chapters 5, 6, and 7. The details of the main memory are examined in Chapter 8. In each of these chapters we will discuss both the general concepts and their application in our example microprocessors.

1.9 REVIEW QUESTIONS

1. Describe the function and the operation of a bus in a microcomputer.

2. What are the advantages and disadvantages of a single-bus structure?

3. What kind of devices can be used as storage units in modern microcomputers?

4. How is the main memory organized?

5. Describe the operation of the CPU when writing a word of data into the main memory.

6. What are the functions and the structure of the CPU?

7. Why are registers included in the CPU?

8. What is the purpose of the program counter?

9. What determines the number of address lines that exist in a microcomputer bus?

10. How can one achieve transfer of data between two devices that operate at substantially different speeds? Where do such situations occur in a computer system?

11. Describe the memory-mapped I/O scheme. What are its advantages and disadvantages?

12. How can the program-controlled I/O technique be used to print a line of English text on a printer that prints one character at a time?

Chapter
2

Machine Programming Fundamentals

Electronic circuits that form a computer are often referred to as *hardware*. The hardware would be of little use without programs that contain detailed instructions which specify the sequences of events that must take place in order to perform useful computational tasks. Since such programs are readily changed they are referred to as *software*.

In this chapter we will discuss the most basic aspects of software, namely, the low-level programs, where each instruction denotes a simple task that can be performed in one well-defined operation of some part of the hardware. Programs consisting of such simple instructions are suitable for direct execution in a computer. However, they are more difficult and time-consuming to write than programs written in high-level languages such as FORTRAN, PASCAL, or PL-1. A program written in a high-level language can be executed only by first being translated into a low-level machine program. Fortunately, the necessary translation process can be automated, and is typically done with the aid of a translating program called a *compiler*. Techniques used for this translation process (compilation) are quite sophisticated and well beyond the scope of this text. We will concentrate only on the low-level programs, to enable the reader to gain an appreciation of the interaction between hardware and software in a general-purpose computer.

In Chapter 1 we espoused a point of view that computers are based on a few well-defined principles. A thorough understanding of these principles will enable one to understand and use effectively most computers, regardless of the peculiarities of detailed implementations found in different machines. This is equally true for both the hardware and software aspects. The best way of developing competence is to study the general principles thoroughly, and then reinforce the gained knowledge with a careful examination of some real example, preferably a successful commercial product.

This chapter deals with programming concepts in a very basic way. An attempt is made to describe the main principles in a general form, using highly descriptive terminology that is not tied to any particular computer. Then, in Chapters 3 and 4, actual machines are considered as examples of how the general principles are implemented in practice.

2.1 INSTRUCTIONS

Computational tasks that the computer hardware performs are specified by means of instructions. The complexity of the task specified in an instruction has a direct bearing on the complexity of the hardware that is to perform this task. In order to keep the hardware reasonably simple, it is prudent to use only relatively simple instructions. We begin by considering simple examples of three kinds of instructions: two-operand, one-operand, and sequence control instructions.

2.1.1 Two-Operand Instructions

Simple instructions have a rather limited scope. Typically, they may specify a basic operation, e.g., addition, comparison, or transfer, and one or two operands involved in the operation. For example, transferring data from one location in the computer to another might be accomplished by an instruction of the type

Move   A,B

where A and B denote the locations involved. In this case the data from location A, the *source,* is copied into location B, the *destination.* The Move instruction does not alter the contents of the source. Note that A and B represent locations that are identified by numerical addresses. For clarity and readability of programs we often use names rather than numbers. Before a program is loaded in the computer memory for execution, the names are replaced with numbers as will be explained in Section 2.8.

The locations specified in an instruction may be main memory locations, central processing unit (CPU) registers, or perhaps, addressable locations in input/output (I/O) units. The number of locations that can be specified in a single instruction is limited by practical considerations. When a program consisting of a sequence of instructions is stored in the main memory the operations required and the locations of the operands must be denoted by means of binary codes, which require considerable space within an instruction. If instructions are not to be unreasonably long, it is necessary to restrict the number of operands that may be included in a given instruction. A practical choice for this number of operands is not greater than two, particularly so for microcomputers.

Limiting the instructions to having at most two operands has some repercussions on the ease of writing programs. Many numerical tasks require operations of the type

$$c = a + b$$

where the sum of a and b is to be computed and the result stored as c. Let a, b, and c be integer variables. In a high-level language, the desired task may be expressed as

$$c := a + b$$

Each of the variables a, b, and c is assigned a location in the main memory. Let a and b be the numbers stored in memory locations A and B, respectively. The sum, c, is to be stored in location C. Then, the computer must perform the operation

$$C \leftarrow [A] + [B]$$

In this notation we are using the square brackets to indicate the "contents of the location specified." Hence, the above expression means that the contents of location A are added to the contents of location B and the result is stored in location C. This operation requires three operands to be specified, which cannot be done in a single instruction that is limited to two operand locations. However, the desired effect may be achieved using more than one instruction. The task may be realized in two steps:

$$A \leftarrow [A] + [B]$$
$$C \leftarrow [A]$$

The first step is readily performed with an instruction

$$\text{Add} \qquad \text{B,A}$$

which adds the source operand B to the destination operand A and leaves the sum as the new contents of location A. The second step is done using a Move instruction.

Execution of the above sequence of two instructions has a possibly undesirable side effect of destroying the original contents of location A. Sometimes, it is important not to destroy the original operands in this way. The problem is easily avoided by first copying the contents of location A into location C, as follows:

$$C \leftarrow [A]$$
$$C \leftarrow [B] + [C]$$

The required operations are performed by the instructions

$$\text{Move} \qquad \text{A,C}$$
$$\text{Add} \qquad \text{B,C}$$

Instructions such as Move A,B require a sufficient number of bits to specify the operation and the addresses of the locations involved. Specification of an arbitrary location in the main memory requires a considerable number of bits, typically 16 or more in microcomputers. Thus, when both operands are specified in this manner, a relatively long instruction is needed. Since long instructions occupy more of the main memory, it is worthwhile to consider alternatives that result in shorter instructions. We may designate a particular location as an *accumulator,* and arrange the addressing scheme so that an operand that is held in this location may be specified using only a few bits. Then, instructions of the type

Move	A,ACCUM
Move	ACCUM,B

and

Add	A,ACCUM

would be considerably shorter than the more general two-operand instructions.

The name "accumulator" is associated with this dedicated location, because it is often used for temporary accumulation of intermediate results. For example, suppose that we want to compute

$$2 ([A] + [B]) + [C]$$

and store the result in location C. A simple way of achieving this is

Move	A,ACCUM
Add	B,ACCUM
Shift_left	ACCUM
Add	C,ACCUM
Move	ACCUM,C

A "shift left" instruction is used to do the required multiplication. When bits of a binary number are shifted left by one bit position, the operation is equivalent to multiplying the number by 2. We should caution that this example is intended to show only how the accumulator may be used to accumulate intermediate results. It is not intended to illustrate how a general expression of the above type should be computed, where the necessary multiplication might be performed with a "Multiply" instruction.

Having introduced the idea of using an accumulator, we should consider the kind of location that may be suitable for this purpose. If the accumulator is a location in the main memory, this location will have to be accessed through a bus transaction as explained in Section 1.5. Such a transaction takes considerable time, typically 0.2 to 1 microsecond for most microcomputers. The instruction Add B,ACCUM requires three such time intervals, because both operands have to be fetched from the main memory and brought into the CPU where the addition operation is carried out. Then, the sum has to be transferred back to the memory. A better alternative is to provide one or more registers in the CPU to serve as accumulators. Using such CPU accumulators reduces the execution time significantly. For this reason most computers have CPU registers that can be used as accumulators. In the remainder of this text we will use the designator ACCUM to refer to a dedicated register in the CPU. When there are several such registers, we will call them ACCUMA, ACCUMB, and so on.

2.1.2 One-Operand Instructions

So far we have considered instructions that specify an operation involving two-operand locations. These locations either contain the two operands involved in the operation, as in the Add instruction, or they merely indicate the source and destination addresses in a transfer operation, as in the Move instruction.

General programming tasks require other types of instructions as well. While it is possible to perform many tasks using relatively few distinct instructions, it is convenient to have a variety of instructions available to make the programming job easier. Consider the simple task of assigning a value 0 to some variable a. The corresponding location, say A, should be cleared, i.e., set to zero. This task could be done using a two-operand instruction, for example,

Subtract   A,A

Clearing memory locations in this manner is rather clumsy. It is much better if there exists an instruction

Clear   A

which will do the job more simply.

The Clear instruction is just one example of a useful one-operand instruction. It is convenient to have other instructions of this type, for example, instructions that can shift or test the value of the specified operand. We encountered a useful application for a shift instruction in the previous subsection.

2.1.3 Sequencing Control Instructions

In addition to instructions that perform functions on operands, it is essential to have some instructions that can be used to control sequencing in the execution of programs. In other words, a mechanism must exist that enables the programmer to specify the order in which the instructions that constitute a program will be executed. Instructions are stored in the main memory in locations having sequential addresses, such as 100, 101, 102, etc. During execution, this natural ordering of instructions is followed most of the time. But, whenever a decision point is reached, two or more options are possible depending upon the results of some computation. An example is shown in Figure 2.1. Let the flowchart in part (a) of the figure depict an action that must be taken at some point in a larger program. Two values, a and b are compared. If a is less than b, then the task 2 must be performed; otherwise task 1 is to be performed. A possible arrangement for the instructions of this program, when loaded in the main memory for execution, is given in Figure 2.1b. The test block in the flow chart is implemented by the two instructions: Compare and Branch_if_<0. These are followed by two program segments that implement the two tasks. We have arbitrarily placed the instructions for task 1 before those of task 2.

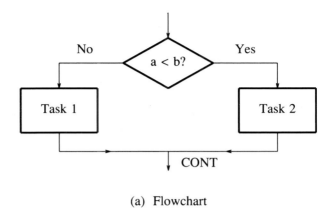

(a) Flowchart

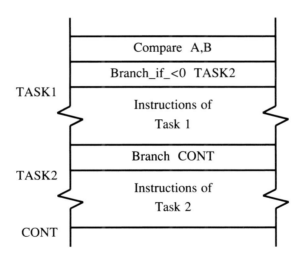

(b) Arrangement in the main memory

Figure 2.1 An example of a decision point in a program.

The Compare instruction compares the values of the two variables, which are assumed to be stored in locations A and B, by performing the subtraction A − B. It does not alter the value of either A or B. If the result of the comparison indicates that task 2 is to be executed, then it is necessary to skip past the instructions of task 1 to get to the first instruction of task 2. The capability to conditionally alter the sequence of execution of instructions in this manner is provided by means of the Branch instruction. This instruction specifies the condition to be tested, namely, that the result of the previous instruction was less than zero. The instruction also specifies the address of the next instruction to be executed if the condition is true. If the condition is false, the next instruction to be executed is the one that immediately follows the branch instruction. Thus, upon executing the instruction

Branch_if_<0   TASK2

a branch to the location TASK2 will be made if the tested condition is true, that is, if the value of a is less than b. Note that if the path taken involves task 1, it is necessary to skip the instructions of task 2 when the end of task 1 is reached. This is accomplished by inserting an unconditional branch instruction after the last instruction of task 1. This instruction always causes a branch to the location indicated within it.

We will discuss the details of the various conditions that are suitable for testing by branch instructions in Chapters 3 and 4, which deal with specific microprocessors. At this point we should note that these conditions are based on the results of executing instructions that immediately precede a branch instruction. Simple conditions that a branch instruction can test are whether the result of a preceding arithmetic operation is positive, negative, or zero. For example, in the two instructions

Add   A,B
Branch_if_=0   NEWLOC

the contents of locations A and B are added. Then, a branch to address NEWLOC takes place if the sum is equal to 0.

Conditions that can be tested by branch instructions are determined by the instructions that perform arithmetic and logic operations. But, they can also be determined by some other instructions. For example, most microprocessors have a Test instruction, which tests the contents of the operand location specified. It checks the sign of the operand and whether or not the operand is equal to 0. Such conditions established by an instruction are usually saved as bits in a register that is often called the "condition code" register. Thus, one bit will be set to 1 if the tested result was 0, and cleared to 0 otherwise. This bit may be referred to as the Z bit, or in commonly accepted jargon as the Z *flag*. Another flag, called N, indicates whether or not the sign of the tested result was negative. Other condition code flags may be defined to reflect a variety of conditions, as will be discussed in conjunction with specific microprocessors in Chapters 3 and 4.

It is important to emphasize one further point about condition code flags. When an instruction sets or clears a given flag, the state of the flag will remain unchanged until another instruction that affects this flag is executed. The flags affected vary among different classes of instructions. Some flags are affected by many instructions, and others by only a few. The state of the flag is often determined by the most recently executed instruction. However, if that instruction does not affect this flag, the state of the flag will remain as set by some previous instruction.

The discussion in this section has shown the need for two basic types of instructions: those that perform operations on either one or two operands, and those that control the flow of execution (sequencing) of a program.

2.2 INSTRUCTION LOOPS

The most obvious way of writing programs is to use straightline sequencing, where each operation is represented by an instruction, and the instructions are stored in the main memory in the order in which the operations are to be performed.

Consider the simple example in Figure 2.2. It depicts the task of copying the contents of three locations A1, A2, and A3 into locations B1, B2, and B3. This task can be performed by a program consisting of the three instructions shown in the figure. Each required transfer is accomplished by a specific instruction, and the instructions are ordered in the sequence in which they should be executed.

Now, suppose that we attempt the same approach with a larger list of numbers, where the numbers from list A are to be copied to list B. Let the numbers in the list

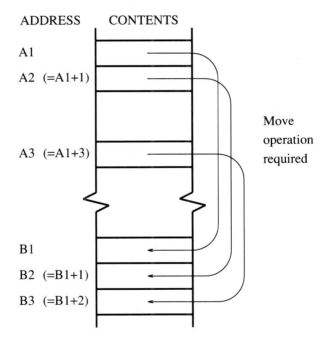

Possible program:

    Move A1,B1

    Move A2,B2

    Move A3,B3

Figure 2.2 A simple straightline program.

occupy consecutive locations in the main memory, as indicated in Figure 2.3. Such lists correspond to the well-known array structures used in high-level programming languages. A straightline program that copies list A into list B is shown in Figure 2.4. It comprises n instructions, where n is the number of entries in the list. Clearly, for a large n this straightline approach is cumbersome.

One of the fundamental principles in computer programming is that large repetitive operations should be handled using the concept of loops. A sequence of

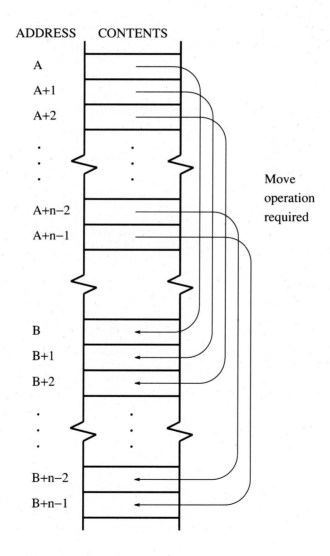

Figure 2.3 An example task that copies a list.

instructions that specify the required operation appear only once in the program, within a loop. This loop is executed as many times as necessary to complete the overall task. In high-level programming languages loops are implemented using specific constructs, such as the familiar DO statements. A low-level programming equivalent involves setting up a counter that keeps track of the number of passes through the loop. The counter is tested during each pass to determine whether or not the required number of passes have been carried out.

Returning to the task shown in Figure 2.3, it is clear that a loop offers a suitable solution. In a high-level language, say PASCAL, the desired copying task can be specified as

$$\text{for i := 1 to n do}$$
$$\text{B[i] := A[i] ;}$$

This exploits the fact that each entry in the list (element in the array) is identified by its index, i. The above statement defines a loop that uses i as a variable index, so that a different entry of the list is specified each time the transfer instruction is executed.

Implementing a loop at the machine-language level is slightly more complicated. Figure 2.5 gives a program that might be used. Location NCOUNT serves as a counter that keeps track of the number of times the loop is executed. It corresponds to the index, i, used above. The loop contains a Move instruction that carries out the desired transfer of one entry from list A to list B. It is written as

$$\text{Move} \quad \text{A(NCOUNT),B(NCOUNT)}$$

The notation A(NCOUNT) is used to denote that the address of the source operand is the sum of the address A and the contents of location NCOUNT. For example, if the first entry of list A is in memory location 500 and the contents of NCOUNT is the value 3, then A(NCOUNT) refers to the contents of location 503, which is the fourth entry in list A. The destination location in list B is defined in the same way. Note that as NCOUNT is incremented during each pass through the loop, a different pair of entries in lists A and B is affected. As the value of NCOUNT increases from 0 to $n-1$, in steps of 1, all of the entries in list A will be moved to list B.

```
Move     A,B
Move     A+1,B+1
Move     A+2,B+2

          .
          .

Move     A+n−2,B+n−2
Move     A+n−1,B+n−1
```

Figure 2.4 A straightline program that performs the task depicted in Figure 2.3.

	Clear	NCOUNT	Use NCOUNT as a counter.
LOOP	Move	A(NCOUNT),B(NCOUNT)	Transfer an entry in list A to list B.
	Increment	NCOUNT	Increment counter.
	Compare	NCOUNT,N	Branch back until the last entry
	Branch_if_<0	LOOP	is copied.
	Halt		

Figure 2.5   A loop program that performs the task shown in Figure 2.3.

The loop also includes instructions that control sequencing. After the counter NCOUNT is incremented, it is compared with the number of entries, n, which is assumed to be stored in location N. As long as the contents of NCOUNT are less than n, the test condition specified in the branch instruction will be true. Hence, the next instruction selected for execution will be the Move instruction that carries the label LOOP. After n passes through the loop, the comparison operation finds the value of the counter to be equal to n, which makes the test condition for branching false. Thus, the instruction that follows the branch instruction is executed next. This is the Halt instruction which indicates the end of the required task.

There are other ways of writing programs of the type in Figure 2.5, such as the option shown in Figure 2.6. Here, the counter is initially set to correspond to the total number of entries in the lists minus 1. This is achieved by loading the value n, from location N, into location NCOUNT and then decrementing NCOUNT before entering the loop. In the loop, the value of NCOUNT is reduced by one each time through the loop. Thus, the entries of the list are copied in reverse order (order of decreasing addresses). The Branch instruction looks at the result of the operation that precedes it (decrementation), and causes branching back to the Move instruction as long as the result is ≥ 0. Thus, the loop will be traversed n times. Observe that this loop has one instruction fewer than the loop in Figure 2.5 because there is no need for a separate Compare instruction. A test for a sign of the result is an integral part of the Decrement instruction.

	Move	N,NCOUNT	Set NCOUNT as a counter.
	Decrement	NCOUNT	
LOOP	Move	A(NCOUNT),B(NCOUNT)	Transfer an entry in list A to list B.
	Decrement	NCOUNT	Decrement counter
	Branch_if_≥0	LOOP	Branch back if NCOUNT≥0.
	Halt		

Figure 2.6   An alternative program for the task shown in Figure 2.3.

2.3 ADDRESSING MODES

In the previous examples a variable name that appears in an instruction is interpreted as referring to the memory location where the operand is to be found. Occasionally, it is desirable to include a specific value to be used directly as an operand of an instruction. For example, if the number of entries in the lists of Figure 2.3 is known at the time the program is written, we do not need to allocate a separate memory location to store the value n. This value may be given directly in the instructions that use it. We will use the symbol # to indicate that a number given in an instruction is to be used directly, rather than being interpreted as the address of an operand. To illustrate this, let $n=50$. Then, the Compare instruction in Figure 2.5 may be replaced with

Compare NCOUNT,#n

This instruction compares the contents of location NCOUNT with the value 50, while the instruction in Figure 2.5 compares the contents of NCOUNT with the contents of location N. Similarly, the first instruction in Figure 2.6 may be replaced with

Move #50,NCOUNT

or better still, the first two instructions can be replaced with

Move #49,NCOUNT

We should consider one other possible way of implementing the task of Figure 2.3. In the two programs of Figures 2.5 and 2.6, the Move instruction required a nontrivial computation (addition) to derive the address location for its source and destination operands. An alternative technique involves the notion of *pointer* locations. These may be either CPU registers or main memory locations. The function of a pointer is to store the address of an operand for use by some instructions. Thus, if a given list is to be traversed in ascending order, the pointer is initially set to the address of the first entry in the list. Successive entries in the list can be accessed by incrementing the pointer accordingly.

Using the notion of pointers, we can write the program given in Figure 2.7. Pointers POINTERA and POINTERB point to entries in lists A and B, respectively. At the start, the pointers are loaded with the addresses of the first entries in the lists. During each pass through the loop, the pointers are incremented to point to successive entries in the lists. The operand addresses used in the Move instruction are the

	Move	#n,NCOUNT	Set NCOUNT as a counter.
	Move	#ADDRESSA,POINTERA	Set pointers to point to first
	Move	#ADDRESSB,POINTERB	entries in lists A and B.
LOOP	Move	[POINTERA],[POINTERB]	Transfer an entry in list A to list B.
	Increment	POINTERA	Increment pointers
	Increment	POINTERB	to point to the next entry.
	Decrement	NCOUNT	Decrement counter.
	Branch_if_>0	LOOP	Branch back if NCOUNT>0.
	Halt		

Figure 2.7 A program for the task in Figure 2.3 that uses pointer addressing.

contents of the pointers. This is indicated in Figure 2.7 by writing the pointer name between square brackets.

It is important to ensure that the square brackets convention as used above is understood clearly. Consider a name NAME that has a value of 1000. Whenever NAME appears as an operand in an instruction, the intended operand is the contents of location 1000 in memory, while [NAME] represents the contents of the memory location whose address is stored in location 1000. Whenever the value 1000 itself is intended as an operand, it will be written as #NAME, or simply as #1000.

The previous examples illustrate the ideas pertinent to instruction sequencing. But, they also hint at the need for having a variety of addressing modes, that is, ways of accessing operands, to simplify the job of writing programs. We will now examine this need and describe some of the techniques commonly found in microprocessors.

The examples in Figures 2.5 through 2.7 have been simple enough not to raise much concern about the characteristics of the operands and the ways in which they are stored in the computer's main memory. Our primary objective was to indicate the nature of machine instructions, in a somewhat informal way. Now, we should take a closer look at the operands to see how they are handled during execution of a program. There are several different ways, called addressing modes, in which operands can be specified in machine instructions.

A given *addressing mode* denotes the steps that must be taken to find the desired operand. If the operand resides in an addressable location, either in the main memory or in a CPU register, the address of that location is called the *effective address,* A_{eff}. In a simple case the effective address may be stated explicitly in the instruction, but more complex modes are useful where some non-trivial computation must be performed to obtain A_{eff}.

In the previous sections, we saw a need for specifying data in several different forms. In particular, we made use of variables, constants and simple lists (arrays). In order to deal with such data, it is common to define four addressing modes:

1. Absolute mode for dealing with variables
2. Register mode for dealing with temporary variables
3. Immediate mode for handling constants
4. Indexed mode for processing simple lists

We also saw, in the program of Figure 2.7, that pointers can be used in an interesting way. This was just one example. The usefulness of pointers in general is such that it is worthwhile providing a suitable addressing mode, called the indirect mode. We will describe these five basic modes first, and then present a few additional modes that facilitate some programming tasks.

Absolute mode. The address of the operand is given in the instruction; this is the addressing mode that we have used in most examples so far. In an instruction such as

Move   A,B

the names A and B stand for numbers that are the addresses of the main memory locations allocated to the two variables. The locations called A and B are represented by numbers that are in fact their addresses.

Register mode. This is the same as the absolute mode, except that the named location is a CPU register. This mode is useful when dealing with temporary variables, such as the intermediate results in a computation. An example of this mode is our previous usage of the accumulator, ACCUM, which is likely to be a register in the CPU.

Consider the instruction

Move   M,ACCUM

This instruction copies the contents of memory location M into the accumulator. Similarly,

Move   ACCUM,M

copies the contents of the accumulator into memory location M. When the CPU contains only one register that can be used as an accumulator, it is not necessary to specify the accumulator explicitly as one of the operands. Instead, it may be specified implicitly in the name of the instruction itself. For example, the two move instructions mentioned above may be written as:

Load_accum   M

and

Store_accum   M

The Load_accum instruction indicates that ACCUM is to be loaded with the contents of location M, an operation that we might state in symbolic notation as

$$ACCUM \leftarrow [M]$$

Similarly, the Store_accum instruction moves the contents of ACCUM into location M, i.e.,

$$M \leftarrow [ACCUM]$$

Because the register is referenced implicitly by the instruction, the register mode is sometimes called the *implied mode* in manufacturers' literature. Modern microprocessors contain several registers that may be used as operand locations. They may be dedicated to certain functions, such as the accumulators or the index registers (see Indexed Mode, below). Their contents can be manipulated by instructions provided specifically for that purpose.

Immediate mode. The operand itself is given in the instruction. This is perhaps the simplest addressing mode. In Figure 2.7 we used the instruction

Move   #n,NCOUNT

with the intent of placing the value *n* into location NCOUNT. This instruction involves two operands. The source is the number *n* which is stated explicitly within the instruction, i.e., it is given in the immediate mode. The destination NCOUNT is specified using the absolute mode. It is customary to use a # sign to specify an immediate operand.

Indexed mode. In this mode, the address of the operand is computed as the sum of the contents of a register, referred to as an *index register*, and a specified constant. This mode is very convenient for accessing data entries in lists. An instruction using the indexed mode must specify a register to be used as the index register, R, as well as an *index value*, X, that is, the required constant. Then,

$$A_{eff} = X + [R]$$

The index register R may be a specific register in the CPU designated for this purpose. However, in many computers, R can be one of several registers that may also be used for other functions.

   An illustration of the indexed mode is given in Figure 2.8. Consider the instruction

Clear   180(R)

Assume that the index register, R, contains the number 300. The index value is equal to 180. Thus, this address specification denotes memory location 480, which contains the desired operand that will be cleared to 0.

   In the example of Figure 2.5 we used the instruction

Move   A(NCOUNT),B(NCOUNT)

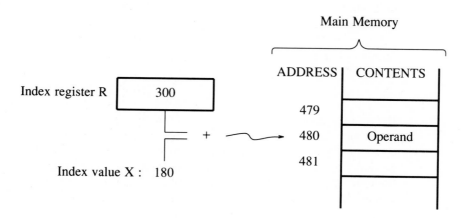

Figure 2.8   An example of indexed addressing mode.

To be consistent with the definition given above, NCOUNT must be an index register. This is convenient because the value of NCOUNT changes each time through the loop, while the addresses of the lists A and B remain constant. Both operands are specified in the indexed mode. For the source operand, A can be specified as the index value X, which is the address of the first entry in list A. The index register, R, should contain the number that corresponds to NCOUNT, which indicates the distance of the source operand from the beginning of the list. The same scheme is applicable to the destination operand. Note that one index register may be used for both operands in this case, because the entries in question are in the same location relative to the beginning of their respective lists. In order to access different entries it is only necessary to change the contents of the index register as required.

Indirect mode. In this mode the effective address of the operand is the contents of a register or a memory location whose address is given in the instruction. This mode is usually denoted by square brackets. For example, in Figure 2.7 we used the notion of pointers to step through the entries of a list. We used the instruction

Move   [POINTERA],[POINTERB]

to move the source operand in the memory location pointed at by POINTERA to the location pointed at by POINTERB. It is important to realize that POINTERA and POINTERB are the names of either registers in the CPU or of locations in the main memory. If POINTERA denotes a register, then the CPU uses the contents of that register as A_{eff}. However, if POINTERA denotes a memory location, then the CPU must access the main memory twice. First, it reads the contents of the location specified by the name POINTERA. Then, it reads the operand, using the contents of location POINTERA as the effective address.

The square brackets used in the indirect mode specify a general addressing mechanism called "indirection." This mechanism can be combined with most other addressing modes, at least conceptually, to generate more powerful modes. When indirection is combined with another mode it means that the effective address computed for that mode should not be used as the address of the operand; instead it should be used as the address of the memory location that contains the address of the operand. For example, indirection may be combined with the indexed mode. Recall that in the indexed mode we have

$$A_{eff\,(indexed)} = X + [R]$$

In the indexed indirect mode the effective address is given by

$$A_{eff\,(indexed\ indirect)} = [X + [R]]$$

As a numerical example consider the values given in Figure 2.9. The index value is X = 180. The index register R contains the number 300. Then, the effective address for the indexed mode is computed as

$$A_{eff\,(indexed)} = 180 + 300 = 480$$

In the indexed indirect mode, the contents of location 480 are interpreted as the address of the operand. Therefore

$$A_{eff\ (indexed\ indirect\,)} = 510$$

That is, the desired operand in this case is the contents of location 510.

In addition to the basic modes described above, one encounters several other addressing modes that perform more specialized addressing functions, which are needed in certain situations. We will now consider three additional addressing modes.

Autoincrement mode. This mode is often useful when a register is used as a pointer to a list. It is particularly important for the implementation of stacks, as will be explained in Section 2.6. Let the operand specification

$$[R]_{autoinc}$$

denote the autoincrement mode using register R. It means that the operand is pointed at by the register R. That is, the contents of register R are the address of the operand. After the operation specified by the instruction is performed, the contents of register R are incremented automatically. Thus, the autoincrement mode may be described symbolically as

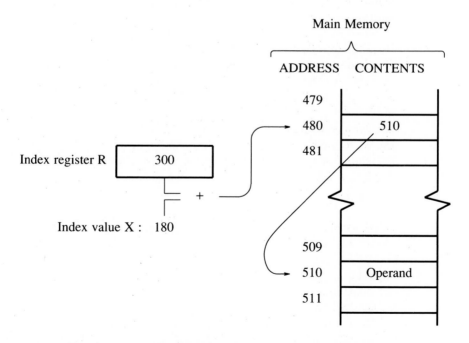

Figure 2.9 An example of indexed indirect addressing mode.

$$A_{eff} = [R]$$
$$R \leftarrow [R] + 1$$

In the example of Figure 2.7, we used pointers to access successive entries. They were incremented explicitly within the instruction loop using the instructions

Move	[POINTERA],[POINTERB]
Increment	POINTERA
Increment	POINTERB

Assuming that POINTERA and POINTERB are CPU registers, the three instructions may be replaced with the single instruction

Move $[POINTERA]_{autoinc}$, $[POINTERB]_{autoinc}$

Figure 2.10 shows how the autoincrement mode can be used to implement the program of Figure 2.7. The starting addresses of lists A and B are assumed to be known as numbers ADDRESSA and ADDRESSB. Hence, they can be specified in the immediate mode. The counter NCOUNT is a memory location and is addressed using the absolute mode. It should be noted that a CPU register could be used for the same purpose, if one is available. The counter is initially set to the value n using the immediate mode.

Autodecrement mode. This is a companion addressing mode to the autoincrement mode. Let us write an operand specified in the autodecrement mode as

$$[R]_{autodec}$$

The pointer register R is first decremented and then its contents are used as the address of the operand, i.e.,

$$R \leftarrow [R] - 1$$
$$A_{eff} = [R]$$

Observe that the contents of the pointer register are updated prior to the determination of A_{eff} in this mode, while they are updated after A_{eff} is determined in the autoincrement mode. The reason for this is related to the way in which these two modes are used to implement a special data structure called a stack, as will be explained in Section 2.6.

Relative mode. In this mode the location of an operand is given relative to the location of the instruction that references the operand. That is, the location of the operand is specified by giving the distance, or offset, between the operand and the instruction.

Sometimes the programmer does not know the location in which a program and its data will be placed in the main memory. In such a case, it is not possible to use the absolute mode to refer to the data items. However, it is possible to use the relative addressing mode, provided that the relative position of the program and the data remains unchanged.

```
              Move          #n,NCOUNT
              Move          #ADDRESSA,POINTERA
              Move          #ADDRESSB,POINTERB
     LOOP     Move          [POINTERA]autoinc,[POINTERB]autoinc
              Decrement     NCOUNT
              Branch_if_≥0  LOOP
              Halt
```

Figure 2.10 A possible program that corresponds to Figure 2.7 if autoincrement addressing is used.

The relative mode is exactly the same as the indexed mode, with the program counter (PC) being used as an index register. In this case, the index value X denotes the distance between the required operand and the current contents of the PC. Symbolically, this means

$$A_{eff} = X + [PC]$$

The addressing modes are summarized in Table 2.1. The indirect mode is included as one of the modes. We should note that the idea of indirection can be applied on a wider scale. Whenever indirection is used, the effective address of an operand is not specified directly in the instruction. Instead, the effective address is the contents of a register or a memory location specified in the instruction. We have discussed the indirect versions of the absolute and the indexed modes, which one may refer to as the "absolute indirect" and the "indexed indirect" modes. Similarly, it is possible to use the "relative indirect" mode. The autoincrement and autodecrement modes are, in fact, indirect modes, because the instruction specifies a register the contents of which become the effective address of the operand. It is also possible to impose a further level of indirection, in which case the resulting addressing modes may be called "autoincrement indirect" and "autodecrement indirect," respectively.

We have considered the addressing modes from a general point of view. The actual modes provided in commercial microprocessors differ from one product to another. In Chapters 3 and 4 we will present the details of the addressing modes in Motorola's 6809 and 68000 microprocessors.

2.4 INPUT AND OUTPUT TASKS

Two of a computer's essential tasks are to acquire data from input devices and to send data to output devices. We introduced the basic ideas for implementation of I/O operations in Section 1.5. One of the suggested possibilities is to provide special registers in the interface of the I/O device, which can be accessed by the CPU as if they were locations in the main memory. A full discussion of I/O techniques will be given in Chapters 5, 6, and 7. However, at this point, having considered some of the essential aspects of machine-level programming, it is useful to see how simple I/O transfers may be realized. Knowledge of such I/O techniques will enable the user to perform interesting exercises on a computer system.

Table 2.1   ADDRESSING MODES

Absolute	Address of the operand is stated explicitly within the instruction.
Register	The operand location is a CPU register specified in the instruction.
Immediate	Value of the operand is stated explicitly within the instruction.
Indexed	Address of the operand is the sum of the contents of an index register and an index value X, where both the register and the value X are specified in the instruction.
Indirect	The effective address of the operand is the contents of a register or a memory location, whose address is given in the instruction.
Autoincrement	Address of the operand is the contents of a register R. After the operand is fetched, the value in R is incremented to point to the next operand location.
Autodecrement	Address of the operand is the contents of a register R. Prior to the fetching of the operand, the value in R is decremented.
Relative	Address of the operand is specified relative to the current contents of the program counter. A value X is given in the instruction, which when added to the contents of the program counter yields the address of the operand.

A video terminal is one of the most useful I/O devices. It allows entering the data into the computer by means of a keyboard, and displays output information on the screen. Let us consider the programming considerations for these functions in more detail. Assume that there exist three registers in the I/O interface that connects the terminal to the computer bus. An input buffer, INBUF, is used to hold a byte of data that corresponds to the key pressed on the keyboard, encoded in the standard ASCII code, which is described in Appendix A. An output buffer, OUTBUF, is used to accept an ASCII-encoded character that is to be displayed on the screen. The third register, called STATUS, indicates the status of the input and output sections of the terminal. Let its least-significant bit, b_0, be set to 1 if there is valid data in INBUF and cleared to 0 if INBUF is empty. Similarly, let b_1 indicate whether or not the display buffer, OUTBUF, is ready to receive a character. Figure 2.11 summarizes the format of these registers. We should note that other registers are usually included in an I/O interface of this kind, but we will not discuss their details until Chapter 5.

Let us associate the names INBUF, OUTBUF, and STATUS with the memory addresses of the I/O registers of the terminal interface mentioned above. Using the program-controlled I/O scheme, I/O transfers can be performed by testing the relevant status bit and transferring the data when the required buffer register is ready. Figure 2.12 shows a routine that transfers one character from the keyboard buffer to the CPU register ACCUM. The program tests the status bit, b_0, repeatedly. While this bit is equal to 0, the input buffer is empty and testing continues. As soon as a key is pressed on the keyboard, b_0 changes to 1. As a result the condition specified in the Branch instruction will be false, and the Move instruction that transfers the character from the buffer to ACCUM will be executed. The status bit is tested after the entire contents of the STATUS register are first loaded into ACCUM. The instruction

$$\text{AND} \qquad \text{\#1,ACCUM}$$

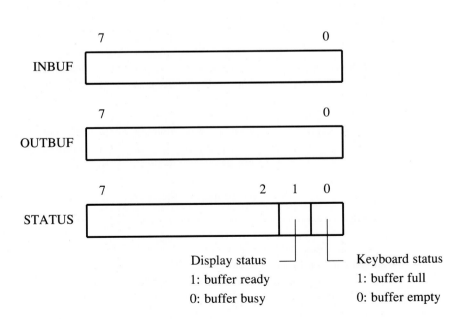

Figure 2.11   Registers in the I/O interface of a video terminal.

performs a logical AND operation on corresponding bits of the two operands. Since the first operand is the constant 1, specified in the immediate addressing mode, the only bit in the result that will not necessarily be equal to 0 is the least-significant bit. This bit will, in fact, be equal to the value of the least-significant bit in ACCUM, which in turn corresponds to the keyboard status.

The display operation is performed in the same manner. A single character can be displayed by transferring it to OUTBUF with a program analogous to that in Figure 2.12. Let us consider a slightly more difficult task of displaying one line of characters. Assume that there is a one-line message stored in the main memory beginning at address MESSAGE. The message is to be transferred to OUTBUF, one character at a time. The end of the message is denoted by the "carriage return" (CR) character.

Figure 2.13 shows a program that displays a one-line message. A pointer register, POINTER, is used to indicate the memory address of the next character to be transferred to the display. This pointer is incremented, using the autoincrement addressing mode, each time a character is moved from the memory to OUTBUF. A wait loop checks the status of the display, as signified by bit b_1 of the STATUS register. Observe that this bit is isolated from the rest of the bits read from the STATUS register by ANDing the status word with the constant 2 (=00000010).

These examples illustrate that the program-controlled I/O scheme is simple to implement. It can be used with a number of different I/O devices. For example, if more than one video terminal is connected to a computer, it is possible to check for the activity on their keyboards simply by implementing a wait loop within which the status registers of all terminals are read in turn, until one of them is found to have data ready for transfer.

```
WAIT    Move            STATUS,ACCUM    Read status.
        AND             #1,ACCUM        Check bit 0.
        Branch_if_=0    WAIT            Loop if no character ready.
        Move            INBUF,ACCUM     Transfer data.
        Halt
```

Figure 2.12 A program that reads one character from a keyboard.

```
        Move            #MESSAGE,POINTER        Initialize pointer.
WAIT    Move            STATUS,ACCUM            Wait for display to
        AND             #2,ACCUM                    become ready.
        Branch_if_=0    WAIT
        Move            [POINTER],OUTBUF        Transfer one character.
        Compare         #CR,[POINTER]autoinc    Continue displaying
        Branch_if_≠0    WAIT                        if not CR.
        Halt
```

Figure 2.13 A program to display a one-line message.

The key concept in program-controlled I/O is to test the status of the I/O device continuously within a "wait" loop, to determine whether or not the data transfer can be made. While the relevant status bit is found to indicate that the device is not yet ready for data transfer, the execution of the wait loop continues. This method of handling I/O transfers occupies the CPU fully with the execution of the wait loop. When dealing with slow devices, such as keyboards whose data transfer rates are limited by the typing speed of the human operator, the instructions in the wait loop will be executed many times for each character transferred from INBUF to the accumulator. We will see in Chapter 7 that there are other techniques, such as the interrupt-driven I/O, where most of the time "wasted" in the wait loop can be put to better use. With such techniques the CPU may perform some other task while waiting for the data in INBUF to become available.

2.5 SUBROUTINES

In digital processing applications it is often necessary to perform a given task several times. If such a task is realized by executing a particular program segment, then this program segment has to be executed whenever the task is to be performed. Thus, the program must include the required segment wherever the corresponding task is needed. It is a poor idea to include identical versions of a given segment in many places of a program, as this unnecessarily wastes the storage space in the main memory. Instead, the segment may be stored in one place in memory. Then, it can be caused to be executed at different points in the program, as needed. A program segment used in this manner is called a *subroutine*.

Since the instructions that constitute a subroutine are not physically inserted in the required places of the main program, it is necessary to provide a mechanism for transferring the execution flow to and from the subroutine. The main program, usually referred to as the *calling program,* includes a special branchlike instruction that invokes execution of the first instruction in the subroutine. This instruction may be appropriately named the Call_subroutine, or simply the Call, instruction. Instructions within the subroutine are executed in the specified sequence, until the last instruction is completed. At this point execution must be returned to the calling program. This is achieved with a Return_from_subroutine instruction.

Figure 2.14 illustrates the desired changes in the flow of execution. It shows a case where one subroutine is needed in two different places of the calling program. The Call instruction is a branchlike instruction, because it transfers execution flow to a unique location, namely, the beginning of the subroutine. But, at the end of the subroutine the return to the calling program cannot be achieved with a simple branchlike instruction, because the location to which the return is made is not unique. This is due to the fact that a subroutine can be called from many different places in the calling program.

In order to return from the subroutine correctly, it is necessary to save the *return address* before the subroutine is executed. This is the address of the instruction that immediately follows the Call instruction in the calling program. Recall that the contents of the program counter (PC) in the CPU always point at the instruction to be

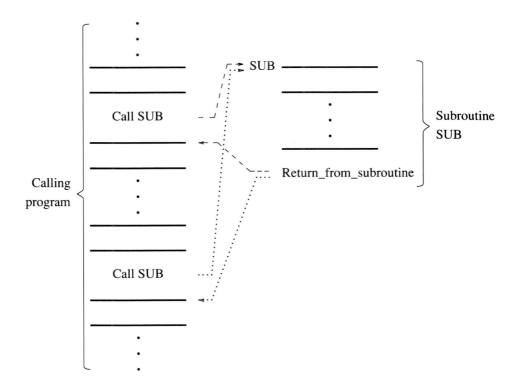

Figure 2.14 Execution flow when a subroutine is used.

executed next. As each instruction is fetched from the main memory and brought into the CPU for execution, the PC is updated to point at the next instruction that will be fetched. Thus, when a Call instruction is being executed the PC contains the address of the next instruction that follows in the calling program. This is the required return address that must be saved before branching to the subroutine. After saving the return address, a branch to the subroutine is made simply by loading the address of the first instruction of the subroutine into the PC.

Where should the return address be stored? One might assume that all that is needed is one register in the CPU which is used exclusively for this purpose. Indeed, this simple scheme will suffice if only one return address is to be saved at any given time. It will be shown later that this is not enough. However, before examining possible alternatives, it is necessary to discuss a related problem.

Any subroutine is likely to need some data to work on. This data, or possibly information about its location, must be passed by the calling program to the subroutine. Data items that pass from the calling program to the subroutine are called *parameters*. When only a few parameters have to be passed, it is possible to do so through CPU registers or some designated memory locations. If the subroutine expects to find the parameters in certain registers and locations, then the calling program has to load the required parameters into these registers and locations before calling the subroutine. The results from the subroutine, if any, can be passed to the calling program in a similar manner.

As a simple example, consider again the program in Figure 2.7. The loop that actually copies the list can be implemented in the form of a subroutine, as shown in Figure 2.15. The calling program must pass three parameters to this subroutine, namely, the length and locations of the two lists. This is done by loading the number of entries into a memory location, NCOUNT, and the parameters into pointer registers POINTERA and POINTERB. These parameters are then used by the instructions in the subroutine to perform the required task.

Consider another simple example—a subroutine that finds the largest number in a given list of numbers. Assume that there are n numbers in the list whose starting location is given by LISTADDRESS. Also, assume that the CPU has three registers that can be used in this task, namely, an accumulator, ACCUM, and two pointer registers, POINTERA and POINTERB. Any parameters that need to be passed between the calling program and the subroutine will be passed via these registers.

Figure 2.16 shows a possible program, consisting of a calling program and a subroutine MAX. The calling program computes and places the addresses of the first and the last entries in the list into registers POINTERA and POINTERB, respectively. The value n is assumed to be available in memory location SIZE. The subroutine uses POINTERA to step through the list. At any stage of the search, it keeps the maximum value found in ACCUM, and the address of the corresponding entry in the list in MAXLOC. Note that MAXLOC may be either another register in the CPU or a main memory location. In either case, MAXLOC is known only to the subroutine, and not to the calling program. Hence, before returning to the calling program, the results must be placed in the registers used for parameter passing. The maximum value is left in ACCUM, while its address is transferred from MAXLOC to POINTERA.

```
                Move            #n,NCOUNT                    Calling
                Move            #ADDRESSA,POINTERA           program
                Move            #ADDRESSB,POINTERB
                Call            SUBMOVE
                next instruction


                                  .
                                  .
                                  .


SUBMOVE         Move            [POINTERA],[POINTERB]        Subroutine
                Increment       POINTERA
                Increment       POINTERB
                Decrement       NCOUNT
                Branch_if_>0    SUBMOVE
                Return_from_subroutine
```

Figure 2.15 *Example of Figure 2.7 using a subroutine.*

Move	#LISTADDRESS,POINTERA	Pointer A points to first entry.
Move	SIZE,POINTERB	Pointer B points to the
Add	POINTERA,POINTERB	location that follows the
Call	MAX	last entry in the list.
next instruction		

$$\vdots$$

MAX	Move	POINTERA,MAXLOC	Address of max value in MAXLOC.
	Move	[POINTERA],ACCUM	Max value is in accumulator.
LOOP	Compare	ACCUM,[POINTERA]	Compare with next entry.
	Branch_if_$\geq$0	NEXT	
	Move	POINTERA,MAXLOC	Update address and value if
	Move	[POINTERA],ACCUM	larger number is found.
NEXT	Increment	POINTERA	Go to next entry.
	Compare	POINTERA,POINTERB	Check for the end of list,
	Branch_if_$<$0	LOOP	and loop back if not done.
	Move	MAXLOC,POINTERA	Address of max number is
	Return_from_subroutine		in pointer A.

Figure 2.16 A subroutine for finding the largest number in a list.

Next, let us consider the more complex task of sorting a list of numbers in descending order. There are many ways of accomplishing this task. Since we wish to examine certain aspects of using multiple subroutines, we will make use of subroutine MAX of Figure 2.16. The list will be traversed from top (first entry) to bottom. In each pass through the list, subroutine MAX will be used to find the largest number in the unsorted sublist. Then, this largest number will be interchanged with the top number in the sublist. Thus, after the first pass, the largest number will be placed at the top of the list while the remaining entries are the unsorted sublist. After the second pass, the top two items are guaranteed to be correctly sorted, and so on. After $n-1$ passes, the entire list will be sorted.

A possible program is given in Figure 2.17. The calling program calls subroutine SORT, and passes to it the location and size of the list as parameters. Subroutine SORT calls subroutine MAX to find the largest number in a sublist. The starting location of the sublist is saved in STARTLOC before subroutine MAX is called, because the contents of POINTERA are altered by the subroutine. Upon return from MAX, the subroutine SORT exchanges the largest number found with the entry at the top of the sublist whose address is in STARTLOC. Note that a temporary location is used in this interchange, because neither entry may be destroyed in the process. Also, observe that no parameters need to be passed from SORT to the calling program, because reordering of the actual list entries is done within the subroutine.

The above example demonstrates the possibility of a subroutine calling another subroutine, a scheme referred to as *subroutine nesting*. This raises some questions about our assumptions with respect to the way in which the return address is saved.

```
              Move           #LISTADDRESS,POINTERA      Initialize pointers to the
              Move           SIZE,POINTERB              beginning and the end
              Add            POINTERA,POINTERB          of the list.
              Call           SORT
              next instruction

                  .
                  .
                  .

SORT          Move           POINTERA,STARTLOC          Store top of sublist.
SLOOP         Call           MAX                        Call subroutine in Fig. 2.16.
              Move           [POINTERA],TEMP            Swap entry found in MAX
              Move           [STARTLOC],[POINTERA]        with entry at the top of
              Move           TEMP,[STARTLOC]            sublist.
              Increment      STARTLOC                   Go to next entry; as top
              Move           STARTLOC,POINTERA            of next sublist.
              Compare        POINTERA,POINTERB          Check for the end of list
              Branch_if_<0   SLOOP                        and loop back if not done.
              Return_from_subroutine
MAX           ...                                       Subroutine in Figure 2.16.
```

Figure 2.17 A sorting program that uses nested subroutines.

If the return address when SORT is called is to be stored in some dedicated CPU register, then where should the return address for MAX be stored? Both return addresses must be remembered. Providing two CPU registers for this purpose is clearly not a good idea, because the number of times subroutines may be nested would be restricted by the number of such registers. A much better solution is to store return addresses in the main memory. Also, it must be possible to store them in such a way that the last return address stored will be the first address retrieved, to correspond to the nesting mechanism. Implementation of such an arrangement will be discussed in the next section.

2.6 STACKS

Whenever a collection of data items is organized in a certain way, the organization is referred to as a *data structure*. A simple list, or an array, is one example of a commonly used data structure. Section 2.5 suggested a need for a data structure for storing data items such that the last item stored will always be the first item to be removed from storage. To depict the nature of such a structure we may draw an analogy with a stack of plates in a cafeteria, where each plate is an item that is stored by placing it on top of an existing stack. The plate removed first is the one at the top of the stack. Appropriately enough, an equivalent data structure is called a *stack*.

How can a stack be implemented in a computer? It is possible to include special hardware, perhaps within the CPU, that realizes the stack directly. However, there is also a less costly alternative which makes use of the main memory. This possibility is exploited in most microcomputers and will be the only one considered here.

A stack may be implemented in the main memory by storing data items in consecutive memory locations, as in the case of a simple list, but with the proviso that new entries can be added only at the end of the list. Moreover, data items can be removed only from the same end of the list. This means that the position of the end of the list will be constantly changing, as new items are added or stored items are deleted. The memory address of the end of such a list is called the *top of the stack*. This address must be known at all times. It is usually kept in a CPU register called a *stack pointer* (SP). The beginning address of the list is sometimes referred to as the bottom of the stack. A stack may be arranged to grow in either the direction of ascending or descending addresses. The latter is the choice adopted in most microprocessors, hence it is the one we will use.

Figure 2.18 illustrates the organization of a stack in the main memory. The stack is assumed to start at memory address 8400. The figure depicts a situation where there are five data items stored in the stack. The stack pointer is pointing at the top item, at location 8396.

The operations of adding items to and deleting items from a stack are known as *push* and *pull* (or *pop*) operations. In many microprocessors there exist specific

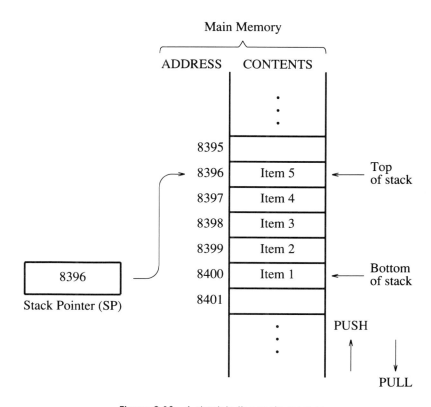

Figure 2.18 A stack in the main memory.

machine instructions that realize these operations. A Push instruction updates the contents of the SP to extend the size of the stack. Then, it places the new item at the top of the stack. For example, an item is pushed on the stack in Figure 2.18 by changing the value in SP to 8395 and storing the item in the corresponding memory location. A Pull instruction removes an item from the stack and updates the contents of the SP. If item 5 is pulled from the stack in Figure 2.18, the value in SP is incremented to 8397 to point at the new top of the stack.

A stack structure may be implemented even when specific Push and Pull instructions are not available. The desired operations can be readily achieved using the autoincrement and autodecrement addressing modes. Assume that the SP is a register with which these modes can be used. Then, the push operation is equivalent to

$$\text{Move} \qquad \text{NEWITEM}, [\text{SP}]_{autodec}$$

Recall that in the autodecrement mode the SP is first decremented, and then its contents are used as the memory address at which NEWITEM is stored. This is exactly the sequence of operations needed to add NEWITEM above existing items in the stack. Similarly, the pull operation becomes

$$\text{Move} \qquad [\text{SP}]_{autoinc}, \text{ITEMDEST}$$

where the item is moved from the stack to its destination location. Then, the SP is incremented to point at the top of the now shortened stack.

In order to ensure correct operation, the stack pointer should not be used for any other purposes that might result in changes in its contents. The stack should be thought of as temporary storage where every data item that is stored is later removed. The order in which items are added and removed is such that the last item added is the first item removed. Hence, the structure is also called a *last-in-first-out* (LIFO) list.

It is essential that no data item be accidentally left forgotten on the stack, as this would ruin the order of access to the remaining items. Note that deletion of one or more items from the top of the stack does not imply that they are physically erased from the main memory. It is only necessary to increment the SP so that the deleted items are no longer included within the stack. For example, items 5 and 4 in the stack of Figure 2.18 can be deleted simply by changing the contents of the SP to 8398, which makes item 3 the top element of the stack. Of course, the values of these items left in the memory will be overwritten when new items are pushed on the stack in the same locations.

2.7 USE OF STACKS FOR SUBROUTINE LINKAGE

In Section 2.5 it was pointed out that the use of subroutines requires the ability to store the return addresses temporarily. Stacks provide a most suitable mechanism for this, which is implemented in most computers. Recall that when a subroutine is called, that is when a Call instruction is executed, the program counter contains the required return address. Hence, the process of executing the Call instruction should consist of

1. Pushing the contents of the PC on the stack, and
2. Loading the address of the first instruction of the subroutine in the PC.

When the execution of the subroutine has been completed, a return to the calling program is made by executing a Return_from_subroutine instruction. This instruction pulls the top entry from the stack and loads it into the PC.

An example involving a calling program and two nested subroutines is given in Figure 2.19. Only the Call and Return instructions are shown explicitly, because they govern the execution flow. Greek letters are used to relate the state of the stack, which is shown near the bottom of the figure, to various places in the program during execution. In each case the position of the top of the stack is shown at the time when the execution of the instruction in question is completed. Just prior to calling subroutine SUB1 the top of the stack is at memory address 8400. This is indicated as point α. After completion of the Call SUB1 instruction—point β—the return address, 1002, is found stored at the top of the stack. Note that the top of the stack is now at location 8399, because we have assumed that the stack grows in the direction of descending addresses. Within SUB1 there is a call for subroutine SUB2, labeled as point γ. The next instruction in SUB1 is at location 1504, which is the return address placed on the stack before the execution of SUB2 is begun. Upon completion of SUB2, the return to SUB1 is achieved by loading the return address from location 8398 on the stack into the PC. This event, labeled δ, brings the top of the stack to address 8399. Finally, the return from SUB1 to the calling program, denoted as ε, restores the stack to its original size. It is clear that any number of nested subroutines can be handled in this manner.

Until now we have used one simple method for passing parameters to subroutines, namely, by means of registers or fixed memory locations. There are two other schemes commonly used for this purpose.

In the first scheme, the parameters are placed in the memory locations that immediately follow the Call instruction. This means that the location of the parameters is denoted by the return address determined when the Call instruction is executed. Using this scheme, the program of Figure 2.15 can be rewritten as shown in Figure 2.20. The calling program is arranged so that all three parameters are placed after the Call instruction. Upon entering the subroutine, the parameters are loaded into CPU registers POINTERA and POINTERB and memory location NCOUNT. Since the parameters are given as a list of entries, they can be accessed easily if the address of the first parameter in the list is available in some pointer register. POINTERC is used for this purpose. Thus, the first instruction of the subroutine loads POINTERC with the "return address" from the top of the stack. Then, POINTERC is used to access the parameters. POINTERC is incremented after each access to point to the next parameter. After the third parameter is dealt with, the updated POINTERC contains the true return address that must replace the value stored on the top of the stack. The rest of the subroutine is the same as in Figure 2.15.

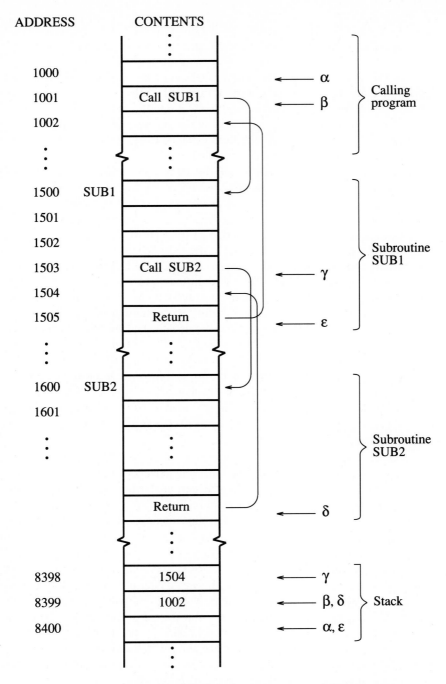

Figure 2.19 Execution flow with nested subroutines.

	Call	SUBMOVE	
	n		Parameters that are
	ADDRESSA		passed to the subroutine.
	ADDRESSB		
	next instruction		
	$\vdots$		
SUBMOVE	Move	[SP],POINTERC	Pointer C points to parameters.
	Move	[POINTERC],NCOUNT	Load parameters
	Increment	POINTERC	
	Move	[POINTERC],POINTERA	into their respective
	Increment	POINTERC	
	Move	[POINTERC],POINTERB	CPU registers.
	Increment	POINTERC	
	Move	POINTERC,[SP]	Correct return address on stack.
LOOP	Move	[POINTERA],[POINTERB]	Copy list entries.
	Increment	POINTERA	
	Increment	POINTERB	
	Decrement	NCOUNT	
	Branch_if_>0	LOOP	
	Return_from_subroutine		

Figure 2.20 Passing parameters after the Call instruction.

For the sake of clarity, separate Increment instructions have been used in the example program in Figure 2.20. The reader should recall that these instructions can be eliminated if we make use of the autoincrement addressing mode, as discussed in Section 2.3.

It is also possible to pass parameters to a subroutine by having the calling program place them on the stack. This can be done as indicated in Figure 2.21(a). The parameters are pushed on the stack before the subroutine is called. Execution of the Call instruction causes the return address to be stored above the parameters. When the subroutine is entered, the top of the stack is as shown in Figure 2.21(b). The parameters may be read from the stack, without disturbing the contents of the SP, and loaded into CPU registers. In this case they can be accessed using the indexed mode, because the subroutine knows the offset for each parameter from the top of the stack. A complication in this scheme is that the parameters cannot be removed from the stack as long as the return address is needed at the top of the stack. Thus, upon return to the calling program, the SP (which is then pointing to the last parameter stored on the stack) must be updated to remove the parameters, which are no longer needed. In our example, this is accomplished by adding 3 to the SP, because there are three parameters and each is assumed to occupy one addressable location in the memory.

A subroutine that copies a list of values from one memory location to another performs the copying function directly in the specified memory locations. It computes no other results that should be passed to the calling program. However, it is

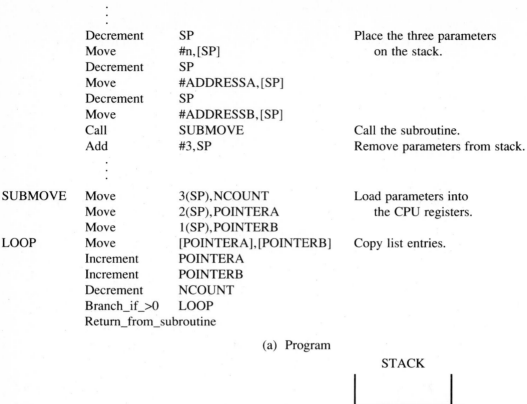

.
.

Decrement	SP	Place the three parameters
Move	#n,[SP]	on the stack.
Decrement	SP	
Move	#ADDRESSA,[SP]	
Decrement	SP	
Move	#ADDRESSB,[SP]	
Call	SUBMOVE	Call the subroutine.
Add	#3,SP	Remove parameters from stack.

.
.

SUBMOVE	Move	3(SP),NCOUNT	Load parameters into
	Move	2(SP),POINTERA	the CPU registers.
	Move	1(SP),POINTERB	
LOOP	Move	[POINTERA],[POINTERB]	Copy list entries.
	Increment	POINTERA	
	Increment	POINTERB	
	Decrement	NCOUNT	
	Branch_if_>0	LOOP	
	Return_from_subroutine		

(a) Program

STACK

SP

Return address

Address of B

Address of A

n

(b) Top of the stack when SUBMOVE is entered.

Figure 2.21   Passing parameters via the stack.

often the case that some results generated in a subroutine have to be passed to the calling program. This can be done by placing the results in CPU registers or in predetermined memory locations. An example of this is found in the program of Figure 2.16. Subroutine MAX determines the address of the largest number in a list, and passes it to the calling program by placing it in POINTERA. Another commonly used alternative is to place the results in the parameter space on the stack, that is, below the top location which contains the return address. Again, the process is similar to the scheme for passing parameters to the subroutine via the stack.

The entire mechanism of controlling the execution flow and parameter passing is referred to as *subroutine linkage*. The details of subroutine linkage vary from one computer to another. However, the mechanisms discussed above are quite general, and are available on most microprocessors.

2.8 EXAMPLE OF A LINKED LIST

The examples in the previous sections have illustrated the concepts of program flow control and the need for a variety of addressing modes. We have assumed that data is structured in the form of arrays or lists. This simple data structure is readily understood and easy to use. In this section we will consider a more complex list structure as another example, to provide further illustration of the previously espoused principles. Simple lists are stored in the main memory in a manner where successive items occupy contiguous memory locations. The simplicity and the ease of use of array structures are obvious advantages of this method of storing data. However, there are cases where the array structure is inconvenient. For example, suppose that items in the array are to be kept in a certain order, say alphabetically. Adding an item at the end of the list is quite simple. But, if an item is to be inserted somewhere in the middle of the list, then the task becomes more complicated. Let the list have n items and assume that the new item is to be placed in location i, where $1 < i < n$. First, it is necessary to move the items in locations i through n into locations $i+1$ through $n+1$. Then, the new item can be placed into the vacated location i. Such tasks can be made simpler if a differently structured list is used.

Instead of storing list items in contiguous locations, let us allow different items to be stored anywhere in the main memory. Let each item in the list include a number that indicates the location of the next item in the list. This number may be simply the address of the next item. Since it provides the "linking information" needed to find the next item, we refer to it as a *link*. Now, each item in the list consists of a link and some data. A list of this type is called a *linked list*. The data component of a list item may consist of one or more numerical or alphanumeric quantities. For example, in a linked list that contains the marks for students in a given course, each item may consist of the link, the name of the student, and four integers corresponding to the student's marks in laboratory work, the term test and the final examination, as well as the overall grade. Figure 2.22 gives a graphical illustration of such a linked list. Part (a) shows a list that is assumed to be sorted alphabetically according to student names. Each link is assumed to be the address of the next item in the list. The last item, in position n, must not point to any more items. This is done by making the link equal to a special "null" code.

A new item to be added to the list may be stored anywhere in the main memory. It is inserted into the linked list as shown in Figure 2.22(b). Here, an entry pertaining to Frank's marks is inserted in a proper alphabetical place. This requires changing the link in the item that corresponds to Dan, which must now point to Frank rather than George. The link of the inserted item should point to George. No other item in the list needs to be changed as a result of this operation. Deletion of an item from a linked list is also very easy. If Dan is to be deleted from the list in

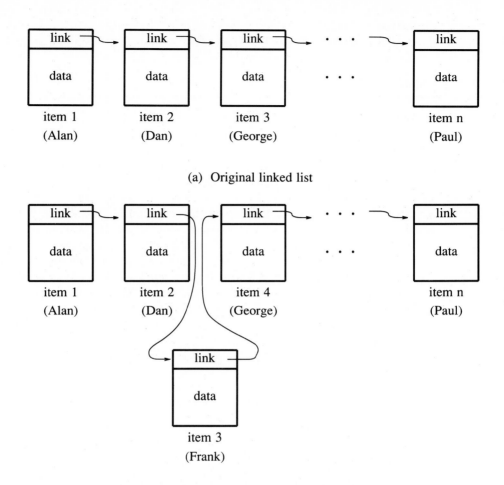

(a) Original linked list

(b) Insertion of an item into a linked list

Figure 2.22   Linked-list structure.

Figure 2.22(a), it is only necessary to change the link in Alan's entry to be the address of George rather than Dan.

The key advantage of the linked list stems from the fact that successive items need not be stored in contiguous memory locations. They can be stored in any available locations. However, all entries belonging to a given item should be stored in consecutive locations, so that they can be identified in relation to the link entry of the item. Figure 2.23 shows a valid arrangement for the example in Figure 2.22(b). The linked list starts at location 100, which is called the "head" of the list.

In order to access any item in a linked list, one must always start at the head of the list. The link value of the first item gives the address of the second item. One then reads the link value of the second item to find the third item, and so on. Thus, an obvious disadvantage of the linked-list structure is that random references to individual items cannot be made as simply as in array structures. In order to find the kth item, it is necessary to start at the head of the list and proceed sequentially through the links until the kth item is reached.

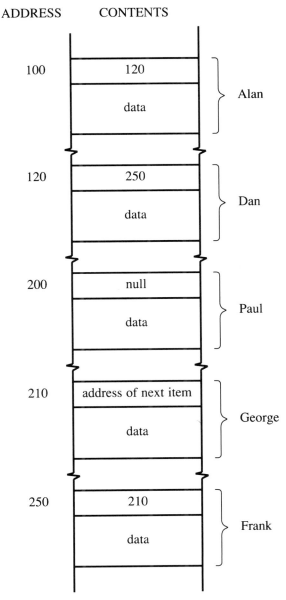

ADDRESS   CONTENTS

100	120	} Alan
	data	
120	250	} Dan
	data	
200	null	} Paul
	data	
210	address of next item	} George
	data	
250	210	} Frank
	data	

Figure 2.23 A possible storage arrangement for the list in Figure 2.22(b).

Let us return to the example of a linked list that contains the marks of students in a given class. Assume that each item in the list occupies 15 consecutive words in the memory, arranged as shown in Figure 2.24.

Suppose that we wish to find the highest final mark attained in the class. We can adopt the same strategy as in the example of Figure 2.16. A possible program is given in Figure 2.25. A pointer register, LINKPOINTR, is used to scan through the items of the list. It is always loaded with the address of the item being considered. At any stage of the search, the accumulator, ACCUM, is used to keep the highest grade found. Zero is taken as the initial maximum value, by clearing the contents of the accumulator. As the list is scanned, the fifteenth word of each item, which contains the student's final grade, is compared with the current maximum value. When a grade higher than the current maximum is found, the value in ACCUM is replaced

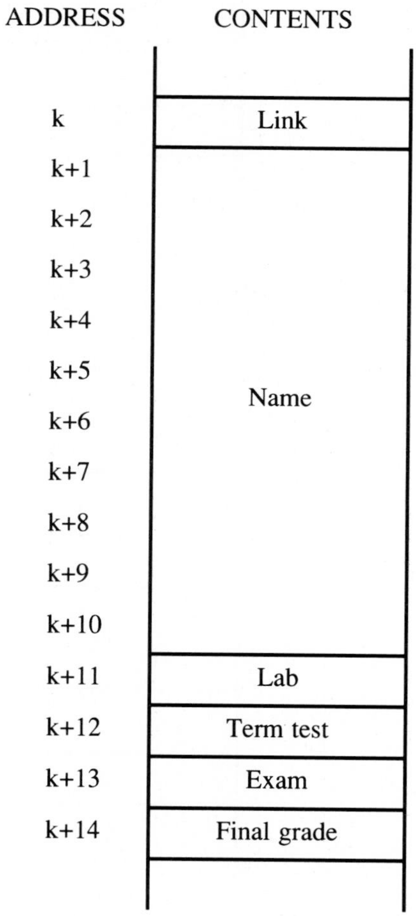

ADDRESS	CONTENTS
k	Link
k+1	Name
k+2	
k+3	
k+4	
k+5	
k+6	
k+7	
k+8	
k+9	
k+10	
k+11	Lab
k+12	Term test
k+13	Exam
k+14	Final grade

Figure 2.24 Arrangement of an item in a linked-list example.

	Move	#LISTADDRESS,LINKPOINTR	Initialize pointer.
	Clear	ACCUM	Initial highest grade.
LOOP	Compare	ACCUM,14(LINKPOINTR)	Compare with next grade.
	Branch_if_$\geq$0	NEXT	Current max still valid.
	Move	14(LINKPOINTR),ACCUM	Update highest grade found.
NEXT	Compare	#0,LINKPOINTR	Check for the end of the
	Branch_if_=0	DONE	list.
	Move	[LINKPOINTR],LINKPOINTR	Go to next student.
	Branch	LOOP	
DONE	next instruction		

Figure 2.25 A program that finds the highest grade in the linked-list example.

with this grade. The last item of the list is identified by its link (the first word) having a value of 0. Hence, a check for the link being equal to 0 is made each time through the loop, in order to stop after the last item has been considered.

In Figure 2.25, the accumulator is initialized to the value of 0, before entering the main loop of the program, to ensure that the existing value in ACCUM will not cause erroneous results. A separate Clear instruction is used for this purpose. It is useful to note that the program can be rewritten so that this instruction is not needed. Consider the alternative program shown in Figure 2.26. In this case, the accumulator is loaded by the first instruction in the main loop. Upon entering the loop, the initial value loaded into ACCUM is the final grade in the first item in the list. It is no longer necessary to initialize the accumulator outside the loop, thus making the program shorter by one instruction.

We have chosen a very simple task to illustrate the use of a linked list. There are many questions that one may ask about the data in the list. For example, it may be of interest to find the name of the best student, the names of the top three students, the median grade of the class, and so on. The reader should contemplate writing programs for such tasks. Some suggested exercises are given at the end of this chapter.

	Move	#LISTADDRESS,LINKPOINTR	Initialize pointer.
LOOP	Move	14(LINKPOINTR),ACCUM	Update highest grade found.
NEXT	Move	[LINKPOINTR],LINKPOINTR	Go to next student.
	Compare	#0,LINKPOINTR	Check for the end of the
	Branch_if_=0	DONE	list.
	Compare	ACCUM,14(LINKPOINTR)	Compare with next grade.
	Branch_if_$\geq$0	NEXT	Current max still valid.
	Branch	LOOP	Better grade found.
DONE	next instruction		

Figure 2.26 A shorter program that finds the highest grade.

2.9 CHARACTER STRINGS

In previous examples, we have made frequent use of names to represent numerical values. However, we have not considered manipulating such names themselves. Manipulation of alphanumeric data is an important function that needs to be understood thoroughly. Such data are handled in encoded form often using the standard ASCII code, described in Appendix A. In the following discussion, we will assume that each character is encoded in the 7-bit ASCII code, with a 0 appended as the most-significant bit.

A collection of characters defined as a meaningful entity is called a *character string*. Character strings can be short, being as small as a single name or even a single letter, or long, as is the case with English text that may be a chapter in a book. Such strings can be manipulated in many ways. Some of the essential operations involve: comparison of two strings, changing a portion of a string, concatenation of two strings, and so forth. We will illustrate the basic ideas in string manipulation by examining the first of these operations, namely, the comparison of two strings.

In the example of Figure 2.24 we set aside a 10-word field to hold the name of a student. For simplicity, let us assume that each word in this field contains one ASCII-encoded character. Thus, "George" would occupy the first 6 words and the remaining 4 words would be filled with the NUL character (represented as an all 0 pattern). Now, suppose that we want to find the marks that George obtained. In order to do this, it is necessary to find the corresponding item in the list. This involves starting at the head of the list and checking the name field of each item to see if it matches "George," until the desired item is found.

Two character strings can be compared to see if they are identical by comparing them one character at a time. Recall that the Compare instruction subtracts the destination operand from the source operand and sets some conditions that can be tested by a branch instruction that follows. When the subtraction involves numerical operands, the result can be positive, negative, or zero, meaning that the first operand is larger than, smaller than, or equal to the second operand. When the operands are ASCII-encoded characters, they can also be subtracted one from another. If the result is 0, then the operands are identical. If the result is greater than 0, then the ASCII code for the source operand is a larger number than the one for the destination operand. If the result is negative, then the destination operand is higher. From the ASCII codes given in Table A.1, it can be seen that letters are encoded so that ascending numerical values represent an ordered sequence from A to Z and from a to z. This means that letters can be sorted alphabetically on the basis of their codes. Treating names as character strings, a given list of names can be sorted alphabetically by comparing the individual characters in sequence.

Figure 2.27 shows a program that searches the linked list of Figure 2.24 to find the marks of a given student. The desired student's name is represented by a character string that starts in memory location STUDENT. Assume that each character occupies one word, to have the same arrangement as assumed in Figure 2.24. The string is terminated by the null character. Three pointer registers are used in the program. LINKPOINTR points to the link entry of the item that is currently being

	Move	#LISTADDRESS,LINKPOINTR	Initialize pointer.
	Branch	NAMEINIT	
NEXTITEM	Move	[LINKPOINTR],LINKPOINTR	Go to next student.
NAMEINIT	Compare	#0,LINKPOINTR	Name not found if
	Branch_if_=0	NOTFOUND	end of list.
	Move	#STUDENT,POINTER1	Get location of name.
	Move	LINKPOINTR,POINTER2	Set pointer to the
	Increment	POINTER2	name field.
MATCH	Compare	[POINTER1]$_{autoinc}$,[POINTER2]$_{autoinc}$	Compare 2 characters.
	Branch_if_≠0	NEXTITEM	No match, next name.
	Compare	#0,[POINTER1]	Continue until end of
	Branch_if_≠0	MATCH	name reached.
	.	Instructions that output the	
	.	name and the marks.	
NOTFOUND	.	Instructions that display a message	
	.	that the student's name is not in the list.	

Figure 2.27 A program that finds the item corresponding to a given student's name.

considered. POINTER1 and POINTER2 are used to scan through the characters of the strings at STUDENT and the name field of the list item, respectively, when the two strings are compared. Note that each time the LINKPOINTR is updated to point to the next item, both POINTER1 and POINTER2 must be initialized to point to the first characters of the respective name strings.

The comparison is done within the loop MATCH. It is successful if all characters in both name strings are identical, at which point the instructions that display the name of the student and the marks are executed. The desired item in the list is identified by the contents of LINKPOINTR.

It is possible that a student's name is not found in the list. This is detected if a link value of 0 is encountered. In such a case, an appropriate message is displayed.

We have made some simplifying assumptions about the strings in this example, to allow us to focus on the most important aspects of character manipulation. Storing strings one character per word is wasteful of memory space if the words are longer than 8 bits. If a word is 16 or 32 bits long, then it can hold 2 or 4 characters, respectively. Another implicit assumption made was that the names of the students do not overlap in terms of their character strings, that is one student's name is not a substring of another name. This means that two names such as Paul and Paula are not expected to be found in the list. What would the program of Figure 2.27 do if both Paul and Paula were students in the class? The reader is encouraged to modify the program to remove this restriction.

Manipulation of character strings is an important function. This section has provided a brief introduction to some of the concepts needed to deal with alphanumeric data.

2.10 ASSEMBLY PROCESS

Until now we have considered some basic ideas about machine programming in fairly general terms. The examples given have been presented in as descriptive a format as possible, without worrying about details that may be pertinent to particular computers. We have also paid little attention to the fact that programs must be encoded in patterns of zeros and ones, before they can be executed on a computer. This section will discuss some aspects of the process needed to produce programs encoded in a way that allows their direct execution on a given machine.

A program encoded in a form that is suitable for execution when loaded in the memory of a computer is called an *object program,* or *object code.* The object code consists of strings of zeros and ones. Hence, it is not easy to work with for a human programmer. It is much easier to write programs in a language that uses names and symbols to represent individual bit patterns of the object code, as we have used in the previous sections. The symbolic notation which we have used is not tied to any specific computer. However, for any given computer a language is defined, which specifies exactly how such names, symbols, and constructs can be used to write programs. Such language is called the *assembly language* for that computer.

Programs written in assembly language are translated into object code by means of another program, called an *assembler.* The process of converting an assembly language program into object code is called the *assembly process.* The main concepts involved in this process will be discussed in this section. These concepts are applicable to most microprocessors. A few specific requirements pertinent to the 6809 and 68000 microprocessors will be treated in Chapters 3 and 4, respectively.

We stated above that object code is the form of a program that can be directly executed on a computer. A program written by a programmer in any language is called a *source program,* or *source code.* Source programs are written either in a high level language or in an assembly language, but almost never in the object code format directly.

In what follows we will consider the basic aspects of assembly languages. The discussion is applicable to a wide range of languages, but the examples will be drawn from assembly languages for the Motorola family of microprocessors.

An assembly language has well-defined rules that must be followed when writing programs. These rules govern the way in which instruction statements are written, names are defined, and data are represented. They may seem to be unnecessarily constraining, but a tight set of rules leads to source programs that are easy to assemble into object code, thus requiring relatively simple assemblers. A set of rules that define the correct format of symbols and their ordering is called the *syntax* of the assembly language.

2.10.1 Names, Numbers, and Characters

From the programmer's point of view, it would be an arduous task to have to deal with data and addresses only in terms of their numerical values. It is much more convenient to associate names with such values, particularly when frequent references are made to any given address or data item. The names must be written according to the rules explained below.

A name can consist of any alphanumeric characters, i.e., letters, numbers, and symbols such as & or %. A usual restriction is that the first character must be a letter. Some examples of valid names are: NEMS, POINT3, and D23&Q. The maximum number of characters in a name may be limited to a particular size. Should the programmer write a name that exceeds the maximum length, a typical assembler will simply ignore the extra characters.

The programmer must be careful to use unique names within the maximum allowable length. For example, if the maximum length is 8, the names DATAITEM2 and DATAITEM37 will be interpreted by the assembler as being the same.

It is advisable to use meaningful names that suggest their purpose. Keywords associated with the assembly language, such as operation descriptors, cannot be used as names.

In this chapter we have used decimal notation to represent numbers. In practice it is often convenient to use other number representations, notably binary and hexadecimal (hex). In assembly languages for Motorola microprocessors, binary and hex numbers are denoted with the prefixes % and $, respectively. A number is assumed to be decimal unless marked otherwise. For example, decimal number 29 can be given as

Decimal:	29
Binary:	%11101
Hex:	$1D

Alphanumeric characters can also be used as data. They are identified with apostrophes. The assembler replaces each character with its ASCII code, e.g.,

A is replaced with $41
5 is replaced with $35

See Appendix A for details of the ASCII codes.

The above discussion indicates how a programmer can write names, numbers, and characters. During the assembly process the assembler produces object code where everything is translated into a binary representation.

2.10.2 Statements

A typical assembly language instruction statement is written in the following four-field format:

Label   Operation   Operand(s)   Comment

The fields are delimited by some characters, usually by one or more blank spaces. This implies that the delimiter may not be used within a field. An exception is the comment field which is ignored by the assembler. Comments are used for documentation purposes only.

Not all of the fields are necessarily used in every statement. Three examples of statements used in the previous sections are

	Move	#ADDRESSA,POINTERA	Set pointer A.
LOOP	Move	[POINTERA],[POINTERB]	Transfer an entry.
	Increment	POINTERA	

A statement starts in column 1 of the source line. Thus, if a label is present it must start in column 1. A statement without a label must have a delimiter, in our case a blank space, in column 1.

Label

A *label* is a programmer defined name that satisfies the constraints given in Subsection 2.10.1. Labels must be unique. A label stands for the address of the memory location where the labeled instruction will be loaded in the main memory. When a label appears as an operand in an instruction, the assembler replaces it with its equivalent address. For example, suppose that the instruction

LOOP   Move   A,B

is loaded in memory location 200. Then, LOOP represents the value 200. If another instruction in the program uses LOOP as a name in the operand field, the assembler will replace this name with 200. Thus, the instruction

Branch   LOOP

will be interpreted by the assembler as

Branch   200

Operation

The *operation* that is to be performed is defined in the operation field of an instruction. The operation is specified by means of capitalized symbolic names, called *mnemonics*. Instruction mnemonics are usually chosen as abbreviations that, as much as possible, indicate the operation represented. Some typical examples of mnemonics for operations in previously used instructions are

Move	==>	MOVE
Increment	==>	INC
Branch_if_>0	==>	BGT

In addition to specifying an operation, a mnemonic may identify a CPU register that is used as an implied operand in the instruction. For example, the instruction

Load_Register_X   #5

may be written as

LDX   #5

It loads the number 5 into register X.

Operands

The operand field contains either an immediate operand or addressing information specifying where the operand is to be found. The addressing modes that can be used depend on a particular computer. The assembler replaces any names that appear in the operand field with their equivalent numerical values. For example, suppose that LOC represents memory address 500. Then, the instruction

LDX   #LOC

is equivalent to

LDX   #500

Other formats that may be used for specifying operand addresses will be presented in Chapters 3 and 4.

Usually, each entity in the specification of an operand is represented by a single name or number. However, many assemblers also allow arithmetic expressions to be used. Such expressions are evaluated by the assembler and replaced with their numerical values. For example, the instruction

Move   #7,LOC+3

is interpreted by the assembler in exactly the same manner as the instruction

Move   #7,503

Either of these two instructions is translated into a machine instruction that moves the value 7 to the memory location whose address is 503. It is important to note that the expression LOC+3 is evaluated at the time the program is assembled. The addition operation does not become a part of the object code program that is later executed. More complex expressions can also be used, the allowed complexity being dependent upon the capabilities of the specific assembler used in the assembly process.

Comments

The comment field is ignored by the assembler. Thus, it can contain any characters, including blank spaces.

Sometimes it is desirable to insert an entire line, or even several lines, of comments to improve the intelligibility of a program. This can usually be done by placing a special character in the first column. We will assume that this character is "*". Then, any line of source code beginning with a "*" will be interpreted as a comment. Readability can also be enhanced with blank lines, which are also ignored by most assemblers.

2.10.3 Assembler Directives

A source program contains the instructions that are to be translated by the assembler into machine code. In addition, it must also contain some auxilliary information needed by the assembler in the translation process. For example, the programmer must specify the starting location in the main memory for various program segments.

Such information is provided to the assembler in the form of statements known as *assembler directives* or *assembler commands*. It should be emphasized that these are commands to the assembler program; they are not assembled into the object program.

Assembler directives perform functions that include defining symbols and names, assigning different sections of a program to specific parts in the memory, assigning locations in memory for temporary data storage, and placing constant data into desired memory locations. The structure of assembler directives is similar to that used for machine instructions. A mnemonic that defines the required function is written in the operation field. If a particular directive requires a label it is placed in the label field, while any address or data is given in the operand field. Following are some commonly used assembler directives.

EQU (Equate)

This directive enables the programmer to equate names to numerical values. It assigns the numerical value in the operand field to the name or symbol in the label field. Some examples are

```
BUFFER     EQU     $2AF0
SEVEN      EQU     7
```

Expressions and names can be used in the operand field, e.g.,

```
NEWBUF     EQU     BUFFER+8
```

Note that this directive is meaningful only if the name BUFFER has been defined earlier in the program.

ORG (Origin)

The ORG directive enables the programmer to specify where in the main memory the program or data should be located. When the programmer writes

```
           ORG     1000
START      LDX     #250
```

the assembler will place the LDX instruction in memory location 1000, and will assign the value 1000 to the name START. Subsequent instructions are placed into locations that follow. The ORG directive can be used several times in a program, if various program segments are to be assembled into different parts of the memory.

DATA (Define data)

Specific data can be placed in the memory by means of the DATA directive. The data can be numbers or characters. A name can be associated with the address of the memory location where the data is stored by writing it in the label field. For example, the statement

```
TWENTY     DATA     20
```

places the number 20 in the memory. The particular location where this number is stored is determined by the preceding statements in the source program. If the programmer wants to store 20 in a specific location, say 1000, this can be accomplished with the sequence

                         ORG   1000
             TWENTY   DATA   20

In this case, the name TWENTY is assigned the value 1000.

Most assemblers allow specification of several data items within one DATA statement. For example, the statement

             LIST   DATA   7,32,17,320,4

loads the numbers given in the operand field into five memory locations, and associates the name LIST with the address of the first of these five locations.

Data can be written in any of the allowed notations for names, numbers, or characters. Thus, a string of characters may be loaded with the statement

             MESG   DATA   "IMPORTANT MESSAGE"

This statement causes successive bytes to be loaded with the ASCII codes for the characters I, M, P, and so forth.

END

When a source program is to be assembled, the assembler must be told where this program begins and ends. The starting point is obviously the first statement in the program. However, the end point must be clearly indicated, and this is accomplished with the END directive. Any source code that follows this directive will be ignored by the assembler.

2.10.4 Example

Figure 2.28 shows an example of source code which illustrates the concepts discussed above. It is the program of Figure 2.20 written in the style of a typical assembly language program. Comments are used liberally to explain what is happening at various points in the program.

Mnemonics used to denote various operations either have obvious meaning, or their function is indicated in the comment field. The notation used in the operand field is consistent with the previous discussion. The program assumes that there exist three CPU registers that can be used as pointers, POINTERA, POINTERB, and POINTERC. In any particular microprocessor these registers would be identified using the names pertinent to that machine.

2.10.5 Assembling a Program

A source program written in assembly language is converted, or assembled, into object code by an assembler program. The assembler must replace all names and symbols with their numerical values represented in binary form.

```
*  This is a part of a program which calls subroutine
*  SUBMOVE to copy 50 items from list A to list B.
*  It is assumed that list A starts at location 2000
*  and list B at location 2500.
```

ADDRESSA	EQU	2000	Define starting
ADDRESSB	EQU	2500	addresses of lists.
	⋮		
	LBSR	SUBMOVE	Branch to subroutine.
	DATA	50, ADDRESSA, ADDRESSB	Define constant data.
	⋮		

```
*  Assume that subroutine SUBMOVE is to be placed in
*  the main memory starting at address 1000.  The subroutine
*  corresponds to the example in Figure 2.20.
*  Let the subroutine use memory location 3000 as
*  the counter NCOUNT.
```

NCOUNT	EQU	3000	
	ORG	1000	
SUBMOVE	MOVE	[SP], POINTERC	Load parameters into counter
	MOVE	[POINTERC], NCOUNT	location NCOUNT and the
	INC	POINTERC	two pointers.
	MOVE	[POINTERC], POINTERA	
	INC	POINTERC	
	MOVE	[POINTERC], POINTERB	
	INC	POINTERC	Place the return address on
	MOVE	POINTERC, [SP]	the stack.
LOOP	MOVE	[POINTERA], [POINTERB]	Copy an item from A to B.
	INC	POINTERA	
	INC	POINTERB	
	DEC	NCOUNT	Continue copying until the
	BGT	LOOP	end of lists.
	RTS		Return from subroutine.
	END		End of source program.

Figure 2.28   An example of source code, based on the program in Figure 2.20.

In order to keep track of memory addresses, the assembler sets up a *location counter*. The location counter contains the memory address of the instruction or data item that is being assembled. It provides the starting address for each instruction, which is associated with the label, if any, of that instruction. When writing the source program, the programmer can refer to the contents of the location counter with a special symbol. In many assembly languages the chosen symbol is "*".

As an example of a possible use of the location counter, consider the sequence of two statements

```
            ORG     1200
            LDX     #*
```

In this case "*" refers to the current contents of the location counter, which is the number 1200. The assembler inserts this number as the desired immediate operand. This means that the LDX instruction will be treated as

```
            LDX     #1200
```

The same effect can be achieved with

```
            ORG     1200
    POINT   LDX     #POINT
```

If register X is to be loaded with the address of a location which is ten locations away in the forward direction, address 1210 in this case, the desired result can be achieved with

```
            ORG     1200
            LDX     *+10
```

A relative offset in either a positive or a negative direction can be specified in this way. Recall that the "*" symbol is also used to begin a line of comments (see Subsection 2.10.2). There is no confusion in this double usage of the symbol. A "*" is interpreted as the location counter only when it appears in the operand field.

During the process of assembling a program, the assembler scans through all instructions in the program looking for names that need to be defined. It constructs a *symbol table,* where each name is entered along with the numerical value associated with it. In the source program, each name must be given a value either in an EQU assembler directive, or by appearing as a label of an instruction. A complete symbol table can be constructed in one pass through the program. Then, the assembler makes a second pass through the program, replacing all names and symbols with their numerical values listed in the symbol table. An assembler which functions in this manner is called a *two-pass assembler.*

A two-pass assembler has no difficulty in resolving forward references, that is, situations where a name in the operand field of an instruction represents a value that is defined in a subsequent instruction in the program. For example, suppose a branch to a forward point is to be made as follows:

```
            BRA     AHEAD
                .
                .
                .
    AHEAD   MOVE    A,B
```

If an assembler tries to assemble each instruction as it is encountered, the name AHEAD in the Branch instruction would cause problems, because the value AHEAD is not known until the Move instruction is assembled. However, during the first pass of a two-pass assembler, the name AHEAD is entered in the symbol table as soon as the Branch instruction is encountered. The value of AHEAD is determined and recorded in the symbol table when the Move instruction is considered. Thus, during the second pass, all information relating names to values is available in the symbol table. Most microprocessor assemblers work in this manner.

We should note that an assembler assembles programs written according to some fairly strict rules. These rules simplify the assembly process. Their existence makes it also possible for the assembler to detect some of the errors that a programmer might make. Of course, the assembler can only detect those errors that cause the source code to be inconsistent with the rules. Most assemblers generate error messages that indicate the rule violations they have detected.

It is important to observe that there exist many different assemblers, and the discussion in this section is applicable to most of them. It is quite usual to see several different assemblers for any given assembly language. They may even differ in some rules for writing source programs. However, they all produce object code that can be executed on the desired computer.

Finally, we should mention that there are two types of assemblers. The first runs on the same computer for which it assembles object programs. Such an assembler is referred to as a *resident assembler*, or a *self-assembler*. Since an assembler is just a program that converts encoded information from one kind of code into another, it can do this on any suitable machine. Hence, there exists a second type of assembler called a *cross-assembler*, which runs on a computer different from the one for which it assembles programs. Usually, cross-assemblers run on computers that are larger and have more extensive software facilities than the machine for which the object code is produced. If a cross-assembler is written in a high-level language, it can run on many different computers.

2.11 TERMINOLOGY

Following is a summary of the new terms introduced in this chapter.

Addressing mode: a scheme for specifying the address of an operand.

- *Absolute*: the address of the operand is given in the instruction.
- *Autodecrement*: a pointer register is decremented and then its contents are used as the address of the operand.
- *Autoincrement*: contents of a pointer register are used as the address of the operand and then the register is incremented.
- *Immediate*: the operand itself is given in the instruction.
- *Indexed*: the address of the operand is the sum of the contents of an index register and a specified constant.
- *Indirect*: the address of the operand is the contents of a register or a memory location whose address is specified in the instruction.

- *Register*: the operand is in a register specified in the instruction.
- *Relative*: the location of the operand is given relative to the location of the instruction that references the operand.

Assembler: a program used to translate source programs written in an assembly language into object programs.

Assembly language: a low-level language intended for a particular microprocessor. It consists of machine instructions represented in a symbolic form.

Character string: a sequence of alphanumeric characters.

Compiler: a program used to translate source programs written in a high-level language into low-level machine programs.

Effective address: the actual address at which a desired operand is found.

Linked list: a list in which each item includes a pointer that indicates the location of the next item in the list.

Object program: a program encoded in a form suitable for execution when loaded in the main memory of a computer.

Source program: a program written by a programmer, in any language.

Stack: a data structure where the last element stored is the first element removed (last-in first-out structure).

Subroutine linkage: a mechanism for controlling the execution flow and parameter passing between a calling program and a subroutine.

Syntax: a set of rules that define the correct format of symbols and their ordering in a given language.

2.12 CONCLUDING REMARKS

The material in this chapter deals with machine programming fundamentals from a general point of view. The discussion is pertinent to most microprocessors, and indeed to most computers. In the next two chapters we will consider two commercially successful microprocessors in some detail, to see how the general principles are applied to specific cases. Chapter 3 describes the Motorola 6809 microprocessor, as an example of a standard inexpensive product that is suitable for a wide range of applications. Chapter 4 deals with the Motorola 68000 family, whose members are larger and more powerful microprocessors, capable of handling much more extensive computing tasks. These two chapters may be read independently, because each is a natural extension of this chapter. The reader may wish to concentrate on Chapter 3, or proceed directly to Chapter 4. However, because the 6809 and 68000 microprocessors are quite different, a study of both chapters would provide the reader with a broader appreciation of the subject.

2.13 REVIEW QUESTIONS AND PROBLEMS

REVIEW QUESTIONS

1. How are the condition code flags used in controlling the flow of execution of a program?

2. Why is it useful to have a number of different addressing modes?

3. Give some examples where it is convenient to use indirect addressing.

4. What mechanism can be used to implement subroutines? Why is a Call_subroutine instruction different from a simple Branch instruction?

5. How can parameters be passed to a subroutine?

6. How can a stack be implemented in the main memory?

7. What is the function of the location counter during the assembly process?

8. What is the difference between machine instructions and assembler directives?

PRIMARY PROBLEMS

9. Rewrite subroutine MAX, of Figure 2.16, so that instead of comparing against the address of the last entry in the list, the loop is controlled by decrementing a counter.

10. Can the Increment instruction in the program of Figure 2.16 be eliminated by using the autoincrement addressing mode?

11. Can the Increment instruction in the program of Figure 2.17 be eliminated by using the autoincrement addressing mode?

12. Improve the speed of execution of the program in Figure 2.17, by not swapping the entry found by MAX and the entry at the top of the sublist if the largest number is already at the top of the sublist.

13. Rewrite the program of Figure 2.13 using a subroutine that displays one character on a video terminal.

14. Write a program that inserts an item for Frank between the items for Dan and George in the linked list of Figure 2.22. Assume that the items have the format given in Figure 2.24, and that the name is stored by placing one ASCII-encoded character per word in the "name" field.

15. Write a program that finds the median final grade in the linked list of Section 2.8.

16. Write a program that finds the names of the top three students in the class, using the linked list of Figure 2.22 and assuming that the list is structured as stated in Problem 14.

17. Modify the program of Figure 2.27, so that it can be used with a linked list structured such that each word has 16 bits and two ASCII characters are stored per word of the name field. Assume that each memory read operation transfers a 16-bit word to the CPU.

ADVANCED PROBLEMS

18. Figure 2.12 shows how a character can be read from the keyboard of a video terminal. Show how this scheme can be used to read characters from four different video terminals connected to the same computer. The received characters should be stored in four separate strings in the main memory. Assume that some specific action will be performed, by executing a subroutine called ACTION, when a CR (carriage return) character is detected from one of the terminals. The number of the terminal should be passed to the subroutine in one of the registers. After returning from the subroutine, reading of input characters should continue. Write a program that can be used for this purpose.

19. In Section 2.5 we considered the task of sorting numbers in a list, using a simple method realized in the program of Figure 2.17. Consider an alternative method. The list is scanned from top to bottom, comparing adjacent entries and swapping them (i.e., interchanging their positions) if they are in the wrong order. After at most $n-1$ passes through the list, where n is the number of entries in the list, all entries will be sorted in the correct numerical order. Because the elements of the list move gradually toward their ultimate positions, this method is known as a "bubble sort." Write a program that sorts a list of numbers using this method. Assume that the execution time of a program is proportional to the number of instructions executed. How does the speed of your bubble sort program compare to that of the program in Figure 2.17?

20. Matrices are two-dimensional arrays. They are stored in the main memory as one-dimensional lists. It is a common practice to store a matrix in a list in column order, i.e., the elements in column $k+1$ are stored in consecutive memory locations, following the elements in column k. This means that access to the elements involves a nontrivial computation. For an $n \times m$ matrix, the address of element (i,j) can be computed as

$$\text{Address}(i,j) = m*(j-1) + (i-1) + \text{Address}(1,1)$$

Write a program that multiplies two matrices, assuming that each element occupies one word in the memory and that the values of the elements are small enough so that when two elements are multiplied the result will fit into one word of memory. Assume that there exists a suitable Multiply instruction.

21. Rewrite the program of Figure 2.27 using two subroutines, one that compares two strings and another that displays the desired messages on a video terminal.

22. (a) Write a program that counts the number of occurrences of the string "the" in a long string consisting of some English text. Assume that the string being searched is terminated by the null character.
    (b) Enhance your program to search for an arbitrary string.

23. Write a program that sorts the linked list of Section 2.8, in descending order of final grades. Assume that the items of the list have the format given in Figure 2.24.

24. Write a program that finds and displays on a video terminal the name and the marks of the best student in the class, using the linked list of Figure 2.22 and assuming that the list is structured as stated in Problem 14.

25. Write a program that can sort the linked list of Section 2.8 in alphabetical order of students' names. Assume that one ASCII-encoded character is stored in each word of the name field.

26. Repeat Problem 25, assuming that two ASCII characters are stored per word, whose length is 16 bits. Assume that each memory read operation transfers a 16-bit word to the CPU.

27. The 10-word name field in Figure 2.24 imposes a restriction on the size of the names that can be handled. A more flexible arrangement is shown in Figure 2.29, where a variable-length name field is used. The second word of the item contains an offset, j, which indicates the distance from the link entry to the location of the lab mark, expressed in terms of the number of words. Repeat Problem 24 for a list structured in this way.

28. Repeat Problem 23, assuming that the items of the list have the format given in Figure 2.29.

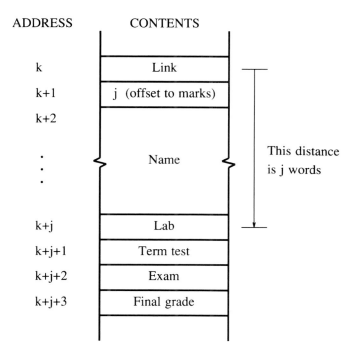

ADDRESS CONTENTS

Address	Contents
k	Link
k+1	j (offset to marks)
k+2	
⋮	Name
k+j	Lab
k+j+1	Term test
k+j+2	Exam
k+j+3	Final grade

This distance is j words

Figure 2.29 An alternative arrangement of an item in a linked-list, using a variable-length name field.

Chapter
3

Motorola 6809 Microprocessor

Chapter 2 dealt with general principles of programming at the machine language level. The concepts discussed are applicable to most commercially available microprocessors. This chapter will focus on one real product—the Motorola 6809 microprocessor—and will show how the ideas and examples of Chapter 2 can be implemented using the 6809, enabling the reader to write application programs in the 6809 assembly language.

Microprocessors made their appearance during the early 1970s. A number of different microprocessors came on the market at about the same time and some became popular choices in a host of diverse applications. The most successful microprocessors have been accompanied by families of supporting chips that facilitate the design of microprocessor systems. One of the most popular families of such chips has been introduced by the Motorola company. It is known as the 6800 family, and has the 6800 microprocessor as its key element. This family includes a number of directly compatible chips that provide I/O functions, controller operations, and storage capability.

The 6800 became one of the most widely used microprocessors, but its relatively simple structure poses certain limits on the performance that can be achieved in systems based on this chip. In order to provide increased performance, Motorola developed the 6809 microprocessor. The 6809 fits into the 6800 family of chips fully, being directly compatible with all the support chips. However, it is sufficiently different from the 6800 microprocessor that the programs written for the 6800 will not run successfully on it without substantial changes, and programs written for the 6809 cannot be run on the 6800 at all. The 6809 is a microprocessor with considerably enhanced computing capability.

In the following sections we will discuss the structure of the 6809, its addressing modes, and instruction set. We will show how the illustrative examples in Chapter 2 can be implemented as 6809 programs. The hardware aspects of this microprocessor, in particular its memory and I/O bus, will be described in Chapters 6 and 7.

3.1 6809 PROGRAMMING MODEL

The 6809 microprocessor handles external data in 8-bit quantities, and most of its instructions operate on 8-bit operands. Therefore, we may say that the word length of the 6809 is 8 bits, or 1 byte.

Addresses generated by the 6809 consist of 16 bits, meaning that a total of 2^{16} (=65,536) distinct addressable locations may be referenced. This number is often referred to as 64K, where 1K is equal to 2^{10} (=1024). Each address denotes a location capable of holding one byte of information. Hence, a 6809 microcomputer system is said to be *byte addressable*.

A 6809 microcomputer may have its main memory as large as 64K bytes. However, in practice, some addressable locations are used for I/O purposes rather than as main memory locations. Input and output operations are implemented using the memory-mapped I/O scheme, discussed in Section 1.5. In this scheme, some of the addressable locations refer to locations (usually registers) in the supporting I/O chips.

Figure 3.1 shows the bus structure of a 6809 microcomputer. The bus consists of 16 address lines, 8 data lines, and 12 control lines. This is the maximum size of the bus. Fewer lines may be used in small systems, where memory and I/O requirements are small.

The 6809 microprocessor contains a number of registers, shown in Figure 3.2. There are two 8-bit *accumulators,* called A and B. Either accumulator may be involved in instructions that use 8-bit operands. But, the 6809 also has some instructions that perform operations on 16-bit operands. They use both registers A and B as a single 16-bit accumulator, called D, where A represents the high-order byte and B the low-order byte of D.

There are four pointer registers, each being 16 bits long to accommodate 16-bit addresses. Two of these are *index registers* X and Y. The other two are stack pointers S and U. The S pointer, also called the *processor stack pointer,* points to a stack in the memory that is used by the microprocessor for functions such as storing return addresses for subroutine linkage. The stack can also be used by the programmer for temporary storage purposes. The U pointer, called the *user stack pointer,* allows the programmer to establish a second stack, to be employed by his

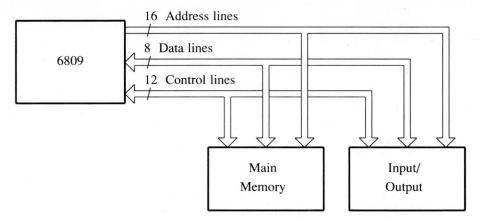

Figure 3.1   6809 bus structure.

programs only. Both the S and U pointers can also be used as additional index registers, but in the case of the S pointer this should be done only with great care.

The 6809 has a 16-bit program counter, an 8-bit page-addressing register, and an 8-bit condition code register. The functions of the latter two will be described in the following subsections.

3.1.1 Direct Page Register

In the absolute addressing mode, the address of the operand is given within the instruction. In this case, the use of 16-bit addresses results in long instructions. It is possible to reduce the requirement for address bits if the addressable space is divided into sections, called *pages,* of suitable size. The 6809 addressable space of 64K locations is divided into 256 pages. Each page consists of 256 locations that can hold 8 bits of information. Figure 3.3 shows this arrangement. Note that addresses are given in hexadecimal notation. The two low-order digits represent the low-order byte of an address. They have a value ranging from 00 to FF (0 to 255_{10}), which defines one of 256 locations on a given page. The high-order byte identifies one of 256 pages.

Instructions in a typical program segment tend to refer to data stored in a relatively small portion of the memory. If a number of data operands are held within one page, and it is known which page is being referenced, the data can be addressed by specifying only their location on that page. With the arrangement of Figure 3.3, 8-bit addresses are needed to refer to any given location within a page. The 6809 addressing mechanism allows the programmer to select a particular page by depositing the corresponding high-order byte of the address in the *direct page register* (DP). Once a particular page is selected, instructions need only provide the locations of their operands within that page. A short-address format is available for this purpose, where the instruction supplies the low byte of the address of the operand, while the high byte is obtained by the microprocessor from the DP register.

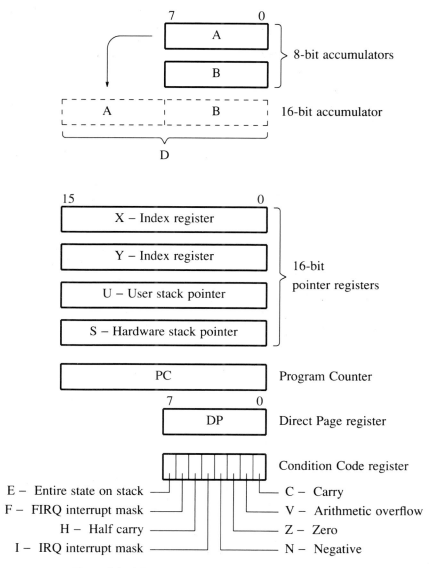

Figure 3.2 Internal registers in the 6809 microprocessor.

When the DP register is used to access memory operands, the address of an operand is specified in an instruction using 8 bits rather than 16. This use of the DP register yields two advantages. Because such instructions are shorter by one byte, less memory space is required. Second, since only one byte of an instruction needs to be read from the memory during one memory access cycle, because the data bus is 8 bits wide, shorter instructions will require fewer memory accesses, thus resulting in faster execution of a program.

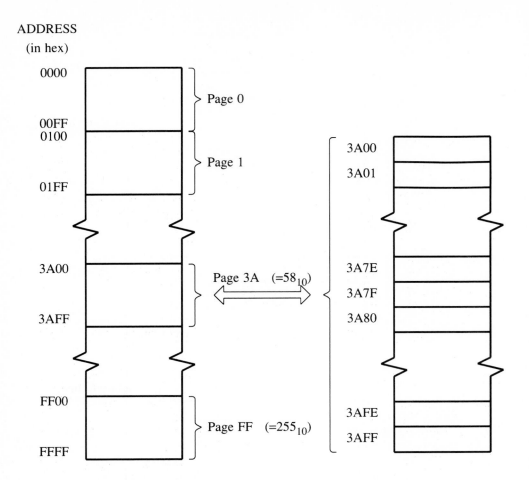

Figure 3.3 Paging arrangement for 6809 addressable space.

3.1.2 Condition Code Register

The last register that must be considered is the *condition code register* (CCR). Whenever an operation is performed, this register is used to retain some information about its result. As we have already mentioned in Section 2.1, the condition codes are used by conditional branch instructions to decide whether or not a branch should be made. If an arithmetic operation, such as addition or subtraction, produces a zero result, this fact is recorded by setting the Z bit in CCR to 1; otherwise, Z is set to 0. Other outcomes are recorded as well, as described below. The action of storing this information in CCR is performed by the microprocessor, as part of executing the instruction in question.

Each of the 8 bits in CCR is associated with one specific condition. The meaning of these bits, called the *flags,* is as follows:

- C: Carry from bit position 7 of the arithmetic and logic unit (ALU)
- V: Arithmetic Overflow
- Z: Zero
- N: Negative
- I: Interrupt Request mask
- H: Half carry, that is, carry from bit position 3 of the ALU
- F: Fast Interrupt Request mask
- E: Entire state stored on the stack

The C, V, Z, N, and H flags are set by the results of arithmetic and logic operations. Some examples of how these flags are set as a result of addition and subtraction operations are given in Figure 3.4. In case of difficulty in interpreting the examples, the reader may find it useful to review the material in Section A.6 of Appendix A.

The C flag is set when a carry from the most-significant bit position occurs. This may happen as a result of adding two operands. If a carry-out occurs then C is set to 1, and it is cleared to 0 otherwise. In Figure 3.4 a carry occurs in parts (c) and (d). Note that in each case the result of the addition is correct, ignoring the carry in accordance with the rules of 2's complement arithmetic. Nevertheless, the C flag is set to 1. The importance of the C flag in arithmetic operations is apparent when larger operands are used. Two 32-bit numbers can be added by summing them in two parts. First, the low-order 16 bits are added. Then, the high-order 16 bits plus the carry that may have resulted from the addition of the low-order 16 bits are added.

In subtraction operations the C flag is used to indicate a "borrow." Subtraction is performed by adding the 2's complement of the subtrahend to the minuend (see Section A.6.3). A carry bit resulting from this operation is actually the complement of the borrow bit. Therefore, upon completion of a subtract instruction, the C flag is set to the complement of the carry from the most-significant bit position. Parts (e) and (f) of Figure 3.4 illustrate the case of subtraction.

The C flag is also affected by some logic operations. For example, an operation that shifts an operand one bit position to the left, will set the C flag to equal the previous value of bit 7 of the operand.

The V flag indicates arithmetic overflow. It is set to 1 when the result of an arithmetic operation is larger in magnitude than can fit in the destination register specified in the instruction. Otherwise, the V flag is cleared to 0. Arithmetic operations involve either 8-bit or 16-bit operands, which are assumed to be represented in 2's complement notation. In each case, the most significant bit (in position 7 or 15) denotes the sign of the operand. The V flag is set if the signed value of the result of the operation exceeds what can be represented in 2's complement, using 8 or 16 bits, respectively. Figure 3.4(b) shows a case where arithmetic overflow occurs. There is a carry from bit position 6, so the V flag is set to 1.

The Z flag is set to 1 if an operation produces a zero result. It is cleared otherwise. Similarly, the N flag is set if a negative result occurs. In addition to being affected by various arithmetic and logic operations, the Z and N flags may be set or cleared by simple data transfer operations. For example if an accumulator is loaded

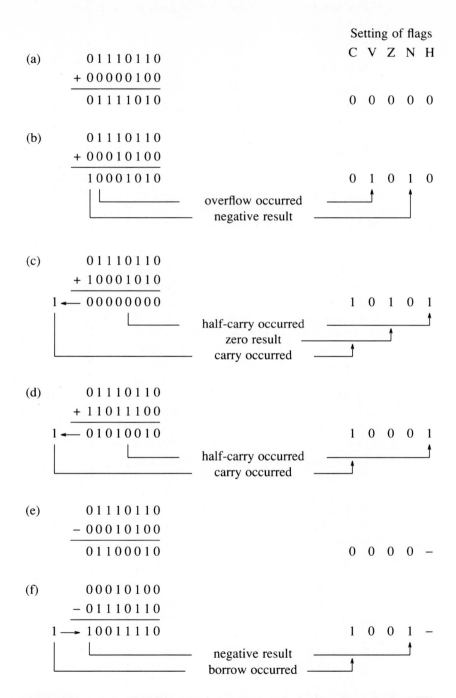

Figure 3.4 Examples of setting the condition code flags, as a result of addition and subtraction operations.

with an operand from the main memory, the Z and N flags are set to indicate whether the operand value is zero or negative.

The H flag is set by arithmetic operations on 8-bit operands when a carry from bit position 3 occurs. This flag is useful in writing programs for handling BCD (Binary Coded Decimal) numbers, where 4 bits are used to represent each decimal digit. Setting of this flag is illustrated in parts (c) and (d) of Figure 3.4. Observe that the H flag is not affected by the subtract operation.

The I, F, and E flags are used in conjunction with the interrupt control mechanism. Their meaning and usage will be discussed in Chapter 7.

The existence of the CCR flags provides effective means for testing the results of operations. Conditions indicated by the flags can be tested by conditional branch instructions. A substantial number of such instructions are available in the 6809. They use various combinations of the values of the CCR flags as conditions upon which branching decisions are made.

Before describing the details of the 6809 instruction set, we will introduce the conventions used in preparing 6809 assembly language programs. This should enable the reader to use various instructions in laboratory exercises while studying the material in the remainder of the chapter.

3.2 ASSEMBLER CONVENTIONS FOR 6809

Section 2.10 discussed the features of a typical assembly process, assembler conventions, and assembler directives. Most of the material presented is directly applicable to the 6809 and will not be repeated here. In this section we will consider only the relevant restrictions and details, as related to the assemblers provided by Motorola.

Names used to represent variables or constants are limited to 6 alphanumeric characters in length. Numbers are assumed to be decimal, unless marked with the prefix % for binary or $ for hexadecimal. The assembler replaces alphanumeric characters, preceded by an apostrophe, with their ASCII codes. For example, 'A is replaced with $41, and '9 is replaced with $39.

Assembly language instruction statements are written in the format described in Subsection 2.10.2. The operand field is normally used to specify one operand only. Operands held in the microprocessor registers are specified in the implied addressing mode, where the name of the register involved is indicated as a part of the mnemonic that denotes the required operation. For example, in the instruction

ADDA   #9

one operand is the contents of accumulator A and the other is the number 9 given as the immediate operand in the instruction. The two operands are added and the result is stored in accumulator A.

A number of assembler directives are provided. The EQU, ORG, and END directives are used as described in Subsection 2.10.3. Instead of the previously discussed DATA directive, the 6809 assembler has three distinct directives: FCB, FDB, and FCC. It also has two more directives, RMB and SETDP. Their use is explained below.

FCB (Form Constant Byte)

Specific data can be placed in the memory by means of the FCB directive. Any number of operands can be specified, separated by commas. They are stored in consecutive memory locations, one byte for each operand. For example, the statements

NUM	EQU	45
	ORG	$100
	FCB	3,28,$A,'A,NUM,NUM+5,9

will load the memory as follows:

Address (hex)	Contents (hex)
100	03
101	1C
102	0A
103	41
104	2D
105	32
106	09

FDB (Form Double Byte)

The FDB directive is similar to FCB. The difference is that each operand results in a 2-byte value being placed in the memory. The directive is particularly useful when address constants need to be defined, because they are 16 bits long. For example, the statements

	ORG	$500
LOC	FDB	$1AF2,23,0,LOC,LOC+6

will load the memory as follows:

Address (hex)	Contents (hex)
500	1A
501	F2
502	00
503	17
504	00
505	00
506	05
507	00
508	05
509	06

Note that the high-order byte of a 16-bit number is stored first, that is, at the lower address, followed by the low-order byte. This convention is used consistently for all 16-bit address or data values in the 6809 microprocessor.

FCC (Form Constant Character)

A string of characters can be placed in the memory with the FCC directive. Each character is represented by its 7-bit ASCII code, with a 0 added in the most-significant bit position to form an 8-bit byte. The string must be delimited (surrounded) by the same character at each end. For example,

> STRING   FCC   /MESSAGE/

loads 7 bytes of memory starting at location STRING with the ASCII codes for M, E, S, S, A, G, and E.

Any single character can be used to delimit the string, instead of "/". Of course, it is necessary to choose a character that is not included within the string.

RMB (Reserve Memory Bytes)

The RMB directive allocates a specified number of memory bytes for purposes such as temporary storage and data tables. It increments the assembly location counter (see Subsection 2.10.5) by the value of the operand. As an example of this directive consider the statements

> ```
>             ORG     $1000
>   TABLE     RMB     50
>   ITEM      FCB     23
> ```

As a result of these statements, the assembler will reserve 50 bytes of memory for a table, starting at address $1000. It will also associate the address $1000 with the name TABLE. The constant 23 ($=17_{16}$) will be placed in the memory location that follows the reserved area, i.e., in location $1000 + 50 = $1032. The value $1032 will be associated with the name ITEM.

SETDP (Set Direct Page Register)

During the assembly process, the assembler assumes that the DP register contains a zero, unless informed otherwise by the SETDP directive. The programmer may use the SETDP directive, at any point in a program, to tell the assembler what value to expect in the DP register. However, this directive does not actually change the contents of the DP register to the specified value. Recall that all assembler directives are used during the assembly process only, and do not appear in the object code at execution time. Having told the assembler what value to expect in the DP register, the programmer must ensure that this value is indeed loaded into the DP register at execution time. This can be accomplished by including machine instructions in the program that place the desired page value into one of the accumulators and then transfer it into the DP register. For example, suppose that the directive

> SETDP $C4

is included in a source program, preceding a routine that makes frequent references to some data on page C400. Then, the DP register may be loaded by including in the routine two instructions such as

LDA   #$C4
TFR   A,DP

The first instruction loads the immediate operand $C4 into accumulator A, and the second transfers the contents of A into the DP register.

3.3 6809 ADDRESSING MODES

Section 2.3 dealt with the addressing modes that are most commonly encountered in computers. Particular machines vary in the specifics of implementation of addressing modes, but the basic concepts are essentially the same. The 6809 supports all addressing modes discussed in Section 2.3. Its addressing modes are described in Table 3.1, which the reader should compare with Table 2.1.

Two aspects of the 6809 organization have a significant impact on the detailed features of its addressing modes: the word length and the CPU register structure. A word length of 8 bits does not allow implementation of a reasonable instruction set if each instruction is limited to a length of one word. The obvious solution is to extend instructions to span several words. In order to provide the desired flexibility, the 6809 instructions vary from 1 to 5 bytes in length, depending upon the nature of each instruction.

The shortest instructions tend to be those that refer to operands in the implied addressing mode. Since these operands are the contents of microprocessor registers, few bits are needed to denote a particular register. The situation is much different when main memory operands are involved. An absolute address of 16 bits requires 2 bytes if it is to be given explicitly within an instruction. An instruction must also specify the operation to be performed and indicate how to interpret the addressing data. This information is encoded in a binary pattern referred to as the *OP-code*. The OP-code constitutes the first byte of an instruction, with addressing data given in subsequent bytes.

As a typical example, consider the LDA (Load accumulator A) instruction, where accumulator A is to be loaded with the 8 bits stored in memory location $1F4C_{16}$. Assume that the memory operand is to be specified using the absolute addressing mode. In assembler notation, this instruction may be written as

LDA $1F4C

The OP-code, which indicates a load operation from the main memory to accumulator A using the absolute addressing mode, is encoded as $B6_{16}$. Two additional bytes are needed to specify the memory address of the operand. At execution time, this instruction appears in the memory as shown in Figure 3.5. We have arbitrarily assumed that the instruction is stored starting at address 100_{16}. Consider now the LDB instruction, which performs an identical operation, except that it uses accumulator B. It may be written as

LDB $1F4C

Its OP-code is encoded as $F6_{16}$, which indicates the LDB operation in the same memory addressing mode. Thus, the OP-code pattern enables the microprocessor to

Table 3.1   6809 ADDRESSING MODES

Absolute long (extended)	A 16-bit address of the operand is stated explicitly in the two bytes that follow the OP-code.
Absolute short (direct)	The 8 least-significant bits of a 16-bit address are given in the byte that follows the OP-code and the 8 most-significant bits are the contents of the Direct Page register.
Register (implied)	Operation specified in the instruction implies an operand in one of the registers in the microprocessor.
Immediate	Value of the operand is given explicitly in one or two bytes that follow the OP-code.
Indexed	Address of the operand is the sum of the contents of a 16-bit register R (R = X,Y,U, or S) and an index value that is contained either in one or two bytes that follow the OP-code or in one of the accumulators.
Autoincrement	Address of the operand is the contents of a 16-bit register R (R = X,Y,U, or S). After the operand is accessed, the value in R is incremented automatically by 1 or 2 as specified in the instruction.
Autodecrement	Address of the operand is the contents of a 16-bit register R (R = X,Y,U, or S). Prior to accessing the operand, the value in R is decremented by 1 or 2 as specified in the instruction.
Relative	Address of the operand is the value in the program counter offset by an 8-bit or a 16-bit signed number specified in the 1 or 2 bytes that follow the OP-code.
Indirect	This mode can be used in conjunction with: extended, indexed, autoincrement, autodecrement and relative modes. In each case, the effective address determined in the primary mode is used to access a memory location that holds the address of the desired operand.

determine the desired implied operand location, either accumulator A or B in this case.

The absolute addressing mode is provided in two versions. A long version, called "extended," provides a full 16-bit address in the 2 bytes that follow the OP-code. In the short version, called "direct," only the low-order byte of the operand

Main Memory

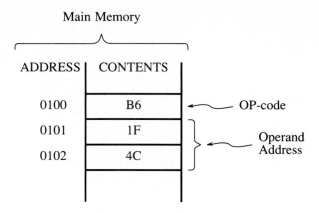

Figure 3.5   The format of LDA $1F4C instruction.

address is given as part of the instruction. In this case, the high-order byte is taken from the direct page register. Figures 3.6 and 3.7 illustrate the use of the extended and the direct modes. In each case an operand in memory location 15CA is to be accessed. This location is on page 15_{16} of the addressable space.

Main Memory

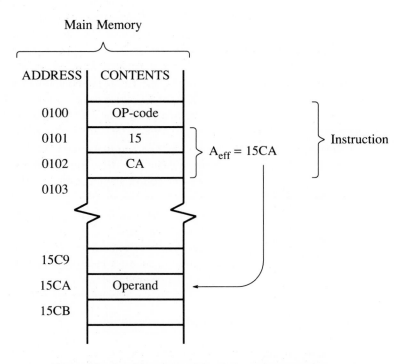

Figure 3.6   An extended addressing mode example.

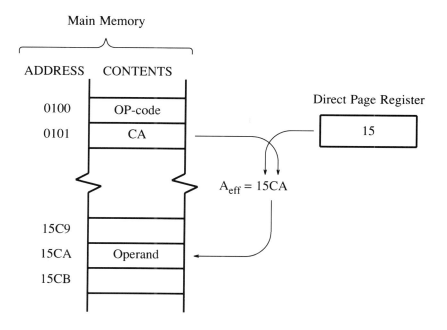

Figure 3.7 A direct addressing mode example.

We have used the term "direct" to describe the short version of the absolute mode to be consistent with the Motorola literature, but this usage is not generally found in the computer literature. In more general use, any addressing mode in which a computed address specifies a memory location where the desired operand is to be found is referred to as a direct addressing mode. This contrasts with indirect addressing modes, in which the computed address specifies a memory location where the effective address of the operand is to be found. In the latter, one extra access to the memory is needed to get the operand itself.

The immediate addressing mode allows for inclusion of either a 1-byte or a 2-byte operand explicitly in the instruction. The operand is given after the OP-code.

The pointer registers X, Y, U, and S can all be used in the indexed, autoincrement, and autodecrement modes. In the indexed mode, the index value can either be included explicitly within the instruction, or it can be given as the contents of one of the accumulators.

The autoincrement and autodecrement modes allow the pointer register to be incremented or decremented by 1 or 2. This depends upon whether a 1-byte or a 2-byte operand is accessed, respectively.

The relative addressing mode uses an offset value that is either a 1-byte or a 2-byte signed number. The register that holds the reference address relative to which the offset is specified is the program counter.

Indirect addressing modes in 6809 are available with extended, indexed, autoincrement, autodecrement, and relative modes. In each case, the contents of the location pointed at by the effective address determined in the primary mode are interpreted as the address of the operand, as was explained in Section 2.3. However, in the indirect version of the autoincrement and autodecrement modes, the pointer register must be incremented or decremented by 2. (It would make little sense to increment a pointer register by 1 when it is pointing at a 2-byte address.)

3.3.1 Assembler Syntax for 6809 Addressing Modes

Table 3.2 shows the assembler notation for the 6809 addressing modes, as well as a description of the addressing function for each mode. We will consider some entries in the table using Clear and Add instructions as examples. Note that, in what follows, instruction mnemonics are always capitalized. Moreover, if a microprocessor register is involved in the instruction as an operand location, it is indicated as a part of the mnemonic. For example, CLR is the mnemonic for a Clear instruction when the operand location is in the memory, while CLRA denotes an instruction that clears the contents of accumulator A.

The absolute mode is specified by writing either a 16-bit address of the operand, or a name representing this address, in the operand field. For example, memory location 452 can be cleared with the instruction

$$\text{CLR} \qquad 452$$

If the name LOC has been associated with the value 452 earlier in the program, then the same effect can be achieved with

$$\text{CLR} \qquad \text{LOC}$$

When an operand is specified in this manner, the assembler will use the direct, that is, 1-byte, version of the absolute mode if the value stored in the DP register is the same as the high-order byte of the address specified. Otherwise, it will use the extended mode. The assembler syntax allows for explicit specification of extended and direct modes with the symbols > and <, respectively. Thus, the instruction

$$\text{CLR} \qquad <452$$

will be assembled using the direct mode, regardless of the value in the DP register. Such explicit specification of either the extended or the direct mode is seldom used in practice. Its usefulness is limited to cases where a forward reference is made. For example, consider the situation where an operand in a given instruction is specified by name, say PLACE, which is defined later in the program. When the assembler comes upon this instruction during the first pass of the assembly process, it does not know the address of PLACE and cannot determine whether the direct addressing mode can be used. While the actual address can be inserted during the second pass, the assembler must allocate either 1 or 2 bytes of space for this address in the assembled instruction during the first pass. Unless the programmer has specified the choice of the mode, the assembler will have to use the extended mode to deal with this forward reference.

The immediate addressing mode is specified by preceding the operand with the # symbol. Thus, an instruction that adds 100 to accumulator A is written as

Table 3.2   ASSEMBLER SYNTAX FOR 6809 ADDRESSING MODES

Mode	Syntax	Addressing Function
Absolute (extended)	>Value or Value	EA = Value
Absolute short (direct)	<Value or Value	EA = [DP] ‖ LOB
Register (implied)	Included in the OP-code	Operand = [Implied register]
Immediate	#Value	Operand = Value
Indexed	,R Value,R SValue,R or ACCUM,R	EA = [R] EA = Value + [R] EA = SValue + [R] EA = [ACCUM] + [R]
Autoincrement	,R+ or ,R++	EA = [R] R ← [R] + 1 EA = [R] R ← [R] + 2
Autodecrement	,-R or ,--R	R ← [R] - 1 EA = [R] R ← [R] - 2 EA = [R]
Relative	Value,PCR	EA = Offset + [PC] EA = SOffset + [PC]
Extended indirect	[Value]	EA = [Value]
Indexed indirect	[,R] [Value,R] or [ACCUM,R]	EA = [EA indexed]
Autoincrement indirect	[,R++]	EA = [EA autoincrement]
Autodecrement indirect	[,--R]	EA = [EA autodecrement]
Relative indirect	[Value,PCR]	EA = [EA relative]

R	= X,Y,U, or S register.
ACCUM	= A,B, or D register.
EA	= effective address.
a ‖ b	= 16-bit number where a is the most significant byte and b is the least significant byte.
Value	= A number definable in up to 16 bits or a name representing it.
SValue	= 8-bit number or a name representing it.
Offset	= 16-bit signed offset.
SOffset	= 8-bit signed offset.
PC	= Program Counter.
DP	= Direct Page Register.
LOB	= Low order 8 bits of Value.

$$\text{ADDA} \qquad \#100$$

In this instruction, one operand—accumulator A—is implied by the instruction mnemonic, and the second operand is given in the immediate mode. Since accumulator A is 8 bits long, the assembler program translates the above instruction into machine code using 1 byte to represent the number 100. If an instruction refers to a 16-bit register, such as D or X, a 2-byte format would be generated. For example, the assembler will use 2 bytes to represent the number 100 in the instruction

$$\text{ADDD} \qquad \#100$$

The first operand in the above instruction is accumulator D, which is the 16-bit register formed by concatenating accumulators A and B.

The indexed mode produces the effective address of the operand by adding an index value to the contents of register R, where R may be one of the registers X, Y, U, or S. The notation for the indexed mode in the 6809 microprocessor is, unfortunately, different from the commonly used notation explained in Chapter 2. Instead of writing

$$\text{Value(R)}$$

the syntax for 6809 requires us to write

$$\text{Value,R}$$

While this appears awkward, it is not ambiguous because the operand field generally contains only one operand, hence a comma is not needed to separate the operands. As will be explained in Section 3.4, there are only a few 6809 instructions that allow specification of more than one operand in the operand field, but the indexed addressing mode cannot be used with these instructions.

The index value may either be given in the instruction or may be the contents of one of the accumulators. For example, suppose that index register X contains 400, while accumulator B contains 52. Then, the task of clearing location 452 can be accomplished with either

$$\text{CLR} \qquad 52,\text{X}$$

or

$$\text{CLR} \qquad \text{B,X}$$

If the index value is equal to 0, it need not be shown explicitly. Thus, location 400 can be cleared with either

$$\text{CLR} \qquad 0,\text{X}$$

or

$$\text{CLR} \qquad ,\text{X}$$

Autoincrement and autodecrement modes are indicated by single or double plus and minus signs as shown in Table 3.2. A 16-bit operand is loaded from memory location 400 into register Y with the instruction

$$\text{LDY} \quad ,X++$$

if the number 400 is the contents of register X. Having accessed the operand, the microprocessor increments the pointer register by 2, thus making 402 the new value in X. As another example, consider the instruction

$$\text{STA} \quad ,-X$$

where the value in register X is 400. In this case, X is first decremented by 1, so that it contains 399. Then, the contents of accumulator A are stored in memory location 399. Incrementing or decrementing a pointer register by 1 or 2 is done depending on whether a 1-byte or a 2-byte operand is involved in the instruction.

The relative mode is just like the indexed mode, except that the register used is the program counter. In the indexed mode, the programmer gives the offset either explicitly in the assembly language instruction or specifies the name of the register that contains the offset. The offset used in the relative mode is computed by the assembler program. The programmer simply gives the address of the desired operand, or a name that represents it, and indicates that the relative mode should be used. The assembler computes the actual offset. For example, the instruction

$$\text{CLR} \quad \text{PLACE,PCR}$$

clears 1 byte at PLACE, and will be assembled in the relative mode. The distance in bytes between the address of PLACE and the location of this instruction [as indicated by the contents of the program counter (PC)] is the offset, which will be computed by the assembler and included in the instruction (object) code it generates. Of course, the name PLACE must be defined elsewhere in the program.

Indirect versions of the addressing modes are indicated by placing square brackets around the direct mode specification, as shown in the table. For example, suppose that register X contains the value $1000, and that memory locations $1000 and $1001 hold one 16-bit entry of a list of addresses. Let this entry be the value $1F00. Then, we can clear memory location $1F00 with the instruction

$$\text{CLR} \quad [,X]$$

which uses the indirect version of the index mode with the index value of 0. This is illustrated in Figure 3.8. Or, if we want to clear a number of locations designated by a list of addresses, we may use the instruction

$$\text{CLR} \quad [,X++]$$

Figure 3.9 (p. 95) depicts a possible situation. Our instruction clears memory location $1F00. Then, it increments register X by 2, making its value $1002, to point to the next address in the list. When the same instruction is executed the second time, it clears memory location $20A4 and increments the value in X to $1004. Executing this instruction in an appropriate loop will clear all the locations desired.

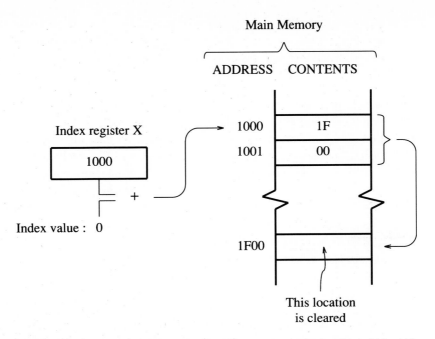

Figure 3.8 An example of indexed indirect addressing, using CLR (,X) instruction.

3.4 6809 INSTRUCTION SET

The 6809 instruction set consists of a variety of instructions that enable reasonably easy implementation of typical programming tasks. Many instructions can be used with several different addressing modes. There is an extensive set of branch instructions, as well as instructions for stack control, subroutine linkage, and interrupt handling.

Table 3.3 shows the complete 6809 instruction set. For each instruction, the table gives the operation mnemonic, the allowable addressing modes, a functional description, and the setting of the CCR flags. There are three entries for each addressing mode. These are the actual hex pattern for the OP-code, the length of the instruction in bytes, and the number of clock cycles needed for its execution.

Two-Operand Instructions

In two-operand instructions, one of the operands must be a register in the microprocessor. From the point of view of the designer of the microprocessor, this choice is motivated by the desire to keep instructions short. If both operands in an instruction could reside in the main memory, the instruction might require specification of two 16-bit addresses.

Since one operand is always a register, the 6809 instruction set does not contain a general Move instruction, where the source and the destination can be specified

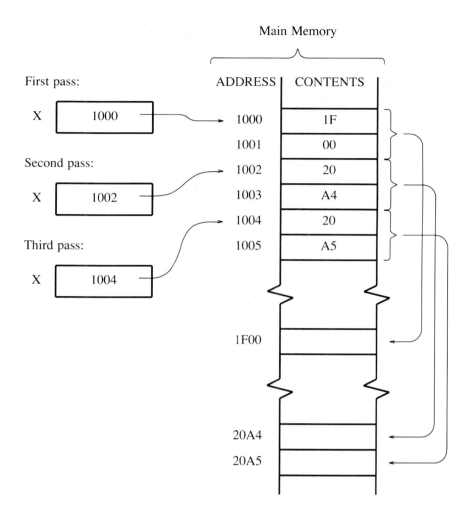

Figure 3.9 An example of autoincrement indirect addressing, using CLR (,X++) instruction.

arbitrarily. Instead, Load and Store instructions are provided. A register, say accumulator A, can be loaded from a memory location LOC with an instruction of the type

LDA   LOC

Conversely, a Store instruction transfers the contents of a processor register into a specified memory location.

Assume that a byte of data in memory location LOCA is to be transferred into memory location LOCB. The desired transfer can be achieved with two instructions

LDA   LOCA
STA   LOCB

Table 3.3   6809 INSTRUCTION SET

Operation	Mnemonic	IMMED			DIRECT			EXTEND		
		OP	#	~	OP	#	~	OP	#	~
Add	ADDA/B	8B/CB	2	2	9B/DB	2	4	BB/FB	3	5
	ADDD	C3	3	4	D3	2	6	F3	3	7
Add with carry	ADCA/B	89/C9	2	2	99/D9	2	4	B9/F9	3	5
Add accumulator B to index register X (unsigned)	ABX									
AND	ANDA/B	84/C4	2	2	94/D4	2	4	B4/F4	3	5
	ANDCC	1C	2	3						
	CWAI	3C	2	20						
Bit test	BITA/B	85/C5	2	2	95/D5	2	4	B5/F5	3	5
Clear	CLR				0F	2	6	7F	3	7
	CLRA/B									
Compare	CMPA/B	81/C1	2	2	91/D1	2	4	B1/F1	3	5
	CMPD	1083	4	5	1093	3	7	10B3	4	8
	CMPS	118C	4	5	119C	3	7	11BC	4	8
	CMPU	1183	4	5	1193	3	7	11B3	4	8
	CMPX	8C	3	4	9C	2	6	BC	3	7
	CMPY	108C	4	5	109C	3	7	10BC	3	8
Complement (1's)	COM				03	2	6	73	3	7
	COMA/B									
Decimal adjust	DAA									
Decrement	DEC				0A	2	6	7A	3	7
	DECA/B									
Exclusive-OR	EORA/B	88/C8	2	2	98/D8	2	4	B8/F8	3	5
Exchange register contents	EXG (R1,R2)									
Increment	INC				0C	2	6	7C	3	7
	INCA/B									

INDEX			IMPLIED			Performed function	Condition code flags affected				
OP	#	~	OP	#	~		H	N	Z	V	C
AB/EB	2+	2+				A ← [A] + [M]	x	x	x	x	x
E3	2+	6+				D ← [D] + [M]	x	x	x	x	x
A9/E9	2+	4+				A ← [A] + [M] + [C]	x	x	x	x	x
			3A	1	3	X ← [X] + [B] (unsigned)					

A4/E4	2+	4+				A ← [A] ∧ [M]
						CCR ← [CCR] ∧ [M]
						CCR ← [CCR] ∧ [M] ;
						  store registers on S-stack;
						  wait for interrupt
							
							x
							x
								x
								x
								x
									x
									x
									x
										0
										x
										x
											
											x
											x

A5/E5	2+	4+				[A] ∧ [M]		x	x	0	
6F	2+	6+				M ← 0	0	1	0	0
			4F/5F	1	2	A ← 0	0	1	0	0
A1/E1	2+	4+				[A] − [M]		x	x	x	x
10A3	3+	7+				[D] − [M]		x	x	x	x
11AC	3+	7+				[S] − [M]		x	x	x	x
11A3	3+	7+				[U] − [M]		x	x	x	x
AC	2+	6+				[X] − [M]		x	x	x	x
10AC	3+	7+				[Y] − [M]		x	x	x	x
63	2+	6+				M ← $\overline{[M]}$		x	x	0	1
			43/53	1	2	A ← $\overline{[A]}$		x	x	0	1
			19	1	2	Converts binary sum of BCD characters into BCD format		x	x	0	x
6A	2+	6+				M ← [M] − 1		x	x	x	
			4A/5A	1	2	A ← [A] − 1		x	x	x	
A8/E8	2+	4+				A ← [A] ⊕ [M]		x	x	0	
			1E	2	7	R1 ↔ R2					
6C	2+	6+				M ← [M] + 1		x	x	x	
			4C/5C	1	2	A ← [A] + 1		x	x	x	

Table 3.3 (continued)   6809 INSTRUCTION SET

Operation	Mnemonic	IMMED			DIRECT			EXTEND		
		OP	#	~	OP	#	~	OP	#	~
Load	LDA/B	86/C6	2	2	96/D6	2	4	B6/F6	3	5
	LDD/U	CC/CE	3	3	DC/DE	2	5	FC/FE	3	6
	LDS	10CE	4	4	10DE	3	6	10FE	4	7
	LDX	8E	3	3	9E	2	5	BE	3	6
	LDY	108E	4	4	109E	3	6	10BE	4	7
Load effective address	LEAS/U									
	LEAX/Y									
Multiply	MUL									
Negate (2's complement)	NEG				00	2	6	70	3	7
	NEGA/B									
No operation	NOP									
OR (inclusive)	ORA/B	8A/CA	2	2	9A/DA	2	4	BA/FA	3	5
	ORCC	1A	2	3						
Push	PSHS/U									
	(reg. list)									
Pull	PULS/U									
	(reg. list)									
Rotate left	ROL				09	2	6	79	3	7
	ROLA/B									
Rotate right	ROR				06	2	6	76	3	7
	RORA/B									
Shift left (arithmetic)	ASL				08	2	6	78	3	7
	ASLA/B									
Shift right (arithmetic)	ASR				07	2	6	77	3	7
	ASRA/B									
Shift left (logic)	LSL				08	2	6	78	3	7
	LSLA/B									
Shift right (logic)	LSR				04	2	6	74	3	7
	LSRA/B									

INDEX			IMPLIED			Performed function	Condition code flags affected				
OP	#	~	OP	#	~		H	N	Z	V	C
A6/E6	2+	4+				$A \leftarrow [M]$		x	x	0	
EC/EE	2+	5+				$D \leftarrow [M]$		x	x	0	
10EE	3+	6+				$S \leftarrow [M]$		x	x	0	
AE	2+	5+				$X \leftarrow [M]$		x	x	0	
10AE	3+	6+				$Y \leftarrow [M]$		x	x	0	
32/33	2+	4+				$S \leftarrow EA$					
30/31	2+	4+				$X \leftarrow EA$		x			
			3D	1	11	$D \leftarrow [A] \times [B]$ (unsigned)			x		x
60	2+	6+				$M \leftarrow 0 - [M]$		x	x	x	x
			40/50	1	2	$A \leftarrow 0 - [A]$		x	x	x	x
			12	1	2	No operation					
AA/EA	2+	4+				$A \leftarrow [A] \vee [M]$		x	x	0	
						$CCR \leftarrow [CCR] \vee [M]$	x	x	x	x	x
			34/36	2	5+	Push registers on S-stack					
			35/37	2	5+	Pull registers from S-stack					
69	2+	6+				M		x	x	x	x
			49/59	1	2	A		x	x	x	x
66	2+	6+				M		x	x		x
			46/56	1	2	A		x	x		x
68	2+	6+				M		x	x	x	x
			48/58	1	2	A		x	x	x	x
67	2+	6+				M		x	x		x
			47/57	1	2	A		x	x		x
68	2+	6+				M		x	x	x	x
			48/58	1	2	A		x	x	x	x
64	2+	6+				M		0	x		x
			44/54	1	2	A		0	x		x

Table 3.3 (continued)   6809 INSTRUCTION SET

Operation	Mnemonic	Addressing modes								
		IMMED			DIRECT			EXTEND		
		OP	#	~	OP	#	~	OP	#	~
Sign extend B into A	SEX									
Store	STA/B				97/D7	2	4	B7/F7	3	5
	STD				DD	2	5	FD	3	6
	STS				10DF	3	6	10FF	4	7
	STU/X				DF/9F	2	5	FF/BF	3	6
	STY				109F	3	6	10BF	4	7
Subtract	SUBA/B	80/C0	2	2	90/D0	2	4	B0/F0	3	5
	SUBD	83	3	4	93	2	6	B3	3	7
Subtract with carry	SBCA/B	82/C2	2	2	92/D2	2	4	B2/F2	3	5
Transfer register to register	TFR (R1,R2)									
Test	TST TSTA/B				0D	2	6	7D	3	7

		RELATIVE			DIRECT			EXTEND		
		OP	#	~	OP	#	~	OP	#	~
Branch always	BRA	20	2	3						
	LBRA	16	3	5						
Branch if carry clear	BCC	24	2	3						
	LBCC	1024	4	5(6)						
Branch if carry set	BCS	25	2	3						
	LBCS	1025	4	5(6)						
Branch if = 0	BEQ	27	2	3						
	LBEQ	1027	4	5(6)						
Branch if ≥ 0	BGE	2C	2	3						
	LBGE	102C	4	5(6)						
Branch if > 0	BGT	2E	2	3						
	LBGT	102E	4	5(6)						
Branch if higher	BHI	22	2	3						
	LBHI	1022	4	5(6)						
Branch if higher or same	BHS	24	2	3						
	LBHS	1024	4	5(6)						

INDEX			IMPLIED			Performed function	Condition code flags affected				
OP	#	~	OP	#	~		H	N	Z	V	C
			1D	1	2	Set all bits in A to equal the sign bit in B		x	x	0	
A7/E7	2+	4+				M ← [A]		x	x	0	
ED	2+	5+				M ← [D]		x	x	0	
10EF	3+	6+				M ← [S]		x	x	0	
EF/AF	2+	5+				M ← [U]		x	x	0	
10AF	3+	6+				M ← [Y]		x	x	0	
A0/E0	2+	4+				A ← [A] − [M]		x	x	x	x
A3	2+	6+				D ← [D] − [M]		x	x	x	x
A2/E2	2+	4+				A ← [A] − [M] − [C]		x	x	x	x
			1F	2	7	R2 ← [R1]					
6D	2+	6+				[M] − 0		x	x	0	
			4D/5D	1	2	[A] − 0		x	x	0	

INDEX			IMPLIED			Branch test
OP	#	~	OP	#	~	
						None
						[C] = 0
						[C] = 1
						[Z] = 1
						[N] ⊕ [V] = 0
						[Z] ∨ ([N] ⊕ [V]) = 0
						[C] ∨ [Z] = 0
						[C] = 0

Table 3.3 (continued) 6809 INSTRUCTION SET

Operation	Mnemonic	RELATIVE			DIRECT			EXTEND		
		OP	#	~	OP	#	~	OP	#	~
Branch if ≤ 0	BLE	2F	2	3						
	LBLE	102F	4	5(6)						
Branch if lower	BLO	25	2	3						
	LBLO	1025	4	5(6)						
Branch if lower or same	BLS	23	2	3						
	LBLS	1023	4	5(6)						
Branch if < 0	BLT	2D	2	3						
	LBLT	102D	4	5(6)						
Branch if minus	BMI	2B	2	3						
	LBMI	102B	4	5(6)						
Branch if ≠ 0	BNE	26	2	3						
	LBNE	1026	4	5(6)						
Branch if plus	BPL	2A	2	3						
	LBPL	102A	4	5(6)						
Branch never	BRN	21	2	3						
	LBRN	1021	4	5						
Branch if overflow clear	BVC	28	2	3						
	LBVC	1028	4	5(6)						
Branch if overflow set	BVS	29	2	3						
	LBVS	1029	4	5(6)						
Branch to subroutine	BSR	8D	2	7						
	LBSR	17	3	9						
Jump	JMP				0D	2	3	7E	3	4
Jump to subroutine	JSR				9D	2	7	BD	3	8
Return from interrupt	RTI									
Return from subroutine	RTS									
Software interrupt	SWI									
	SWI2									
	SWI3									
Wait for interrupt	SYNC									

INDEX			IMPLIED			Branch test
OP	#	~	OP	#	~	
						$[Z] \vee ([N] \oplus [V]) = 1$
						$[C] = 1$
						$[C] \vee [Z] = 1$
						$[N] \vee [V] = 1$
						$[N] = 1$
						$[Z] = 0$
						$[N] = 0$
						None
						$[V] = 0$
						$[V] = 1$
6E	2+	3+				
AD	2+	7+				
			3B	1	6/15	
			39	1	5	
			3F	1	19	
			103F	2	20	
			113F	2	20	
			13	1	2	

Table 3.3 (continued)   6809 INSTRUCTION SET

Note: OP = OP code

   # = number of bytes in the instruction

   EA = effective address

   For 2-byte operands: $R \leftarrow [M]$ implies $R_H \leftarrow [M]$ and $R_L \leftarrow [M + 1]$

   5(6) means: 5 cycles if branch is not taken, 6 cycles if taken

   In EXG and TFR instructions, the R1 and R2 registers are specified
   in the second byte as follows:

   ```
   7     4 3     0
   ┌───────┬───────┐
   │  R1   │  R2   │
   └───────┴───────┘
   ```
   where bit patterns for the registers that
   can be used as R1 and R2 are

D :	0000	PC :	0101
X :	0001	A :	1000
Y :	0010	B :	1001
U :	0011	CCR :	1010
S :	0100	DP :	1011

Sixteen-bit quantities can be moved conveniently using 16-bit registers. For example, assume that 2 bytes of data in memory locations LOCA and LOCA+1 are to be transferred into memory locations LOCB and LOCB+1. Using register Y, this may be done as

   LDY   LOCA
   STY   LOCB

In two-operand instructions, one operand is specified in the operand field, while the other operand is usually indicated within the mnemonic for the instruction. There are six exceptions to this rule. The instruction

   TFR   R1,R2

transfers the contents of the register specified as R1 into the register specified as R2. Both registers are stated in the operand field. The same format is used to indicate the registers in the EXG instruction, which exchanges the contents of the specified registers. In both these instructions, the R1 and R2 registers can be any of the registers shown in Figure 3.2. The other four exceptions are the PSHS, PSHU, PULS, and PULU instructions which push or pull the processor registers onto or off either the S or U stack. The registers involved are written in the operand field. For example, the instruction

   PSHS   A,B,Y

pushes registers A, B, and Y onto the S stack. These instructions will be discussed in detail in Section 3.8. Observe that in these six instructions, commas are used to separate the registers specified in the operand field. Only the microprocessor registers can be the operands in these instructions, and no memory location can be specified instead. This use of commas should not be confused with that in the indexed addressing mode, which was explained in the previous section.

One-Operand Instructions

In one-operand instructions the operand may be either in the memory or in a processor register. The memory operands are specified in the operand field, as in

CLR   LOC

which clears memory location LOC.

Microprocessor registers are indicated within the operation mnemonic, as in

CLRA

which clears accumulator A.

Branch Instructions

A substantial number of branch instructions are included in the 6809 instruction set. Branch test conditions are evaluated on the basis of the status of flags in the condition code register, CCR. The branch address, which is the address of the next instruction to be executed if a test condition is true, is specified relative to the current contents of the program counter. An offset value in the form of either an 8-bit or a 16-bit signed number is given in the instruction. The branch address is derived as

Address of next instruction = [PC] + offset

It should be pointed out that at the time the branch address is computed, the contents of the program counter will have been incremented to point to the next byte in memory following the branch instruction. That is, the PC will be pointing to the first byte of the instruction that will be executed if the branch test condition is not true. Since the branch address is calculated by adding an offset value to the contents of the PC, it is quite common to say that branch instructions specify branch addresses in the relative mode.

Note that the value of an 8-bit offset, which must be interpreted as a 2's complement number, is in the range

$$-128 \ (=10000000_2) \leq \ \text{Offset} \ \leq +127 \ (=01111111_2)$$

This range is sufficient for most branches that are needed in short program loops. In cases where a branch is to be made to an instruction far away from the branch instruction, it is necessary to use a 16-bit offset. For this reason, two types of branch instructions are provided, one with short and one with long offset. For example, the BEQ (Branch if Equal to 0) instruction uses an 8-bit offset and causes a branch if the Z flag in the CCR is equal to 0. The long version of this instruction has the mnemonic LBEQ.

The flags in the condition code register are examined by the branch instructions to determine if a branch is to be performed. Many instructions affect the status of these flags. All arithmetic and logic operations set the flags based on the result produced. Instructions of the Load, Store, and Test type set the flags according to the numerical value of the operand involved. The only flags affected by such instructions are the N, Z, and V flags. The first two are set to 1 if the operand is negative or zero, respectively, while V is always cleared to 0. It is important to note that a flag in the CCR always reflects the status produced by the most recently executed instruction that has an effect on this flag.

The functional meaning for most branch instructions is obvious from either their names or the logic expressions that describe the branch test condition. However, there are some subtle cases that should be recognized. There are two instructions that test if the result of a previous arithmetic operation was greater than zero. These are the BGT (Branch if Greater Than zero) and BHI (Branch if HIgher) instructions. Typically, these instructions are preceded in a program by a Compare instruction, which subtracts the values of its two operands. While the BGT and BHI instructions seemingly perform the same branch test, they differ in the way the operands are interpreted. The BGT instruction assumes that the previous comparison operation was done on signed operands, while the BHI instruction assumes that the operands were unsigned. There exists a similar difference between the branch instructions in each of the instruction pairs (BGE, BHS), (BLT, BLO), and (BLE, BLS). Thus, when dealing with data consisting of signed operands, one of the instructions BGT, BGE, BLT, or BLE should be used. The other four instructions are needed for numbers that represent unsigned quantities, such as addresses and counts.

Branch instructions use the relative mode in determining the address of the next instruction. However, there is another instruction—the Jump instruction (JMP)— where this address is given explicitly in the instruction, by including a 16-bit value that is the required address to which a branch is made. The JMP instruction is unconditional; that is, the branch is always performed.

Finally, we should mention the 6809 instructions available for writing subroutines. There are three Call instructions. Two are of the branch type, named BSR (Branch to Subroutine) and LBSR (Long Branch to Subroutine), which use the relative mode in determining the branch address. The third is JSR (Jump to Subroutine), which includes the address of the subroutine as an absolute 16-bit number. Return from subroutine is carried out with the RTS instruction.

Instruction Length

As pointed out in the previous section, the 6809 instructions can be from 1 to 5 bytes in length, depending upon the type of instruction and the addressing mode used. The reader should note that for the indexed mode, the length shown is the minimum number of bytes for this mode. The indexed mode incorporates a number of options, to be discussed in Section 3.6. It can have 1 byte or 2 bytes more than the number given in Table 3.3. This fact is indicated with a plus sign that follows the minimum number of bytes.

Speed of Execution

The number of clock cycles, given in the column marked "~" in Table 3.3, is indicative of the speed with which instructions can be executed. One clock cycle is required for each access to memory. Additional clock cycles are needed in some instructions for processing within the microprocessor. For example, the instruction ADDA #12 can be executed in two cycles; one is needed to fetch the OP-code byte, and another to fetch the immediate operand byte that is to be added to accumulator A. The necessary addition is carried out while the OP-code of the next instruction is being fetched. On the other hand, the instruction ADDD #12 , which adds the value 12 to the 16-bit accumulator D, requires four cycles. Fetching of the OP-code and the two bytes of the immediate operand takes three cycles. One more cycle is needed to add the 16-bit operand to accumulator D.

3.5 EXAMPLES

At this point it is worthwhile to reconsider some of the programming examples given in Chapter 2, to see how they can be implemented in 6809 assembly language.

List Transfer

The program of Figure 2.5 can be realized as shown in Figure 3.10. The program is largely self-explanatory, although some instructions warrant a few comments. Index register X is initially cleared by loading a zero as an immediate operand into it. The reason for this somewhat awkward way of clearing a register is the fact that a simpler CLRX instruction is not provided in the 6809 instruction set. The two Load and Store instructions are given in the indexed mode, where LISTA and LISTB are the index values that correspond to the starting addresses of the two lists.

The only unfamiliar instruction in the program of Figure 3.10 is

$$LEAX \qquad 1,X$$

Its function is merely to increment the contents of register X by 1. It is used because of the absence of a simpler INCX instruction. The mnemonic LEAX means Load Effective Address into register X. The operand in the instruction is specified in the indexed mode, so that its effective address is

$$A_{eff} = 1 + [X]$$

Thus, the LEAX instruction used results in

$$X \leftarrow 1 + [X]$$

which increments the index register as required.

```
          LDX    #0          Clear index register X.
   LOOP   LDA    LISTA,X     Transfer an entry from list A
          STA    LISTB,X        to list B.
          LEAX   1,X         Increment index register X by 1.
          CMPX   N           Branch back until the last
          BLO    LOOP           entry is copied.
          next instruction
```

Figure 3.10   A 6809 program corresponding to Figure 2.5.

```
          CLRA                  Clear accumulator A.
          LDX       #LISTA      Initialize index registers X and Y
          LDY       #LISTB          to point to lists A and B.
   LOOP   LDB       A,X         Transfer an entry from list A to
          STB       A,Y             list B, by means of accumulator B.
          INCA                  Increment the counter.
          CMPA      N           Branch back until the last
          BLO       LOOP            entry is copied.
          next instruction
```

Figure 3.11   An alternative 6809 program corresponding to Figure 2.5.

Figure 3.11 shows an alternative 6809 program that corresponds to Figure 2.5. In this case two index registers are used. They are loaded with the starting addresses of lists A and B. These addresses are provided as immediate operands in the LDX and LDY instructions. The reader should note that the instruction

$$\text{LDX} \qquad \text{\#LISTA}$$

loads the X register with the address of the first entry in list A. This is to be contrasted with the instruction

$$\text{LDX} \qquad \text{LISTA}$$

which loads the X register with the 2 bytes of data that are the contents of memory locations LISTA and LISTA + 1.

The Load and Store instructions use the indexed addressing mode, with the index value being the contents of accumulator A. The accumulator also serves as the loop counter.

The program of Figure 2.10 may be implemented as indicated in Figure 3.12. This particular program will copy 50 bytes from list A to list B. Note that accumulator A is used as the counter NCOUNT. This makes the program shorter and faster to execute than if a main memory location is used as the counter.

The program of Figure 3.12 may be written in the form of a subroutine as shown in Figure 3.13. The resulting subroutine corresponds to the program of Figure 2.15, with a particular number, 50, being chosen as the value n.

```
          LDA       #50         Set the counter to 50.
          LDX       #LISTA      Initialize pointer registers X and Y
          LDY       #LISTB          to point to lists A and B.
   LOOP   LDB       ,X+         Transfer an entry from list A to
          STB       ,Y+             list B, by means of accumulator B.
          DECA                  Branch back until the last
          BHI       LOOP            entry is copied.
          next instruction
```

Figure 3.12   A 6809 program corresponding to Figure 2.10.

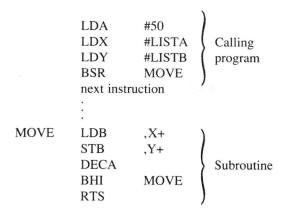

```
          LDA      #50      ⎫
          LDX      #LISTA   ⎬  Calling
          LDY      #LISTB   ⎪  program
          BSR      MOVE     ⎭
          next instruction
             ⋮

MOVE      LDB      ,X+      ⎫
          STB      ,Y+      ⎪
          DECA              ⎬  Subroutine
          BHI      MOVE     ⎪
          RTS               ⎭
```

Figure 3.13 A 6809 program for the example in Figure 2.15.

In all of the examples above, the address distance between each branch instruction and its branch destination is small. Hence, short-offset branch instructions have been used. A possible exception could be the subroutine call, BSR, in Figure 3.13. If the subroutine MOVE happens to be located farther away from the BSR instruction than can be specified with an 8-bit offset, the LBSR instruction must be used instead.

Input/Output

The input and output programs discussed in Section 2.4 can be implemented in straightforward fashion. Figure 3.14 shows a program that reads one character from a keyboard, which corresponds to the program of Figure 2.12. Note that the accumulator A is used as ACCUM. Also, we have assumed that the names STATUS and INBUF have been assigned numerical values that are the actual addresses of these registers, by means of EQU assembler directives. Figure 3.15 gives a program that displays a line of characters on a video screen, which corresponds to Figure 2.13. Register X is used as a pointer register that indicates the memory location of the character that is being transferred to the display buffer. Because the 6809 microprocessor does not have a general Move instruction, each character is first loaded into accumulator A and then transferred to the OUTBUF register. The accumulator A is also used for testing the display status bit, b_1 in Figure 2.11, within the WAIT loop. The last character transferred is the carriage return character, CR. Its ASCII code, $0D_{16}$, is assigned to the name CR by an EQU directive. Other EQU directives are needed to specify the addresses represented by the names MESG, STATUS, and OUTBUF. These are not shown in Figure 3.15.

```
WAIT      LDA      STATUS    Read status.
          ANDA     #1        Check bit 0.
          BEQ      WAIT      Loop if no character ready.
          LDA      INBUF     Transfer data.
          next instruction
```

Figure 3.14 A 6809 program that reads one character from a keyboard.

```
CR      EQU     $0D             ASCII code for carriage return.
        LDX     #MESG           Initialize pointer.
WAIT    LDA     STATUS          Wait for the display
        ANDA    #2                  to become ready.
        BEQ     WAIT
        LDA     ,X+             Transfer a character to the
        STA     OUTBUF              output buffer.
        CMPA    #CR             Continue displaying
        BNE     WAIT                if not CR.
        next instruction
```

Figure 3.15   A 6809 program that displays a one-line message.

3.6 MACHINE ENCODING OF 6809 INSTRUCTIONS

So far we have considered the 6809 instructions in their assembly language format. In order to execute a program, each instruction must be encoded into machine code, which is the job performed by the assembler program. The programmer can write a source program in assembly language, or more preferably in a high-level language, without any knowledge of the actual machine code that this program will be translated into. However, there are times when it is very useful to understand the machine code, particularly when difficulties are encountered in having a program executed correctly. In order to "debug" a program, that is, to correct all errors in it, the programmer may find it necessary to resort to an examination of the object program directly. This section will discuss some of the relevant aspects of the 6809 machine code.

An instruction may be represented by a single byte. This is the case with instructions where, in addition to the required operation, the OP-code byte specifies one or two implied operands. Some examples are INCA (Increment Accumulator A), DECB (Decrement Accumulator B) and TSTA (Test Accumulator A). Table 3.3 shows that the OP-codes of these instructions are 4C for INCA, 5A for DECB, and 4D for TSTA. An example of a two-operand instruction that can be represented with a single byte is MUL (Multiply), which multiplies the contents of accumulators A and B and places the result in the 16-bit accumulator D. Recall that accumulator D consists of accumulator B as its low-order byte and accumulator A as its high-order byte, which means that as a result of this instruction the original contents of accumulators A and B are lost. The OP-code for the MUL instruction is 3D.

In instructions that specify operands in the memory, the OP-code is followed by the addressing information in the subsequent bytes, as explained in Section 3.3. Such instructions may require 1 or 2 bytes for the OP-code. Examples of 1-byte OP-codes are LDA, STB, ADDA, CMPB, and LDX. Ideally, all instructions should have a 1-byte OP-code. However, the number of distinct codes that can be defined with 8 bits is 256. This number is insufficient to define all OP-codes for 6809. Recall that the OP-codes not only indicate the operations to be performed, but they also designate various registers that are involved in these operations and the address mode to be used. The solution is to use 2-byte OP-codes for some instructions. Such is the case, for example, with LDY, STY, and CMPY instructions. Table 3.3 gives the OP-code for LDY in the immediate addressing mode as 108E. This means that the instruction

LDY   #$2FA0

will occupy 4 bytes of main memory, as shown in Figure 3.16. The OP-code is followed by the immediate operand to be loaded into index register Y. The operand must have 16 bits, because this is the size of register Y. It is interesting to note that these instructions have 2-byte OP-codes when they involve register Y, whereas the equivalent instructions using register X, i.e., LDX, STX, and CMPX, have only a 1-byte OP-code. This may be explained by the fact that the 6809 is an enhancement of the 6800 microprocessor, which had only one index register, X, and all its instructions were encoded in 1-byte OP-codes. When writing 6809 programs, it is advisable to use register X if only one index register is needed, in order to take advantage of the shorter OP-code.

Post-byte

An extra byte to extend the number of possible OP-codes is found in the "indexed" addressing mode. Table 3.3 shows either a 1-byte or a 2-byte OP-code for each instruction in the indexed mode. However, the same OP-code is also used to designate the autoincrement, the autodecrement, and the relative mode, as well as some indirect addressing modes. The specification of which of these is intended is provided by an additional byte, called the *post-byte* in the manufacturer's literature.

The format of the post-byte is shown in Table 3.4. Six different versions of the indexed mode are available. The simplest version includes no index value at all (referred to as zero offset in the table). Thus the effective address is just the contents of register R, where R can be any of the 16-bit pointer registers X, Y, S, or U. The shortest index value consists of 5 bits, included within the post-byte, which specify a signed number in the range −16 to +15. This number is added to the contents of register R to derive the effective address. Index values with lengths of 8 or 16 bits can also be specified in an instruction as 1 or 2 bytes that follow the post-byte. The

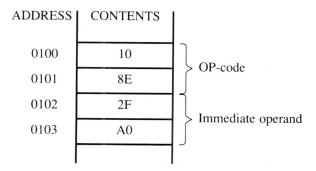

Figure 3.16 Memory layout for the instruction LDY #$2FA0.

Table 3.4 POST-BYTE FORMAT IN 6809

Post-byte bits 7 6 5 4 3 2 1 0	Assembler Syntax	Addressing Mode	#	Extra cycles
0 r r q q q q q	Q,R	Indexed with 5-bit signed offset	0	1
1 r r 0 0 0 0 0	,R+	Autoincrement	0	2
1 r r i 0 0 0 1	,R++		0	3
1 r r 0 0 0 1 0	,-R	Autodecrement	0	2
1 r r i 0 0 1 1	,--R		0	3
1 r r i 0 1 0 0	,R	Indexed with zero offset	0	0
1 r r i 0 1 0 1	B,R	Indexed with accumulator offset	0	1
1 r r i 0 1 1 0	A,R		0	1
1 r r i 1 0 0 0	SValue,R	Indexed with 8-bit signed offset	1	1
1 r r i 1 0 0 1	Value,R	Indexed with 16-bit signed offset	2	4
1 r r i 1 0 1 1	D,R	Indexed with register D offset	0	4
1 x x i 1 1 0 0	Value,PCR	Relative with 8-bit signed offset	1	1
1 x x i 1 1 0 1	Value,PCR	Relative with 16-bit signed offset	2	5
1 x x 1 1 1 1 1	[Value]	Extended indirect	2	5

Notes: rr specifies register R where 0 0 => R = X
                                       0 1 => R = Y
                                         1 0 => R = U
                                         1 1 => R = S

Q = (q q q q q) is a 5-bit signed number ($-16 \leq Q \leq 15$)
i = 1 specifies indirect address mode
x x = undefined

The number of cycles in the last column corresponds to the case where i = 0. This number is increased by 3 when i = 1.

remaining versions of the indexed mode use the 8-bit contents of accumulators A and B, or their combined 16-bit value in register D, as the index value. Note that the term "offset" is used in Table 3.4 for the index value, to be consistent with the terminology in Motorola literature.

Two forms of the relative addressing mode can be specified in the post-byte. These use either an 8-bit or a 16-bit signed offset from the contents of the program counter.

The autoincrement and autodecrement modes are also designated by the post-byte. Each comes in two forms, allowing register R to be incremented by either 1 or 2.

Finally, all indirect addressing modes are defined through the post-byte. As shown in Table 3.4, bit position 4 is used in most cases to denote whether a direct or an indirect version of a given addressing mode is required. This bit is set to 1 to specify the indirect mode.

To illustrate the use of the post-byte we will consider the LDA instruction with various addressing modes. According to Table 3.3, the OP-code for LDA in the "indexed" mode is A6. The post-byte provides the remainder of the specification. Figure 3.17 shows several forms of the LDA instruction, with detailed indication of how the final code is obtained. The post-byte codes are derived from the information in Table 3.4.

Figure 3.17(a) is a simple case where no index value is required. The post-byte consists of the code for this case and an indication that register X is used as the index register R.

Part (b) of the figure describes a case where a negative index value is specified. The value, -9, is small enough to fit into a 5-bit field that can be specified within the post-byte. A 2's-complement representation of this number, 10111, is placed in bits b_{4-0} of the post-byte, where b_4 is interpreted as the sign bit. The index register used in this example is Y, indicated by setting $b_6 = 0$ and $b_5 = 1$.

Indexed indirect mode is illustrated by the instruction

LDA   [$1A3F,U]

whose code is given in Figure 3.17(c). The indirect mode is denoted by setting b_5 to 1. The 16-bit index value, 1A3F, occupies the two bytes that follow the post-byte.

An autoincrement mode example is given in part (d) of the figure. While this is obviously not a version of the indexed mode, it is specified through the post-byte.

The final example, in Figure 3.17(e), illustrates the relative mode. The instruction is written as

LDA   LOC,PCR

where LOC is the symbolic name for the address of the required operand. We have assumed that the value of LOC is 02F0. The symbol PCR denotes that the effective address of the operand is to be derived relative to the current contents of the program counter. At the time the effective address is computed, the contents of the PC are 0104, pointing to the first byte of the instruction that follows. The offset (distance) between this value and the operand address, 02F0, is equal to 01EC. This is the index value that is specified in the 2 bytes that follow the OP-code. It is important to note that the programmer can specify LOC either as a name or as an explicit numerical value. During the assembly of the program, the assembler computes the required offset and encodes the instruction as shown in Figure 3.17(e).

From the discussion above, it is clear that the number of bytes required to specify an instruction that uses the "indexed" mode is variable. The OP-code is either 1 or 2 bytes long. The post-byte specifies the exact addressing mode to be

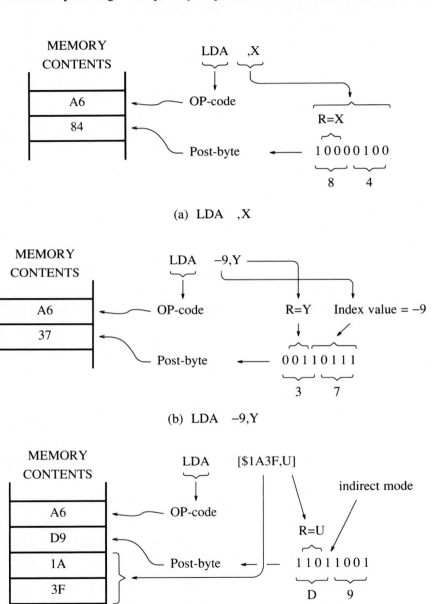

(a) LDA   ,X

(b) LDA   −9,Y

(c) LDA   [$1A3F,U]

Figure 3.17   (continued on next page)

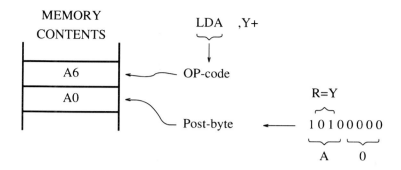

(d) LDA   ,Y+

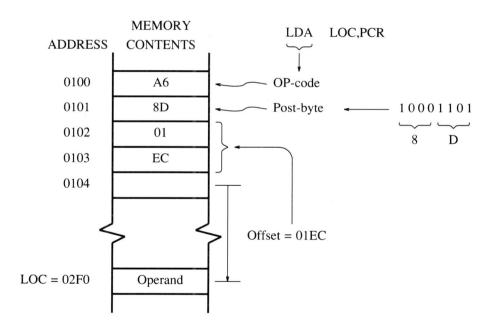

(e) LDA   LOC,PCR

Figure 3.17   Examples of machine code for various versions of the LDA instruction.

used, and may be followed by additional information in another 1 or 2 bytes. Thus, the length of these instructions ranges from 2 to 5 bytes. In Table 3.3 this is denoted by showing the minimum number of bytes and a plus sign to indicate that extra bytes are needed in some modes. Table 3.4 shows how many extra bytes are needed in each case. The number of clock cycles needed to execute a given instruction is dependent upon its length and is shown in the tables in a similar fashion. The additional number of clock cycles in the "indexed" mode is given in Table 3.4. The numbers shown correspond to the cases where the indirect bit, bit 4 of the post-byte, is equal to 0. If the indirect version of these addressing modes is specified, that is when bit 4 is equal to 1, another three cycles are needed. For example, the instruction

$$\text{STA} \qquad 1000, \text{X}$$

is encoded in 4 bytes and takes 8 cycles to execute. Its indirect version

$$\text{STA} \qquad [1000, \text{X}]$$

is also encoded in 4 bytes, but it takes 11 cycles to execute.

Branch Instructions

Branch instructions are specified with 1- or 2-byte OP-codes. Typically, short branch instructions use 1-byte OP-codes and long branch instructions have 2-byte OP-codes. The OP-code is followed by a signed offset, 8 bits for short branches, and 16 bits for long ones. A positive offset causes a branch in the forward direction, while a negative offset causes a branch back as encountered in loops.

Figure 3.18 shows the machine code for the loop segment in the program of Figure 3.12. Recall that a branch is made by loading the branch address, i.e., the address of the instruction to be executed next into the program counter (PC). This address is derived by adding the offset value to the current contents of the PC. Because the contents of the PC are updated as each instruction byte is fetched, the current contents at the time of computation of the branch address correspond to the first byte of the instruction that follows the branch instruction. In our example, the contents of the PC will be 0107 when a branch is to be made back to address 0100. Therefore, the required offset is −7, which in 2's-complement representation using 1 byte becomes 11111001 ($=\text{F}9_{16}$).

An example of a forward branch is given in Figure 3.19. Here, a branch from the instruction at address 0100 is to be made to an instruction at address 12FB. The required offset is large enough to require 16 bits to represent it. Thus, a long branch instruction, LBEQ, is used which is 4 bytes in length, 2 for the OP-code, and 2 for the offset. This means that the contents of the PC will be 0104 when the offset is calculated. The difference between 12FB and 0104 is 11F7, which is the offset placed after the OP-code in the LBEQ instruction.

In this section we have considered some detailed examples of how the 6809 instructions are encoded. The reader can derive the machine code for any instruction with the aid of Tables 3.3 and 3.4.

3.7 REVIEW AND EXERCISES

We have discussed the register structure, the addressing modes, and the basic characteristics of the instruction set of the 6809 microprocessor. This material, coupled with an understanding of the 6809 assembly language rules, will allow the reader to

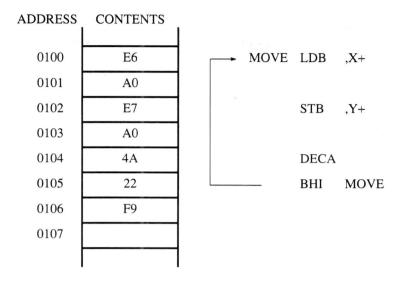

Figure 3.18 Machine code for the loop in the program of Figure 3.12.

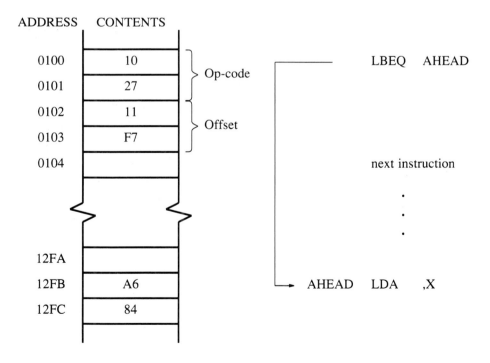

Figure 3.19 An example of a forward branch.

write a variety of nontrivial programs. It is particularly useful to run such programs on a real machine. The I/O approach presented in Sections 2.4 and 3.5 is applicable to most 6809 microcomputers that involve a video terminal. But, it will be necessary to use the actual assignment of addresses and status bits of the registers in the interface of the particular terminal.

In order to consolidate the appreciation of the material covered in the previous sections, the reader may try the exercises that follow.

1. Assume that accumulator A contains the value $6E_{16}$. Find the setting of condition code flags H, N, Z, V, and C, after executing each of the following instructions, assuming that the instructions are executed in the sequence shown:

   (a)   ADDA   #$17
   (b)   ADDA   #$D8
   (c)   SUBA   #$90
   (d)   SUBA   #$92
   (e)   SUBA   #8
   (f)   STA $2000

2. Consider the following 6809 assembler statements:

	ORG	$1000
CONST	FCB	$3B,$B,21,'B
	FDB	500,CONST+10,$AF24

   What memory locations will be affected by these statements? What will be the contents of these locations?

3. Identify all errors in the following 6809 assembly-language statements:

NUMLIST	FCB	23,$30,#3A
ST1	CMPA	#$100
	LDX	,Y+
	ANDA	X,Y

4. Indicate all possible ways of placing the value 0 into accumulator B, using just one 6809 instruction.

5. Assume that the contents of registers A, B, and X are $10, $20, and $300, respectively. Determine the effective addresses for the following instructions:

   (a)   CLR   A,X
   (b)   LDA   ,X+
   (c)   CMPY   D,X
   (d)   LEAU   −5,X
   (e)   ADDA $15

6. In a given object program, the instruction

   LBHI   LOC

   is assembled to be loaded in the main memory at address $500. Give the assembled code for this instruction if LOC corresponds to address

(a) $2000

(b) $150

7. In Subsection 3.1.2 it was stated that the C flag is set to the complement of the carry from the most-significant bit position during a subtraction operation, in order to indicate a borrow. Show, using a few numerical examples, that this is the required action if the subtraction is performed by adding a 2's complement of the subtrahend to the minuend.

8. Assume that there is an ordered list of 500 one-byte entries, beginning at address $1000. Write a program that inserts a new entry, whose value is stored in location NEW, into the 125th position of the list, thus making the list 501 entries long.

9. Assume that there is an ordered list of 200 2-byte entries, beginning at address $1000. Write a program that removes the 75th entry in the list. The remaining entries must be moved to occupy 398 contiguous locations starting at address $1000.

10. Assume that there is a list, called LIST, of 400 1-byte entries, beginning at address $1000. Write a program to create two new lists, LISTEV and LISTOD, such that LISTEV contains all entries in LIST that have even addresses and LISTOD contains all entries at odd addresses. The two lists LISTEV and LISTOD are to be stored in the main memory starting at addresses $1000 and $10C8, respectively.

11. Assemble the program in Figure 3.12 by hand. How many bytes of memory are needed to hold it? How many clock cycles are needed to execute it?

12. Assemble by hand the program in Figure 3.15. How many bytes of memory are needed to hold it? How many clock cycles are needed to execute the WAIT loop when a character is being transferred to the output buffer?

3.8 SUBROUTINE LINKAGE

As mentioned in Section 3.4, there are four instructions used to deal with subroutines. These are BSR, LBSR, JSR, and RTS. When a subroutine is called, the return address is stored on the processor stack, S. Also, several facilities are available for passing parameters to and from a subroutine. Before considering the details of parameter passing, we will describe the implementation of stacks in the 6809 microprocessor.

3.8.1 Stacks in 6809

The 6809 microprocessor has two stack pointers, S and U, which allow simple realization of system and user stacks. The system stack is used by the microprocessor for functions such as storing the return addresses when subroutines are called.

While a stack can be realized using Load and Store instructions in the autoincrement and the autodecrement addressing modes, as discussed in Section 2.6, it is convenient to have specific machine instructions to do the same job. There are four such instructions in the 6809, two for each stack. Register contents can be stored (pushed) on the S-stack using the PSHS instruction. The registers can be loaded with data from the S-stack using the PULS instruction. Similarly, the contents of a register can be pushed onto or pulled off the U-stack with PSHU or PULU instructions, respectively. The programmer can specify which registers are to be pushed, or pulled, by listing them in the operand field of the PSH and PUL instructions.

Since the PSH and PUL instructions need to specify the microprocessor registers affected, they consist of 2 bytes—the first byte being the OP-code and the second byte indicating the registers involved. The format of the second byte is given in Figure 3.20a. Each register is assigned 1 bit as shown. If this bit is set to 1, the corresponding register is affected. The order in which the registers are assigned to bit locations in the second byte, starting with bit position 7, is also the order in which they are stored on the stack. Thus, in PSH instructions, the program counter is stored first and the condition code register last. In PUL instructions the order is reversed. If all registers are stored, then the top of the stack would be as depicted in Figure 3.20b. Note that either register S or U, but not both, is stored on the stack. The S register can be stored only on the U-stack, and the U register on the S-stack. It would be of no value to store the contents of a stack pointer on its own stack.

The assembler syntax for the PSH and PUL instructions allows the specification of registers in the operand field in any order, but they will be stored in the standard order. For example, when the instruction

$$\text{PSHS} \qquad \text{A,B,CC,U,X}$$

is assembled, the second byte contains 01010111. From Figure 3.20 it is apparent that the order in which these registers will be stored on the S-stack is U, X, B, A, CC.

3.8.2 Parameter-Passing

Parameter-passing was discussed from a general point of view in Sections 2.5 and 2.7. A calling program may exchange parameters with a subroutine by placing them in processor registers or in fixed memory locations, by listing them immediately after the call instruction, or by placing them on the stack.

PARAMETERS IN PROCESSOR REGISTERS

The most straightforward way of passing parameters is to load them into one or more registers in the microprocessor. Figure 3.21 shows a 6809 version for the program in Figure 2.16. The subroutine finds the largest number in a list of 500 eight-bit numbers. The calling program places the address of the first entry in the list and the total number of entries into registers Y and D, respectively, to be passed as parameters to subroutine MAX.

Subroutine MAX uses register X as POINTERA, the list pointer, for accessing successive entries in the list. Register Y is used as MAXLOC, which holds the address of the largest number found at any stage while the list is being scanned. The corresponding value of the largest number is held in accumulator A. This value is compared with successive entries in the list and whenever a larger value is encountered the latter is placed in accumulator A. Also, the address of the new largest entry is placed in register Y. Note that this is achieved with the instruction

$$\text{LEAY} \qquad -1,X$$

which loads register Y with the contents of register X minus 1. The reason for subtracting 1 is that the comparison of successive entries in the list is carried out in

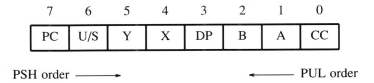

7	6	5	4	3	2	1	0
PC	U/S	Y	X	DP	B	A	CC

PSH order ⟶   ⟵ PUL order

(a) Format of the second byte

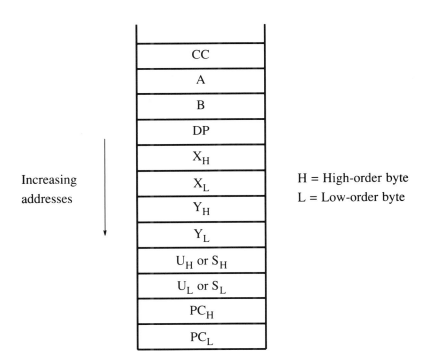

Increasing
addresses

| CC |
| A |
| B |
| DP |
| X_H |
| X_L |
| Y_H |
| Y_L |
| U_H or S_H |
| U_L or S_L |
| PC_H |
| PC_L |

H = High-order byte
L = Low-order byte

(b) Top of the stack if all registers are stored

Figure 3.20 Register ordering in PSH and PUL instructions.

```
              LDY        #LIST        Starting address of LIST and the number of entries
              LDD        #500            are loaded as parameters in Y and D.
              BSR        MAX          Call subroutine MAX.
              next instruction
                .
                .
                .

MAX           LEAX       D,Y          Generate the address that follows the last entry
              STX        LSTEND          in the list and store it.
              TFR        Y,X          Register X will be the list pointer. Registers
*                                        X and Y now point to the first number.
              LDA        ,X           Accum A holds the first number.
LOOP          CMPA       ,X+          Check if A is ≥ the next number.
              BGE        NEXT         If yes, go to next entry.
              LEAY       −1,X         If no, update pointer to the new largest number.
              LDA        −1,X         New largest number is in accum A.
NEXT          CMPX       LSTEND       Check if the end of the list is reached;
              BLO        LOOP            branch back if not.
              RTS
```

Figure 3.21   A 6809 subroutine for finding the largest number in a list.

the autoincrement mode. The contents of X are incremented by 1 during the execution of the CMPA instruction. Thus, when a larger number is found in the list, register X will already have been incremented to point at the next entry.

At the start of the subroutine, the address of the first memory location after the last entry in the list is placed into memory location LSTEND. During each pass through the loop, a check is made to determine if the end of the list has been reached. This is done by comparing the value in the list pointer, register X, with the address in LSTEND.

PARAMETERS LISTED AFTER A CALL INSTRUCTION

The possibility of listing parameters after a call instruction was discussed in Section 2.7. The program in Figure 2.20 illustrates this technique. A straightforward implementation of this program for the 6809 is given in Figure 3.22. A list of 50 one-byte entries is copied from one area in memory to another. The length and addresses of the list are the parameters placed after the BSR instruction, by means of FCB and FDB assembler directives. Subroutine MOVE uses register U, which corresponds to POINTERC in Figure 2.20, to access the parameters. After the last parameter is loaded into a register, register U contains the address of the next instruction in the calling program. This is the correct return address that should replace the original return address pointed at by register S. This is done by the STU instruction. The loop in the subroutine is the same as in previous examples.

```
        BSR     MOVE
        FCB     50              Parameters are the length of the lists
        FDB     LISTA,LISTB        and their addresses.
        next instruction
        .
        .
        .

MOVE    LDU     ,S              Register U is used to access parameters.
        LDA     ,U+             Load n into accum A.
        LDX     ,U++            Load address of LISTA into X.
        LDY     ,U++            Load address of LISTB into Y.
        STU     ,S              Adjust the return address.
LOOP    LDB     ,X+             Transfer one entry from LISTA
        STB     ,Y+                to LISTB.
        DECA                    Decrement the counter and loop
        BHI     LOOP               back until done.
        RTS
```

Figure 3.22 A 6809 realization of the program of Figure 2.20.

The existence of the PULU instruction allows a more compact version of the same program to be written, as shown in Figure 3.23. We can pretend that the parameter list constitutes the user stack, pointed to by the stack pointer U. Thus, the PULU instruction loads the three parameters into registers A, X, and Y. The rest of the program is unchanged.

If a subroutine does not use a particular register, the contents of that register remain undisturbed. However, any registers used by the subroutine lose their original contents. The programmer can ensure that the original contents of a register are not lost by storing them on the stack at the start of the subroutine. Then, they should be restored upon completion of the subroutine, before returning to the calling program, that is, prior to executing the RTS instruction.

The program in Figure 3.24 is a modification of the program in Figure 3.23, where all registers affected by the subroutine are saved on the stack. The five registers in question are stored in 8 bytes (recall that X, Y, and U are 16-bit registers) at the top of the stack, that is, above the return address. The return address stored on the stack is actually the address of the first parameter in the calling program. Hence, this address is loaded into register U, as before, to provide access to the parameters. The indexed mode, with an offset of 8, is used to read the return address from the stack, because it is now 8 bytes below the top of the stack. Then, after reading all the parameters, the return address on the stack is updated to become the true return address; that is, it is adjusted to point to the first instruction following the subroutine parameters in the calling program. The last instruction in the subroutine restores the registers from the stack. Note that the PC is included as one of the registers restored.

```
          BSR        MOVE
          FCB        50
          FDB        LISTA,LISTB
          next instruction

            .
            .
            .

MOVE      LDU        ,S
          PULU       A,X,Y          Load parameters into processor registers.
          STU        ,S
LOOP      LDB        ,X+
          STB        ,Y+
          DECA
          BHI        LOOP
          RTS
```

Figure 3.23   A more compact realization of the program of Figure 2.20.

```
          BSR        MOVE
          FCB        50
          FDB        LISTA,LISTB
          next instruction

            .
            .
            .

MOVE      PSHS       A,B,X,Y,U        Save registers.
          LDU        8,S              Set up U as a pointer to the parameter
          PULU       A,X,Y               list, and load parameters.
          STU        8,S              Adjust the return address.
LOOP      LDB        ,X+
          STB        ,Y+
          DECA
          BHI        LOOP
          PULS       A,B,X,Y,U,PC     Restore registers and return.
```

Figure 3.24   A 6809 realization of the program of Figure 2.20, where the CPU registers are saved on the stack.

Thus, the PULS instruction also loads the return address into the PC, obviating the need for a separate RTS instruction.

PARAMETERS ON THE STACK

Parameters can be passed to a subroutine via the stack. In this case the calling program places the parameters on the stack before calling the subroutine. The calling program is also responsible for removing the parameters from the stack upon return from the subroutine.

Figure 3.25 shows how the program of Figure 2.21 can be implemented. The calling program pushes the parameters onto the stack, where they occupy 5 bytes. Then, the subroutine is called. Note that the LBSR instruction is used instead of BSR, because the address of the subroutine requires a branch offset that is too large to be specified in an 8-bit offset.

The subroutine saves the registers on the stack. At this point the top entries in the stack are as shown in Figure 3.26. The parameter list begins 10 bytes from the top of the stack. The address of the first parameter is loaded into register U. Next, the parameters are copied from the stack, as if they were on top of a U stack, and loaded into the processor registers. This operation does not affect the S stack. The

```
* This is a 6809 implementation of the program in Figure 2.21.
* It illustrates passing of parameters via the stack.
*
LISTA      EQU        $1000              Define starting address for
LISTB      EQU        $2000                  lists A and B.
*
* Calling program
*
           ORG        $150               Start the program at location $150.
           LDA        #50
           LDX        #LISTA             Place the parameters
           LDY        #LISTB
           PSHS       A,X,Y                 on the stack.
           LBSR       MOVE
           LEAS       5,S                Remove parameters from the stack.
*
* The rest of the calling program is entered here.
*
*
* Subroutine MOVE
*
           ORG        $500               Subroutine starts at location $500.
MOVE       PSHS       A,B,X,Y,U          Save registers on the stack.
           LEAU       10,S               Set U as a pointer to parameters on stack.
           PULU       A,X,Y             Load parameters into registers.
LOOP       LDB        ,X+                Transfer one entry from list A
           STB        ,Y+                   to list B.
           DECA                          Decrement counter and loop
           BHI        LOOP                  back until done.
           PULS       A,B,X,Y,U,PC       Restore registers from the stack and return.
```

Figure 3.25   An example of passing parameters on the stack.

rest of the subroutine is the same as in the previous example. Again, no RTS instruction is used, as the return address is loaded into the PC with the PULS instruction that restores the contents of the registers from the stack.

Upon return from the subroutine the parameters on the stack must be removed. At this point in the program the stack pointer S is pointing at the first parameter in the stack. The removal of parameters is achieved simply by changing the contents of the stack pointer to make it point past the parameter list, namely, to the top entry on the stack before the parameters were pushed. Hence, the LEAS instruction adds 5, which is the number of bytes occupied by the parameters, to register S.

The example of Figure 3.25 is presented in a more complete style than the previous examples, in order to illustrate the use of assembler directives and to indicate

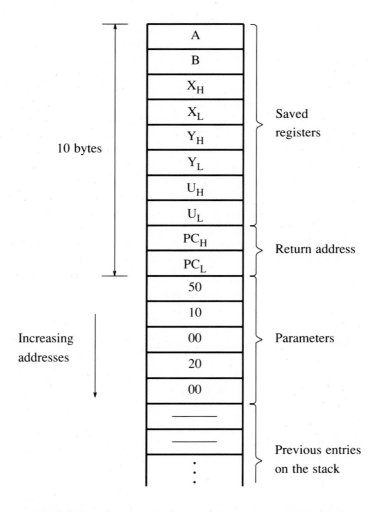

Figure 3.26 Top entries in the stack for example in Figure 3.25.

how a program may be documented with appropriate comments. A vital element in good programming practice is proper documentation. Its importance grows rapidly as programs become larger, particularly when several people are involved in writing various sections of a large program. Moreover, making changes to a poorly documented assembly language program written by another person is likely to be a most trying experience.

3.9 VERSATILITY OF THE 6809 INSTRUCTION SET

The 6809 has a powerful and versatile set of machine instructions. In the previous examples we made use of only a few of the available instructions. In order to give the reader some indication of the versatility of the instruction set we will present some further examples in this section.

The instruction mnemonics suggest fairly clearly the operation performed by a given instruction. The reader should have little difficulty in understanding the operations performed by most instructions listed in Table 3.3. However, the usefulness of some instructions is not likely to be apparent to an inexperienced programmer. We will consider some of these in the discussion below.

ADDRESS COMPUTATIONS

We have already encountered the LEA instructions, which load the effective address of the operand into a specified register. They can be used to perform addition and subtraction on registers X, Y, S, and U, as discussed in Section 3.5. They also provide a simple means for dealing with addresses of operands, rather than with the operands themselves. For example, suppose that register X points to the first entry in a character list. Then, the address of the tenth entry in this list can be loaded into register Y with the instruction

$$\text{LEAY} \qquad 9,\text{X}$$

BINARY CODED DECIMAL ADDITION

Another instruction whose purpose may not be apparent at first glance is DAA (Decimal Adjust Accumulator A). It is intended for dealing with binary-coded-decimal (BCD) data. In BCD representation, each of the digits of a decimal number is encoded in 4 bits. Since the main memory of a 6809 microcomputer is organized in 8-bit words, we may store two BCD digits in each byte, in order to conserve memory space. A BCD number may be represented using several bytes. We will assume that a number with $2n$ digits is stored so that the most significant two digits are in location LOC, followed by the next two digits in LOC+1, and so on. The least significant two digits are thus in LOC+n−1.

Let us now consider the problem of adding two BCD numbers. Suppose that 89 is a digit pair of one number, in accumulator A. It is to be added to a digit pair of the second number, say 38, fetched from the main memory. Using the ADDA instruction, this operation becomes

$$10001001 \quad (89)$$
$$+ \quad 00111000 \quad (38)$$
$$11000001$$

The result is the correct sum of the two 8-bit binary numbers, but it is not the desired BCD sum which is

$$1\ 0010\ 0111 \quad (127)$$

The leftmost 1 of this sum is the carry-out to be added to the next more significant digit pair. The binary sum produced in accumulator A can be converted into the correct result with the DAA instruction. This instruction uses the half-carry bit, H, in the condition code register to generate the correct BCD sum. We will not pursue the details of how the DAA instruction achieves this correction.

An example of BCD addition is given in Figure 3.27. The desired operation is

$$P \leftarrow P + Q$$

The operands P and Q are assumed to be 20 digits long. They occupy 10 bytes of memory each, starting at locations NUMP and NUMQ, respectively. The two locations NUMP and NUMQ are assumed to contain the most significant digits of the two numbers. Registers X and Y are loaded with the starting addresses. Accumulator B is used as a byte counter, while the numbers are added 1 byte at a time. The addition of any digit pair may generate a carry-out that must be added to the sum of the next digit pair. Therefore, the addition is done with the ADCA (Add with Carry to Accumulator A), which adds the memory operand to the contents of accumulator A plus the value of the carry flag C. Then, the result of this binary addition is converted into a BCD format with the DAA instruction. The corrected result replaces the previous contents of the corresponding digit pair of P.

The loop in Figure 3.27 includes the ADCA instruction. The first time through the loop the carry input, that is, the C flag, must be cleared to 0. This is accomplished by the instruction

ANDCC   #%11111110

which performs a logical AND operation on the contents of the condition code register and the immediate operand specified. This instruction clears the C flag, which occupies bit position 0 in CCR, leaving the remaining flags undisturbed. Note that the immediate operand of the ANDCC instruction is specified in the binary format, indicated by the % prefix. In this case, the binary pattern is easier to visualize than its hex equivalent, FE.

The last comment about Figure 3.27 concerns the addressing mode in which the BCD operands are accessed. As can be easily appreciated from the figure, it is convenient to use the indexed mode with the contents of accumulator B being the desired index value. This enables both operands to be scanned, two digits at a time, by decrementing accumulator B.

```
          LDX      #NUMP           Load addresses of summands into
          LDY      #NUMQ             registers X and Y.
          LDB      #9              Initialize B as a byte counter.
          ANDCC    #%11111110      Clear the C flag.
LOOP      LDA      B,X             Form a binary sum for one
          ADCA     B,Y               digit pair.
          DAA                      Decimal adjust.
          STA      B,X             Store the sum into location of number P.
          DECB                     Decrement the counter and
          BHS      LOOP              loop back until done.
          next instruction
```

Figure 3.27   Program for addition of 20-digit BCD operands.

LOGIC OPERATIONS

The 6809 has several instructions for performing logic operations on 8-bit operands. Included are instructions for the logic functions: AND, OR, and Exclusive-OR. There is also a class of instructions that shift or rotate the operand value. The shift operations involve the C flag in the condition code register. The bit shifted out of the operand register is placed into the C flag. This facility enables bits to be shifted from one register to another. It also provides a convenient way for testing the value of the bit that is shifted out, because the status of the C flag can be used as a test condition in branch instructions. As shown in Table 3.3, the rotate instructions include the C flag in the rotation pattern. The new value of C is the bit shifted out of the operand register, while the old value of C fills the 1-bit space created by shifting the operand.

To illustrate the use of the logic instructions, the shift instructions in particular, we will consider a program that determines the parity of an 8-bit operand. The *parity* indicates whether there is an even or an odd number of 1s in a given item of data. Thus, we can talk about even and odd parity. The notion of parity is useful for error checking. In order to improve the reliability of digital systems, data is often encoded to have either even or odd parity. For example, if even parity is used, then every unit of data (either a byte or some larger quantity) has one extra bit appended to it. This bit, called the *parity bit* , is set to 0 or 1, depending upon the number of 1s in the data unit, so that the parity of the complete item is even. If a parity encoded data item is subjected to some noise or interference while it is being transmitted from one point to another, some of its bits may be changed to erroneous values. Checking the parity of the received data will reveal such errors if only one bit, or any odd number of bits, are wrong. Of course, if an even number of bits are in error, then the parity check will not detect the fault. The derivation of parity and its checking is a simple process, which yields a useful degree of error protection that is commonly used in digital systems.

The program in Figure 3.28 computes the parity of an 8-bit word in memory location DATA. It does this by counting the total number of 1s in the word. Then, the least-significant bit of this total will show if the number of 1s is even or odd.

```
                LDA      DATA           Load test data in accum A.
                CLRB                    Accum B to indicate the number of 1s.
      LOOP      LSRA                    Shift right to test a bit.
                ADCB     #0             If the bit is a 1, increment B.
                TSTA                    Test the contents of A, and
                BNE      LOOP              loop back if not zero.
                LSRB                    Check the least significant bit in B.
                BCS      ODD            If 1, then parity is odd,
                process even parity        otherwise it is even.
                  .
                  .
                  .

      ODD       process odd parity
```

Figure 3.28 Program for determining the parity of an 8-bit data item.

Accumulator A holds the data under test, while accumulator B holds the number of 1s. The LSRA (Logical Shift Right Accumulator A) instruction shifts the least-significant bit in A into the C flag and places a 0 into the most-significant bit of A. The C bit is then added to the contents of B with the ADCB instruction. Instead of using a counter to keep track of eight passes through the loop, the program tests for all zeros in accumulator A. Since the LSRA instruction fills the accumulator with 0s from the most significant end, accumulator A will contain all 0s after at most eight shifts. It is important to note that the TSTA instruction sets the N and Z flags in the condition code register, but it does not change the contents of A. Having completed the loop, the least-significant bit of accumulator B is checked to see if the even or the odd parity program segment is to be executed next.

3.10 EXAMPLE OF A LINKED LIST

So far, we have discussed fairly simple examples, which require only a few instructions to implement. Our primary motivation has been to show how various addressing modes and instructions can be used. At this point, it will be beneficial to consider a more complex example that illustrates a number of different concepts.

The linked-list structure, discussed in Section 2.8, provides a convenient vehicle for illustrating many ideas. Let us begin by considering a 6809 implementation of the program in Figure 2.25, which finds the highest final mark in the linked list containing the marks of the students in a given class. Let each item in the list have the format shown in Figure 3.29. The 6809 microprocessor transfers data in 8-bit words, but uses 16-bit addresses. Thus, 2 bytes are needed to store each link pointer to the next item. Otherwise, the format is the same as that of Figure 2.24. The name field is 10 bytes long, restricting the names to 10 characters in length. Each mark occupies 1 byte, allowing unsigned integers in the range 0 to 255 to be stored. We will assume that the marks are stored as binary integers, rather than BCD numbers.

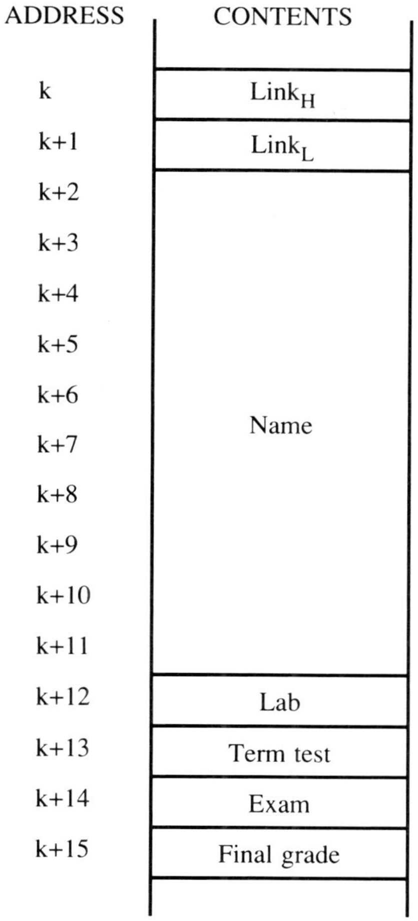

ADDRESS | CONTENTS

Address	Contents
k	$Link_H$
k+1	$Link_L$
k+2	
k+3	
k+4	
k+5	
k+6	
k+7	Name
k+8	
k+9	
k+10	
k+11	
k+12	Lab
k+13	Term test
k+14	Exam
k+15	Final grade

Figure 3.29 Arrangement of one student item in a linked-list example, that corresponds to Figure 2.24.

The program in Figure 3.30 uses register X as a pointer to successive items in the list. The final grade is located 15 bytes away from the address of the link entry of each item. The grade in successive items is compared against the largest value that has been found, which is held in accumulator A. Upon changing the value in register X, to point to the next item, it is necessary to check if the new link is equal to 0 indicating the end of the list. A separate Compare instruction, to correspond to the one in the program of Figure 2.25, is not needed because the LDX instruction loads the X register and sets the Z flag to 1 if the new contents of the register are equal to 0. The execution of the program will leave the highest grade in accumulator A.

```
            LDX     #LSTADR    Initialize register X as a list pointer.
LOOP        LDA     15,X       Update higher grade found.
NEXT        LDX     ,X         Go to next student.
            BEQ     DONE       Check if end of the list.
            CMPA    15,X       Compare with next grade.
            BGE     NEXT       Current max still valid.
            BRA     LOOP       Better grade found.
DONE        next instruction
```

Figure 3.30 A 6809 program that finds the highest grade in a linked list.

An unpleasant, and unnecessary, consequence of the fixed-length name field is that either long names cannot be used, or many bytes of storage will be wasted if the field is long enough to accommodate the longest name. A more flexible scheme is to use a variable-length format, as suggested in Figure 2.29. An adaptation of this format using 8-bit words is shown in Figure 3.31. The third entry of an item indicates the offset in bytes, j, from the address of the link entry and the location

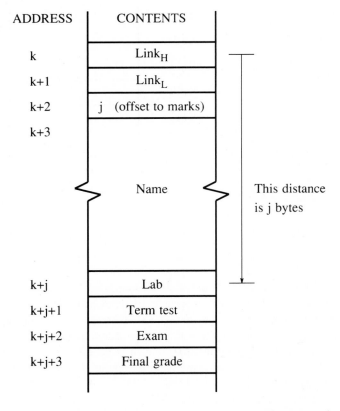

Figure 3.31 Arrangement of an item with a variable-length name field.

where the laboratory mark is stored. We will assume that the character string in the name field is terminated by a null character. While this is not necessary for accessing the marks entries, it is useful in tasks such as sorting the list in alphabetical order.

Consider now the task of locating the item that corresponds to a given student and displaying the marks on a video terminal. Figure 3.32 depicts the tabular form in which the marks should appear on the screen. The required task involves three subtasks. First, the desired item has to be located by comparing the target name with the names in the list. Having found the item, the marks have to be converted from binary form into ASCII-encoded decimal numbers. Finally, appropriately formatted messages that include the marks have to be displayed on the screen.

A possible program is shown in Figure 3.33. It is a more complete version of the approach outlined in Figure 2.26, including the routines needed to display the information. The program is written in a style that illustrates several important aspects of good programming practice. It is modular, using subroutines to perform all natural subtasks. Parameters are passed to subroutines in a consistent manner, via the stack. Comments are used liberally, to facilitate understanding of the program. Each subroutine is preceded by a comment that states its function and the method used to pass the parameters. Symbolic names are used in instructions, and their actual numerical values are specified in assembler directives.

The code consists of a main program and four subroutines that implement the subtasks of: searching for the target name, converting a binary integer into an ASCII-encoded decimal number, displaying a one-line message, and transferring one character from the microprocessor to the display buffer. The subroutines are written in a way that makes them suitable for use in other programs that need such general subtasks. To ensure their generality, the contents of all registers used within a subroutine are saved on the stack and restored prior to returning to the calling program. The name of the student whose record is to be found is assumed to be stored as a character string beginning at location STUDNT and ending with a null character. The first item in the linked list is stored beginning at location LSTADR.

Subroutine MATCH compares two character strings. One is the target name of the student whose record is sought, and the other is in the name field of an item in the linked list. The starting addresses of both strings are passed as parameters via the stack. Note that the name field starts at an address that is three bytes higher than the address of the link entry of the item. This requires an adjustment in the value of

Paul Jones
Laboratory   =   85
Term test   =   70
Examination   =   80
Final grade   =   78

Figure 3.32 Desired format for marks display.

```
*
* Main program; to find the item in the linked list that corresponds
* to a student identified by a character string that starts in location
* STUDNT, and to display the student's name and marks.
*
              ORG      $100
              LDX      #STUDNT      Get the location of student's name.
              LDY      #LSTADR      Initialize Y to point to the name
NEXT          LEAY     3,Y              field in the list.
              PSHS     X,Y          Put parameters on the stack and call
              BSR      MATCH            the string match subroutine.
              LEAS     4,S          Remove parameters from the stack.
              BEQ      FOUND        Strings matched, display the marks.
              LDY      [-3,Y]       Strings did not match, go to
              BNE      NEXT             next item; or
              LDX      #NOTFND      Last item was reached without
              PSHS     X                finding the name, print an
              BSR      MSGOUT           error message.
              LEAS     2,S          Remove parameters from the stack.
              BRA      CONT         Finished.
FOUND         PSHS     X            Display the name of
              BSR      MSGOUT           the student.
              LEAS     2,S
              LDB      #4           There are four marks to be displayed.
              LDA      -1,Y
              SUBA     #3
              LEAY     A,Y          Register Y points to the lab mark.
              LDX      #MARKS       Point to the marks titles.
LOOP          PSHS     X,Y          Display the next mark title
              BSR      MSGOUT
              PULS     X                and the corresponding
              BSR      ASCOUT           mark in decimal form.
              PULS     Y            Prepare the top of the stack and
              LEAY     1,Y              reg Y for subsequent mark.
              DECB                  Repeat the loop four times.
              BNE      LOOP
CONT          next instruction      End of task.
*
*
* Subroutine that compares two character strings whose starting
* addresses, to be loaded into registers X and Y, are passed as
* parameters on the stack. The subroutine returns the Z flag set
* to 1 if the two names matched, otherwise Z is cleared to 0.
*
```

Figure 3.33 (continued next page)

```
MATCH     PSHS     A,B,X,Y,U       Save registers.
          CLRB                     Use accum B to step through the strings.
          LEAU     10,S            Use reg U to pull the parameters off
          PULU     X,Y                the stack.
LP1       LDA      B,X             Compare a pair of
          CMPA     B,Y                characters.
          BNE      LEAVE           No match if different.
          INCB                     They match, go to next pair until
          TSTA                        a NUL character is
          BNE      LP1                reached.
LEAVE     PULS     A,B,X,Y,U,PC    Restore registers and return.
*
*
```

* This subroutine takes an 8-bit unsigned number in location whose
* address is passed as a parameter on the stack, and converts the
* number into two ASCII-encoded decimal digits to be displayed on
* the screen.
*

```
ASCOUT    PSHS     A,B             Save registers.
          CLRA                     Use accum A to count the 10s.
          LDB      [4,S]           Load the 8-bit binary number.
LP2       SUBB     #10             Subtract 10 until the result
          BMI      RESTOR             is negative.
          INCA                     Count the 10s.
          BRA      LP2
RESTOR    ADDA     #$30            Make ASCII code for the tens digit.
          BSR      CHOUT           Output the tens digit.
          ADDB     #$3A            Restore B and make ASCII.
          TFR      B,A             Output the ones
          BSR      CHOUT              digit.
          PULS     A,B,PC          Restore registers and return.
*
*
```

* This subroutine sends one character to the display. The character
* is passed as a parameter in accumulator A.
*

```
CHOUT     PSHS     B               Save register.
WAIT      LDB      STATUS          Check the status of
          ANDB     #2                 the display.
          BEQ      WAIT
          STA      OUTBUF          Transfer the character.
          PULS     B,PC            Restore register and return.
*
*
```

Figure 3.33   (continued next page)

```
* This subroutine displays a one-line message. The starting address
* of the message is passed as a parameter on the stack. The address
* of the first character that follows the end of the message is
* returned to the calling program via the stack.
*
MSGOUT    PSHS    A,X              Save registers.
          LDX     5,S              Use X to scan through the message.
MORE      LDA     ,X+              Load the next character.
          BEQ     DONE             Check if NUL.
          BSR     CHOUT            Display the character.
          BRA     MORE
DONE      STX     5,S              Place address on the stack.
          PULS    A,X,PC           Restore registers and return.
*
*
* Student's name and messages to be displayed.
*
CR        EQU     $0D
LF        EQU     $0A
          ORG     $1000
STUDNT    RMB     30
MARKS     FCB     CR,LF
          FCC     /Laboratory   = /
          FCB     0,CR,LF
          FCC     /Term test    = /
          FCB     0,CR,LF
          FCC     /Examination  = /
          FCB     0,CR,LF
          FCC     /Final grade = /
          FCB     0,CR,LF,0
NOTFND    FCC     /The name not found./
          FCB     CR,LF,0
          END
```

Figure 3.33 A program that finds a student's record and displays the marks.

register Y, in the main program, which keeps track of the link addresses as the list is scanned. If the two strings are identical, the exit from subroutine MATCH will be caused by the TSTA instruction detecting a null character. This results in setting the Z flag to 1. Upon return from the subroutine, the Z flag is tested to see if the two strings matched. (The value of the Z flag is not affected by the LEAS instruction.) In the case of a match, the desired item has been located. Otherwise, the next item in the list is considered. If the last item has been reached without finding a match, indicated by observing a null character in the link entry, an error message is displayed stating that the student's name is not in the list.

All messages are displayed by calling the subroutine MSGOUT. It sends a one-line message to the display by calling subroutine CHOUT repeatedly. The latter waits for the display to become ready and then transfers one character to OUTBUF register. Note that the character to be displayed is passed to CHOUT as a parameter in accumulator A, for simplicity. This is the only exception to the general strategy of passing the parameters via the stack. Because several messages are to be displayed, and they are stored in consecutive memory locations, it is useful to return the last address derived within subroutine MSGOUT to the calling program. Observe that this address is in fact the starting address of the next message, because the autoincrement addressing mode is used to scan through the message. The address is returned to the calling program via the stack.

The subroutine ASCOUT converts an 8-bit unsigned binary integer into a two-digit ASCII-encoded decimal number, and displays it on the screen by calling subroutine CHOUT. It performs the conversion by subtracting 10 from the binary number repeatedly, until a negative result is reached. The number of times that 10 can be subtracted without leaving a negative result is the "tens" digit of the decimal number. The remaining value is the "ones" digit. In order to finish with the correct ones digit, it is necessary to restore the result in accumulator B after it has been observed to be negative, by adding 10 back to it.

From Table A.1 in Appendix A, it can be seen that a BCD digit can be converted into an 8-bit ASCII-encoded character by using the four bits of the BCD code as the least-significant four bits and the pattern 0011 as the most-significant four bits. For example, the digit 7 is encoded as 00110111 $(=37_{16})$. Thus, the tens digit held in accumulator A is converted into ASCII by the instruction

$$\text{ADDA} \qquad \text{\#\$30}$$

For the ones digit, which is equal to the contents of accumulator B, the ASCII conversion can be combined with the necessary correction of the result that became negative during the previous subtraction. The correction requires adding 10 $(=A_{16})$ to accumulator B, hence the combined addition uses the value $30_{16} + A_{16} = 3A_{16}$ as the immediate operand.

The character strings that constitute the messages to be displayed are given at the end of the program. Each is preceded by a Carriage Return (CR) and a Line Feed (LF), and terminated by a null character (0). The CR and the LF are needed to produce the tabular format depicted in Figure 3.32. When the display receives a CR character, it returns the cursor to the column 1 position. When it receives a LF character, it moves the cursor to the line below. The MSGOUT subroutine starts displaying from location MESGS and continues until it encounters the null character. At this point, the cursor on the screen is at the second character position to the right of the equal sign (because a blank follows the = sign in the message). Now, when the ASCOUT subroutine sends the two digits that represent the mark, these digits will appear in the proper place on the displayed line. Since the next message is preceded by CR and LF characters, it will be displayed on the line that follows.

The reader may have noticed a flaw in this program. The ASCOUT subroutine converts the binary integer into a two-digit decimal number. This limits the range to

0 to 99, not allowing the mark 100 to be displayed. We have left the simple modification of the subroutine to handle 100, and possibly all numbers in the range 0 to 255, as an exercise for the reader.

In order to run the program in Figure 3.33 on a 6809 microcomputer, it is necessary to specify the actual addresses of the buffer registers STATUS and OUTBUF, as well as the position of the "display ready" bit in the STATUS register (which is not necessarily b_1 as we have assumed in Figure 2.11). Of course, it is also necessary to have the linked list available in the main memory, starting at location LSTADR.

3.11 PERFORMANCE CONSIDERATIONS

The primary requirement for any program is that it work properly. It is advisable that programs be written in a clear and logical manner, breaking large tasks into smaller subtasks that can be implemented in modular fashion. Tricky pieces of code are not easy to debug and are likely to cause problems. In many applications the size of a program and its speed of execution are not of major concern, however, there are cases where performance requirements place severe constraints on either the size or the speed, or both.

Program size is seldom a constraint in fully configured systems that include large amounts of main memory and a secondary storage device. The price of memory chips has been falling dramatically, allowing the implementation of large memories at a reasonably low cost. A personal computer based on the 6809 microprocessor will almost certainly have full 64K bytes of main memory provided (less a few locations used for I/O devices). It will also have some secondary storage, probably a magnetic disk. For programs written in assembly language, few constraints on their size are likely to be experienced by a typical user. But, consider an application where a commercial product is to be designed using the 6809 microprocessor—perhaps a measuring instrument, an appliance, a toy, or an electronic game. Such products should be inexpensive and fit into small physical packages. This precludes the use of secondary storage devices and limits the size of the programs to whatever can be handled in the main memory provided, some of it in the form of a ROM. Hence, it may become important to optimize the code to use as few bytes as possible.

In other applications, the speed of execution may be of critical importance. Suppose that a microcomputer is used in industrial control, where the response to a given input condition must be provided in a short time. Some computation is needed to determine the required response. It may be necessary to optimize the software with respect to the speed of execution, in order to do this computation in the time available. When such responses must meet the time constraints imposed by other equipment, the application is said to be of the *real-time* type. Of course, even in non-real-time applications the speed of execution may be of interest. For example, if a microcomputer supports several video terminals, the users of the terminals will expect a reasonably fast response to their requests. A user who is doing text editing will become annoyed if the computer takes more than a second or two to perform an editing function. While nothing disastrous happens if longer delays are experienced, the user will quickly become disenchanted with the slow response.

What can be done to reduce the size of a program and improve the speed of execution? Clearly, it is beneficial to use efficient algorithms that can be implemented with small programs. Then, there are the opportunities to remove functionally redundant steps in the execution of a program, although such attempts may make the program less general.

Consider the recommended guideline of saving the contents of the registers that are used in a subroutine on the stack and restoring them just before returning to the calling program. This takes time and requires additional instructions. If the programmer is sure that the registers affected will not contain any information that must be preserved, then it is not necessary to save their contents. Of course, this will reduce the flexibility with which the subroutine can be used in different places of a calling program.

The method chosen to pass parameters to subroutines has an effect on both the size and the speed of execution of a program. If the parameters are passed via the stack, they will have to be pushed onto the stack and pulled from it, requiring extra transfers. Moreover, additional instructions are needed to push and load the parameters and to remove them from the stack (by adjusting the value of the stack pointer S) upon return to the calling program. Passing the parameters through registers results in shorter and faster executing programs, but this can be done only when few parameters are involved and sufficient registers are available.

It is always prudent to make good use of combined operations. For example, when a pointer register is used to traverse a list of items, the program can usually be written such that the pointer is updated when an item is accessed, thus exploiting the possibility of using the autoincrement addressing mode. We encountered another example of combining operations in eliminating the need for the RTS instruction in Figures 3.24 and 3.33. When a return from a subroutine is preceded by restoring the register contents from the stack, both tasks can be performed by a single instruction. The transfer of the return address from the stack to the program counter takes two cycles and no extra code space, as a part of the PULS instruction. In contrast, a separate RTS instruction requires 1 byte of code and five cycles to execute.

Branch instructions come in two formats—short or long. The long branch instructions need two extra bytes of storage and take two or three cycles longer to execute than short branch instructions. It is worthwhile to ensure that short branch instructions are used whenever the required branch offset can be specified in 1 byte.

Another possibility for optimizing programs is provided by the direct page register feature of the 6809. Organizing the data in pages, such that many accesses to a given page are made before moving to another page, makes it possible to set the contents of the DP register accordingly. In this case, the direct addressing mode, which requires only a 1-byte address, can be used instead of the extended mode. This shortens the program and thus speeds up its execution.

The X and Y registers in the 6809 microprocessor have similar functions. Yet, as pointed out in Section 3.6, some instructions need an extra byte when register Y is used. These instructions, in addition to being longer, take one more cycle to execute. When space and speed are important, it is wise to use register X instead of Y as much as possible.

Finally, we should note that the order in which operations are performed can produce interesting effects. In Figure 3.14, we implemented the WAIT loop with the instructions

```
WAIT    LDA     STATUS
        ANDA    #1
        BEQ     WAIT
```

These three instructions occupy 7 bytes and take 10 cycles to execute. We can accomplish the same task with the following instructions:

```
        LDA     #1
WAIT    BITA    STATUS
        BEQ     WAIT
```

The BITA instruction performs a bit test operation by ANDing the contents in accumulator A and location STATUS, without changing either of them. It sets the Z flag that is used as the required condition by the branch instruction. This arrangement of the WAIT loop also needs 7 bytes. However, it takes only eight cycles to execute. All of this is of little consequence in the implementation of a loop that waits for a single I/O device. But, suppose that a microprocessor is connected to many I/O devices, that must be polled to see if they require service. The reader should contemplate such a larger polling task and write a suitable loop that examines the status of all devices, assuming that the status bit to be tested is in the same bit position in all status registers. How are the space and time requirements affected by the arrangement of the loop?

In this section we discussed a number of aspects that have an impact on the size and speed of execution of assembly language programs. We cannot end without reiterating that, whenever possible, programs should be written in a high-level language. Also, clarity and ease of understanding should be primary goals in the preparation of microprocessor software. Optimization of code at the assembly language level should be undertaken only when dictated by performance constraints.

3.12 CONCLUDING REMARKS

This chapter has dealt with the details of the 6809 microprocessor, as seen from the programmer's perspective. Its register structure, addressing modes, and instruction set have been presented. Knowledge of this material is sufficient for writing many application programs.

We have not discussed the 6809 interrupt mechanism for dealing with I/O operations. Interrupt techniques will be introduced in Chapter 7, including those features that are found in the 6809. We have also not considered the physical structure of the 6809. Understanding of various control signals, along with the address and data connections, is essential for a designer who wishes to build a microcomputer. Such physical characteristics and related hardware system aspects will be the subject of Chapters 5, 6, 7, and 10.

The instruction set of the 6809 is summarized in Table 3.3. For space reasons, the description of each instruction is brief. More detailed description can be found in the manufacturer's literature [1] or in specialized texts, such as reference [2] which includes numerous illustrative examples. A discussion of the 6809 in the framework of more general computer organization books can be found in references [3–5].

The reader will notice in Table 3.3 an instruction, MUL, for multiplication of unsigned operands. However, the 6809 does not have an instruction for multiplication of signed operands, and has no instructions for division. These operations can be performed using routines based on algorithms that make use of simple instructions for addition and subtraction. Such algorithms are discussed in references [2–4].

3.13 REVIEW QUESTIONS AND PROBLEMS

REVIEW QUESTIONS

1. Describe possible uses for each register in the 6809 microprocessor.

2. When should one attempt to make extensive use of the direct page register?

3. How many addressing modes can be used to load the contents of memory location $200 into accumulator A? Give examples of both the assembler format and the corresponding object code for this LDA instruction, for each possible mode.

4. In Section 3.4 it was stated that there exist pairs of branch instructions, such as BGT and BHI, that perform the same branch test on signed and unsigned operands, respectively. Show that the branch tests given in Table 3.3, which involve the condition code flags, make the desired distinction for all such pairs.

5. Give examples of instructions that occupy 1, 2, 3, 4, and 5 bytes.

6. What are the relative advantages and disadvantages of passing parameters to a subroutine by listing them after a Call instruction versus placing them on the stack?

7. Can pointer registers in the 6809 microprocessor be used to do arithmetic operations? Give an example.

PRIMARY PROBLEMS

8. Write a 6809 program that sorts a list of 8-bit unsigned integers, to correspond to the program in Figure 2.17.

9. Write a program that finds the largest value in a list of 16-bit unsigned integers. Assume that the list contains 100 numbers, and is stored starting at location LSTADR.

10. Write a program that sorts a list of 16-bit positive integers in ascending order. Let the list have 100 entries, starting at address LSTADR.

11. The 6809 load and store instructions set the condition code flags N and Z according to the value of the operand moved. They also clear the overflow flag,

V, while leaving the carry flags, C and H, unaffected. If these instructions did not clear the V flag, when will the branch occur in the following case:

```
LDA    LOC
BGT    LOOP
```

How would you change these instructions to ensure that the branch will always take place for numbers greater than 0?

12. Assemble by hand the subroutine MAX in Figure 3.21. How many bytes of memory are needed to hold it? How many clock cycles are needed to execute this subroutine for a list of 500 entries?

13. How many bytes of main memory are needed for the program in Figure 3.25? How many clock cycles are needed to execute this program?

14. What is the value of the return address in Figure 3.26?

15. Repeat Problem 13 for the program in Figure 3.27.

16. Rewrite the program in Figure 3.15, using a subroutine CHOUT that displays one character on the screen.

17. Write a program that interchanges (swaps) the high- and low-order bytes in index register X.

18. Write a program that reverses the order of bits in index register X. For example, if the starting pattern in register X is 1001011101100110, the result left in X should be 0110011011101001.

19. Consider the following program:

```
        ORG    $1000
LOC     FDB    1234
        ORG    $100
        LDD    LOC
        LDX    #8
LOOP    RORA
        RORB
        LEAX   -1,X
        BNE    LOOP
        RORA
        STD    LOC
        END
```

(a) What does this program achieve?

(b) Is it an efficient way of accomplishing this task? If not, then give a better program that accomplishes the same task. (Better means shorter and/or faster.)

20. The following subroutine is intended to add two 3-byte integers P and Q. The 3 bytes of P and Q are in memory locations P3, P2, and P1, and Q3, Q2, and Q1, respectively.

```
SUMSUB    LDA     P1          Add low-order bytes.
          ADDA    Q1
          STA     S1          Save low-order sum.
          LDA     P2          Add middle bytes.
          BCC     FINE
          INCA
FINE      ADDA    Q2
          STA     S2          Save middle byte of sum.
          LDA     P3          Add high-order bytes.
          BCC     NOFIX
          INCA
NOFIX     ADDA    Q3
          STA     S3          Save high-order byte.
          RTS
```

Does this routine work properly for all values of P and Q? If not, explain the problem and show any necessary corrections.

21. Write two subroutines that can be used to implement a stack which is independent of the S and U stacks.

PUSHA places the contents of accumulator A on the stack.
PULLA loads accumulator A with the top item on the stack.

Use the 2-byte variable STKPTR as the stack pointer. (STKPTR is a RAM variable, not a 6809 register.) STKPTR must increase with each PUSHA operation, and must decrease with each PULLA.

22. Write a program that finds the names of the three worst students in the linked list of Section 2.8, where each item has the format shown in Figure 3.29.

23. Write a program that finds the names of the top three students in the class, using the linked list of Section 2.8, and assuming that each item is structured as shown in Figure 3.31.

ADVANCED PROBLEMS

24. Write a 6809 program for the task of Problem 2.18.

25. Write a 6809 program using the bubble sort scheme explained in Problem 2.19.

26. Write a program that reads two consecutive ASCII-encoded characters typed on the keyboard of a video terminal. The characters represent a two-digit hexadecimal number. Having read the characters, your program should convert them into an 8-bit number and load this number into accumulator A.

27. Figure 3.27 shows a program that adds two 20-digit BCD numbers. Write a similar program that subtracts such numbers. (*Hint*: use 10's complement BCD number representation.)

28. Write a program that adds two 32-bit numbers, represented in 2's complement notation and stored in memory locations NUM1 and NUM2.

29. The 6809 microprocessor has an instruction MUL for unsigned 8-bit multiplication. However, MUL does not produce correct answers if the multiplicand or the multiplier is negative. To perform signed 2's complement multiplication, you need to convert negative numbers into unsigned notation, multiply, and convert the product back to the 2's complement notation.

Write a subroutine NMUL to multiply two 8-bit numbers, one of which is positive and the other is negative. To simplify the task, assume that the positive number is in accumulator A and the negative number is in B. The product, in 2's complement notation, should be placed in accumulator D.

30. Write a 6809 subroutine that examines a string of 32 bits as a continuous sequence of 0s and 1s. The result returned by the subroutine is the number of times two 1s are found in adjacent bit positions. You must account for overlapped occurrences of the pairs of 1s. For example, in the sequence

$$0110\ 0111\ 0111\ 1010$$

there are six such pairs of 1s.

31. Write a program that can sort the linked-list of Section 2.8 in alphabetical order of students' names. Assume that each item is structured as shown in Figure 3.31.

32. In many microprocessor applications it is necessary to be able to delay a certain action by a predetermined amount of time. The required delay can be realized by executing a subroutine that does nothing except "wasting" the desired time on execution of its instructions. Assume that a given 6809 microprocessor is driven by a 1 MHz clock, so that each machine cycle takes 1 microsecond to perform. It is intended to use the following subroutine to produce a 1 ms delay:

```
DELAY    PSHS    B,CC
         LDB     #Value
LOOP     DECB
         BNE     LOOP
         PULS    B,CC,PC
```

(a) What number should be used as Value to produce the required delay?

(b) Modify the subroutine so that it can be used by a calling program to generate delays in the range from 1 to 200 ms.

(c) Write the shortest program that you can to produce a delay of 1 minute.

33. A *disassembler* is a program that examines memory and produces an assembly language source program, that is, it tries to undo the work of the assembler program. How can this be done? Would the output look like the original source program that was assembled by the assembler? Why? What would happen if data were in a part of the main memory where the disassembler expected a program? Is there a way to get around this problem? What other problems could arise?

LABORATORY EXERCISES

34. Write a 6809 program to multiply two matrices, as described in Problem 2.20.

35. Write a 6809 program that implements the task of Problem 2.22.

36. Write a program that reads two decimal integers from a video terminal, multiplies them, and displays their product. The program should include the following subroutines:

(a) RNUM: Read in decimal digits to form a 16-bit binary integer. Store the value in 2 bytes of memory labeled NUM.

(b) MULT: Multiply two 16-bit integers stored in memory, say in locations NUM1 and NUM2, and store their product in location PROD (note that PROD may occupy 32 bits).

(c) DNUM: Display the contents of PROD as a decimal number.

3.14 REFERENCES

1. *M6809 Microprocessor Programming Manual*. Austin, TX: Motorola Inc., 1979.

2. Leventhal, L.A., *6809 Assembly Language Programming*. Berkeley, CA: Osborne/McGraw-Hill, 1981.

3. Wakerly, J.F., *Microcomputer Architecture and Programming*. New York: Wiley, 1981.

4. Hamacher, V.C., Z.G. Vranesic, and S.G. Zaky, *Computer Organization*. 2nd ed., New York: McGraw-Hill, 1984.

5. Wagner, T.J., and G.J. Lipovski, *Fundamentals of Microcomputer Programming*. New York: Macmillan, 1984.

Chapter
4

Motorola 68000 Microprocessor Family

In the previous chapter we considered the Motorola 6809 microprocessor as a representative of 8-bit processors. Eight-bit microprocessors are suitable for use in numerous applications, which include computer terminals, word processors, personal computers, video games, and a variety of digitally controlled instruments. Yet, the 8-bit word length is rather restrictive in applications where high performance of the microprocessor is of critical importance.

VLSI technology has progressed to the point where it is possible to manufacture 16- and 32-bit microprocessors on a single chip. These are powerful processors, capable of performance that used to be associated with rather large computers. There are a number of such microprocessors available commercially. They include Intel's iAPX 86 [1] through iAPX 386 [2], Motorola's MC68000 [3], MC68020 [4] and MC68030 [5], Zilog's Z8000 [6], and National's 32016 and 32032 [7]. In this chapter we will discuss the Motorola chips. The MC68000 is an example of a 16-bit microprocessor, while the MC68020 and MC68030 represent the 32-bit chips. They have the same basic structure, hence we will deal first with the MC68000 and then describe the enhancements found in the MC68020 and MC68030. In the remainder of the text we will refer to these microprocessors simply as 68000, 68020, and 68030, omitting the prefix MC.

Usefulness of a microprocessor depends on several factors. While some specific application may be handled well with a particular microprocessor, the commercial viability of any microprocessor is contingent upon its suitability for use in a host of different applications. This implies that microprocessors should have certain generally useful features. Typically, it is convenient to have a large addressable space and flexible addressing modes. The former allows sizeable programs to be written without having to impose artificial size constraints, and the latter permits fast and easy accessing of data that may be stored in the main memory in a variety of data structures (lists, arrays, and so on). It is also useful to have a reasonably large number of registers in the processor. The availability of registers reduces the number of times that a given data item has to be read from or written into the main memory, thus speeding up overall processing. Another important feature is the size and versatility of the instruction set provided.

Using a high-level language is the easiest and fastest way of writing software for any computer. It is particularly useful if the instruction set of a microprocessor lends itself easily to the implementation of high-level language constructs, and allows efficient processing of programs written in such languages. Also, the microprocessor should have adequate input/output capability. It is essential that the basic I/O schemes be supported, which include polled-status I/O, interrupts, and direct memory access.

The 68000 family of microprocessors has been designed to provide all of these features. These are powerful processors, capable of supporting sophisticated applications.

This chapter presents the programmer's view of the 68000 microprocessor family at the assembly language level. The discussion builds on the general concepts described in Chapter 2. Sufficient details are given to enable the reader to write programs in the 68000 assembly language.

4.1 BASIC STRUCTURE OF THE 68000 MICROPROCESSOR

The 68000 handles external data in 16-bit words. Internally, it can deal with both data and addresses in up to 32-bit quantities, which is the size of most registers in the processor. External addresses are 24 bits long, meaning that 2^{24} (16M) locations can be addressed. The 68000 microprocessor is byte addressable; that is, each byte in the main memory has a unique address associated with it. The reader may wonder why only 24-bit addresses are used if the processor registers are 32 bits long. The reason is the desire to keep the number of pins on the microprocessor integrated circuit chip within certain limits. The 68000 has a total of 64 pins, which provide separate address, data, and control lines. A larger addressable space would have required a larger number of pins, resulting in a package that would have exceeded the maximum size that was available at the time the 68000 was introduced. Restricting addresses to 24 bits is not necessarily detrimental, because the 16-Mbyte addressable space is large enough for many applications. However, as we will see in Section 4.11, the 68020 and 68030 microprocessors have no such restriction; they are implemented in larger packages and use all 32 address bits.

The bus structure of a 68000 microcomputer is shown in Figure 4.1. Input and output operations are implemented using the memory-mapped I/O scheme discussed in Section 1.5. That is, some locations in the memory address space are assigned to I/O devices. Note that there are only 23 lines in the address bus, because bit 0 of an address is not transmitted on the data lines. Instead, the control lines provide an indication of whether a high or a low byte is needed, as will be explained in Chapter 6.

The register arrangement of the 68000 as seen by the programmer is given in Figure 4.2. Most of the registers are 32 bits long. There are eight data registers, D0 through D7, which can deal with operands in three sizes: byte, word, and long word, as shown in the figure. *Byte* operands involve only the 8 least-significant bits of a register. *Word* operands involve the 16 least-significant bits. All 32 bits are used for *long word* operands. The shorter operands can be used by instructions without affecting the remaining high-order bits in a register. Alternatively, an instruction may specify that the sign bit (i.e., bit 7 for byte and bit 15 for word operands) should be extended. In this case, the most-significant bit of the result of any operation is duplicated into all higher-order bits. The data registers can be used as general-purpose accumulators and counters. They can also be used as index registers in some addressing modes.

There are eight address registers, A0 through A7. The first seven of these registers are used to hold addressing information that is either 16 or 32 bits long. This information may be addresses of memory or I/O locations or index values to be used in indexed addressing modes. Address register A7 serves as a stack pointer.

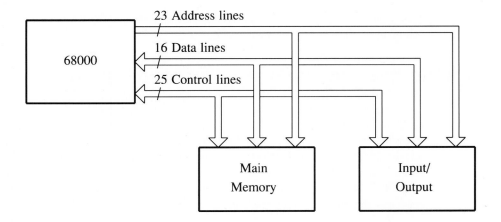

Figure 4.1 The 68000 bus structure.

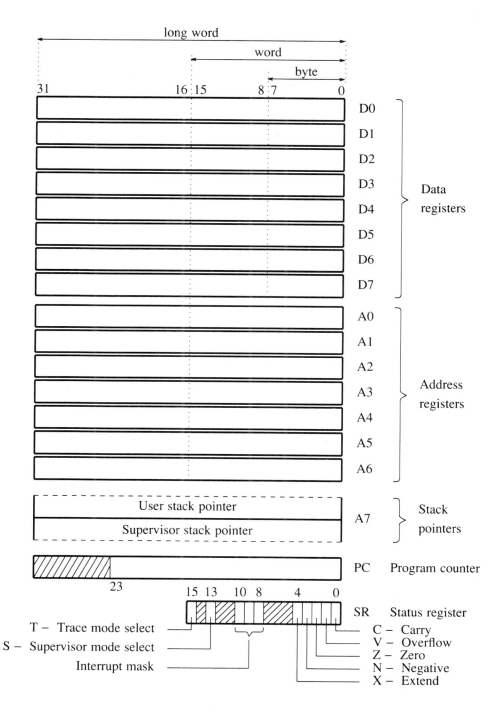

Figure 4.2 The 68000 register structure.

Actually, there are two stack pointers addressable as A7. They are associated with the two basic modes of operation of the 68000. In the first mode, called the *Supervisor* mode, all machine instructions can be executed, and A7 refers to the *supervisor stack pointer*. The second mode is the *User* mode. In this mode some instructions (called *privileged* instructions) cannot be executed and A7 refers to the *user stack pointer*. When a machine instruction refers to register A7, the register selected is determined by the information held in the processor's status register, SR. Bit 13 of this register determines the mode of operation of the processor, as will be explained later.

The program counter is a 32-bit register, but only its 24 least-significant bits are used as the address placed on the address lines of the bus. We should note that using 32-bit internal registers for computation of addresses allows for future modifications to the 68000 so that more address lines can be provided. The number of address lines may be increased by having a larger package, by replacing some of the existing control lines, or by time sharing the use of some pins. In the case of 68020, which will be discussed in Section 4.10, the address bus has 32 dedicated lines.

The SR has only 16 bits. Its low-order byte is also called the condition code register (CCR), because it contains the condition-code flags. The S bit (in position 13) of the SR register indicates whether the processor is in the Supervisor (S=1) or the User (S=0) mode. We will consider this feature in Section 4.5.6. Bits 8 to 10 are used for interrupts, as will be described in Chapter 7. Bit 15, called the *trace* bit T, is useful in debugging of programs, which will be discussed in Chapter 12.

4.1.1 Condition Code Flags

As already discussed in Section 2.1, it is useful to retain some information about the results of the most recently performed operations. This information is kept in the form of condition code flags, which are used by conditional branch instructions and for several other purposes such as in multiple-precision arithmetic and in the manipulation of logical operands. A flag is set or cleared whenever an instruction that affects it is executed.

There are five flags defined, as indicated in Figure 4.2. They are:

- C: Carry from the most-significant bit position.
- V: Arithmetic Overflow
- Z: Zero
- N: Negative
- X: Extend

Figure 4.3 gives some examples of how these flags are set as a result of addition and subtraction operations on signed operands of different lengths. The arithmetic operations are performed using the 2's complement rules (see Section A.6 in Appendix A).

The Z flag is set to 1 when the result is 0 and it is cleared otherwise. The N flag is set to 1 if the most-significant bit position of the result is equal to 1. This is bit position 7, 15, or 31 when the result is a byte, a word, or a long-word, respectively.

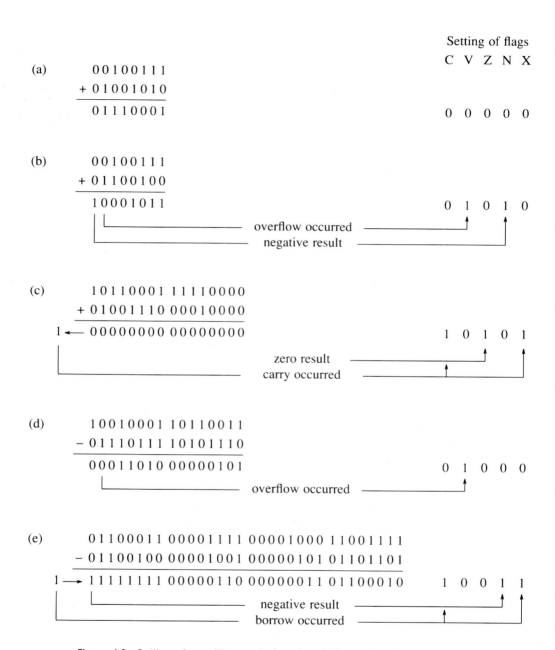

Figure 4.3 Setting of condition code flags in addition and subtraction operations.

The V flag is set to 1 if the signed value of the result of an arithmetic operation cannot fit within the operand length specified. Parts (b) and (d) of Figure 4.3 illustrate cases where overflow occurs, using 8- and 16-bit operands.

The C flag is set to 1 when a carry from the most-significant bit position occurs as a result of an addition operation. This flag is also used to indicate a borrow after a subtraction operation. Subtraction is done by adding the 2's complement of the subtrahend to the minuend (see Section A.6.3). This addition process results in a carry from the most-significant bit position when a borrow does not occur in the subtraction in question. Hence, in order to represent the borrow bit correctly, the C flag is set to the complement of the carry out of the most-significant bit position when a subtraction operation is executed. This is illustrated in parts (d) and (e) of Figure 4.3.

The X flag is set in the same way as the C flag, but it is not affected by as many instructions as the C flag. It is useful in the implementation of multiple-precision arithmetic operations, e.g., addition and subtraction of 64-bit numbers.

The examples in Figure 4.3 show how the condition code flags are manipulated during addition and subtraction operations. These flags are also affected by a variety of instructions that move operands and perform logic operations. In general, they are used to represent the nature of the result of a given operation. The details will be discussed shortly, when the complete instruction set is presented.

4.1.2 Memory Addresses

The 68000 is a byte-addressable microprocessor, with the byte locations having addresses in the range 0 to $2^{24}-1$. However, since the 68000 uses a 16-bit data bus, it is convenient to think of its memory as being composed of 16-bit words, where each word has an even address.

Figure 4.4 illustrates the memory addressing scheme. The high-order byte of each word, comprising bits 8 to 15, is in an even address location, and the low-order byte, bits 0 to 7, is in an odd address location. The address used for a word or a long-word operand is that of its most-significant byte. Word and long-word operands must be aligned on even-numbered boundaries; that is, they must start at an even address. All data pushed on or pulled from the system stack, which is implemented in the main memory using register A7 as the stack pointer, must be word

Address of left byte	CONTENTS (hex) 15 8,7 0		Address of right byte
1000	3F	20	1001
1002	00	4A	1003
1004	23	08	1005

Figure 4.4   Memory addresses in the 68000.

aligned. Thus, even the byte operands must be stored in stack locations with even-numbered addresses.

In the example of Figure 4.4, the byte operands in locations 1000 to 1005 are 3F, 20, 00, 4A, 23, and 08. The same memory contents can be interpreted as the word operands 3F20, 004A, and 2308 in locations 1000, 1002, and 1004, respectively, or a long-word operand 004A2308 at address 1002.

A special type of operands that are handled in 8-bit quantities are BCD (Binary Coded Decimal) numbers. They are represented in a packed format, such that two BCD digits occupy 1 byte. The BCD instructions use the operands 1 byte at a time.

4.2 ASSEMBLER CONVENTIONS FOR THE 68000 MICROPROCESSOR

Before introducing the instruction set of the 68000 microprocessor, we will give a brief description of the conventions and assembler commands used in preparing programs. This information should enable the reader to do laboratory exercises while studying the material in the remainder of the chapter.

A typical assembly process, general assembler conventions, and assembler directives were discussed in Section 2.10. That discussion is fully applicable to the 68000 microprocessor. In this section we will describe only the differences pertinent to the 68000 assembler supplied by the manufacturer.

The maximum length allowed for operand and label names is 8 characters. Numbers are taken to be decimal, unless marked with the prefix % for binary or $ for hexadecimal. Alphanumeric characters enclosed between single quotes are replaced by their ASCII codes. One or more characters can be specified within a string delimited by two single quotes. For example, the specification 'NUM2' is interpreted as the ASCII codes corresponding to N, U, M, and 2, respectively.

The 68000 assembler makes use of several assembler directives. The EQU, ORG, and END directives are the same as those described in Section 2.10.3. The DATA directive is called DC. This and other 68000 assembler directives are described below.

DC (Define Constant)

The DC directive is useful for placing specific data into the main memory. A number of constants can be defined, separated by commas, in a single DC statement. The constants defined may be 8, 16, or 32 bits long, where the size is indicated by appending one of the letters B, W, or L to the DC designator. For example, the directive

$$\text{LOC} \quad \text{DC.W} \quad \$1250,28,\%11011001$$

will interpret all three constants as being 16 bits long and load the memory as follows:

Address	Contents (hex)
LOC	1250
LOC+2	001C
LOC+4	00D9

Note that since 16-bit constants are being defined in this example, the address LOC must be an even number.

Alphanumeric characters are stored in individual bytes. The directive

LOC   DC.B   'MESSAGE'

will load the 7 bytes of memory starting at location LOC with the ASCII code for M, E, S, S, A, G, and E. In this case, the address LOC may be either even or odd.

DS (Define Storage)

Memory locations can be reserved for some specific purpose by means of the DS directive. The size of the reserved area can be specified as a number of bytes, words, or long words. For example, the statement

TABLE   DS.L   100

reserves 100 long words (400 bytes) of memory starting at the address identified as TABLE.

It is possible to use expressions to specify operands in the DC and DS directives. As an example, consider the following statements:

NUM	EQU	50
	ORG	$1000
TABLE	DS.B	NUM−10
VRBL	DC.W	$240,VRBL+6

These statements will cause the assembler to reserve 40 bytes of memory for a table that starts at address $1000. The value 1000 will be associated with the name TABLE. The name VRBL will be associated with the address $1028, and the constants $240 and $102E will be placed in locations $1028 and $102A, respectively.

SET

The SET directive is used to define names in the same manner as the EQU directive. The difference between them is that a name defined in an EQU statement cannot be redefined later in the program, while a name defined in a SET statement can be subsequently redefined using another SET statement.

4.3 68000 ADDRESSING MODES

A powerful microprocessor has a variety of instructions and addressing modes. In order to encode the instructions, many distinct bit patterns must be available. When the length of such patterns exceeds the word length of the main memory, several words have to be used to encode a given instruction. In the 68000 microprocessor, the first word of an instruction is a 16-bit *OP-code* that specifies the operation to be performed and indicates how to interpret the addressing data. Additional words are used as required by the desired addressing modes. For example, if an instruction includes a 32-bit immediate operand, this operand is given in the two words that follow the OP-code.

The 68000 has addressing modes that offer considerable flexibility in accessing operands. All of the addressing modes described in Section 2.3 are included, as well as some others. Table 4.1 gives a summary of the available modes. The first three columns of the table show the name, the assembler syntax, and the addressing function of each mode. The fourth column gives a short, descriptive notational name for each mode, which will be used in later tables for labeling purposes. The entries in the table are mostly self-explanatory. The following discussion elaborates further on some important features.

Table 4.1 THE 68000 ADDRESSING MODES

Name	Assembler Syntax	Addressing Function	Notational Name
Absolute long	Value	EA = Value	Abs.L
Absolute short	Value	EA = Sign extended WValue	Abs.W
Register	Rn	EA = R_n that is, Operand = $[R_n]$	Rn
Register indirect	(An)	EA = $[A_n]$	(An)
Immediate	#Value	Operand = Value	Immed
Indexed basic	WValue(An)	EA = WValue+$[A_n]$	d(An)
Indexed full	BValue(An,Rk.S)	EA = BValue+$[A_n]$+$[R_k]$	d(An,Xi)
Autoincrement	(An)+	EA = $[A_n]$; Increment A_n	(An)+
Autodecrement	−(An)	Decrement A_n ; EA = $[A_n]$	−(An)
Relative basic	WValue(PC) or Label	EA = WValue+[PC]	d(PC)
Relative full	BValue(PC,Rk.S) or Label(Rk)	EA = BValue+[PC]+$[R_k]$	d(PC,Xi)

Notes: EA = effective address
  Value = a number given either explicitly or represented by a label
  BValue = an 8-bit Value
  WValue = a 16-bit Value
  A_n = an address register
  R_n, R_k = an address or a data register
  S = a size indicator: W for sign-extended 16 bit word and L for 32-bit long word.

Absolute mode

The absolute address of an operand is given in the instruction, following the OP-code word. Two versions of this mode are provided. In the absolute long mode, a 32-bit address is given explicitly in the instruction, while in the absolute short mode, only a 16-bit value is given. The high-order 8 bits of the address are obtained by replicating the MSB (i.e., bit 15) of this value. Since the MSB can be either 0 or 1, the short mode permits accessing of only two pages of the addressable space, namely, the 0 page and the $FF8 page. Each page consists of 32K bytes.

Register mode

The operand location is either a data or an address register indicated in the instruction.

Register indirect mode

The contents of an address register indicated in the instruction is the effective address of the operand.

Immediate mode

The value of the operand is given explicitly in the instruction. In assembler notation, the operand value is either stated explicitly, or it may be represented by a name. Four sizes of operands are allowed. Byte, word, and long-word operands are placed in the memory locations that follow the OP-code word. A fourth type of immediate operand uses only small numbers (typically having 3 bits) and is included within the OP-code word. A description of the way in which operand lengths are specified in the assembler syntax can be found at the end of Section 4.4.1.

Basic indexed mode

The effective address of the operand is the sum of a 16-bit signed index value (displacement) and the contents of an address register, An. The index value and An are both specified in the instruction.

Full indexed mode

In this mode two index registers are used, one being an address register An and the other either an address register or a data register, Rk. The effective address of the operand is the sum of an 8-bit signed index value and the contents of registers An and Rk. The index value and registers An and Rk are specified in the instruction. All 32 bits of the An register are used in the derivation of the address. However, either all 32 bits, or the sign-extended low-order 16 bits of Rk are used. The former possibility is indicated by Rk.L in the assembler syntax, while the latter is the default case that can also be shown as Rk.W.

Autoincrement mode

The effective address of the operand is the contents of an address register, An, which is specified in the instruction. After accessing the operand, the contents of An are incremented automatically by 1, 2, or 4, for a byte, word, or long-word operand, respectively.

Autodecrement mode

The effective address of the operand is the contents of an address register, An, specified in the instruction. Before accessing the operand, the contents of An are decremented by 1, 2, or 4, for a byte, word, or long-word operand, respectively.

Basic relative mode

This mode is the same as the basic indexed mode, except that the program counter is used in place of an address register An.

Full relative mode

This mode is the same as the full indexed mode, except that the program counter is used in place of an address register An.

Finally, let us consider briefly the choice of notation in the last column of Table 4.1, which is intended to be as descriptive as possible. For example, Abs.W denotes the absolute short mode, where a 16-bit word in the instruction gives the desired effective address. The indexed full mode is indicated as d(An,Xi), because it involves a displacement (index value) and the contents of one address register and one index register. The index register can be either an address or a data register. It should be emphasized that this is not an assembler notation to be used in programming. The entries in the fourth column of the table are simply abbreviations introduced for ease of reference later in this chapter. The correct assembler syntax is indicated in column 2 of the table.

The above discussion provided a simple description of the 68000 addressing modes, using generally accepted terminology. Some of the names used for various addressing modes are different from those found in Motorola literature. In order to enable the reader to readily consult the manufacturer's data sheets and user's manuals, we have summarized the differences in terminology in Table 4.2. While the Motorola terminology is very descriptive, it is rather unwieldy for discussion purposes.

Comparing the 68000 addressing modes with the general addressing modes described in Section 2.3, it is apparent that only the direct versions of most modes are provided in the 68000. The only exception is the register indirect mode. However, considerably more powerful indexed and relative modes are available, with increased flexibility gained by making use of two index registers in the derivation of an effective address.

Table 4.2   THE 68000 TERMINOLOGY DIFFERENCES

Terminology used in this text	Motorola terminology
Autoincrement	Address register indirect with postincrement
Autodecrement	Address register indirect with predecrement
Indexed basic	Address register indirect with displacement
Indexed full	Address register indirect with index
Relative basic	Program counter with displacement
Relative full	Program counter with index

4.4 68000 INSTRUCTION FORMATS

The 68000 microprocessor has a powerful instruction set, intended to provide good support for implementation of programs written in high-level programming languages. Most instructions can use any of the three operand sizes and any addressing mode in a uniform way. In what follows, we will describe the format in which machine instructions are written when preparing assembly language programs. In Section 4.5 we will consider the details of the entire instruction set, including the encoding needed to produce executable object code.

4.4.1 Two-Operand Instructions

The assembler format for 2-operand instructions is

OP   src,dst

which states that an operation OP is to be performed using the source and destination operands, with the result placed in the destination location. The operands can be specified using a variety of addressing modes, but not all addressing modes can be used with all the instructions. One notable difference between the 68000 and 6809 microprocessors is that the 68000 has several instructions that allow both the source and destination operands to be in the main memory. In the case of the 6809, at least one operand must be in a processor register.

The most general specification of operands is allowed in the MOVE instruction, where any pertinent addressing mode may be used for both operands. Such wide scope in specifying addresses is not provided for other instructions, because there would not be enough codes available within a 16-bit OP-code word. The choice is to provide few instructions with many possible addressing modes, or many instructions with fewer addressing modes. The latter choice governs the design of the 68000 microprocessor.

Several instructions, other than MOVE, allow specification of two memory operands, but in a more restricted manner. For example, there exists an instruction for comparing lists of data items. Its format is

CMPM   (Ai)+,(Aj)+

using address registers Ai and Aj as pointers to the operands. This instruction can be used in the autoincrement mode only, which is a reasonable choice when lists of items are scanned.

In a number of two-operand instructions, one operand must be in a processor register, while the second operand may be specified using any addressing mode. For instance, in the ADD instruction

ADD   src,dst

if the source is a data register, then the destination may be stated in absolute, register, register indirect, autoincrement, autodecrement, or indexed modes. However, if the source is specified in any other mode, the destination must be a data register.

In assembler syntax the size of the operands used in an instruction is indicated by appending one of the letters B, W, or L to the operation mnemonic, to denote byte, word, or long-word operands, respectively. As an example, two 32-bit numbers in data registers D1 and D2 can be added with the instruction

ADD.L   D1,D2

On the other hand, if the instruction is specified as

ADD.W   D1,D2

the addition is performed on only the least-significant 16 bits of the operands, leaving the high-order 16 bits of register D2 unchanged. If the size is not specified explicitly in an instruction, the default case is the W specification. Thus, 16-bit addition can also be achieved with

ADD   D1,D2

As mentioned in Section 4.3, the immediate operands in some instructions can be small numbers defined within the OP-code word. Such operands are called "quick" and are denoted by appending the letter Q to the operation mnemonic. An example of this format is the instruction

SUBQ.W   #2,D4

which causes the value 2 to be subtracted from the 16-bit operand in register D4. The quick immediate operand, 2, is specified in a 3-bit field within the OP-code word. It should be noted that the use of the quick operand is the only case in which separate length specifications are given for the source and destination operands. In our example, the source operand is a 3-bit number, while the destination operand has 16 bits.

4.4.2 One-Operand Instructions

In these instructions the operand field specifies an operand and possibly some additional information. As in the case of the two-operand instructions, there are some restrictions on the addressing modes allowed. For example, the contents of a data register can be shifted left by means of an LSL instruction. The contents involved may be either all 32 bits, or the least-significant 16 or 8 bits of the register. Moreover, it is possible to specify the number of bit positions by which the contents should be shifted. This number, called the *count*, can be given either in another data register, or as a quick immediate operand within the LSL instruction. As an illustration, suppose that the contents of register D2 are to be shifted left by five positions, and that register D1 has the number 5 stored in it. Then, the desired shift operation can be realized with either

LSL.L   D1,D2

or

LSL.L   #5,D2

Table 4.3   68000 INSTRUCTION SET

Mnemonic (Name)	Size	Addressing mode	Dn	An	(An)	(An)+	-(An)	d(An)	d(An,Xi)	Abs.W	Abs.L	d(PC)	d(PC,Xi)	Immed	SR or CCR
ABCD (Add BCD)	B	s = Dn d =	x												
		s = −(An) d =					x								
ADD (Add)	B, W, L	s = Dn d =			x	x	x	x	x	x	x				
		d = Dn s =	x	x	x	x	x	x	x	x	x	x	x	x	
ADDA (Add address)	W	d = An s =	x	x	x	x	x	x	x	x	x	x	x	x	
	L	d = An s =	x	x	x	x	x	x	x	x	x	x	x	x	
ADDI (Add immediate)	B, W, L	s = Immed d =	x		x	x	x	x	x	x	x				
ADDQ (Add quick)	B, W, L	s = Immed3 d =	x	x	x	x	x	x	x	x	x				
ADDX (Add extended)	B, W, L	s = Dn d =	x												
		s = −(An) d =					x								
AND (Logical AND)	B, W, L	s = Dn d =			x	x	x	x	x	x	x				
		d = Dn s =	x		x	x	x	x	x	x	x	x	x	x	
ANDI (AND immediate)	B, W, L	s = Immed d =	x		x	x	x	x	x	x	x				x
ASL (Arithmetic shift left)	B, W, L	count = [Dn] d =	x												
		count = QQQ d =	x												
		count = 1 d =			x	x	x	x	x	x					
ASR (Arithmetic shift right)	B, W, L	count = [Dn] d =	x												
		count = QQQ d =	x												
		count = 1 d =			x	x	x	x	x	x					
BCHG * (Test a bit and change it)	B	bit# = [Dn] d =			x	x	x	x	x	x	x				
		bit# = Immed d =			x	x	x	x	x	x	x				
	L	bit# = [Dn] d =	x												
		bit# = Immed d =	x												
BCLR * (Test a bit and clear it)	B	bit# = [Dn] d =			x	x	x	x	x	x	x				
		bit# = Immed d =			x	x	x	x	x	x	x				
	L	bit# = [Dn] d =	x												
		bit# = Immed d =	x												

OP code b_{15} $\cdots$ b_0	Operation performed	Condition flags				
		X	N	Z	V	C
1100 RRR1 0000 0rrr 1100 RRR1 0000 1rrr	$d \leftarrow [s] + [d] + [X]$ Binary coded decimal addition	x	u	x	u	x
1101 DDD1 SSEE EEEE 1101 DDD0 SSee eeee	$d \leftarrow [Dn] + [d]$ $Dn \leftarrow [s] + [Dn]$	x x	x x	x x	x x	x x
1101 AAA0 11ee eeee 1101 AAA1 11ee eeee	$An \leftarrow [s] + [An]$					
0000 0110 SSEE EEEE	$d \leftarrow s + [d]$	x	x	x	x	x
0101 QQQ0 SSEE EEEE	$d \leftarrow QQQ + [d]$	x	x	x	x	x
1101 RRR1 SS00 0rrr 1101 RRR1 SS00 1rrr	$d \leftarrow [s] + [d] + [X]$ Multiprecision addition	x	x	x	x	x
1100 DDD1 SSEE EEEE 1100 DDD0 SSee eeee	$d \leftarrow [Dn] \wedge [d]$		x	x	0	0
0000 0010 SSEE EEEE	$d \leftarrow s \wedge [d]$		x	x	0	0
1110 rrr1 SS10 0DDD 1110 QQQ1 SS00 0DDD 1110 0001 11EE EEEE	C ← [operand] ← 0 X ←	x	x	x	x	x
1110 rrr0 SS10 0DDD 1110 QQQ0 SS00 0DDD 1110 0000 11EE EEEE	→ [operand] → C → X	x	x	x	x	x
0000 rrr1 01EE EEEE 0000 1000 01EE EEEE 0000 rrr1 01EE EEEE 0000 1000 01EE EEEE	$Z \leftarrow \overline{(\text{bit\# of } d)}$; then complement the tested bit in d			x		
0000 rrr1 10EE EEEE 0000 1000 10EE EEEE 0000 rrr1 10EE EEEE 0000 1000 10EE EEEE	$Z \leftarrow \overline{(\text{bit\# of } d)}$; then clear the tested bit in d			x		

Table 4.3 (continued)   68000 INSTRUCTION SET

Mnemonic (Name)	Size	Addressing mode		Dn	An	(An)	(An)+	-(An)	d(An)	d(An,Xi)	Abs.W	Abs.L	d(PC)	d(PC,Xi)	Immed	SR or CCR
BSET * (Test a bit and set it)	B	bit# = [Dn]	d =			x	x	x	x	x	x	x				
		bit# = Immed	d =			x	x	x	x	x	x	x				
	L	bit# = [Dn]	d =	x												
		bit# = Immed	d =	x												
BTST * (Test a bit)	B	bit# = [Dn]	d =			x	x	x	x	x	x	x				
		bit# = Immed	d =			x	x	x	x	x	x	x				
	L	bit# = [Dn]	d =	x												
		bit# = Immed	d =	x												
CHK (Check register against bounds)	W	d = Dn	s =	x		x	x	x	x	x	x	x	x	x	x	
CLR (Clear)	B, W, L		d =	x		x	x	x	x	x	x	x				
CMP * (Compare)	B, W, L	d = Dn	s =	x	x	x	x	x	x	x	x	x	x	x	x	
CMPA * (Compare address)	W	d = An	s =	x	x	x	x	x	x	x	x	x	x	x	x	
	L	d = An	s =	x	x	x	x	x	x	x	x	x	x	x	x	
CMPI * (Compare immediate)	B, W, L	s = Immed	d =	x		x	x	x	x	x	x	x				
CMPM * (Compare memory)	B, W, L	s = (An)+	d =				x									
DIVS * (Divide signed)	W	d = Dn	s =	x		x	x	x	x	x	x	x	x	x	x	
DIVU * (Divide unsigned)	W	d = Dn	s =	x		x	x	x	x	x	x	x	x	x	x	
EOR (Exclusive OR)	B, W, L	s = Dn	d =	x		x	x	x	x	x	x	x				
EORI (Exclusive OR immediate)	B, W, L	s = Immed	d =	x		x	x	x	x	x	x	x				x

OP code $b_{15} \quad \cdots \quad b_0$	Operation performed	Condition flags				
		X	N	Z	V	C
0000 rrr1 11EE EEEE 0000 1000 11EE EEEE 0000 rrr1 11EE EEEE 0000 1000 11EE EEEE	$Z \leftarrow \overline{(\text{bit\# of d})}$; then set to 1 the tested bit in d			x		
0000 rrr1 00EE EEEE 0000 1000 00EE EEEE 0000 rrr1 00EE EEEE 0000 1000 00EE EEEE	$Z \leftarrow \overline{(\text{bit\# of d})}$;			x		
0100 DDD1 10ee eeee	If $[Dn] < 0$ or $[Dn] > [s]$ then raise an interrupt		x	u	u	u
0100 0010 SSEE EEEE	$d \leftarrow 0$		0	1	0	0
1011 DDD0 SSee eeee	$[d] - [s]$		x	x	x	x
1011 AAA0 11ee eeee 1011 AAA1 11ee eeee	$[An] - [s]$		x	x	x	x
0000 1100 SSEE EEEE	$[d] - [s]$		x	x	x	x
1011 RRR1 SS00 1rrr	$[d] - s$		x	x	x	x
1000 DDD1 11ee eeee	$d \leftarrow [d] \div [s]$, using 32 bits of s and 16 bits of d		x	x	x	0
1000 DDD0 11ee eeee	$d \leftarrow [d] \div [s]$, using 32 bits of s and 16 bits of d		x	x	x	0
1011 rrr1 SSEE EEEE	$d \leftarrow [Dn] \oplus [d]$		x	x	0	0
0000 1010 SSEE EEEE	$d \leftarrow s \oplus [d]$		x	x	0	0

Table 4.3 (continued)   68000 INSTRUCTION SET

Mnemonic (Name)	Size	Addressing mode		Dn	An	(An)	(An)+	-(An)	d(An)	d(An,Xi)	Abs.W	Abs.L	d(PC)	d(PC,Xi)	Immed	SR or CCR
EXG (Exchange)	L	s = Dn	d =	x	x											
		s = An	d =	x	x											
EXT (Sign extend)	W		d =	x												
	L		d =	x												
JMP (Jump)			d =			x			x	x	x	x	x	x		
JSR (Jump to subroutine)			d =			x			x	x	x	x	x	x		
LEA (Load effective address	L	d = An	s =			x			x	x	x	x	x	x		
LINK (Link and allocate)		disp = Immed	s =		x											
LSL (Logical shift left)	B, W, L	count = [Dn]	d =	x												
		count = QQQ	d =	x												
	W	count = 1	d =			x	x	x	x	x	x	x				
LSR (Logical shift right)	B, W, L	count = [Dn]	d =	x												
		count = QQQ	d =	x												
	W	count = 1	d =			x	x	x	x	x	x	x				
MOVE (Move)	W	d = CCR	s =	x		x	x	x	x	x	x	x	x	x	x	
		d = SR	s =	x		x	x	x	x	x	x	x	x	x	x	
		s = SR	d =	x		x	x	x	x	x	x	x				
	L	s = SP	d =		x											
		d = SP	s =		x											

OP code b_{15} $\cdots$ b_0	Operation performed	X	N	Z	V	C
1100 DDD1 0100 0DDD 1100 AAA1 0100 1AAA 1100 DDD1 1000 1AAA	$[s] \leftrightarrow [d]$					
0100 1000 1000 0DDD 0100 1000 1100 0DDD	(bits 8-15 of d) ← (bit 7 of d) (bits 16-31 of d) ← (bit 15 of d)	x x	x x	0 0	0 0	
0100 1110 11EE EEEE	PC ← [d]					
0100 1110 10EE EEEE	SP ← [SP] – 2 ; [SP] ← [PC] ; PC ← [d]					
0100 AAA1 11ee eeee	An ← effective address of s					
0100 1110 0101 0AAA	SP ← [SP] – 4 ; [SP] ← [An] ; An ← [SP] ; SP ← [SP] + disp					
1110 rrr1 SS10 1DDD 1110 QQQ1 SS00 1DDD 1110 0011 11EE EEEE	C ← operand ← 0 ; X ←	x	x	x	0	x
1110 rrr0 SS10 1DDD 1110 QQQ0 SS00 1DDD 1110 0010 11EE EEEE	0 → operand → C ; → X	x	x	x	0	x
0100 0100 11ee eeee 0100 0110 11ee eeee 0100 0000 11EE EEEE 0100 1110 0110 1AAA 0100 1110 0110 0AAA	CCR ← [s] SR ← [s] d ← [SR] d ← [SP] SP ← [d]	x x	x x	x x	x x	x x

Table 4.3 (continued)   68000 INSTRUCTION SET

Mnemonic (Name)	Size	Addressing mode		Dn	An	(An)	(An)+	-(An)	d(An)	d(An,Xi)	Abs.W	Abs.L	d(PC)	d(PC,Xi)	Immed	SR or CCR
MOVE (Move)	B, W, L	s = Dn	d =	x		x	x	x	x	x	x	x				
		s = An	d =	x		x	x	x	x	x	x	x				
		s = (An)	d =	x		x	x	x	x	x	x	x				
		s = (An)+	d =	x		x	x	x	x	x	x	x				
		s = −(An)	d =	x		x	x	x	x	x	x	x				
		s = d(An)	d =	x		x	x	x	x	x	x	x				
		s = d(An,Xi)	d =	x		x	x	x	x	x	x	x				
		s = Abs.W	d =	x		x	x	x	x	x	x	x				
		s = Abs.L	d =	x		x	x	x	x	x	x	x				
		s = d(PC)	d =	x		x	x	x	x	x	x	x				
		s = d(PC,Xi)	d =	x		x	x	x	x	x	x	x				
		s = Immed	d =	x		x	x	x	x	x	x	x				
MOVEA (Move address)	W, L	d = An	s =	x	x	x	x	x	x	x	x	x	x	x	x	
MOVEM * (Move multiple registers)	W	s = Xn	d =			x			x	x	x	x				
		d = Xn	s =			x	x		x	x	x	x	x	x		
	L	s = Xn	d =			x		x	x	x	x	x				
		d = Xn	s =			x	x		x	x	x	x	x	x		
MOVEP * (Move peripheral data)	W	s = Dn	d =						x							
	L	s = Dn	d =						x							
	W	s = d(An)	d =	x												
	L	s = d(An)	d =	x												
MOVEQ (Move quick)	L	s = Immed8	d =	x												
MULS * (Multiply signed)	W	d = Dn	s =	x		x	x	x	x	x	x	x	x	x	x	
MULU * (Multiply unsigned)	W	d = Dn	s =	x		x	x	x	x	x	x	x	x	x	x	
NBCD (Negate BCD)	B		d =	x		x	x	x	x	x	x	x				

OP code b_{15} $\cdots$ b_0	Operation performed	Condition flags				
		X	N	Z	V	C
00SS RRRM MMee eeee	$d \leftarrow [s]$		x	x	0	0
00SS AAA0 01ee eeee	$An \leftarrow [s]$					
0100 1000 10EE EEEE 0100 1100 10ee eeee 0100 1000 11EE EEEE 0100 1100 11ee eeee	$d \leftarrow [Xn]$ ⎫ A second word is $Xn \leftarrow [s]$ ⎬ used to specify $d \leftarrow [Xn]$ ⎪ the registers $Xn \leftarrow [s]$ ⎭ involved					
0000 DDD1 1000 1AAA 0000 DDD1 1100 1AAA 0000 DDD1 0000 1AAA 0000 DDD1 0100 1AAA	Alternate bytes of $d \leftarrow [Dn]$ $Dn \leftarrow$ Alternate bytes of d					
0111 DDD0 QQQQ QQQQ	$Dn \leftarrow QQQQQQQQ$		x	x	0	0
1100 DDD1 11ee eeee	$Dn \leftarrow [s] \times [Dn]$		x	x	0	0
1100 DDD0 11ee eeee	$Dn \leftarrow [s] \times [Dn]$		x	x	0	0
0100 1000 00EE EEEE	$d \leftarrow 0 - [d] - [X]$ using BCD arithmetic	x	u	x	u	x

Table 4.3 (continued)   68000 INSTRUCTION SET

Mnemonic (Name)	Size	Addressing mode		Dn	An	(An)	(An)+	-(An)	d(An)	d(An,Xi)	Abs.W	Abs.L	d(PC)	d(PC,Xi)	Immed	SR or CCR
NEG (Negate)	B, W, L		d =	x		x	x	x	x	x	x	x				
NEGX (Negate extended)	B, W, L		d =	x		x	x	x	x	x	x	x				
NOP (No operation)																
NOT (Complement)	B, W, L		d =	x		x	x	x	x	x	x	x				
OR (Logical OR)	B, W, L	s = Dn	d =			x	x	x	x	x	x	x				
		d = Dn	s =	x		x	x	x	x	x	x	x	x	x	x	
ORI (OR immediate)	B, W, L	s = Immed	d =	x		x	x	x	x	x	x	x				x
PEA (Push effective address)	L		s =			x			x	x	x	x	x	x		
RESET																
ROL (Rotate left without X)	B, W, L · W	count = [Dn] · count = QQQ · count = 1	d = · d = · d =	x · x		x	x	x	x	x	x	x				
ROR (Rotate right without X)	B, W, L · W	count = [Dn] · count = QQQ · count = 1	d = · d = · d =	x · x		x	x	x	x	x	x	x				
ROXL (Rotate left with X)	B, W, L · W	count = [Dn] · count = QQQ · count = 1	d = · d = · d =	x · x		x	x	x	x	x	x	x				
ROXR (Rotate right with X)	B, W, L · W	count = [Dn] · count = QQQ · count = 1	d = · d = · d =	x · x		x	x	x	x	x	x	x				

OP code b_{15} $\cdots$ b_0	Operation performed	X	N	Z	V	C
0100 0100 SSEE EEEE	$d \leftarrow 0 - [d]$	x	x	x	x	x
0100 0000 SSEE EEEE	$d \leftarrow 0 - [d] - [X]$	x	x	x	x	x
0100 1110 0111 0001	None					
0100 0110 SSEE EEEE	$d \leftarrow \overline{[d]}$		x	x	0	0
1000 DDD1 SSEE EEEE 1000 DDD0 SSee eeee	$d \leftarrow [s] \vee [d]$		x	x	0	0
0000 0000 SSEE EEEE	$d \leftarrow s \vee [d]$		x	x	0	0
0100 1000 01ee eeee	$SP \leftarrow [SP] - 2$; $[SP] \leftarrow$ Effective address of s					
0100 1110 0111 0000	Assert RESET output line					
1110 rrr1 SS11 1DDD 1110 QQQ1 SS01 1DDD 1110 0111 11EE EEEE	[C] ← [operand] ←↺ (rotate left)		x	x	0	x
1110 rrr0 SS11 1DDD 1110 QQQ0 SS01 1DDD 1110 0110 11EE EEEE	↻→ [operand] → [C] (rotate right)		x	x	0	x
1110 rrr1 SS11 0DDD 1110 QQQ1 SS01 0DDD 1110 0101 11EE EEEE	[C] ← [operand] ← [X] (rotate left through X)	x	x	x	0	x
1110 rrr0 SS11 0DDD 1110 QQQ0 SS01 0DDD 1110 0100 11EE EEEE	[X] → [operand] → [C] (rotate right through X)	x	x	x	0	x

Table 4.3 (continued)   68000 INSTRUCTION SET

Mnemonic (Name)	Size	Addressing mode	Dn	An	(An)	(An)+	-(An)	d(An)	d(An,Xi)	Abs.W	Abs.L	d(PC)	d(PC,Xi)	Immed	SR or CCR
RTE (Return from exception)															
RTR (Return and restore CCR)															
RTS (Return from subroutine)															
SBCD (Subtract BCD)	B	s = Dn d = s = −(An) d =	x				x								
Scc (Set on condition)	B	d =	x		x	x	x	x	x	x	x				
STOP (Load SR and stop)		s =												x	
SUB (Subtract)	B, W, L	s = Dn d = d = Dn s =	x	x	x x	x x	x x	x x	x x	x x	x x	x	x	x	
SUBA (Subtract address)	W L	d = An s = d = An s =	x x	x x	x x	x x	x x	x x	x x	x x	x x	x x	x x	x x	
SUBI (Subtract immediate)	B, W, L	s = Immed d =	x		x	x	x	x	x	x	x				
SUBQ (Subtract quick)	B, W, L	s = Immed3 d =	x	x	x	x	x	x	x	x	x				
SUBX (Subtract extended)	B, W, L	s = Dn d = s = −(An) d =	x				x								
SWAP (Swap register halves)	W	d =	x												

OP code	Operation performed	Condition flags				
b_{15} • • • b_0		X	N	Z	V	C
0100 1110 0111 0011	SR ← [[SP]] ; SP ← [SP] + 2 ; PC ← [[SP]] ; SP ← [SP] + 2	x	x	x	x	x
0100 1110 0111 0111	CCR ← [[SP]] ; SP ← [SP] + 2 ; PC ← [[SP]] ; SP ← [SP] + 2	x	x	x	x	x
0100 1110 0111 0101	PC ← [[SP]] ; SP ← [SP] + 2					
1000 RRR1 0000 0rrr 1000 RRR1 0000 1rrr	d ← [d] − [s] − [X] Binary coded decimal subtraction	x	u	x	u	x
0101 CCCC 11EE EEEE	Set all 8 bits of d to 1 if cc is true, otherwise clear them to 0					
0100 1110 0111 0010	SR ← s ; Wait for interrupt	x	x	x	x	x
1001 DDD1 SSEE EEEE 1001 DDD0 SSee eeee	d ← [d] − [s]	x	x	x	x	x
1001 AAA0 11ee eeee 1001 AAA1 11ee eeee	An ← [An] − [s]					
0000 0100 SSEE EEEE	d ← [d] − s	x	x	x	x	x
0101 QQQ1 SSEE EEEE	d ← [d] − QQQ	x	x	x	x	x
1001 RRR1 SS00 0rrr 1001 RRR1 SS00 1rrr	d ← [d] − [s] − [X]	x	x	x	x	x
0100 1000 0100 0DDD	$[Dn]_{31-16} \leftrightarrow [Dn]_{15-0}$		x	x	0	0

Table 4.3 (continued)   68000 INSTRUCTION SET

Mnemonic (Name)	Size	Addressing mode	Addressing mode												
			Dn	An	(An)	(An)+	-(An)	d(An)	d(An,Xi)	Abs.W	Abs.L	d(PC)	d(PC,Xi)	Immed	SR or CCR
TAS (Test and set)	B	d =	x		x	x	x	x	x	x	x				
TRAP (Trap)															
TRAPV (Trap on overflow)															
TST (Test)	B, W, L	d =	x		x	x	x	x	x	x	x				
UNLK (Unlink)				x											

OP code		Operation performed	Condition flags				
b_{15} · · · b_0			X	N	Z	V	C
0100 1010 11EE EEEE		Test d and set N and Z flags; set bit 7 of d to 1		x	x	0	0
0100 1110 0100 VVVV		SP ← [SP] − 2 ; [SP] ← [PC] ; SP ← [SP] − 2 ; [SP] ← [SR] ; PC ← vector					
0100 1110 0111 0110		If V = 1, then SP ← [SP] − 2 ; [SP] ← [PC] ; SP ← [SP] − 2 ; [SP] ← [SR] ; PC ← TRAPV vector					
0100 1010 SSEE EEEE		Test d and set N and Z flags		x	x	0	0
0100 1110 0101 1AAA		SP ← [An] ; An ← [[SP]] ; SP ← [SP] + 4					

Table 4.3 (continued)   68000 INSTRUCTION SET

Branch instructions

Mnemonic (Name)	Displacement size	OP code	Operation performed
BRA (Branch always)	8	0110 0000 PPPP PPPP	PC ← [PC] + disp
	16	0110 0000 0000 0000 PPPP PPPP PPPP PPPP	
Bcc (Branch conditionally)	8	0110 CCCC PPPP PPPP	If cc is true, then PC ← [PC] + disp
	16	0110 CCCC 0000 0000 PPPP PPPP PPPP PPPP	
BSR (Branch to subroutine)	8	0110 0001 PPPP PPPP	SP ← [SP] − 2 ; [SP] ← [PC] ; PC ← [PC] + disp
	16	0110 0001 0000 0000 PPPP PPPP PPPP PPPP	
DBcc (Decrement and branch conditionally)	16	0101 CCCC 1100 1DDD PPPP PPPP PPPP PPPP	If cc is false, then Dn ← [Dn] − 1 ; If [Dn] ≠ −1, then PC ← [PC] + disp
DBRA (Decrement and branch)	The assembler interprets this instruction as DBF (see the DBcc entry)		

Note that while two entries are specified in the operand field, only the second defines the operand involved; the first entry indicates the count.

In contrast to this version of the LSL instruction, when a memory operand is to be shifted the count is restricted to 1 and the operand length can only be 16 bits. However, in this case a wider range of addressing modes may be used, namely, the absolute, register indirect, autoincrement, autodecrement, or indexed modes.

Various instructions and the addressing modes that can be used with them are summarized in Table 4.3. We will consider the details of this table in Section 4.6.

4.4.3 Branch Instructions

A wide range of branch instructions is available, and they use 8- or 16-bit signed offset values. An 8-bit offset value is given within the OP-code word, requiring a total of 16 bits for the instruction. With 16-bit offsets, a second word is used.

In addition to the usual branch instructions, there are a number of instructions that first decrement a counter and then perform a branching function. These instructions are convenient for implementation of program loops. They will be discussed in more detail in the following section.

The branch instructions specify the branch address in terms of a signed offset. Since at most 16 bits can be used to indicate an offset, the branch addresses must be within ±32K of the current value in the program counter. Transfers of control outside this range can be made with the unconditional Jump instruction, JMP, in which the destination address can be specified using the general addressing modes.

4.5 THE 68000 INSTRUCTION SET

The instruction set of the 68000 microprocessor is shown in Table 4.3. The instructions are listed in alphabetical order, except for the branch instructions which are given at the end of the table. For each instruction the table shows the size of the operands allowed, the addressing modes provided, the OP-code pattern, the operation performed, and the condition flags that are affected. In order to present the table in a reasonably compact form, a considerable number of notational symbols are used, which are defined in Table 4.4.

In Table 4.3, a two-operand instruction occupies one or more rows, depending on the number of allowable addressing-mode combinations for the source and

Table 4.4 ABBREVIATIONS AND SYMBOLS USED IN TABLE 4.3

Symbol	Meaning
s	Source operand
d	Destination operand
An	Address register n
Dn	Data register n
Xn	An address or data register, used as index register
PC	Program counter
SP	Stack pointer
SR	Status register
CCR	Condition code flags in SR
AAA	Address register number
DDD	Data register number
rrr	Source register number
RRR	Destination register number
eeeee	Effective address of the source operand
EEEEE	Effective address of the destination operand
MMM	Effective address mode of destination
CCCC	Specification for a condition code test
P...P	Displacement
Q...Q	Quick immediate data
SS	Size: 00≡byte, 01≡word, 10≡long word
VVVV	Trap vector number
u	Condition code flag state is undefined (meaningless)
d(An)	Indexed basic addressing mode
d(An,Xi)	Indexed full addressing mode
d(PC)	Relative basic addressing mode
d(PC,Xi)	Relative full addressing mode

destination operands. Each row corresponds to a given addressing mode, either for the source operand or for the destination operand. For instance, in a row associated with a given source addressing mode, an x is shown in the columns of the addressing modes that can be used to specify the destination operand for this source mode. As an example, in the ADD instruction, if the source is a data register, Dn, then the destination can be specified in any of the modes: (An), (An)+, −(An), d(An), d(An,Xi), Abs.W, or Abs.L. But, if the source is given in any other addressing mode, the destination can only be a data register. For some instructions, one of the operand locations can be the status register (SR) or the condition-code register (CCR). Recall that CCR refers to the least-significant byte of the status register. Hence, only 8 bits are affected in the operations specifying the CCR contents as an operand.

The operations performed by the instructions and their impact on the condition codes are also given in Table 4.3. In most cases, the information shown suffices for proper understanding of the relevant details. However, there are some instructions that warrant further comments; they will be given at the end of this section. These instructions are denoted by an asterisk in the mnemonic column of the table. A good appreciation of various instructions can be attained from the example programs given in the subsequent sections.

4.5.1 OP-Code Patterns

To make the table useful for work with the assembled object code, the OP-code patterns are given for all instructions. Most OP-codes consist of one 16-bit word. Included in the OP-codes are bits that specify the nature of the operation to be performed, the registers involved, data for short immediate operands, the length of the operands, offsets for branch instructions, and so on. In order to utilize the available code space efficiently, the bits that specify this information are not necessarily partitioned into uniform fields that always occupy the same position within an OP-code.

For most instructions only one operand can be specified in a general format, where several addressing modes can be used. This may be either a source or a destination operand. The format of this address field, which is 6 bits long, is shown in Table 4.5. A 3-bit field denotes the addressing mode, and the remaining 3 bits indicate which of the eight registers is involved. For those addressing modes where no register is involved (i.e., absolute and immediate) or where a specific register is implied (e.g., the program counter or status register) all 6 bits of the address field are used as the mode indicators. The 6 bits of the general address field are indicated in Table 4.3 as eeeee for the source operand and as EEEEEE for the destination operand.

The OP-code specifies the size of the operands to be used in the instruction. Typically, 2 bits are assigned to denote operand length. These bits are shown as SS, with the patterns 00, 01, and 10 indicating byte, word, and long word, respectively.

As an example, consider a version of the AND instruction that has the OP-code 1100 DDD1 SSEE EEEE. This instruction performs a logical AND operation on the contents of the source and destination operands and leaves the result in the destination location. The source operand is the contents of a data register identified by the

Table 4.5 ADDRESS FIELD ENCODING FOR 68000 MICROPROCESSOR

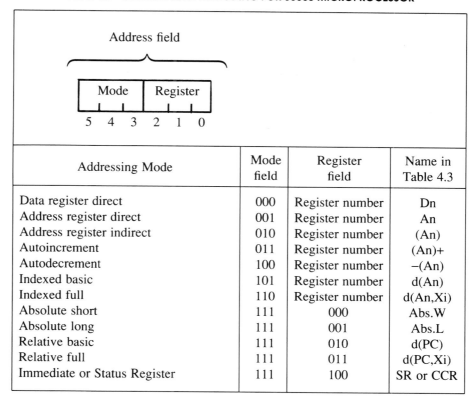

Addressing Mode	Mode field	Register field	Name in Table 4.3
Data register direct	000	Register number	Dn
Address register direct	001	Register number	An
Address register indirect	010	Register number	(An)
Autoincrement	011	Register number	(An)+
Autodecrement	100	Register number	−(An)
Indexed basic	101	Register number	d(An)
Indexed full	110	Register number	d(An,Xi)
Absolute short	111	000	Abs.W
Absolute long	111	001	Abs.L
Relative basic	111	010	d(PC)
Relative full	111	011	d(PC,Xi)
Immediate or Status Register	111	100	SR or CCR

3 bits labeled DDD. The destination location is specified by the 6 bits of the address field denoted as EEEEEE. The size of the operand is determined by the pattern in the SS bits. The remaining 5 bits indicate the AND operation and that the source operand is in a data register. The next row in the table gives the second basic version of the AND instruction, where the destination location can only be a data register and the source may be specified in any of several different modes. The OP-codes for the two versions of the AND instruction differ only in bit position 8, if the same addressing mode is used in each case. The reader may find it interesting to check the OP-codes of some similar instructions, such as ADD, OR, and SUB.

When a short immediate operand is used in the quick immediate mode, the operand is given explicitly within the 16-bit OP-code word. In the ADDQ and SUBQ instructions this immediate datum consists of three bits, denoted as QQQ in the OP-codes of Table 4.3. In the MOVEQ instruction, 8 bits are included as the operand. Immediate datum is also used in shift and rotate instructions. Here, a 3-bit number is included in the OP-code word to indicate the number of bit positions (the count) by which an operand is to be shifted or rotated. The count can also be given as the contents of a data register, as described in Section 4.4.2.

4.5.2 Implementation of Branch Instructions

The 68000 instruction set includes many branch instructions, shown at the end of Table 4.3. As explained in Section 4.4.3, the branch instructions can use either 8- or 16-bit offsets. The offset is a signed 2's-complement number that specifies the distance in bytes between the current value of the program counter and the address of the instruction to which a branch is made.

A number of conditional branch instructions are available. They all have the format indicated under Bcc in Table 4.3. The condition-code suffix, cc, can be any of the 16 possibilities given in Table 4.6. For example, the BNE (Branch if Not Equal) instruction causes a branch if the value of the Z flag in the CCR register is equal to zero. Recall that the status of the condition tested, the value of Z in this case, is assumed to have been set by a previous instruction, perhaps a Compare or a Subtract instruction. Table 4.6 shows the test conditions for all conditional branch instructions.

Table 4.6   CONDITION CODES FOR Bcc, DBcc AND Scc INSTRUCTIONS

Machine Code CCCC	Condition Suffix cc	Name	Test Condition
*0000	T	True	Always true
*0001	F	False	Always false
0010	HI	High	$C \vee Z = 0$
0011	LS	Lower or same	$C \vee Z = 1$
0100	CC	Carry clear	$C = 0$
0101	CS	Carry set	$C = 1$
0110	NE	Not equal	$Z = 0$
0111	EQ	Equal	$Z = 1$
1000	VC	Overflow clear	$V = 0$
1001	VS	Overflow set	$V = 1$
1010	PL	Plus	$N = 0$
1011	MI	Minus	$N = 1$
1100	GE	Greater or equal	$N \oplus V = 0$
1101	LT	Less than	$N \oplus V = 1$
1110	GT	Greater than	$Z \vee (N \oplus V) = 0$
1111	LE	Less or equal	$Z \vee (N \oplus V) = 1$

*T and F suffixes cannot be used in the Bcc instruction.

A useful extension of conditional branching is provided by the DBcc instructions, which have the assembler form

$$DBcc \qquad Dn, LABEL$$

The condition specified by the cc suffix is tested. If it is satisfied, then the instruction that immediately follows the DBcc instruction is executed next. If this condition is not met, the least-significant 16 bits of register Dn are decremented by one. If the result is equal to −1, the instruction that follows the DBcc instruction is executed, as in the above case where the branch condition was satisfied. If the result is not equal to −1, then a branch is made to the instruction at location LABEL. The branch address is specified by means of a 16-bit signed offset that is added to the current value in the program counter.

To illustrate the functioning of DBcc instructions, consider the example in Figure 4.5. The program given scans a list of 1000 eight-bit signed numbers, until it finds the first positive number with a non-zero value. It determines the location of this number in terms of the distance in bytes from the beginning of the list. Each number is tested by the TST.B instruction, which sets the condition-code flags N and/or Z to 1 if the tested number is negative and/or zero, respectively. The instruction also clears the C and V flags. The state of the flags is used by the DBGT instruction in determining whether a branch should be made. From Table 4.6 it can be seen that the branch condition is satisfied if N or Z is set to 1 (remember that V=0). Thus, the loop will be executed until the desired number is found or the end of the list is reached. At this point, the contents of register D0 will have been decremented by the number of passes through the loop, corresponding to the distance sought.

The reader will have noted that the logic of the DBcc instructions is opposite to that of the Bcc instructions. In the case of DBcc, control is transferred to the branch address only if the branch condition is not met and if the decremented contents of the specified register, Dn, are not equal to −1. On the other hand, a Bcc instruction causes a branch when the branch condition is satisfied. The power of the DBcc instructions comes from the inclusion of the counter mechanism within them. They are particularly useful in termination of loops, as will be seen in the examples given in Section 4.6.

```
          MOVEA.L   #LIST,A0      Initialize the list pointer.
          MOVE      #999,D0       List has 1000 entries.
          MOVE      D0,D1
LOOP      TST.B     (A0)+         Test the next entry.
          DBGT      D0,LOOP       Branch back if negative or zero.
          SUB       D0,D1         Find the distance from the
          MOVE      D1,HOWFAR       beginning of the list.
          next instruction
```

Figure 4.5   A program that locates the first positive nonzero 8-bit number in a list.

A few of the instructions in Table 4.3, those labeled with an asterisk, are discussed more in the subsections below. The information given in Tables 4.1 to 4.6 should suffice for writing and debugging programs in the 68000 assembly language. The structure and size of assembled instructions can be deduced easily from the given OP codes and addressing modes. We should note that no timing information is shown in Table 4.3, because its inclusion would have increased the size of the table greatly. The reader who needs to know the number of machine cycles involved in the execution of a given instruction should consult the manufacturer's literature [3].

4.5.3 Compare Instructions

The purpose of compare instructions is to set the condition code flags to reflect the relative values of the two operands being compared. The flags are subsequently tested by a conditional branch instruction. Table 4.3 indicates that the 68000 has four compare instructions, specified in the format

$$\text{CMP} \qquad \text{src,dst}$$

The comparison is performed by subtracting the value of the source operand from the destination operand, without changing the value of either operand. Note that this is different from our interpretation of the Compare instruction in Chapter 2, where we assumed that the comparison involves subtracting the destination operand from the source operand. Thus, if a 68000 compare instruction is followed by the BGT instruction, branch will take place if the destination operand is larger than the source operand.

4.5.4 BCHG, BCLR, BSET, and BTST Instructions

These instructions operate on a single bit. The location of this bit is identified by means of a number (bit #), which is either given as an immediate value within the instruction, or in one of the data registers. The complement of the designated bit is loaded into the condition flag, Z. Then the specified operation is performed on the bit. In the BCHG instruction the bit is changed to the complement of its previous value. The BCLR and BSET instructions force the specified bit to 0 and 1, respectively. The BTST instruction performs the test only, which results in setting the Z flag, without affecting the state of the bit tested.

As an example, consider the instruction

$$\text{BCLR} \qquad \text{D2,D3}$$

assuming that register D2 contains the value 6. This instruction tests bit b_6 of register D3 by loading its complement into the Z flag. Then it clears b_6 to 0. The same effect can be achieved with the instruction

$$\text{BCLR} \qquad \text{\#6,D3}$$

It is important to note that in these instructions the state of the Z flag is determined before the actual operation specified by the instruction takes place, because the flag is used to indicate the original value of the bit tested. In all other instructions, the state of the condition code flags is determined after the specified operation has been performed.

4.5.5 MOVEM Instruction

This instruction transfers the contents of one or more registers to or from consecutive memory locations. A second word in the instruction, called a *register list mask*, is used to indicate which registers are to be involved in the transfer. Bits b_0 through b_7 of this word correspond to data registers D0 through D7 and bits b_8 through b_{15} to address registers A0 through A7, respectively. The registers moved are those whose corresponding bits in the register list mask are equal to 1.

Figure 4.6 gives an example of the use of the MOVEM instruction. It shows the effect of executing two such instructions. The first is

$$\text{MOVEM.L} \qquad \text{(A7)+,D0-D2/A1/A3}$$

Five registers are involved in this instruction. They are specified in the instruction statement in symbolic form, separated by a slash character "/". A range of contiguous registers may be denoted by a dash, as in D0–D2. The instruction above loads five long words from the memory into registers D0, D1, D2, A1, and A3. When the instruction is assembled, the five registers are indicated by setting bits b_0, b_1, b_2, b_9, and b_{11} in the register list mask to 1. Thus, the second word of the instruction has the hex pattern 0A07. The memory locations that contain the source operands are accessed in the autoincrement mode, according to the source operand specification. The first long word, at address 20000_{16}, is loaded into D0, and the fifth long word into A3.

The second instruction in Figure 4.6 is

$$\text{MOVEM} \qquad \text{(A7)+,D3/D5}$$

Since the length of the operands is not stated explicitly, a word operand is assumed by default. The two 16-bit words, pointed at by register A7, are transferred into registers D3 and D5. The register list mask for this case is 0028.

When word-sized operands are used in a MOVEM instruction that loads registers from memory, the most-significant bit of an operand is sign-extended into the high-order bits of a 32-bit register. In our example, when the word 0F52 in location 200144 is loaded into D3, the contents of D3 become 00000F52. But, the word 8A00 in the next location has its most-significant bit equal to 1. Therefore, the pattern FFFF8A00 is loaded into D5.

One final comment about the MOVEM instruction should be made. The order in which the registers are accessed is as explained above, namely, D0 through D7 followed by A0 through A7. However, there is one exception when the autodecrement addressing mode is used. Since this mode is the converse of the autoincrement mode, the order in which the registers are accessed is reversed. This is accomplished by reversing the association between registers and the bits in the register mask, so that bit b_0 corresponds to A7 and b_{15} to D0. Note, from Table 4.3, that the autoincrement mode can be used in the MOVEM instruction only for the source operands.

MEMORY		REGISTERS		REGISTERS	
Address	Contents	before		after	
⋮	⋮	001000	PC	001008	PC
001000	4CDF		A1	0008FC30	A1
001002	0A07				
001004	4C9F		A3	0003AB4D	A3
001006	0028				
⋮	⋮	00020000	A7	00020018	A7
020000	10BA				
020002	0024		D0	10BA0024	D0
020004	0000				
020006	01F2		D1	000001F2	D1
020008	3F00				
02000A	2240		D2	3F002240	D2
02000C	0008				
02000E	FC30		D3	00000F52	D3
020010	0003				
020012	AB4D				
020014	0F52		D5	FFFF8A00	D5
020016	8A00				

Figure 4.6   Effect of executing the instructions: MOVEM.L   (A7)+,D0–D2/A1/A3
MOVEM   (A7)+,D3/D5.

The destination must be the microprocessor registers. In contrast, the autodecrement mode can be used to specify the destination only. The source is the set of registers. Thus, if data are transferred from the memory into the registers with a MOVEM instruction that uses the autoincrement mode, the data can be copied back into the memory with a MOVEM instruction that uses the autodecrement mode.

4.5.6 MOVEP Instruction

The MOVEP instruction can be used only for transfers between a data register and memory locations specified in the basic indexed addressing mode. Either 32- or 16-bit operands can be transferred, 1 byte at a time. The high-order byte of the register is transferred first and the low-order byte last, and the memory address is incremented by 2 after each byte transfer. For example, assume that registers D3 and A5 contain 3B56C174 and 1F0000, respectively. The instruction

$$MOVEP.L \qquad D3,\$200(A5)$$

transfers 3B, 56, C1, and 74 to byte locations 1F0200, 1F0202, 1F0204, and 1F0206. Note that all byte addresses will be even or odd depending upon the address of the first location, which is specified in the instruction.

The MOVEP instruction is intended to facilitate data transfer to and from 8-bit peripheral devices. Consider such a device connected to the eight high-order lines of the data bus, D_{15-8}. According to Figure 4.4, all internal registers of this device will appear at even addresses. If some of the parameters needed by the device are 16 or 32 bits long, each parameter may be stored in two or four register locations, whose addresses differ by an increment of 2. Thus, a single MOVEP instruction can be used to transfer one 16- or 32-bit item from a microprocessor register to the appropriate register locations in the peripheral device. Without the MOVEP instruction, several instructions would be needed to perform the same task by shifting data in a microprocessor register appropriately and transferring one byte at a time to the peripheral device, using MOVE.B instructions. A typical example is a 16-bit timer counter found in a general purpose parallel interface chip, that will be discussed in Chapter 7. The MOVEP instruction provides convenient means for both loading and reading the contents of such a counter.

4.5.7 Multiply and Divide Instructions

Multiplication and division operations can be performed using either signed or unsigned operands, and the 68000 has separate instructions to deal with each type of operand. When signed operands are involved, the operations are performed using 2's complement arithmetic. Unsigned operands are treated as positive numbers where all bit positions are used to represent the magnitude of the number. Let us consider multiplication first. One multiply instruction is available for each type of numbers, as follows:

- MULS: multiplies two signed 16-bit operands and leaves a 32-bit signed product in the destination location, which must be a data register.

- MULU: multiplies two unsigned 16-bit operands and leaves a 32-bit unsigned product in the destination location, which must be a data register.

These instructions set condition code flags Z and N to 1 if the result is 0 and the most-significant bit is equal to 1, respectively. They clear flags C and V, while the state of the extend flag, X, is not affected. Figure 4.7 illustrates the use of these instructions. In part (a) of the figure, two examples of signed multiplication of an immediate operand, 5000_{16}, and the contents of register D0 are given. The figure shows the contents of D0 before and after the multiplication. Note that the high-order 16 bits of the original contents of D0 are not used as a part of the operand value, but are overwritten by the 32-bit product left in D0. Part (b) of the figure shows similar examples of unsigned multiplication. All condition code flags are cleared to 0 in these examples, except for case *ii* where the N flag would be set to 1. The reader should verify the values of the products given in the figure.

		Before	**After**
(i)	D0 :	A27B0006	0001E000
(ii)	D0 :	A27BFFFC	FFFEC000

(a) MULS   #$5000,D0

		Before	**After**
(iii)	D0 :	A27B0006	0001E000
(iv)	D0 :	A27BFFFC	4FFEC000

(b) MULU   #$5000,D0

Figure 4.7   Examples of multiplication.

There are two instructions for division, again one for signed and one for unsigned operands

- DIVS: divides a 32-bit dividend by a 16-bit divisor. The dividend is the destination operand, which must be in a data register. The divisor is the source operand. Division is performed using 2's complement arithmetic, and a 32-bit result is left in the destination data register. The result consists of a signed quotient, which occupies the 16 least-significant bit positions in the register. A signed remainder is left in the most-significant 16 bits of the register. The sign of the remainder is the same as the sign of the dividend.
- DIVU: performs the same function as DIVS, except that all operand values are treated as unsigned numbers.

		Before	**After**
(i)	D0 :	0000002C	0002000E
(ii)	D0 :	FFFFFFDE	FFFEFFF5

(a) DIVS   #3,D0

		Before	**After**
(iii)	D0 :	0A38003F	003F028E
	D0 :	FFF03FDE	3FDE3FFC

(b) DIVU   #$40000,D0

Figure 4.8   Examples of division.

The DIVS and DIVU instructions set the Z, N, and V flags if the quotient is zero, negative, or cannot fit in the 16-bit destination field, respectively. If overflow occurs, that is, if V is set, the state of Z and N flags is undefined. The C flag is always cleared and the X flag is not affected. If the divisor is equal to 0, the division is not performed, and an "exception" is raised, as will be discussed in Section 7.4.2. Figure 4.8 shows some examples of division, where the divisor is specified as an immediate operand in the divide instruction.

4.5.8 Privileged Instructions

The 68000 microprocessor has two modes of operation, one called the Supervisor mode and the other the User mode. The mode in which the processor is operating at any given time is defined by the S bit of the status register (see Fig. 4.2), which is equal to 1 in the Supervisor mode and equal to 0 otherwise. The main difference between the two modes is that certain instructions, called *privileged instructions,* can only be executed in the Supervisor mode.

The existence of User and Supervisor modes allows for a certain amount of protection in a microcomputer system. In the User mode, a program cannot overwrite the contents of the status register or the stack pointers. It cannot perform drastic actions such as resetting or stopping the system. Such functions can only be performed in the Supervisor mode, which is normally reserved for system software (see Chapter 12). Thus, there exists a simple mechanism where an application program, running in the User mode, is prevented from inadvertently disrupting the system.

The privileged instructions in the 68000 are

- ANDI, EORI, ORI, and MOVE instructions when the destination is the status register SR,
- a special MOVE USP instruction that transfers the contents of the user stack pointer to or from an address register,
- RESET, RTE, and STOP instructions.

4.6 EXAMPLES

Let us now return to some examples in Chapter 2 to see how they might be implemented using the 68000 assembly language.

LIST TRANSFER

Figure 4.9 shows a program that corresponds to the program in Figure 2.5, which copies data from list A to list B. The lists are assumed to contain n long-word entries. Thus, the total number of bytes transferred is $4n$. Address registers A1 and A2 are initialized to point at the beginning of the two lists, and data register D1 is used as an index register to scan along the lists during the copying process. Register D2 holds the value $4n$, used in terminating execution of the loop. Note that the multiplication by 4 is achieved with the LSL instruction, by shifting the value n

```
        MOVEA.L      #LISTA,A1           Initialize register A1 and A2 to point to the
        MOVEA.L      #LISTB,A2               start of lists A and B.
        MOVE.L       N,D2                Register D2 indicates the number of bytes
        LSL.L        #2,D2                   in the lists.
        CLR.L        D1                  Clear index register D1.
LOOP    MOVE.L       (A1,D1),(A2,D1)     Transfer an entry in list A to list B.
        ADD          #4,D1               Increment index register.
        CMP.L        D2,D1               Branch back until the last entry is copied.
        BLT          LOOP
        next instruction
```

Figure 4.9   A 68000 program corresponding to Figure 2.5.

two bit positions to the left. Since one long word is transferred at a time, the index register, D1, is incremented by 4 to access the next entry. The first two instructions, which initialize the address pointers, use the MOVEA (Move Address) instruction, which is a special version of the general MOVE instruction. It transfers the contents of the source operand to one of the address registers. An important difference between these two instructions is that MOVE sets the N and Z flags according to the value of the operand involved and clears the V and C flags, while MOVEA has no effect on the flags. The MOVE instruction is used to deal with signed data operands, while MOVEA is used for addresses which are always unsigned numbers.

The next example is given in Figure 4.10. It corresponds to Figure 2.10, where the autoincrement mode is used in the MOVE instruction. Again, address registers A1 and A2 are initialized to point to the start of the lists. In this case, we have assumed the lists to consist of fifty 16-bit entries. Hence, the autoincrement mode will cause the address registers to be incremented by 2 each time through the loop. The example illustrates the use of DBRA (Decrement and Branch) instruction to terminate a loop. This is actually the DBF instruction, which is used so frequently that it is given a more descriptive name. Note that the test for the branch condition (cc=F) in DBF is always false (see Table 4.6), so that the only test made to decide whether or not a branch is to be made is whether the contents of the specified counter are equal to -1. In our example register D1 is the counter specified. Since its initial value is 49, the loop will be traversed 50 times before control passes to the instruction that follows DBRA.

```
        MOVE.L       #49,D1              Use register D1 as a counter.
        MOVEA.L      #LISTA,A1           Initialize registers A1 and
        MOVEA.L      #LISTB,A2               A2 as pointers to lists.
LOOP    MOVE         (A1)+,(A2)+         Transfer 16 bits from list A to list B.
        DBRA         D1,LOOP             Branch back until the last entry
                                            is transferred.
        next instruction
```

Figure 4.10   A 68000 program for copying a list of 16-bit entries.

```
        MOVEA.L    LIST,A1      Starting address and the address
        MOVE.L     SIZE,A2          that follows the last entry in the
        ADDA.L     A1,A2            list are passed as parameters.
        BSR        MAX          Call subroutine MAX.

              ⋮
              ⋮

MAX     MOVEA.L    A1,A3        Register A3 points to the largest number.
        MOVE.B     (A1)+,D1     Register D1 holds the largest number.
LOOP    CMP.B      (A1)+,D1     Check if the next number is larger.
        BGE.S      NEXT         If no, go to next entry.
        LEA        −1(A1),A3    Update pointer to the largest number.
        MOVE.B     (A3),D1      New largest number is in D1.
NEXT    CMPA.L     A2,A1        Check if the end of the list is reached;
        BLT        LOOP         Branch back if not.
        RTS
```

Figure 4.11   A 68000 program for finding the largest number in a list.

SORTING

A program for finding the largest number in a list of 8-bit numbers is presented in Figure 4.11, following the strategy outlined in Figure 2.16. The program is written as a subroutine to which the addresses that denote the beginning and the end of a list are passed as parameters. Address register A1 is used to scan through the list. Register A2 contains the address that follows the last entry in the list, which is used in a test to terminate the loop. The largest number found at any given time is held in D1, and its address in A3. We should note that at the time register A3 is updated by the LEA (Load Effective Address) instruction, register A1 is already pointing at the next entry in the list, because the autoincrement mode is used when comparing the list entries. Therefore, the LEA instruction subtracts 1 from the number in A1, before loading it into A3. Of course, the contents of A1 are not changed by this instruction.

Observe that the first branch instruction in the subroutine is written as

BGE.S   NEXT

The suffix S is appended to the mnemonic BGE to indicate that a short (8-bit) displacement can be used in computing the branch address, because location NEXT is not far away. When the assembler encounters a forward branch, during the process of assembling a program, it does not know the address of the location specified by a name until this name is encountered as a label. Hence, it cannot determine if a short displacement can be used. Yet, it must allocate either one word for a short branch or two words for a long branch, to store the branch instruction in the main memory. (See Table 4.3 for the format of the branch instructions.) The suffix S tells the assembler to use the short version. No such indication is needed in the case of backward branches, because the address corresponding to the name of the branch location is already known to the assembler.

	MOVEA.L	LIST, A4	Starting address and the address
	MOVE.L	SIZE, A2	that follows the last entry in the
	ADD.L	A4, A2	list are passed as parameters.
	BSR	SORT	Call subroutine SORT.
	next instruction		

$$\vdots$$

SORT	MOVE.L	A4, A1	Pass the starting address of sublist to MAX.
	BSR	MAX	Call subroutine MAX (see Figure 4.11).
	MOVE.B	(A4),(A3)	Exchange the first entry of the sublist with the
	MOVE.B	D1,(A4)+	largest value and increment the sublist pointer.
	CMPA.L	A4, A2	Check if the end of the list is
	BLT	SORT	reached; branch back if not.
	RTS		

Figure 4.12 A 68000 program that sorts a list of numbers in decreasing order (corresponds to Figure 2.17).

Subroutine MAX is used in Figure 4.12, which is a routine for sorting a list of 8-bit entries corresponding to the program in Figure 2.17. Again, the parameters are passed to the subroutines in the same way as before. Register A4 points to the start of a smaller sublist, during each pass through the sorting subroutine, SORT. Note that the first entry in the sublist and the maximum value found are exchanged without using a temporary storage location, because a copy of the maximum value is already available in a processor register.

INPUT/OUTPUT

The 68000 microprocessor can interact with a variety of I/O devices. While the detailed discussion of its I/O capabilities will be given in Chapters 6 and 7, in this section we will consider the simple example of program-controlled I/O introduced in Section 2.4. The example involves communication with a video terminal, which has interface registers as depicted in Figure 2.11.

A program that reads one character from the keyboard is shown in Figure 4.13. It is a straightforward implementation of the scheme suggested in Figure 2.12, using data register D0 in place of the accumulator. Note that the same scheme was used for the 6809 microprocessor, in Figure 3.14. The only significant difference is that the 68000 has a general Move instruction that can be used instead of the more specific Load and Store instructions of the 6809. Observe also that the instruction that performs the logical AND operation using the immediate operand 1 and the least-significant byte of register D0 is written as

$$\text{ANDI.B} \quad \text{\#1,D0}$$

The immediate nature of the source operand is indicated twice, by including the I in the operation mnemonic and by the # sign in the operand field. The same effect could be achieved with the instruction

AND.B   #1,D0

While the inclusion of I appears to be redundant, there exists a subtle difference between the AND and ANDI instructions. From Table 4.3 it can be seen that if an immediate operand is used in the AND instruction, then the destination operand must be in a data register. However, in the ANDI instruction, where only an immediate source operand is allowed, the destination operand can be specified in several different addressing modes. There are five other 68000 instructions that exhibit a similar difference between the general and the immediate versions. These are the arithmetic instructions ADDI, SUBI, and CMPI as well as the logic instructions ORI and EORI. In each case, there exists a general version that does not include the suffix I, but has a limited range of addressing modes for the destination operand.

The program of Figure 4.13 is not very efficient. The 68000 microprocessor has a number of powerful instructions that we have not discussed in Chapters 2 or 3, because they would have obscured the presentation of general concepts or because they are simply not available in the 6809. The purpose of the first two instructions in Figure 4.13 is to determine whether or not bit 0 of location STATUS is equal to 1. The 68000 includes a Bit Test instruction, BTST, which sets the value of the Z flag to the complement of the value of the bit specified in the instruction. Hence, this instruction is sufficient to perform the desired test. The state of the Z flag can then be used as the condition that decides if a branch to WAIT should be made. The improved program is shown in Figure 4.14. The BTST instruction is representative of three other instructions that, in addition to testing the specified bit, either complement this bit or set it to a constant value. These are the BCHG, BCLR, and BSET instructions, which were introduced in Section 4.5.3.

A one-line message, terminated by the carriage-return character, can be displayed on a video terminal with a program such as the one given in Figure 4.15. This is an implementation of the scheme presented in Figure 2.13. Address register A0 is used as the pointer to scan through the characters of the message stored in the main memory. The loop that waits for the display to become ready is based on the BTST instruction, as discussed above. As each character is transferred to the display

```
WAIT    MOVE.B      STATUS,D0     Read status.
        ANDI.B      #1,D0         Check bit 0.
        BEQ         WAIT          Loop if no character ready.
        MOVE.B      INBUF,D0      Transfer data.
        next instruction
```

Figure 4.13 A program that reads one character from a keyboard.

```
WAIT    BTST.B      #0,STATUS     Check status bit 0.
        BEQ         WAIT          Loop if no character ready.
        MOVE.B      INBUF,D0      Transfer data.
        next instruction
```

Figure 4.14 An improved program that reads one character from a keyboard.

CR	EQU	$0D	ASCII code for carriage return.
MESG	DC.B	'Hello, this is a 1-line message'	
	DC.B	CR	
	MOVEA.L	#MESG,A0	Initialize pointer.
WAIT	BTST.B	#1,STATUS	Wait for the display
	BEQ	WAIT	to become ready.
	MOVE.B	(A0)+,OUTBUF	Transfer a character to display.
	CMPI.B	#CR,−1(A0)	Continue displaying if not CR.
	BNE	WAIT	
	next instruction		

Figure 4.15 A program that displays a one-line message.

buffer, A0 is autoincremented to point at the next character. Thus, in order to check if the transmitted character was CR, the Compare instruction must access this character again, which is done in the indexed mode with an offset of −1. Note that in this case the CMPI instruction cannot be replaced with the CMP instruction, because the destination location is not a data register.

4.7 REVIEW EXERCISES

The 68000 is a powerful microprocessor. So far we have seen that it has a sizeable register set, a variety of addressing modes, and a substantial repertoire of instructions. While its overall complexity is considerable, the reader can learn to use the 68000 reasonably quickly, starting with simple programming tasks and then delving into more intricate possibilities. An understanding of the basic types of instructions, coupled with an access to a machine for hands-on experience, will enable the reader to do many worthwhile exercises. The more complex and specialized instructions usually provide a more convenient and more efficient means for implementing tasks that can also be realized with simple instructions.

In previous examples, we have used a number of simple instructions. Table 4.3 gives much of the information about the entire instruction set. However, due to space restrictions it is not complete. Detailed explanations for all instructions, a full specification of all bit-patterns, and the number of clock cycles needed to execute the instructions are not included. The table contains sufficient details to gain an appreciation of the scope of the instruction set. The reader who will do extensive work with a 68000 microprocessor should acquire a programmer's manual [7] or a book that gives all of the details [8].

The following exercises are suggested as a quick test of the grasp of the preceding material in this chapter.

1. Let the contents of registers A0 and D1 be 103A0 and 4C, respectively, when each of the following instructions is executed. What will be the effective addresses of the source and destination operands?

   (a) MOVE.B   (A0)+,D1
   (b) MOVE $4A0,10(A0,D1.W)

(c)	MOVEM.L	D1–D4,–(A0)
(d)	MOVEQ.L	#5,D1

2. Which of the following 68000 assembly-language instructions have syntax errors? Correct the errors.

	ADD.B	(A2)+,(A3)+
POINT23	MOVE	34(A0,D1),LOC
	SUB.L	#38,(D2)
	ABCD	–(A3),D0
	ANDI.L	#$FFA0,3(A0,D0.L)
	BEQ	INCREMENT
	MOVEQ	#300,D1

3. Consider the 68000 assembler statements.

	ORG	$1000
TEXT	DC.B	'Title'
	DC.B	23,$4A,$5
	DC.L	52,TEXT+20

What memory locations will be affected by these statements? What will be the contents of these locations? What will be the long word starting at location $1004?

4. What is the main difference between the CMP and CMPI instructions?

5. Let the contents of registers A0, D0, D1, D2, and D3 be 1A00, 2, 6, 1A00C0FF, and 20300A01, respectively. Also, let the long word in memory location 1A00 be FCA800C7. What will the contents of the destination operand and the state of the condition code flags be after executing the instruction

(a)	ADD.L	D3,(A0)
(b)	ADD.B	D2,D3
(c)	SUB.L	D3,D2
(d)	ASR	2(A0)
(e)	ASR.B	3(A0)
(f)	LSR.B	3(A0)
(g)	BCHG	D1,$1A00
(h)	BTST	#2,2(A0)
(i)	MOVE.L	$1A00,D0
(j)	MOVEM	(A0)+,D0/D1

6. Assume that the contents of registers D1 and D2 are 5 and A3AB, respectively, when each of the following instructions is executed. What will be their contents as a result of executing the instruction

(a)	LSL.W	D1,D2
(b)	LSL.L	D1,D2
(c)	ROR.B	D1,D2

    (d)   ROXL.W   D1,D2

    (e)   ROR.L   #7,D1

7. Show two different ways of clearing the high-order 16 bits of register D2, without disturbing its low-order bits.

8. Consider the following instructions:

                BTST.L   #18,D0

                BCLR.L   #18,D0

                BSET.L   #18,D0

                BCHG.L   #18,D0

How could the tasks performed by these instructions be carried out if this class of instructions was not available?

9. Rewrite the program of Figure 4.11, so that the subroutine MAX controls the loop by decrementing a counter, instead of comparing against the address of the end of the list.

10. Assume that there is an ordered list of 200 long-word entries, starting at address $1000. Write a program that inserts a new entry, whose value is found in location NEW, into the 85th position of the list, thus making the list 201 entries long.

11. Assume that there is a list of 500 two-byte entries, beginning at address $1000.

(a) Write a program that removes the 210th entry, thus making the list 499 entries long.

(b) Write another program that removes the entry whose value is 2357.

12. Assume that there is a list of 200 long-word entries, starting at address $1000. Write a program that finds the sum of all 8-bit numbers found in the least-significant byte position of each entry in the list.

13. Assemble by hand the subroutine MAX in Figure 4.11. How many bytes of memory are needed to store it?

14. Assemble by hand the program in Figure 4.15. How many bytes of memory are needed to store it?

4.8 SUBROUTINE LINKAGE

The 68000 microprocessor deals with subroutines in a manner consistent with the discussion in Sections 2.7 and 3.8. There are two instructions that can be used to call a subroutine. The BSR (Branch to Subroutine) instruction uses the relative mode to determine the branch address, while JSR (Jump to Subroutine) includes the address of the subroutine as an absolute 32-bit number. The BSR instruction can have a 16-bit (short) or a 32-bit (long) offset, depending upon the distance between its location in a program and that of the called subroutine. Both BSR and JSR instructions save the return address on the system stack, using address register A7 as a stack pointer. The return from a subroutine is performed with the RTS instruction.

There exist two other instructions, LINK and UNLK, which will be described later in this section.

4.8.1 Parameter-Passing

Parameters can be passed between a calling program and a subroutine as discussed in Sections 2.5, 2.7, and 3.8. They can be placed in registers, in some predetermined memory locations or on the stack.

Parameters in Processor Registers

The simplest and fastest way of passing parameters is via processor registers. We have used this method in the examples of Figures 4.11 and 4.12. The main drawback of the scheme is that there may not be a sufficient number of registers available. This is less of a problem in the 68000 than in the 6809 microprocessor, because the former has a much larger register set. Of the 16 data and address registers available, only the stack pointer, A7, may not be used for this purpose.

Parameters Listed After a Call Instruction

Another possibility is to place the parameters immediately after a Call instruction, as illustrated in Figure 2.20. A 68000 implementation of this example is given in Figure 4.16. Subroutine SUBMOVE copies a list of entries from one part of the memory to another. Fifty 32-bit entries are to be copied. The calling program gives the length and the addresses of the lists after the BSR instruction, using the DC (Define Constant) assembler directive.

The subroutine first saves the contents of the registers that are used within it. These are the registers D1, A0, A1, and A2. Thus, the return address placed on the

```
             BSR        SUBMOVE
             DC.W       49              Parameters are the length of the
             DC.L       LISTA,LISTB        lists and their addresses.
             next instruction

                 ⋮
                 ⋮

SUBMOVE      MOVEM.L    D1/A0−A2,−(A7)  Save registers.
             MOVEA.L    16(A7),A0       Reg A0 points to parameters.
             MOVE.W     (A0)+,D1        Load length into D1.
             MOVEM.L    (A0)+,A1−A2     Load addresses of lists.
             MOVEA.L    A0,16(A7)       Adjust the return address.
LOOP         MOVE.L     (A1)+,(A2)+     Transfer one entry.
             DBRA       D1,LOOP         Loop back until done.
             MOVEM.L    (A7)+,D1/A0−A2  Restore registers.
             RTS
```

Figure 4.16 A 68000 realization of the program that copies 50 long words.

stack when the subroutine was called is 16 bytes below the current top of the stack. It is loaded into A0 and used to access the parameters passed from the calling program. After the parameters have been loaded into the desired registers, the return address is adjusted to point to the next instruction of the calling program. Then, the main body of the subroutine is executed. The registers are restored prior to leaving the subroutine, using another MOVEM instruction.

Parameters On the Stack

The most flexible way of passing parameters is to place them on the stack. Any number of parameters can be handled in this way. Implementation of this technique is illustrated in Figure 4.17, using the list-copying example of Figure 2.21(a). The length parameter, which is one less than the number of items in the list to be copied, is pushed on the stack. Next, the starting addresses for the lists are placed on the stack using the PEA (Push Effective Address) instructions. Note that instead of using the PEA instruction, the desired function can be achieved with the MOVE instruction, e.g.,

$$\text{MOVE.L} \qquad \text{\#LISTA},-(\text{A7})$$

The difference is that the MOVE instruction affects the condition-code flags, while PEA does not. Of course, in our example this difference is of no consequence.

```
              ORG         $1000
LISTA         DS.L        50
              ORG         $2000
LISTB         DS.L        50
              ORG         $4000
              MOVE.W      #49,-(A7)          Size and addresses of lists A and B
              PEA         LISTA                 are placed on the stack.
              PEA         LISTB
              BSR         SUBMOVE           Call the subroutine.
              ADDI.L      #10,A7            Remove parameters from stack.
              next instruction
                   .
                   .
                   .
SUBMOVE       MOVEM.L     D1/A1-A2,-(A7)    Save registers.
              MOVE.W      24(A7),D1         Size of lists into D1.
              MOVEA.L     20(A7),A1         Address of list A into A1.
              MOVEA.L     16(A7),A2         Address of list B into A2.
LOOP          MOVE.L      (A1)+,(A2)+       Copy list entries.
              DBRA        D1,LOOP
              MOVEM.L     (A7)+,D1/A1-A2    Restore registers.
              RTS
```

Figure 4.17   Passing parameters on the stack—a 68000 realization of the program in Figure 2.21(a).

Subroutine SUBMOVE begins by saving the contents of the registers it uses. The contents of the stack after the registers are saved are shown in Figure 4.18. The distance from the top entry on the stack to the first parameter is 8 words, or 16 bytes. The parameters are accessed in the basic indexed mode and loaded into the respective registers. Then, the copying task is performed and the registers are restored to their previous values. Upon return from the subroutine, the parameters are removed from the stack by adjusting the stack pointer. Since the parameters occupy 10 bytes, register A7 is incremented by this amount.

4.8.2 LINK and UNLK Instructions

We have already seen that the stack plays a central role in subroutine linkage. It is used for storing return addresses, as well as passing parameters to subroutines and results to calling programs. It may also be used by a subroutine to store local

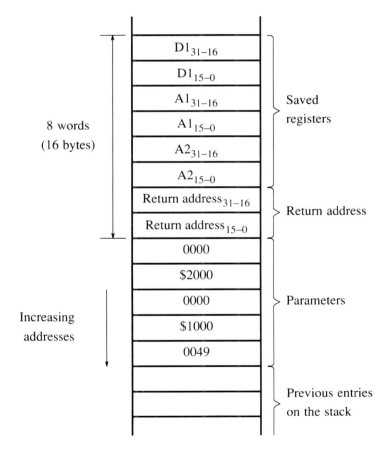

Figure 4.18 Top entries in the stack after the first instruction of the subroutine in Figure 4.17 has been executed.

variables, temporary results, and so forth. Let us consider this latter possibility in more detail.

A variable used only within a particular subroutine is called a *local* variable. If it is used outside the subroutine as well, it is said to be *global*. Local variables can be held and manipulated in the processor registers, as we have done in previous examples. For instance, in subroutine MAX of Figure 4.11, the location and the value of the largest number are local variables held in registers A3 and D1. It is not essential to keep local variables in registers, as they can also be stored in the main memory, which is clearly necessary when there are more local variables than registers available.

If the local variables are to be stored in the main memory, then a convenient place to keep them is on the stack. The stack space occupied while a subroutine is being executed can be recovered prior to the return to the calling program. It is important that such data be easily accessible within the subroutine. The address of the top item on the stack is available in the stack pointer, A7. When an item other than the top item on the stack is needed, it can be accessed using the indexed addressing mode, with A7 as the index register (as we have done in Fig. 4.17). However, because the top of the stack may change as the execution of the subroutine progresses, the technique of accessing local variables through the stack pointer may be somewhat awkward to use. It is easier if the items can be accessed relative to a fixed position in the stack, perhaps the location that holds the first item placed on the stack by the subroutine.

The 68000 microprocessor provides two instructions that facilitate the use of the stack in this manner. These are the LINK (Link and Allocate) and UNLK (Unlink) instructions. Their purpose is to allow the creation of a workspace on the stack, that can be used by a subroutine in an easily accessible manner. Let us consider the LINK instruction first. It names an address register (other than A7), which is used as a pointer to the workspace on the stack. It also specifies the size of the workspace, by means of a displacement operand. Figure 4.19 gives a simple example, showing the effect of executing the instruction

LINK   A6,#-8

Execution of this instruction causes the contents of the specified register, A6, to be saved on the stack. Then, the contents of the stack pointer (A7) are copied into register A6, and the stack pointer is augmented by the displacement value. Because the stack grows in the direction of decreasing addresses, the displacement is given as a negative number. By adding the (negative) displacement to the stack pointer, the desired workspace is created on the stack. In our example, a space of 8 bytes is allocated. Thus, the new value of the stack pointer sets the top of the stack at address 1FFF4. Now, the locations within the allocated workspace are easily accessible using the indexed addressing mode with A6 as the index register.

The UNLK instruction

UNLK   A6

reverses the action of the above LINK instruction. It restores the original contents of the address register specified and the stack pointer, thus removing the allocated workspace from the stack.

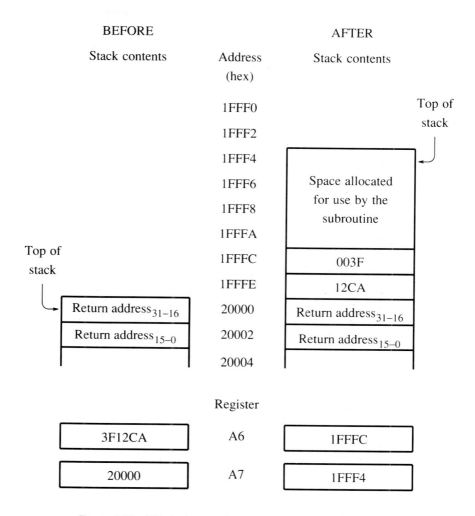

Figure 4.19   Effect of executing the instruction: LINK   A6,# −8

Workspace created on the stack is sometimes referred to as a *stack frame.* The register used to access the information, A6 in our example, is called a *frame pointer.*

4.9 FLOATING-POINT ADDITION EXAMPLE

The examples in the preceding sections are rather simple, intended to illustrate the basic features of the 68000 instruction set and addressing modes. We should now consider a more complex example that uses a variety of instructions and illustrates the flexibility provided by a large register set.

Many scientific and engineering applications require the use of the floating-point number representation. This representation is explained in Section A.7 of Appendix A, which discusses the format recommended by the IEEE Standards Committee. Its single-precision (32-bit) version is depicted in Figure A.18. The most-significant bit is the sign, followed by an 8-bit exponent, which is a signed integer expressed in excess-127 notation. The remaining bits denote a 24-bit mantissa, which is normalized such that its most-significant bit is always equal to 1. Hence, this bit is not recorded explicitly, and only the least-significant 23 bits are stored, in bit positions 22–0. Of course, the implied 1 to the left of these 23 bits must be included in all arithmetic operations.

Arithmetic operations with floating-point numbers are considerably more difficult than those dealing with fixed-point numbers. Also, the addition and subtraction are more involved than multiplication and division. We will discuss the addition of two single-precision numbers in this section. As explained in Section A.7, the numbers must first be aligned so that their exponents are equal. Then, the mantissas are either added or subtracted, depending upon the signs of the numbers. If the signs are the same, the mantissas are added. If they are different, then the smaller number is subtracted from the larger one. The resultant sum is normalized and the sign is chosen to correspond to that of the larger number.

A possible program that implements this procedure is shown in Figure 4.20. It is presented in a complete form that includes sufficient comments and the assembler directives required. The entire program could serve as a subroutine for use in a more general program that deals with floating-point numbers. In such a case it would be prudent to save the contents of all registers used, which we have not done in the example. The program includes two subroutines, where both the parameters and the results are passed through processor registers. We should emphasize that this program is intended to illustrate the use of 68000 instructions, rather than give a precise implementation of floating-point addition. Many details related to maintaining the accuracy of the computation have been omitted for clarity.

The program adds two numbers, A and B, which are found in long words at addresses $4000 and $4004, and places the sum in location $4008. Since the sign, exponent, and mantissa parts of a number must often be considered individually, it is convenient to keep each of them in a separate data register. Our choice is to place the sign in D0, the exponent in D1, and the mantissa in D2, for number A, and the corresponding parts of B in registers D3, D4, and D5. The required information is placed in the least-significant bit positions of each register. Subroutine SETUP is called twice to perform this task. The subroutine demonstrates the use of several logic instructions. It receives a floating-point number in register D3 and returns the mantissa, exponent, and sign in registers D3, D4, and D5, respectively. The subroutine begins by copying the floating-point number into D4. Then, the sign bit, which is originally in bit position 31, is loaded into the X (Extend) and C condition-code flags, by shifting the contents of register D4 one position to the left. Next, the value of the X flag is rotated into bit 0 of register D5, using the ROXL (Rotate Left With X) instruction. Note, from Table 4.3, that there exist two types of rotate instructions. In the first type, namely, ROL and ROR, the contents of a specified register are rotated directly, without including the condition-code flags in the feedback loop. The

```
* Program that adds two floating-point numbers.
*
MASKM       EQU         $FFFFFF             Mask for mantissa.
NUMA        EQU         $4000
NUMB        EQU         $4004
SUM         EQU         $4008
*
            ORG         $1000
            MOVE.L      NUMA,D3             Represent number A
            BSR.S       SETUP                 so that its
            MOVE.L      D3,D0               mantissa is in D0,
            MOVE.L      D4,D1               exponent in D1,
            MOVE.L      D5,D2               and sign in D2.
            MOVE.L      NUMB,D3             Represent number B the same
            BSR.S       SETUP                 way in D3, D4 and D5.
*
* Align the summands, if necessary, by shifting the mantissa
* of one of them until their exponents are equal.
*
ALIGN       CLR.L       D6
            MOVE.B      D1,D6               Assume exponent of A.
            SUB.B       D4,D1               Find the extent of alignment.
            BEQ         SIGN                No alignment needed.
            BHI         ADJUSTB             E_A > E_B.
            MOVE.B      D4,D6               Use exponent of B.
            NEG.B       D1                  Need positive difference.
            CMPI.B      #23,D1              If have to shift more
            BLS.S       FIXA                  than 23 bit positions,
            MOVE.L      NUMB,SUM              then sum is equal
            BRA.S       DONE                  to number B.
FIXA        LSR.L       D1,D0               Otherwise, change A to
            BRA.S       SIGN                  align with B.
ADJUSTB     CMPI.B      #23,D1              If have to shift more
            BLS.S       FIXB                  than 23 bit positions,
            MOVE.L      NUMA,SUM              then sum is equal
            BRA.S       DONE                  to number A.
FIXB        LSR.L       D1,D3               Otherwise, change B.
*
* If the signs are equal, then just add the mantissas. If they are opposite,
* then compare mantissas and subtract the smaller one from the larger one
* and use the sign of the latter.
*
SIGN        MOVE.B      D2,D7               Assume sign of A.
            CMP.B       D5,D2               Compare signs.
            BNE.S       COMPM               If not equal, compare mantissas.
            ADD.L       D3,D0               Otherwise, it is only necessary
            BRA.S       NORMAL                to add mantissas.
```

Figure 4.20   (continued next page)

```
COMPM     CMP.L     D0,D3          Compare mantissas.
          BLS.S     SUBTM          If M_B > M_A, then use
          MOVE.B    D5,D7            the sign of B and
          EXG       D0,D3            generate M_B - M_A
SUBTM     SUB.L     D3,D0            in register D0.
*
```

* Normalize the sum, put it into proper format and store it in the memory.
*

```
NORMAL    BSR.S     NORM           Normalize the sum.
          ROR.L     #8,D6          Assemble the exponent,
          MOVE.B    D7,D6            the sign and
          ROR.L     #1,D6            the mantissa of the
          BCLR      #23,D0           sum into a single
          OR.L      D0,D6            long word, and
          MOVE.L    D6,SUM           store in memory.
DONE      next instruction
*
```

* Subroutine that breaks up the 32-bit floating-point representation of a
* number into 3 pieces. The 24-bit mantissa, which includes the implied 1
* in bit-position 23, is in register D3. The exponent is in D4, and the
* sign in D5. The original number is passed to the subroutine in register D3.
*

```
SETUP     CLR.L     D5
          MOVE.L    D3,D4
          LSL.L     #1,D4          Put the sign bit into
          ROXL.L    #1,D5            bit 0 position of D5.
          ROL.L     #8,D4          Put exponent in D4_{7-0}.
          ANDI.L    #MASKM,D3      Put mantissa in D3_{23-0},
          BSET      #23,D3           including the implied 1.
          RTS
*
```

* Subroutine that normalizes the sum. The mantissa and the exponent
* are passed between the calling program and the subroutine in
* registers D0 and D6, respectively.
*

```
NORM      TST.L     D0             First check if mantissa
          BEQ.S     ZEROM            is equal to 0.
          BTST      #24,D0         Should the mantissa be reduced?
          BEQ.S     BIGGER
          LSR.L     #1,D0          Mantissa is too big, reduce it
          ADDQ.B    #1,D6            and increase the exponent.
          CMPI.B    #255,D6        If there is no exponent overflow,
          BNE.S     RETURN           then finished.
          CLR.L     D0             Otherwise, set mantissa to 0.
          BRA.S     RETURN
BIGGER    BTST      #23,D0         Should mantissa be increased?
```

Figure 4.20 (continued next page)

```
          BNE.S     RETURN
          LSL.L     #1,D0          Mantissa too small, increase it
          SUBQ.B    #1,D6             and decrease the exponent.
          BNE.S     BIGGER
          CLR.L     D0             Exponent underflow.
ZEROM     CLR.L     D6             Set exponent to 0, and make
          CLR.B     D7                the sign positive.
RETURN    RTS
          END
```

Figure 4.20   A program that adds two floating-point numbers.

other type comprises the ROXL and ROXR instructions, where the X flag is included in the feedback loop. The latter instructions correspond to the rotate instructions found in the 6809 microprocessor, which does not have rotate instructions of the first type.

While shifting the sign bit out of register D4, the eight exponent bits are also shifted to the left, so as to occupy the most-significant byte. In order to place these bits into the least-significant byte, we have used the instruction

$$\text{ROL.L} \qquad \text{\#8,D4}$$

The source field indicates that the rotation across eight bit positions is to take place. Note that we could not have used

$$\text{ROR.L} \qquad \text{\#24,D4}$$

because the rotate count is limited to a 3-bit number when specified explicitly. If a count exceeding 8 is required, it can be given in another register, as explained in Section 4.4.2.

The last step in subroutine SETUP is to prepare the 24-bit mantissa. Since the original number is still in D3, the least-significant 23 bits are isolated simply by masking out the remaining high-order bits. The 24th bit, in position 23, is the implied 1, which is generated by the BSET instruction.

After obtaining the component parts of the two numbers, it is necessary to align them so that their exponents are equal. The larger of the two exponents will be the exponent of the resultant unnormalized sum. This exponent is placed in register D6, where it will ultimately be used for assembling the sum in the 32-bit single-precision format. The scheme used chooses the exponent of A and replaces it with that of B if necessary. The two numbers are aligned by shifting the mantissa of the smaller number by a number of bit positions equal to the difference in the exponents. Bits shifted out of the right-hand side are ignored. Prior to shifting a mantissa, a check is made to see if the shift count exceeds 23. In this case, the sum is simply set equal to the larger summand, because all of the meaningful bits of the smaller number would be shifted out of the register. (An exact implementation of the IEEE Standard would not ignore the bits shifted out, even if the exponents differ by more than 23.) It is interesting to note what could happen without making this check. The required shifting is done with instructions of the type

LSR.L   #D1,D0

The count in D1 is the difference in the exponents, which is a number in the range 0 to 255. But, the 68000 microprocessor executes this instruction by using the contents of register D1 as the shift count modulo 64. Hence, if the difference in exponents exceeds 63, an erroneous shifting operation may occur. For example, if the difference is 66, then the effective shift count will be 2 (=66 mod 64), which will obviously produce a wrong result.

Having aligned the summands, we proceed to add the two mantissas. Because the mantissa of a floating-point number is represented in the sign-and-magnitude representation, the mantissas should be added if their signs are equal and subtracted if they are not. This is done in the program segment labeled SIGN. For unequal signs, the smaller mantissa, assumed to be in register D3, is subtracted from the larger one, assumed to be in register D0. This is true when B ≤ A. The case where B > A is handled by exchanging the contents of registers D0 and D3.

The EXG (Exchange Registers) instruction swaps the entire contents of the registers specified. The suffix L, to denote the long word size, is not needed because no other size is allowed.

The sum, whose components are in registers D7, D6, and D0, is normalized in subroutine NORM. Several tests must be made to recognize some special cases. As explained in Section A.7, the exponent values 0 and 255 have a special meaning, representing the numbers 0 and infinity. Zero is represented with all 32 bits of the number equal to 0. Infinity is denoted by an exponent of 255 and mantissa of 0. The normalization is done such that the most-significant bit of the mantissa, in bit position 23, becomes equal to 1. This bit is the implied 1, which is later omitted when the three parts of the number are assembled into a single 32-bit format.

The final step is to assemble the sign, the exponent, and the mantissa into a long word and place it at location SUM.

This example indicates how floating-point numbers can be handled with a microprocessor that has no particular design features intended to deal with such numbers. It is apparent that a reasonably complex software package is needed to implement all the operations involved. Moreover, it is clear that the speed with which floating-point arithmetic can be performed in this manner is low compared to the speed of handling fixed-point numbers. The obvious conclusion is that a fast computer for scientific computation cannot be realized if the floating-point capability is provided through software.

An attractive approach for building computers with good floating-point capability is to implement special circuits, often within a single integrated circuit package, to deal with such operations. One such chip is the MC68881, which operates as a coprocessor with the 68000 family of microprocessors [8]. In particular, it was designed for use with the 68020 microprocessor, which will be discussed in the next section. The coprocessor executes a set of special floating-point instructions in the same way as the microprocessor executes its own instructions. Both chips have the same bus connection requirements, so that they can be incorporated easily into a central processing unit of a powerful microcomputer.

4.10 SPEED OF EXECUTION

The effectiveness of a given microprocessor is closely related to the speed of execution of its instructions. This speed depends on the time needed to access memory locations and to perform the desired operations. The 68000 microprocessor functions under the control of an externally provided clock signal. As will be explained in Section 6.10, data transfers over the 68000 bus require a minimum of four clock cycles. More time is needed when slow devices are involved. In what follows, we will discuss the execution time for various instructions, assuming that all data transfers can be completed in the minimum time of four cycles. When more cycles are needed the total execution time will increase accordingly.

Table 4.7   THE NUMBER OF CLOCK CYCLES NEEDED FOR ADDRESS COMPUTATION AND OPERAND ACCESS

Mode	Byte or Word		Long Word	
	src c[r/w]	dst c[r/w]	src c[r/w]	dst c[r/w]
Dn	0[0/0]	0[0/0]	0[0/0]	0[0/0]
An	0[0/0]	0[0/0]	0[0/0]	0[0/0]
(An)	4[1/0]	4[0/1]	8[2/0]	8[0/2]
(An)+	4[1/0]	4[0/1]	8[2/0]	8[0/2]
−(An)	6[1/0]	4[0/1]	10[2/0]	8[0/2]
d(An)	8[2/0]	8[1/1]	12[3/0]	12[1/2]
d(An,Xi)	10[2/0]	10[1/1]	14[3/0]	14[1/2]
Abs.W	8[2/0]	8[1/1]	12[3/0]	12[1/2]
Abs.L	12[3/0]	12[2/1]	16[4/0]	16[2/2]
d(PC)	8[2/0]	–	12[3/0]	–
d(PC,Xi)	10[2/0]	–	14[3/0]	–
immed	4[1/0]	–	8[2/0]	–

Notes: c = number of clock cycles
r = number of read accesses
w = number of write accesses

In order to execute a given instruction, the OP-code word must be fetched first, followed by fetching any addressing information that is needed, and then accessing the operand locations. Most of the execution time is spent on accessing the main memory. The rest is used to perform various arithmetic and logic operations, inside the microprocessor. Some of these steps can be overlapped. From the user's point of view, the detailed arrangement and timing of internal operations is not of great significance. Only the total time needed to execute a given instruction matters.

When a memory operand is involved, it is first necessary to determine its effective address. Table 4.7 shows the numbers of clock cycles and memory accesses for each addressing mode. Each entry is presented in the form

$$c[r/w]$$

where c, r, and w are the numbers of clock cycles, read accesses, and write accesses, respectively. The memory accesses include reading of addressing information as well as accessing the operand locations themselves. The table shows the different requirements for source and destination operands. Note that byte and word operands have the same requirements, while long-word operands usually involve additional clock cycles and memory accesses.

The simplest case is that of an operand in a microprocessor register. In this case, no extra time is needed either for the determination of the effective address or for accessing the operand. In other cases, several clock cycles and one or more memory accesses are needed. For example, for a source operand in the register indirect mode, (An), one memory read access is needed to fetch the desired operand, if the operand is 8 or 16 bits long. This takes four clock cycles. If the operand is 32 bits long, two read accesses are needed, which occupy eight clock cycles. In the case of a destination operand, the same number of clock cycles are needed, but access to the main memory will be for a write instead of a read operation. As another example, consider the indexed modes for a byte or word destination operand. In both cases it is necessary to access the memory twice. First, the index value is fetched. Then, the operand is stored in the location denoted by the effective address. These two accesses require eight clock cycles. This is the number of cycles indicated in the table for the basic indexed mode, d(An). However, in the full indexed mode, d(An,Xi), a total of 10 cycles are needed. The extra time is required for the more complex computation of the effective address, which involves adding the contents of two registers and the index value.

Timing requirements for the complete 68000 instruction set are rather complex. We will consider the details for only a few instructions, which are given in Table 4.8. The table gives the numbers of clock cycles and memory accesses needed when all operands are in microprocessor registers. The plus sign in the last column is used to indicate that additional clock cycles and memory accesses are needed for each memory operand as specified in Table 4.7.

Consider the instruction

MOVE.W   8(A2),\$4A300CB20

Table 4.8 states that four clock cycles and one read access (to fetch the OP-code word) are needed, plus whatever is required for the operands. According to Table

Table 4.8   INSTRUCTION EXECUTION TIMES

Instruction	Size	Addressing Modes	c[r/w]
MOVE	B, W, L	src,dst	4[1/0]+
ADD, SUB	B, W	src,dst	4[1/0]+
	L	Rn,Dn	8[1/0]
		immed,Dn	8[1/0]+
		Msrc,Dn	6[1/0]+
		Dn,Mdst	4[1/0]+
CMP	B, W	src,Dn	4[1/0]+
	L	src,Dn	6[1/0]+
CMPA	W, L	src,An	6[1/0]+
LEA	L	src,An	4[1/0]+
		Branch taken	Branch not taken
Bcc	disp = 8	10[1/0]	8[1/0]
	disp = 16	10[2/0]	12[2/0]
DBcc	cc true	–	12[2/0]
	cc false	10[2/0]	14[3/0]

Notes: Msrc = memory source (except immediate)
Mdst = memory destination

4.7, the source operand, which is specified in the basic indexed mode, is dealt within eight clock cycles and two read accesses (one to read the index value and one to read the operand itself). The destination location is given in the absolute long mode, which requires 12 cycles—two read accesses (to fetch the 32-bit address) and one write access (to store the operand). Thus, a total of 24 cycles, five read accesses and one write access, are needed to execute this instruction.

As another example, consider the instruction

$$\text{ADD.L} \qquad \text{(A2),D3}$$

A memory source operand and a data register destination are involved. Table 4.8 indicates that, for this case, six clock cycles and one read access are needed in addition to the requirements for operand addressing. From Table 4.7, it follows that the register indirect mode, (An), for a source operand requires eight cycles and two read accesses (to read the long-word operand). The destination operand, being in a microprocessor register, requires no additional time. Thus, 14 clock cycles and three read accesses are needed to execute the instruction.

Table 4.8 includes the "conditional branch" and "decrement and branch" instructions. In all of these instructions, the execution time depends on whether or not the branch is taken. For example, the DBF instruction (which is also known as DBRA, as explained in Section 4.6) has the branch condition always false. If decrementing the counter register, Dn, specified in this instruction causes a branch to be taken, then 10 cycles and two read accesses are needed. But, if the branch is not taken, then 14 cycles and three read accesses are used. It may seem paradoxical to the reader that it takes longer not to take a branch, which presumably requires less computational effort. The reason is that the 68000 microprocessor prefetches the next instruction it expects to be executed, while the execution of the current instruction is still in progress. The objective is to improve the overall performance. The DBcc instructions are mainly used to control loops. They are intended to be placed at the end of a loop, to cause a branch to the beginning of the loop. Hence, most of the time the branch will be taken. While executing a DBcc instruction, the microprocessor prefetches the instruction that is most likely to be needed next, which is the instruction at the branch destination. When the branch is not taken, that is when the instruction that follows the DBcc instruction is to be executed next, the prefetched instruction is of no use and must be thrown away. One more read access is needed to fetch the OP-code of the desired instruction, increasing the total number of read operations to three.

The timing requirements of the complete instruction set are not included in this book for space reasons. We have shown a few representative examples in Table 4.8, to give the reader a feeling for the kind of execution speed that can be achieved. The complete details can be found in the manufacturer's literature [9].

4.11 THE 68020 AND 68030 MICROPROCESSORS

Microprocessors, like most computing devices, evolve over a period of time. When a certain microprocessor becomes accepted and widely used, it is usually followed by several versions of chips that exhibit the same basic structure but differ in some application-oriented aspects. The 68000 has attained considerable popularity. In order to allow its use in low-cost applications, where 8-bit data transfers are acceptable, a less powerful version was developed, known as the MC68008 microprocessor. In the other direction, more powerful chips have also been designed. The

MC68010 is an enhanced version that allows implementation of the virtual memory concept, which will be discussed in Chapter 8. Further enhancements are the MC68020 and MC68030, which are truly 32-bit microprocessors having many features that lead to excellent performance. We will discuss some important aspects of these chips in this section. The MC68020 and MC68030 are source-code compatible chips, hence to a programmer they appear to be essentially the same. In view of this, we will present the following discussion in terms of the MC68020. As we have done with the 68000, we will refer to it simply as the 68020, omitting the prefix MC.

The 68020 has a full 32-bit bus, which includes 32 address and 32 data lines. Thus, the addressable space is 4 gigabytes (=4,294,967,296 bytes). Only a small fraction of this space is likely to be occupied by the physical memory in any given microcomputer, but its large size is useful in conjunction with the idea of a virtual memory, as will be seen in Chapter 8. While the data bus is 32 bits wide, the 68020 can be used with devices that transfer 8, 16, or 32 bits at a time. We will see in Chapter 6 how the microprocessor dynamically adjusts to the data bus needs of a particular device.

4.11.1 Register Set

When the 68020 is operating in the User mode, its register set, as seen by the programmer, is essentially the same as that of Figure 4.2. All 32 bits of the program counter are used. However, several additional control registers exist, which are accessible only in the supervisor state. They facilitate implementation of sophisticated operating system software.

4.11.2 Data Types

Data can be handled in bit, byte, word, and long-word sizes, as well as in 4-bit BCD digits, as in the 68000. The 68020 has three other data types. BCD digits can be represented in the unpacked format, where 1 byte is used for each digit. Another longer size is provided for representing integers, consisting of 64 bits, called a *quad word*. Unlike the 68000, the word, long-word, and quad-word operands need not be aligned on even-address boundaries. Finally, there is a bit-field format, illustrated in Figure 4.21. The bit field can span from 1 to 32 bits. It is specified by the location of its most-significant (leftmost) bit and the field width. The effective address of an operand is used as the base address. An offset from this address specifies the beginning of the bit field. In our example, the base address is k, the offset is 11, and the width of the field is 15 bits.

4.11.3 Addressing Modes

All of the addressing modes discussed in Section 4.3 are available. The suitability of the 68020 microprocessor for running programs written in high-level languages has been enhanced through additional addressing modes.

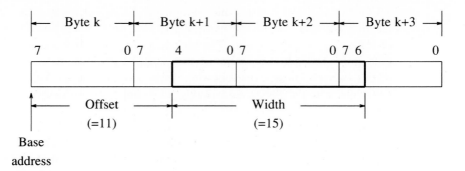

Figure 4.21 Bit-field format.

The full indexed mode has been made more flexible by allowing different sizes of displacement and by scaling the contents of the index register. Figure 4.22 shows how the effective address of an operand is determined. The displacement is an 8-, 16-, or 32-bit signed integer. The 8- and 16-bit values are sign extended, that is, their sign is replicated into high-order bit positions to obtain a 32-bit value. Another change, in comparison with the 68000, is the use of scaling with the index register.

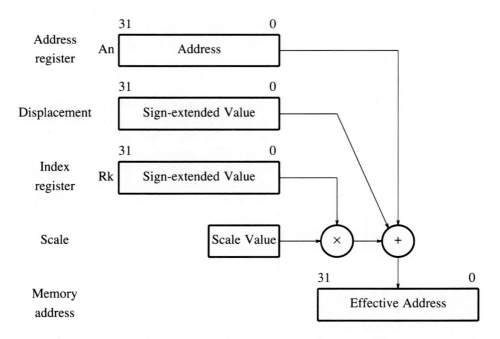

Figure 4.22 Full indexed addressing mode of 68020.

The contents of this register are multiplied by 1, 2, 4, or 8, depending on the value of the scale. The assembler syntax for the mode is

$$(disp, An, Rk.size*scale)$$

Note that the displacement is indicated within the brackets (it is written outside in the 68000). Also, recall that the "size" is either W (word) or L (long word). The effective address is determined as

$$EA = disp + [An] + [Rk] \times scale$$

It is easy to appreciate the usefulness of scale in this mode. If the index register is used to scan through the entries of a list, then byte, word, long-word, or quad-word entries can be accessed by incrementing the index register by 1 for each entry and setting the scale value to denote the number of bytes in the entry. Incrementing an index variable by 1 to reach the next entry is a natural way of dealing with arrays in high-level languages.

Another powerful extension of indexed addressing involves an address component called the address operand, which is obtained indirectly from the main memory. There are two versions of this mode: *memory indirect post-indexed* and *memory indirect pre-indexed* modes. The first of these is illustrated in Figure 4.23. Two displacements are involved in the computation of the effective address. First, a "base displacement," having 16 or 32 bits, is added to the contents of the address register specified, producing a memory address at which an address operand is found. The value of the address operand is, then, used in the full indexed format (as in Fig. 4.22) to determine the effective address of the desired operand. In order to distinguish the second displacement from the first, the name "outer displacement" is associated with it. The assembler syntax for this mode is

$$([basedisp, An], Rk.size*scale, outdisp)$$

It follows the general indexed-mode format, with the memory indirect fetching of the address operand indicated in square brackets.

The memory indirect preindexed mode uses the same constituent parts in determining the effective address. However, in this case, the indexing modification is done before the address operand is fetched from the memory. The syntax for this mode is

$$([basedisp, An, Rk.size*scale], outdisp)$$

The address operand is modified only through the addition of the outer displacement, to yield the effective address.

The indexed modes that we have discussed make use of four user-specified values: address register, index register, base displacement and outer displacement. All of these values are optional and if not specified are treated as having the value 0 in the computation of the effective address.

The usefulness of the memory indirect indexed modes is that they provide easy access to lists where contiguous memory locations are used not to store the actual data items but to hold the addresses of data items. The data items themselves need

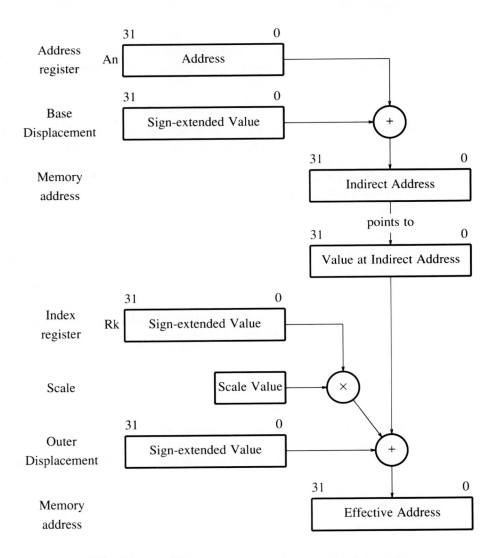

Figure 4.23   Memory indirect post-indexed addressing mode.

not be in contiguous locations. The memory indirect post-indexed mode is particularly useful for accessing the elements of an array whose base address is passed to a subroutine via a stack frame. (See Subsection 4.8.2 for a discussion of how a stack frame may be implemented.)

It is usual to find a special case of indexed addressing modes, referred to as the relative addressing modes, where the program counter contents are used in the computation of the effective address. The 68020 provides a relative mode version of all the indexed modes discussed above, with the program counter used instead of an address register.

4.11.4 Instruction Set

The 68020 instruction set features several enhancements of the basic 68000 instructions, as well as a number of new instructions. Typical enhancements involve the option of using longer operands in some instructions and 32-bit displacements in branch instructions. The new instructions are intended to facilitate dealing with bit fields, BCD numbers, coprocessor chips, and operating system software.

Several instructions use bit-field operands, discussed in conjunction with Figure 4.21. Some are direct extensions of the bit instructions, such as BCLR, BSET, and others presented in Subsection 4.5.4. Instead of operating on a single bit, these instructions perform the specified function on all bits of the field. Other instructions of this class perform tasks such as extracting a bit field and loading it into a data register, inserting a bit field, or finding the first bit equal to 1 in a bit field. The bit field instructions are particularly useful in applications where bits are manipulated in sizes other than the standard byte, word, or long-word formats.

In the previous section we mentioned the existence of the MC68881 floating-point coprocessor, intended for use with the 68020. Other coprocessor chips have been designed to handle some specific tasks that would otherwise occupy much of the time of the main microprocessor. We will see in Chapter 10 how coprocessors can be incorporated into a microcomputer system. The 68020 contains instructions that can be used to interact with such coprocessors.

Despite its relatively small physical size, the 68020 is a very powerful processor, capable of supporting sophisticated multiuser applications. Such applications require considerable system software. The microprocessor has specialized instructions that facilitate the implementation of this software.

4.11.5 Performance

Performance of a computer system is usually assessed in terms of the time needed to perform various computational tasks. Much of the time is spent on accessing instructions and data in the main memory. Having a 32-bit data bus reduces the total time spent on bus transfers, in comparison with the 68000 and its 16-bit data bus.

A powerful technique for reducing the number of bus transfers is to prefetch a block of instructions that are executed in a loop repeatedly and store them in a *cache* within the microprocessor chip. The 68020 contains a 256-byte cache. We will discuss the cache concept in more detail in Chapter 8.

Another way of increasing the performance is to perform the operations within the microprocessor in as parallel a manner as possible. Several suboperations of a given operation can often be overlapped if they are carried out by different circuits. This is done in the 68020. Moreover, all internal data paths are 32 bits wide, providing a further improvement over the 68000 which handles data in 16-bit quantities.

In summary, the 68020 is superior to the 68000 in the versatility and flexibility of addressing modes and the instruction set, as well as in the performance resulting from full exploitation of the 32-bit data paths and parallelism within the microprocessor chip.

4.11.6 68030 Enhancements

From the programmer's point of view, there is little difference between the 68030 and 68020 microprocessors. The 68030 offers better performance and some hardware interfacing advantages. Its better performance stems largely from the inclusion of extra memory circuits within the microprocessor chip. It has a 256-byte data cache, as well as an instruction cache of the same size. The cache details will be given in Section 8.9. The 68030 chip also contains circuits that implement the memory management function, which will be discussed in Section 8.10.

The interfacing flexibility of the 68030 is the result of a more versatile bus structure, to be presented in Section 6.10.

4.12 CONCLUDING REMARKS

In this chapter we have considered the 68000 structure as seen by the programmer. We have not dealt with the I/O aspects in detail, because they will be treated comprehensively in Chapters 5, 6, and 7. Moreover, the reader will find a discussion of cache storage and virtual memory, which are relevant to this family of microprocessors, in Chapter 8.

The 68000 instruction set has been summarized in tabular form. The information given should enable the reader to write many application programs. For space reasons, we have not included all details, such as the number of clock cycles needed to execute a given instruction. For additional information the reader may consult either the manufacturer's literature [9] or a specialized book such as [10], which gives numerous illustrative examples. A discussion of the 68000 microprocessor can also be found in several books on computer organization [11–13].

This chapter concludes the introduction of microprocessors and basic principles of assembly-language programming. In the following chapters we consider other major aspects of a microprocessor system and their impact on the programmer and on the hardware designer.

4.13 REVIEW QUESTIONS AND PROBLEMS

REVIEW QUESTIONS

1. What type of operands can be used with the 68000 instructions?

2. For what common purposes can address and data registers be used?

3. What are the restrictions on the alignment of operands in the main memory in general and on the stack in particular?

4. How many different addressing modes can be used to clear a long word at

memory address $1000? Give examples of both the assembler format and the corresponding object code for this CLR instruction, for each possible mode.

5. The 68000 does not have the "absolute indirect" addressing mode. How can one access an operand whose address is given as the contents of memory location ADRLOC?

6. Give an example that illustrates the usefulness of the full indexed addressing mode. How would you deal with your example if this mode was not available?

7. How is the operand length specified in 68000 assembly-language instructions?

8. Why is it impractical to allow the use of all addressing modes in all two-operand instructions?

9. The 68000 microprocessor has many instructions that can be specified in the format

OPER   ENTRY1,ENTRY2

where OPER denotes the operation to be performed while ENTRY1 and ENTRY2 are the entries in the operand fields. Give examples of various types of information that can be specified as ENTRY1 and ENTRY2.

10. Explain the differences between the BMI and DBMI instructions.

11. Condition code flags are affected by many instructions. Give examples of the possible ways of determining the state of these flags.

12. What is the difference between the Supervisor and User modes of operation?

13. What is the purpose of privileged instructions?

14. What are the three ways of passing parameters to a subroutine? Discuss their relative advantages and disadvantages in terms of the 68000 microprocessor.

PRIMARY PROBLEMS

15. Write a program that reverses the order of bytes in register D0. For example, if the starting contents of D0 are 3F017DEA, then the result left in D0 should be EA7D013F.

16. Write a program that takes the low-order 4 bits in registers D0 to D7 and concatenates them into a 32-bit long word in register D5, with the bits in D0 forming the least-significant bits of the result.

17. When data are transmittted between two microcomputers or a microcomputer and a peripheral device such as a magnetic cassette recorder, it is important to ensure that the received data contains no errors caused by the transmission process. A common technique is to append a *checksum* byte at the end of each record (i.e., fixed number of bytes of data) transmitted. The checksum may be simply the arithmetic sum of all the bytes of module 256. It can be tested by the receiver to improve the confidence in the validity of the data received. Write a subroutine that computes the checksum of an n-byte record.

18. Write a subroutine that converts an ASCII string consisting of both upper- and lower-case letters into a string that has only the upper-case letters.

19. A word or a phrase that is spelled the same forward and backward is called a *palindrome*. Some examples are: "dad," "toot," and "level." In a phrase, all blanks and punctuation marks are ignored, as in "race car." Write a program that determines if a given character string is a palindrome.

20. Write a program that finds all prime numbers in the range 1 to 1000 and displays them on a video terminal.

21. Suppose that we wish to pass two 32-bit numbers as parameters from a calling program to a subroutine. Also, suppose that the subroutine is to pass five 32-bit numbers, which are the results of a computation that it performs, to the calling program. Show how can this be done, assuming that all of these numbers are to be passed via the stack. Write the instructions that perform the subroutine linkage desired.

22. The factorial of a given number is defined as

$$n! = n(n-1)(n-2)\times...\times2\times1$$

It can also be defined *recursively* as

$$n! = n(n-1)!$$

This recursive definition suggests that the factorial can be computed with a subroutine that calls itself. Figure 4.24 gives a program that might be suitable. Does it compute the factorial correctly? Show the contents of the stack as the computation progresses. What range of numbers can this program handle? Can

```
                    MOVE.W      N,D1
                    BSR         FACT
                    next instruction

                        ⋮

    FACT    MOVE.W      D1,-(A7)
            SUBQ.W      #1,D1
            BNE.S       NEXT
            MOVE.W      (A7)+,D1
            BRA.S       DONE
    NEXT    BSR         FACT
            MULU        (A7)+,D1
    DONE    RTS
```

Figure 4.24   A recursive program for computing the factorial.

you write a simpler program to compute the factorial? (If you can, don't jump to a conclusion that recursion is not a useful concept!)

23. Consider a linked list containing the names of students in a given class and their marks, as discussed in Section 2.8. Assume that each item in the list has the format shown in Figure 4.25, with a variable-length name field. Two ASCII-encoded characters are stored in each word of the name field, and the last character of the name is followed by the "null" character. Write a program that finds the name of the student(s) who obtained the highest final grade.

24. Write a program that finds the item corresponding to a given student's name in the linked list of Problem 23.

25. Write a program that finds the median mark on the term test in the linked list of Problem 23.

26. Estimate the time it takes to transfer a list of 1000 entries using the program of Figure 4.9, assuming that the microprocessor is driven by a 10 MHz clock.

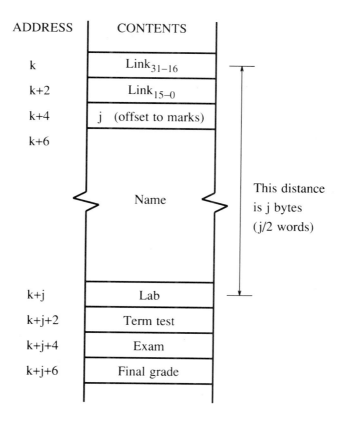

Figure 4.25 Arrangement of an item in a linked list.

How would this time compare with the time needed if one used the autoincrement addressing as suggested in Figure 4.16?

27. What is the longest time that it may take to find the largest number in a list of 1000 items using the program in Figure 4.11, if the microprocessor is driven by a 10 MHz clock?

ADVANCED PROBLEMS

28. Write a program to implement the bubble sort scheme explained in Problem 2.19.

29. Write a program, using SWAP, shift, and MOVE.B instructions, to perform an equivalent task to the instruction

$$\text{MOVEP.L} \qquad \text{D3,\$200(A5)}$$

described in Section 4.5.5.

30. Write a program that converts an eight-digit BCD-encoded number into a binary number. Assume that two BCD digits are packed into each byte.

31. Write a subroutine that adds two n-digit BCD numbers, P and Q, stored in the packed format that contains two digits per byte. Assume that the two most-significant digits of each number are stored in memory locations whose addresses are passed to the subroutine as parameters. The sum is to replace the number P. Note that there exists a special instruction, ABCD, for this purpose, which performs BCD addition on two digits packed in 1 byte. How does your program compare with the 6809 equivalent given in Figure 3.27?

32. Write a subroutine that subtracts the BCD numbers of Problem 31.

33. Modify the program of Figure 4.20 to add two double-precision (i.e., 64-bit format) floating-point numbers.

34. Write a program that multiplies two single-precision floating-point numbers.

35. Write a program that divides two single-precision floating-point numbers.

36. Write a program that sorts the linked list of Problem 23 in alphabetical order of students' names.

37. Write a program that sorts the linked list of Problem 23 in descending order of final grades.

38. Write a program that displays on a video terminal the names and the final grades of the three top students in the class represented by the linked list of Problem 23.

39. Write a subroutine that produces a delay of 1 ms on a microprocessor driven by a 10 MHz clock. How would you use this subroutine in a delay routine that is supposed to take as close as possible to 1 second to execute?

LABORATORY EXERCISES

40. Write a program that implements the task of Problem 2.22.

41. Write a program to multiply two matrices, as described in Problem 2.20.

42. Write a program that subtracts two single-precision floating-point numbers.

43. Write a program that reads a decimal number from the keyboard of a video terminal and converts it into the IEEE Standard single-precision floating-point format. Assume that the number is entered in the exponential notation used in most high-level programming languages. For example, 2.31E6 denotes 2.3×10^6 and 35.24E−3 denotes 35.24×10^{-3}. Each character of the input number is read in ASCII code.

4.14 REFERENCES

1. *iAPX 86,88 User's Manual*. Santa Clara, CA: Intel Corp., 1981.
2. *iAPX 386 Programmer's Reference Manual*. Intel Corp., 1986.
3. *MC68000 16-Bit Microprocessor User's Manual*. Austin, TX: Motorola Inc., 1983.
4. *MC68020 32-Bit Microprocessor User's Manual*. Motorola Inc., 1985.
5. *MC68030 Second Generation 32-Bit Enhanced Microprocessor*. Motorola Inc., 1987.
6. *Z8000 PLZ/ASM—Assembly Language Programming Manual*. Campbell, CA: Zilog Inc., 1979.
7. *Series 32000 Databook*. Santa Clara, CA: National Semiconductor Corp., 1986.
8. *MC68881 User's Manual*. Motorola Inc., 1985.
9. *M68000 16/32-Bit Microprocessor—Programmer's Reference Manual*. Motorola Inc., 1984.
10. Kane, G., D. Hawkins, and L. Leventhal, *68000 Assembly Language Programming*. Berkeley, CA: Osborne/McGraw-Hill, 1981.
11. Wakerly, J.F., *Microcomputer Architecture and Programming*. New York: Wiley, 1981.
12. Hamacher, V.C., Z.G. Vranesic, and S.G. Zaky, *Computer Organization*. 2nd ed., New York: McGraw-Hill, 1984.
13. Eccles, W.J., *Microprocessor Systems—A 16-Bit Approach*. Reading, MA: Addison-Wesley, 1985.

Chapter
5

Input/Output Programming

An overview of the building blocks of a computer and the ways in which they are interconnected was given in Chapter 1. Fundamentals of programming at the machine level were introduced in Chapter 2, and specific examples of commercial microprocessors were discussed in Chapters 3 and 4. In this and the following two chapters we will consider the problem of transferring data between the computer and the outside world. Mechanisms for performing such transfers are essential in a computing system of any kind or size. They are needed to feed data into the computer and to make the results of a computation available to the user.

The way in which machine instructions can be used to perform input and output operations has already been explained in the previous chapters. In this chapter, we will discuss these operations in more detail, but still from the programmer's point of view. In order to understand fully the handling of input and output tasks the programmer must not only be familiar with the software aspects, but must also appreciate the hardware features that influence the design of the software. The main objective of this chapter is to explain the interaction between the software and hardware during input/output tasks. Hardware issues will be discussed in this chapter at a functional level. Circuit details will be discussed in Chapter 6.

Input/output techniques that involve the use of interrupts and DMA will be introduced in Chapter 7, and interconnection standards will be described in Chapter 11.

Commercial integrated circuit (IC) chips used for connecting I/O devices to a microprocessor will be introduced as examples. They will be described in sufficient detail to enable the reader to carry out laboratory exercises. Because such chips are intended for use in a broad range of applications, they incorporate many features. In this chapter, we will concentrate on the features with which the programmer must be familiar. Other details, such as signal timing and connection to the microprocessor bus, will be discussed in Chapter 6.

5.1 SERIAL AND PARALLEL I/O

Some I/O devices, such as video terminals and printers, may be located far away from the microprocessor to which they are connected. Others, particularly high-speed devices such as magnetic disks, should be physically close to the microprocessor. In fact, small disk drives are often housed within the same cabinet as the microprocessor. But, in all cases, an I/O device is a distinct unit, which is connected to the microprocessor via an *interface circuit*, as shown in Figure 5.1. The main purpose of this circuit is to receive commands and data from the microprocessor and to

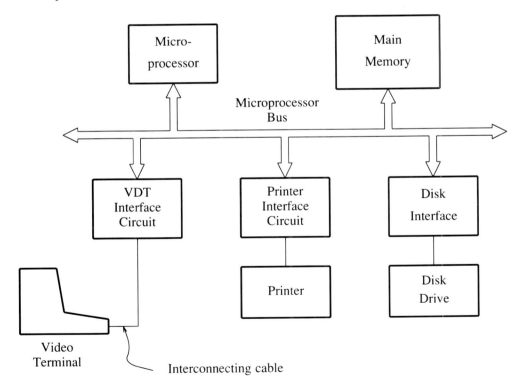

Figure 5.1 Use of I/O interface circuits.

translate them into a format that suits the requirements of the I/O device and the interconnecting cables.

In most microprocessor systems, data are handled in multiples of 8 bits. Eight, 16, or 32 bits are transferred in parallel over the microprocessor bus, depending upon the width of the bus and the capabilities of the devices involved. Data may also be transferred in parallel between an I/O device and its interface circuit. In this case, the device is said to have a *parallel interface*. Many I/O devices, particularly those requiring high data-transfer rates, use this arrangement. Figure 5.2 shows a printer connected to a microprocessor bus via a parallel interface. There are eight data lines and several control lines. Of course, there must also exist one or more ground connections to provide a path for the return current. These connections are not shown in the figure. The control lines serve to coordinate the transfer of data over the data lines. A widely used commercial standard for connecting printers in this manner is called the Centronix interface, which will be described in Chapter 11.

When I/O devices are physically far away from the computer, a connection that requires a large number of wires is expensive to implement. Instead, serial transmission may be used, where bits are transmitted one at a time. An interface that connects a device in this manner is called a *serial interface*. Figure 5.3 depicts a data terminal consisting of a keyboard and a video display, connected to a microprocessor bus via a serial interface. Here, one signal path is provided for input and one for

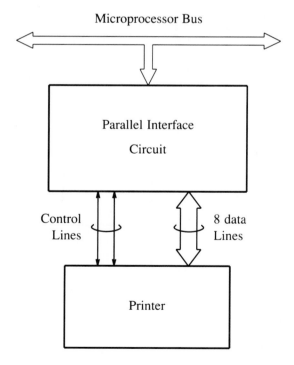

Figure 5.2 A printer connected to a microprocessor via a parallel interface circuit.

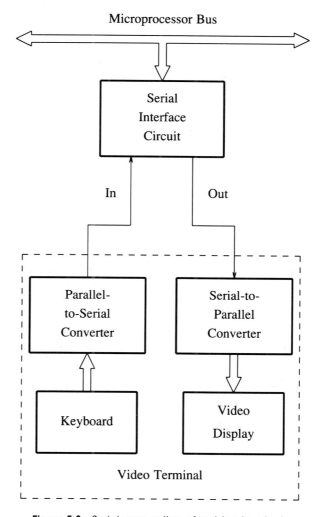

Figure 5.3 Serial connection of a video terminal.

output. The circuitry in the interface is responsible for translating the parallel data on the bus into the serial format, and vice versa. For example, 8 bits of data may be loaded from the bus into a shift register, and shifted out one bit at a time for transmission to the terminal. At the terminal, the received bits are assembled into their original 8-bit format. The reverse process takes place when transferring data from the terminal to the interface circuit. Data bits received from the terminal are loaded serially into a shift register, then placed on the microprocessor bus in parallel.

Typically, the speed at which data are transmitted over a serial link ranges from 300 to 9600 bits per second. Because separate wires are used for transmission in each direction, characters may be sent in both directions simultaneously. Such a connection is called a *full-duplex* connection. A widely used scheme for connecting serial lines to a computer is defined by a commercial standard called RS-232-C. The standard ensures both electrical and mechanical compatibility among devices, and defines the functions of the signals they exchange. Its details will be presented in Chapter 11.

An interface circuit creates a *port* to which external devices may be connected. A port is the collection of lines needed to carry data and control signals between the external device and the microprocessor bus. It is called a *serial port* if data are transferred one bit at a time, and a *parallel port* if several data bits are transferred in parallel.

In the following two sections we will discuss the basic serial I/O interface, concentrating on those aspects that are pertinent to the connection of video terminals to microcomputers. This will enable the reader to perform laboratory exercises on a microprocessor system at an early stage in studying the subject. Later, we will examine a number of issues relating to I/O tasks in general, and develop the essential concepts of parallel I/O interfaces.

5.2 COMMUNICATION WITH I/O DEVICES

Many peripheral devices are intended to handle textual information. Perhaps, the most commonly used such device is the video terminal. We will use data transfers to and from a video terminal to illustrate I/O programming. However, the techniques discussed are applicable to most devices.

Typically, a video terminal is connected to a microprocessor via a serial interface. It deals with characters encoded in an alphanumeric code, with ASCII being the most commonly used (see Appendix A). Since computers usually handle information in multiples of 8 bits, characters are represented using 8-bit quantities, as shown in Figure 5.4. The low-order 7 bits contain the ASCII code for a character, and the most-significant bit is set to 0.

The information exchanged between a microprocessor and an I/O interface circuit consists of the input or output data and status and control information. The

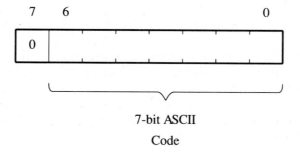

Figure 5.4 Commonly used format for storing a character in an 8-bit byte.

status information tells the microprocessor when the I/O device is ready to send or receive data. It may also enable the microprocessor to monitor the device, to ensure that it is in good operating order. For instance, a printer may indicate whether it has run out of paper or ink or whether the paper is jammed. Control information comprises the commands sent by the microprocessor to change the mode of operation of an I/O device or to cause it to take some specific action. For example, in the case of a device that can operate at different speeds, the microprocessor may send a command to select a particular speed of operation.

All communication between the microprocessor and an I/O interface circuit takes place by transferring information over the microprocessor bus. The locations in the interface where data, status, or control information is stored are called *I/O registers*, and are identified by unique addresses, as explained in Section 1.5. Sometimes, the terms *input buffer* and *output buffer* are also used to refer to the registers that hold the input and output data. In order to send data to a particular I/O device, the microprocessor writes the data into the output data register of the interface. From there, the data are automatically sent to the device. Data received from the device are read by the microprocessor from the input data register in the interface.

It was pointed out in Section 1.5 that I/O devices may be assigned locations in the same address space as the main memory — an arrangement known as *memory-mapped I/O*. In this case, any machine instruction can operate on data either in the main memory or in an I/O interface register, depending upon the address used. This is the approach used with Motorola's microprocessors, as described in Chapters 3 and 4. However, other arrangements are also possible. For example, the microprocessor may have instructions dedicated to I/O transfers. Special signals on the microprocessor bus select the I/O registers, thus providing a separate address space for the I/O devices. Since I/O instructions cannot be used to refer to the main memory, the addresses they use need only distinguish between different I/O registers. This approach has the advantage of providing some flexibility. I/O devices can be assigned addresses using a small number of bits, and longer addresses can be used to access the main memory. On the other hand, the need for specialized instructions increases the size of the instruction set of the machine and limits programming flexibility. A separate I/O address space is used in the Intel family of microprocessors. Unless stated otherwise, in the remainder of this book we will assume a memory-mapped organization.

Simple programs for performing input and output operations were given in previous chapters. Before examining specific I/O devices in more detail, we will review the general organization of an I/O routine.

5.2.1 A Simple I/O Program

With memory-mapped I/O, any instruction that moves data to or from a memory location can be used to read from or write into an I/O device register. For example, if DATAIN is the name of a device register that contains input data, the instruction

$$\text{Move} \qquad \text{DATAIN,ACCUMA}$$

will read the contents of DATAIN and deposit them into a microprocessor register called ACCUMA. For the 6809 microprocessor, the instruction

LDA   DATAIN

transfers the input data to accumulator A. In the case of the 68000, the instruction

MOVE   DATAIN,D0

stores the input data in data register D0. Similar instructions can be used to transfer data to an output register.

Let us now examine the question of coordination, or synchronization, of the activities of the microprocessor and the I/O device. Consider a video terminal connected to a microprocessor, and assume that the I/O registers in the terminal's interface circuit are organized as shown in Figure 5.5. When a character is typed at the keyboard, its ASCII code is sent to the interface, where it is stored in the Input Data register. Bit 0 of the Status register is set to 1 to indicate that an input character is available. As soon as the microprocessor reads the contents of the Input Data register this bit returns to 0, to ensure that the character is read only once.

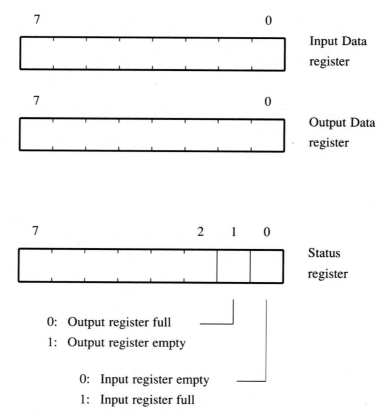

Figure 5.5   I/O registers in the interface circuit of a video terminal.

Bit 1 of the Status register is associated with output operations. When it is in the 1 state, it signifies that the interface circuit is ready to accept data. To display a character on the screen, the microprocessor deposits the ASCII code for this character into the Output Data register. Transferring this information to the terminal involves some delay, which depends upon the transmission speed used. The microprocessor should not send another character until transmission of the previous character has been completed. Hence, as soon as the microprocessor deposits new information in the Output Data register, the output status bit changes to 0. It returns to 1 when the character has been transmitted to the terminal and the interface circuit is ready to receive another character.

A program that receives one character from the keyboard and sends it back to the screen is given in Figure 5.6. The assembler commands at the beginning of the program assign values to the address constants STATUS, DATAIN, and DATAOUT equal to the addresses of the I/O registers they represent. The three-instruction loop starting at WAIT1 is executed repeatedly until bit 0 of the Status register becomes equal to 1. At this time, the input character is read and placed in accumulator B.

During normal operation, a video terminal does not automatically display the characters typed at its keyboard. Hence, the microprocessor should send each character it receives back to the terminal to have it displayed on the screen. In computer jargon, the typed characters are said to be *echoed back* to the terminal. Before sending a character, the microprocessor must ensure that the interface circuit is ready to accept it. The required check is performed in the loop starting at WAIT2, which

STATUS	EQU	$\cdots$	Declare addresses of
DATAIN	EQU	$\cdots$	Status and Data
DATAOUT	EQU	$\cdots$	registers.

* Read input character from the keyboard.

WAIT1	Move	STATUS, ACCUMA	Read Status.
	AND	#1, ACCUMA	Check bit 0.
	Branch_if_=0	WAIT1	Loop if input register empty.
	Move	DATAIN, ACCUMB	Read character.

* Send character to display screen.

WAIT2	Move	STATUS, ACCUMA	Read Status.
	AND	#2, ACCUMA	Check bit 1.
	Branch_if_=0	WAIT2	Loop if output register full.
	Move	ACCUMB, DATAOUT	Send output character.

Figure 5.6 A program to read a character from the keyboard and send it to the display.

causes the microprocessor to wait until the output register in the interface becomes empty. The last Move instruction writes the character into the output register, DATAOUT.

5.3 SERIAL INTERFACES

There exist two basic serial transmission schemes, known as synchronous and asynchronous transmission. The vast majority of video terminals use asynchronous transmission, which is simple and inexpensive to implement, and is suitable for low to medium data transfer rates. In asynchronous transmission, data are transmitted in a standard arrangement known as the start-stop format.

Serial transmission schemes in general and the start-stop format in particular will be discussed in Chapter 6. For this discussion it suffices to know that in the start-stop format each transmitted character is preceded by one bit called the *start bit*, and succeded by one or two bits called the *stop bits*. The number of bits transmitted per character, exclusive of the start and stop bits, may be anywhere from 5 to 8, depending upon the character code used. In the case of ASCII characters, at least 7 bits are needed. The 8-bit format shown in Figure 5.4 is often used, because it matches the way in which characters are stored in the main memory of a microprocessor system.

We will now examine a specific example of an interface circuit that uses asynchronous transmission.

5.3.1 An Interface Example—the ACIA

Most functions of the interface circuit needed for asynchronous transmission can be implemented on a single integrated-circuit chip, called a *Universal Asynchronous Receiver Transmitter* (UART). Several such chips are available from different IC manufacturers. A well known example is Motorola's ACIA (Asynchronous Communications Interface Adapter), chip number 6551 [1,2]. It incorporates data, status, and control registers, as well as all the circuitry needed for parallel-to-serial and serial-to-parallel conversion and for handling the start-stop format used in asynchronous transmission. It was pointed out in Section 5.1 that devices connected to a serial port often use a connector defined by the RS-232-C Standard. The ACIA contains provisions for dealing with some of the RS-232-C signals.

The internal organization of the ACIA is depicted in Figure 5.7. The ACIA is connected to the microprocessor bus using the signals on the left side of the figure, which will be described in Chapter 6 in conjunction with the discussion of interface-circuit hardware details. The remaining signals are used to connect the I/O device and to provide the ACIA with the clock signals needed in the transmission and reception operations. There are separate paths for serial input data and serial output data. Data received at the serial input (RxD) are shifted into a shift register, IN-SR. When a complete character is received, it is transferred to the Receiver Data register (RX) where it is stored until it is read by the microprocessor. Meanwhile, if another character arrives, it can be shifted into the input shift register. The two registers

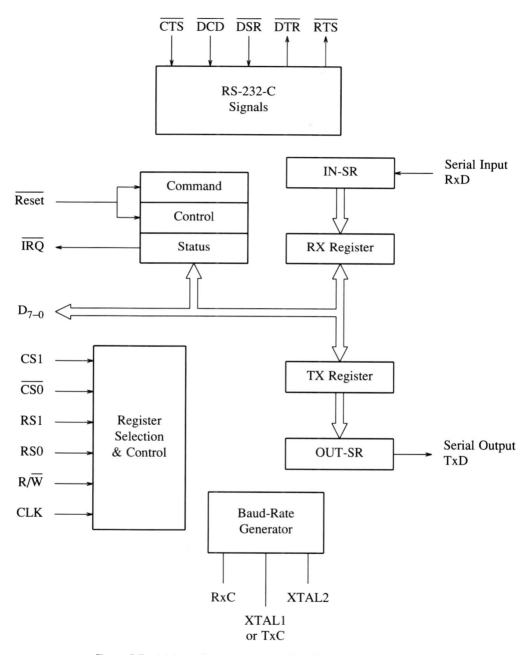

Figure 5.7 A block diagram representing the ACIA registers and signals.

IN-SR and RX constitute a double buffer. They enable the microprocessor to read the input characters correctly, even if they are being received one immediately after the other. If the character transmission time is T seconds, each character will be available in the RX register for T seconds before it is overwritten by the next character. Hence, characters will be received correctly, provided the microprocessor is able to read each character within T seconds from its arrival. The transmitter section of the interface comprises similar components. A double buffer consisting of the Transmitter Data register, TX, and an output shift register, OUT-SR, allows one character to be deposited in TX while another is being transmitted on the serial output line, TxD.

STATUS REGISTER

Status information for the receiver and the transmitter sections is available in the Status register, whose organization is shown in Figure 5.8. The Receiver-Full bit, RF, is set to 1 as soon as a character has been received and is available in the RX register. It is cleared when this character is read by the microprocessor. The state of the transmitter buffer is described by the Transmitter-Empty bit, TE, which is equal to 1 whenever register TX is empty. Note that the RF and TE bits are in bit positions 3 and 4 of the Status register. In the general discussion given earlier, bit positions 0 and 1 were used for similar functions (see Figure 5.5).

Three status bits, PE, FE, and VR, are set to 1 when certain error conditions are detected by the receiver hardware. The start-stop transmission format has a provision for using parity checking to detect transmission errors. The idea of parity and its use for detecting errors were introduced briefly in Section 3.9, and will be elaborated further in Section 6.7.1. When parity checking is used, bit PE indicates whether a parity error has been detected by the receiver section of the ACIA. It is set to 1 if parity checking is enabled and the parity of an incoming character is opposite to that expected. Another type of error that can be detected by the ACIA is called a *framing error*. This error means that the received character does not conform to the start-stop format. Bit FE is set to 1 when such an error is detected by

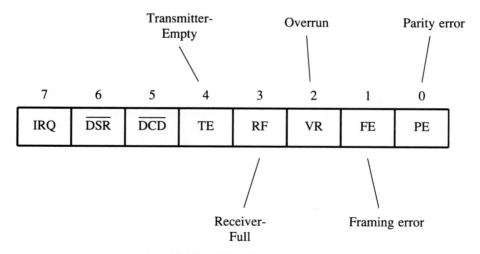

Figure 5.8   Organization of the Status register of the ACIA.

the receiver. The third error bit, VR, indicates an overrun condition. The receiver is said to have been overrun when a character is lost as a result of a new input character entering the RX buffer before the previous contents of this register have been read by the microprocessor. Note that because the receiver is double buffered, it may store up to two characters, one in RX and one in IN-SR, without an overrun condition.

Bit 7 of the Status register, IRQ, is set to 1 when the ACIA has an outstanding interrupt request. Interrupts are an alternative mechanism to the wait loops in Figure 5.6 for controlling input/output operations. They will be discussed in Chapter 7.

Finally, bits 5 and 6 of the Status register are used in conjunction with the signals of the RS-232-C Standard. They represent two input lines to the ACIA called Data Carrier Detect ($\overline{\text{DCD}}$) and Data Set Ready ($\overline{\text{DSR}}$). The use of these two lines, as well as other RS-232-C signals, will be discussed in Section 11.4.1. In the case of a simple connection to a video terminal, bits 5 and 6 may be ignored.

CONTROL REGISTER

In addition to the data buffers and the status register, the ACIA has two registers, called the Control register and the Command register, which serve to select the desired mode of operation. The speed with which the ACIA can send or receive data covers a wide range. The transmission speed, which is also called the *baud rate* (see Section 9.4.2), is determined by the frequency of the clock signal used in the parallel-to-serial or serial-to-parallel conversion process. It may be different for the sender and receiver sections. Some of the available options are summarized in Figure 5.9. The send and receive clock signals may be supplied externally, or may be derived from a circuit inside the ACIA called the *baud-rate generator*. When an external clock signal is used, its frequency should be 16 times the desired bit rate. The send clock should be connected to input TxC and the receive clock to input RxC in Figure 5.7. The baud-rate generator incorporates an oscillator whose frequency is determined by a crystal connected externally between points XTAL1 and XTAL2. The crystal causes the frequency of the oscillator to be accurate and stable over a wide range of operating conditions. Normally, a 1.8432 MHz crystal is used. The output signal of the oscillator is divided down in frequency to obtain the desired transmission baud rate, as specified in bits 0 to 3 of the Control register. When these bits are all set to 0, the baud-rate generator is ignored, and the transmission clock is derived from the external clock instead. Similarly, the receiver clock is selected by bit 4 of the Control register. It may be derived either from the baud-rate generator, in which case it is the same as the transmitter clock, or from the external clock connected to input RxC.

Bits 5 and 6 of the Control register select the number of data bits in each character transmitted or received, where a character may contain from 5 to 8 bits. This field should be used to select an appropriate word length, to match the characteristics of the I/O device connected to the ACIA. For ASCII devices, at least 7 bits are needed per character. Most devices use a word length of 8 bits. As will be explained in Section 6.7.1, each character is followed by one or two stop bits. The number of stop bits used is determined by bit 7 of the Control register.

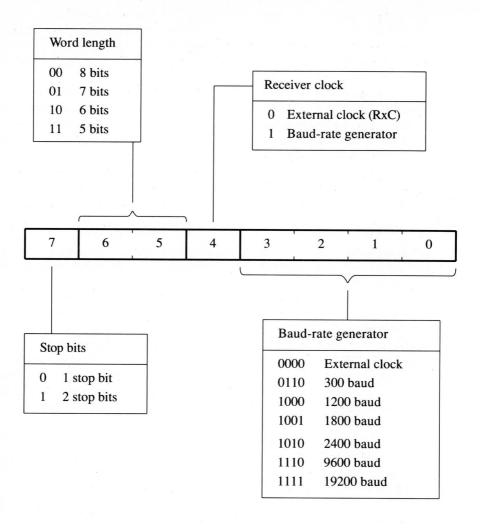

Figure 5.9   A partial description of the fields of the Control register of the ACIA.

COMMAND REGISTER

The Command register is shown in Figure 5.10. The receiver and transmitter sections of the ACIA are disabled when bit 0 of this register is equal to 0. This condition is indicated externally by an inactive state on an output signal called $\overline{DTR}$ (Data Terminal Ready), which is one of the RS-232-C signals. Bit 0 of the Command register must be set to 1 before the ACIA can be used to receive or transmit data. The next three bits, 1 to 3, may be used to enable interrupts, as will be discussed in Chapter 7. They also control the state of another RS-232-C signal called $\overline{RTS}$ (Request To Send). When interrupts are not used, bit 1 should be set to 1 and bits 2 and 3 to 00 or 10. Note that the two signals $\overline{DTR}$ and $\overline{RTS}$ have an overbar as

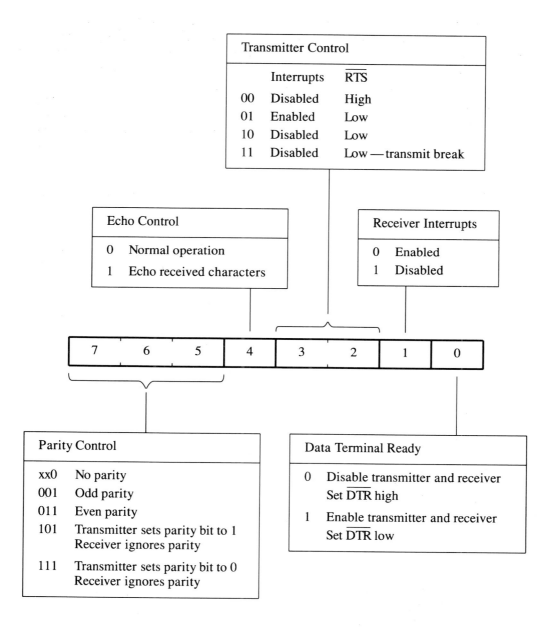

Figure 5.10 Command register of the ACIA.

part of their names. This is a commonly used convention, which indicates that the signal is active when in the low-voltage state.

The echo control bit, bit 4, of the Command register is used for testing purposes. When set to 1, it causes the ACIA to echo back any received data by copying the contents of the RX register into the TX register. As a result, data received on the RxD input will be transmitted on the TxD output. Bit 4 should be set to 0 during normal operation. Finally, bits 5 to 7 determine the way in which the parity of the transmitted and received data is handled by the ACIA. Bit 5 can be regarded as the parity-enable bit. When it is equal to 0, the ACIA does not generate parity bits for the characters it transmits, and does not expect parity bits in the characters it receives. Otherwise, parity bits are treated by the ACIA as specified by bits 6 and 7. The available options are described in the figure.

REGISTER SELECTION

Whenever the ACIA is addressed by the microprocessor, two register-select lines, RS0 and RS1 in Figure 5.7, determine which of the internal registers is to be accessed. These lines are usually connected to the least-significant address bits of the microprocessor bus. Thus, the internal registers appear as four adjacent locations in the address space of the microprocessor. Individual registers are selected during read and write operations according to Table 5.1. The two data registers are assigned the same address, 00; the receiver register is selected during a read operation and the transmitter register during a write operation. A read operation at location 01 reads the contents of the Status register. A write operation at this address is said to *reset* the ACIA. It loads the pattern 00000010 into the Command register, irrespective of the actual data sent by the microprocessor as part of the write operation. Examination of Figure 5.10 indicates that this pattern causes both the transmitter and the receiver to be disabled, and the $\overline{\text{DTR}}$ and $\overline{\text{RTS}}$ lines to become inactive. Activating the Reset input line of the ACIA (see Figure 5.7) has the same effect on the Command register. But, it also resets the Control register to an all-0 pattern. This is the starting state of the ACIA when power is turned on, because the Reset input is usually connected to a line by the same name on the microprocessor bus. This line is activated when power is turned on or when the microprocessor executes a reset instruction. It may also be activated manually by a pushbutton provided for this purpose. Before using the ACIA to transmit or receive data, the Command and Control

Table 5.1   REGISTER SELECTION IN ACIA

RS1	RS0	Read	Write
0	0	Read Receiver Data register, RX.	Write into Transmit Data register, TX.
0	1	Read Status register.	Reset the chip.
1	0	Read Command register.	Write into Command register.
1	1	Read Control register.	Write into Control register.

registers must be loaded with patterns that select the desired mode of operation. These two registers are assigned addresses 10 and 11, respectively, as shown in Table 5.1.

PROGRAMMING EXAMPLES

Figure 5.11 gives a programming example for a typical connection in which an ACIA is used as the interface between a video terminal and a microprocessor. The

```
RX          EQU         $CB00           Declare addresses of Data, Status,
TX          EQU         $CB00               Command and Control registers
STATUS      EQU         $CB01           of the ACIA.
CMND        EQU         $CB02
CNTRL       EQU         $CB03
*
* Initialization
*
            Move        #$1E,CNTRL      Initialize Control and
            Move        #$B,CMND            Command registers.
            .
            .
            .
            Call        RCVCHR          Get input character.
            Call        TRMCHR          Echo it back.
            .
            .
            .
*
* Subroutine to receive one character and place it in ACCUMA.
*
RCVCHR      Move        STATUS,ACCUMB   Check bit 3 of Status register.
            AND         #8,ACCUMB
            Branch_if_=0  RCVCHR        Loop if not ready.
            Move        RX,ACCUMA       Read data.
            Return_from_subroutine
*
* Subroutine to transmit the character in ACCUMA.
*
TRMCHR      Move        STATUS,ACCUMB   Check bit 4 of Status register.
            AND         #$10,ACCUMB
            Branch_if_=0  TRMCHR        Loop if not ready.
            Move        ACCUMA,TX       Send data.
            Return_from_subroutine
```

Figure 5.11 Sending and receiving characters through the ACIA.

ACIA is assumed to occupy the four address locations $CB00_{16}$ to $CB03_{16}$. Hence, its RX and TX registers can be accessed at address CB00, its Status register at address CB01, and so on. In the initialization section of the program, the first Move instruction selects the internal baud-rate generator as the clock source, and sets the baud rate to 9600 for both the transmitter and the receiver. It also selects a transmission format consisting of an 8-bit word followed by one stop bit. These choices are specified by the pattern $1E_{16}$, which is deposited in the Control register. By loading the pattern $0B_{16}$ into the Command register, the second Move instruction enables the receiver and the transmitter, disables interrupts, activates $\overline{RTS}$, and sets the ACIA's mode of operation to no echo and no parity.

The program uses two subroutines, one to receive and one to transmit a character. Each begins with a wait loop to test the respective status bit, followed by an instruction to transfer a byte of data. The main program calls subroutine RCVCHR whenever it wishes to receive input data from the terminal. The subroutine places the received character in ACCUMA. The second subroutine, TRMCHR, is then called to send the contents of ACCUMA to the output data register of the ACIA, thus causing the received character to be displayed on the terminal's screen.

Consider now an ACIA connected to a 6809 microprocessor, and assigned the same addresses as in the example of Figure 5.11. A program that reads one input line from the keyboard of a terminal connected to the ACIA is given in Figure 5.12. Input characters are deposited in the main memory starting at location LINE. They are also echoed back to the terminal, so that they may be displayed on the screen. When a Carriage Return character is encountered, indicating the end of the line, the program calls subroutine INTPRT, which is assumed to interpret the input line. The starting address, LINE, is passed to the subroutine as a parameter in index register X.

A similar program is given in Figure 5.13 for the 68000 microprocessor. Since the ACIA has 8 data lines, it may be connected to either the low- or the high-order byte of the 68000 bus. In either case, the register-select lines, RS0 and RS1, will be connected to address lines corresponding to address bits A_1 and A_2, respectively. Hence, these two bits should be set to select the desired register. If the ACIA is connected to the high-order byte of the bus, its internal registers will appear at successive even addresses, that is, addresses having $A_0 = 0$. If the ACIA is connected to the low-order byte, its registers will appear at successive odd addresses, having $A_0 = 1$. The addresses used in the program of Figure 5.13 assume that the ACIA occupies even address locations in the range D60000 to D60006. After initializing the ACIA Control and Command registers, the program reads input characters, deposits them in the main memory, and echoes them back to the terminal, as before.

5.4 PARALLEL INTERFACE

In its simplest form, a parallel interface does little more than provide a buffer between the microprocessor bus and an I/O device. In general, the word "buffer" refers to a system component that isolates one subsystem from another. It is needed

```
RX         EQU    $CB00      Receiver register.
TX         EQU    $CB00      Transmitter register.
STATUS     EQU    $CB01      Status register.
CMND       EQU    $CB02      Command register.
CNTRL      EQU    $CB03      Control register.
LINE       EQU    · · ·      Starting address for input data.
CR         EQU    $0D        ASCII code for Carriage Return.
*
* Initialization:
*
           ORG    $1000
           LDA    #$1E       Select transmission
           STA    CNTRL          format and speed.
           LDA    #$B
           STA    CMND       Enable ACIA.
           LDX    #LINE      Initialize pointer.
NEXT       LBSR   RCVCHR     Get character.
           STA    ,X+        Store it in memory.
           LBSR   TRMCHR     Echo it back.
           CMPA   #CR        Check for Carriage Return.
           BNE    NEXT
           LDX    #LINE      Pass parameter to subroutine.
           LBSR   INTPRT
           · · ·
*
* Subroutine to receive one character:
*
RCVCHR     LDB    STATUS     Check RF bit.
           ANDB   #8
           BEQ    RCVCHR     Loop if no character ready.
           LDA    RX         Read input character.
           RTS
*
* Subroutine to transmit one character:
*
TRMCHR     LDB    STATUS     Check TE bit.
           ANDB   #$10
           BEQ    TRMCHR     Loop if display not ready.
           STA    TX         Transmit character.
           RTS
           END
```

Figure 5.12 A 6809 program that reads one line from the keyboard and echoes it to the screen.

```
RX          EQU         $D60000             Receiver register.
TX          EQU         $D60000             Transmitter register.
STATUS      EQU         $D60002             Status register.
CMND        EQU         $D60004             Command register.
CNTRL       EQU         $D60006             Control register.
LINE        EQU         . . .               Starting address for input data.
CR          EQU         $D                  ASCII code for Carriage Return.
*
* Initialization:
*
            ORG         $1000
            MOVE.B      #$1E,CNTRL          Select transmission
*                                               format and speed.
            MOVE.B      #$B,CMND            Enable ACIA.
            MOVEA       #LINE,A0            Initialize pointer.
NEXT        BSR         RCVCHR             Get character.
            MOVE.B      D0,(A0)+            Store it in memory.
            BSR         TRMCHR             Echo it back.
            CMP.B       #CR,D0             Check for Carriage Return.
            BNE         NEXT
            BSR         INTPRT
            . . .

*
* Subroutine to receive one character:
*
RCVCHR      BTST.B      #3,STATUS           Check RF bit.
            BEQ         RCVCHR             Loop if no character ready.
            MOVE.B      RX,D0              Read input character.
            RTS
*
* Subroutine to transmit one character:
*
TRMCHR      BTST.B      #4,STATUS           Check TE bit.
            BEQ         TRMCHR             Loop if display not ready.
            MOVE.B      D0,TX              Transmit character.
            RTS
            END
```

Figure 5.13 A 68000 program that reads one line from the keyboard and echoes it to the screen.

when the two subsystems have different electrical or timing characteristics. For example, they may use different voltages to represent 0s and 1s, or one subsystem may require higher current than can be supplied by the other. A buffer may also store information temporarily, while the information is being transferred from one subsystem to the other. The presence of a storage buffer enables the two subsystems to operate independently. In particular, they may operate at different speeds.

Consider a simple input device consisting of eight switches connected to a parallel interface circuit, as shown in Figure 5.14. These switches may be on the control panel of a measuring instrument or a microwave oven, or they may represent the output of sensors that detect overheating, excessive pressure, open or closed doors, and so forth, at various locations in a plant. The state of each switch, ON or OFF, is translated into a logic signal using resistors R_0 to R_7, which are connected to a +5 volt power supply. The voltage at each of the eight points P_0 to P_7 is equal to either 0, which represents logic 0, or +5 V, which represents logic 1, depending upon whether the corresponding switch is closed or open, respectively.

The combined state of the eight switches represents 8 bits of information, which may be read by the microprocessor as an 8-bit byte. For example, if SWITCHES is the address assigned to this interface, the instruction:

$$\text{Move} \qquad \text{SWITCHES, ACCUMA}$$

will read the input byte and deposit it in accumulator A. A 0 will be read at any bit position where the switch is closed and a 1 where the switch is open. For example, if switches SW_0, SW_3 and SW_5 are closed and the rest are open, the data byte read

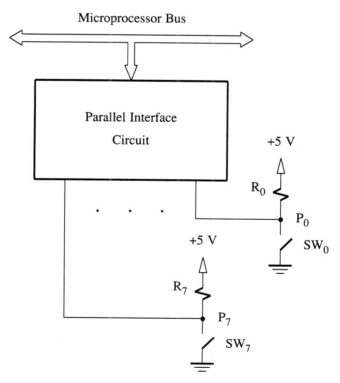

Figure 5.14 A parallel interface connecting 8 switches to a microprocessor.

by the microprocessor will be $D6_{16}$ ($=11010110_2$). Let us assume that the interface circuit does not contain a storage register. It merely provides a path between the switches and the bus, and a mechanism to recognize the address assigned to them. Hence, the value read depends on the setting of the switches at the time the Move instruction is executed.

Consider now a parallel output interface. Again we start by examining a very simple, yet realistic, example. Figure 5.15(a) shows a familiar seven-segment display, found in calculators, cash registers, measuring instruments, and numerous other devices. Many such displays use a light-emitting diode (LED) as a source of light. An LED is a semiconductor component that emits light when electrical current passes through it. Assume that eight LEDs provide the seven segments of the display and a decimal point, and that the LEDs are connected as shown in Figure 5.15(b). Part (c) of the figure shows how switches may be used to control the state of the display. Closing a switch causes the corresponding segment to be lit. The magnitude of the current that flows through the LEDs is determined by resistors R_0 to R_7, whose values should be chosen according to the required current level of the display.

The seven-segment display may be connected to a microprocessor via a parallel interface, as shown in Figure 5.16. In this case a storage buffer is needed to enable a given pattern to be displayed for a long time, independent of the activity on the microprocessor bus. This buffer constitutes the output data register of the interface. Its outputs, Q_0 to Q_7, replace the switches of Figure 5.15(c). Thus, segment i will light up when the voltage at Q_i is near ground, that is, when the corresponding bit in the register contains a 0. For example, if the value 49_{16} ($=01001001_2$), is deposited in the output register, the digit 2 will be displayed.

Only a single instruction is needed to transfer data to the seven-segment display. The instruction

$$\text{Move} \qquad \#\$49,\text{DISPLAY}$$

initiates a write operation to transfer the binary pattern 49_{16} to the interface. It will cause the digit 2 to be displayed. If later in the program the instruction

$$\text{Move} \qquad \#\$F1,\text{DISPLAY}$$

is executed, the digit 2 will be replaced by 7, and so on.

The discussion above deals with very simple devices. In the following subsections, we will introduce some features needed to support more complex I/O transfers. We begin with the important question of synchronizing the activities of the microprocessor with external events.

5.4.1 Synchronization of I/O Transfers

In Section 5.2, it was pointed out that it is often necessary to coordinate the time at which input or output instructions are executed with the occurrence of some external events. Consider, for example, a simple keyboard consisting of eight switches similar to those in Figure 5.14, where each switch is of the pushbutton type. In order

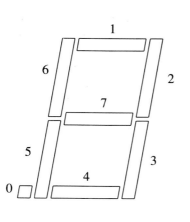

(a) Arrangement of segments

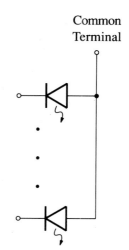

(b) Equivalent circuit

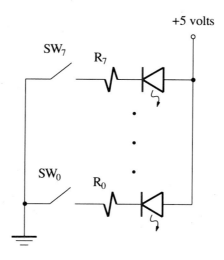

(c) Connection example

Figure 5.15 A seven-segment display.

to use this keyboard as an input device, the microprocessor should be able to detect that a key has been activated. The required indication may be obtained by observing

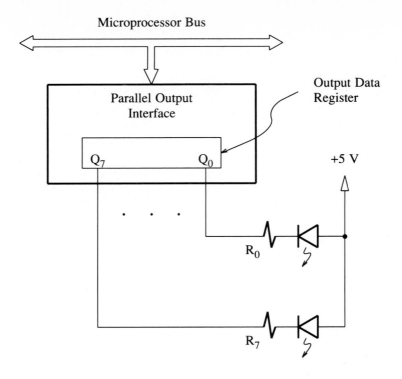

Figure 5.16 A seven-segment display (8 LEDs) driven by a parallel interface circuit.

that all eight bits are equal to 1 when none of the keys is depressed. Hence, the microprocessor may repeatedly read the state of the input port until it finds that one of the eight bits is equal to 0.

The next consideration is to ensure that when a key is activated, it is read only once. In the program of Figure 5.6, which uses the interface defined in Figure 5.5, this goal was easily achieved using the input status bit. The input status bit is set to 1 when a character is received in the input data register, and it is automatically cleared as soon as the contents of the input register are read by the microprocessor. The microprocessor will not read this register again until a new character is received and the status bit is set to 1 once more. Without such a facility, the job of ensuring that an input character is read only once must be done in software. For example, after one of the keys is activated and the input byte is read, we may repeatedly check the input byte until it returns to all 1s, indicating that the key has been released.

Another problem arises from the nature of the operation of mechanical switches. Almost all mechanical contacts are subject to vibrations, or bouncing, while the switch is being opened or closed. Typically, a switch requires about 2 ms to settle to a fully open or a fully closed position. While this may seem like a short time, it is sufficiently long for a microprocessor to execute several hundred instructions. Thus, a mechanism is required to enable the microprocessor to ignore the keyboard until switch bouncing has subsided. This may be done by checking that the input byte remains the same for a few milliseconds before accepting it as valid. Switch bouncing may also be dealt with in hardware, as will be discussed in Section 6.4. The mechanism used, whether in software or hardware, is referred to as *switch debouncing*.

Figure 5.17 shows an example of a program that reads the input byte corresponding to one key being depressed and then released. The loop starting at WAIT is executed repeatedly until one of the input bits becomes equal to 0. The next section of the program places the input byte in accumulator A and initializes a counter location, COUNTER, to some value, N. Then it repeatedly reads the input byte and checks that it is still equal to the contents of accumulator A. At the same time it keeps track of time by decrementing COUNTER. If due to bouncing the input byte returns to FF_{16}, a branch is made to WAIT, and the process is repeated. When bouncing ends, the CHECK1 loop will continue until the counter reaches 0. At this point the contents of accumulator A are accepted as a valid input byte and deposited into memory location INPUT.

The program ends with a loop that waits for the key to be released. Since switch bouncing may also occur while the switch is being released, the RELEASE loop ends only when the input byte is equal to FF_{16} for M successive passes through the CHECK2 loop. The values of N and M should be determined based on the time

WAIT	Compare	#$FF,KEYBRD	Wait for a key
	Branch_if_=0	WAIT	to be activated.
	Move	KEYBRD,ACCUMA	
	Move	#N,COUNTER	
CHECK1	Compare	ACCUMA,KEYBRD	If input changes
	Branch_if_≠0	WAIT	start again.
	Decrement	COUNTER	Repeat N times.
	Branch_if_≠0	CHECK1	
	Move	ACCUMA,INPUT	Accept input.

* Wait for key to be released.

RELEASE	Compare	#$FF,KEYBRD	Wait for key to
	Branch_if_≠0	RELEASE	be released.
	Move	#M,COUNTER	
CHECK2	Compare	#$FF,KEYBRD	Check for bouncing.
	Branch_if_≠0	RELEASE	
	Decrement	COUNTER	Repeat M times.
	Branch_if_≠0	CHECK2	

Next instruction

Figure 5.17 A program to read one switch activation from a simple keyboard, allowing for switch bouncing.

taken by the microprocessor to execute the respective loops once and on the delay needed for debouncing. For example, if one pass through the CHECK1 loop takes 10 μs and a delay of 4 ms is needed, N should be set to 4000.

I/O synchronization in this example is implemented totally in software. Clearly, the existence of a status bit, as in the case of the serial interface of Section 5.3, makes it possible to use much simpler I/O routines. A status bit may be incorporated in a parallel interface circuit if the I/O device is capable of generating a control signal that indicates when it is ready for data transfer. This signal can then be used to set the status flag in the interface. An example of this approach is given below, in conjunction with the discussion of input latching.

5.4.2 Input Latching

When the microprocessor reads an input byte from the interface of Figure 5.14, it obtains a value that corresponds to the setting of the switches at the time the input instruction is executed. With some devices, the information to be read by the microprocessor may be available only for a short time. In such a case, the input data should be stored, or *latched*, into an input buffer, so that it may be read later by the microprocessor.

To illustrate the use of input latching, consider a simple microprocessor-based instrument. Suppose that we desire to measure the speed of a car by counting the number of turns one of the wheels makes in one second, and have the microprocessor record and display the result. We will discuss only the measuring circuit here. A suitable sensor, perhaps optical or magnetic, near the wheel generates an electrical pulse each time the wheel completes one turn. The arrangement shown in Figure 5.18 may be used to repeatedly count the number of pulses generated in one-second intervals. The sensor's output is fed to a counter on the line labeled INC, such that each pulse increments the counter by one.

At the beginning of a measurement cycle, the counter is reset to 0 by a pulse from the Timer circuit. From that moment on, the contents of the counter are proportional to the distance traveled. After exactly one second the Timer sends a pulse on the Data-Valid line, indicating that the contents of the counter should be read at that instant. This is done by using the Data-Valid pulse to load the input data into a register in the interface circuit. At the same time, a status bit is set to 1 to indicate that the desired count is available to be read by the microprocessor. Since the contents of the counter have been saved, the Timer may now reset the counter and start another one-second measurement period.

The I/O routine for this simple speedometer, or frequency meter, waits for the status bit to become equal to 1, reads the contents of the input data register, then computes and displays the speed of the car. The interface circuit hardware should be designed such that the status bit is reset to 0 as soon as the input register is read, thus avoiding the possibility that the microprocessor may read the same input data more than once.

The difference between the parallel interface circuits of Figure 5.14 and Figure 5.18 should now be clear. In Figure 5.14, the microprocessor reads the information present on the input lines of the interface at the time the read operation takes place.

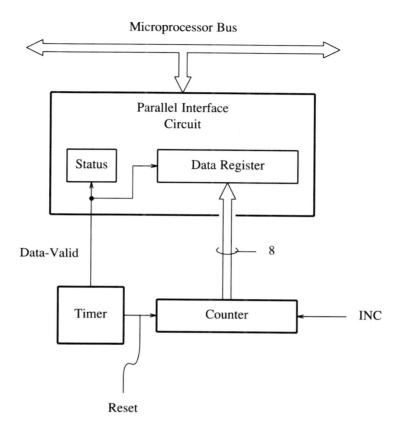

Figure 5.18 A parallel interface used for pulse-rate measurement.

In Figure 5.18, the microprocessor reads the contents of the input data register, which is the information that was available on the input lines at the time a pulse was received on the Data-Valid line. The data-latching feature used in Figure 5.18 is needed whenever the input data may change before being read by the microprocessor.

5.4.3 Handshake Signaling

The interface circuit of Figure 5.18 has one control input, Data-Valid, which indicates when the input data is valid and ready for transfer to the microprocessor. After issuing a Data-Valid pulse, the Timer resets the counter and starts another measurement cycle. It has been assumed implicitly that when the result of one measurement is ready, the microprocessor will have read the previous result. However, if the speed calculations are lengthy, or if the microprocessor is involved in other tasks, it may be necessary to delay a measurement cycle until the results of the previous one have been read. This means that another control signal is needed, to indicate to the

Timer that the contents of the input data register have been read by the microprocessor. The new control line, which we will call Data-Taken, is shown in Figure 5.19, and the sequence of events is illustrated in Figure 5.20. A measurement cycle ends at time t_1, and a pulse is sent on the Data-Valid line by the timer circuit. Arrows in timing diagrams of this kind describe the cause-effect relationship among various signals. The arrow starting at the rising edge of Data-Valid indicates that this transition causes the status bit to change from 0 to 1. At the same time the input data, which are not shown in the figure, are loaded into the data register. When the microprocessor completes reading the input data, at time t_2, the status bit is cleared and a pulse is sent by the interface circuit on the Data-Taken line. This pulse is a response indicating to the Timer that the input register is free. Upon receiving the Data-Taken pulse, the Timer issues a Reset pulse to the counter. A new measurement cycle begins at t_0, following the reset pulse.

As a result of the signaling exchange that takes place over the Data-Valid and Data-Taken lines, the microprocessor and the input device are interlocked in such a way that correct data transfer will take place, irrespective of the delays on either side.

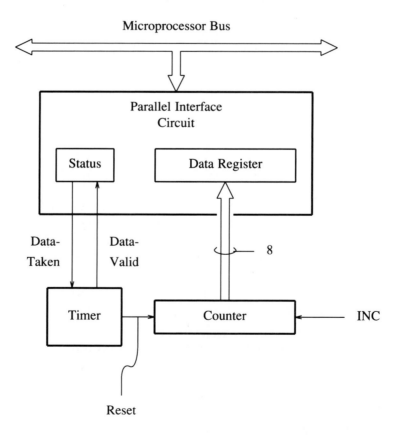

Figure 5.19 Addition of the response signal, Data-Taken.

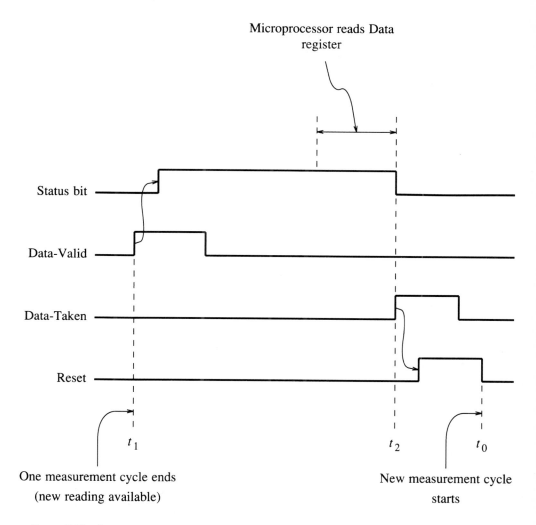

Figure 5.20 Sequence of events while transferring one reading from the counter in Figure 5.19.

Neither device will outrun the other. The Data-Valid and Data-Taken signals are said to perform a *handshake*. Similar command/response exchanges are frequently encountered in microprocessor systems whenever it is desired to coordinate the operation of two devices that are otherwise independent.

The control signal exchange depicted in Figure 5.20 consists of one pulse in each direction. Each pulse must be wide enough to be recognized by the receiving hardware. A similar but more reliable exchange that serves the same purpose is illustrated in Figure 5.21. In this case, after setting Data-Valid to 1 at time t_1, the Timer maintains the signal in that state until Data-Taken becomes equal to 1 at time t_2. The latter transition confirms that the Data-Valid signal has been recognized by

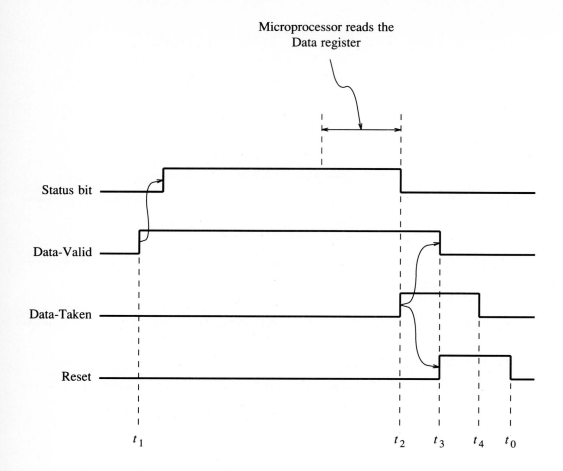

Figure 5.21   A full handshake between Data-Valid and Data-Taken signals.

the interface circuit. Hence, the Timer generates a reset pulse, and returns Data-Valid to 0 (t_3), and in response the interface circuit returns Data-Taken to 0 (t_4). A new measurement cycle begins at t_0, as before. As a result of the interlock between the Data-Valid and Data-Taken signals, correct data transfer will take place irrespective of circuit or transmission delays. This scheme is often called a *full handshake*. When pulses of fixed width are used, as in Figure 5.20, the exchange is sometimes referred to as a *pulse-mode handshake* or a *partial handshake*.

OUTPUT HANDSHAKE

To illustrate the use of handshake signaling in output operations, we will examine the case of a printer connected to a microprocessor via a parallel interface. Printers vary considerably in their speed of operation, from as low as 10 to upward of 10,000 characters per second. Therefore, it is essential to provide a mechanism that indicates to the microprocessor, or more accurately to the program controlling the printing operation, when the next character may be sent to the printer. The handshake is a simple and reliable method for synchronizing the operation of the two devices. A possible arrangement for connecting the printer to the microprocessor is shown in Figure 5.22. Since we are dealing with an output operation, the Data-Valid signal is generated by the interface circuit, and the response, Data-Taken, by the printer. A status bit, called Printer-Ready, is incorporated in the interface circuit to indicate the status of the printer. While the printer is idle, that is when it is ready to accept a new character, Printer-Ready should be equal to 1.

The sequence of actions needed to print one character is indicated in the timing diagram in Figure 5.23. When the microprocessor writes new data into the output data register, the interface clears its status bit and sends a pulse on the Data-Valid line. This happens at time t_1. Upon receiving this pulse, the printer starts the printing operation, which is completed at time t_2. The printer sends a pulse on the Data-Taken line to indicate that it has completed the requested operation, thus causing the status bit to be set to 1.

The Data-Valid and Data-Taken signals in Figure 5.23 perform a handshake similar to that of Figure 5.20, with the roles of the input and output control lines

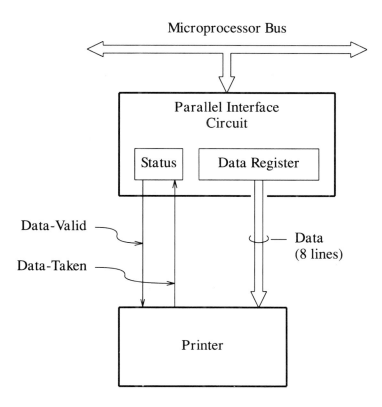

Figure 5.22 A printer connected via a parallel interface.

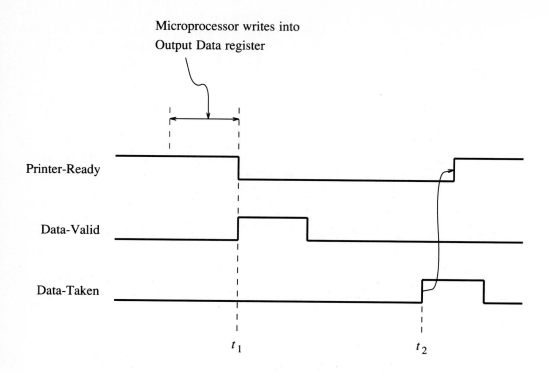

Figure 5.23 Pulse-mode output handshake between a parallel interface circuit and a printer.

interchanged. A full handshake equivalent to that of Figure 5.21 may also be used, as shown in Figure 5.24. As before, the main advantage of a full handshake relative to the pulse-mode handshake is increased reliability in the signal exchange between the interface circuit and the I/O device.

With the introduction of a status flag, I/O programming for a device connected to a parallel interface becomes the same as for a serial interface. A program to print a message on a printer would be essentially the same as a program that displays a message on a video screen. We should note that from the software point of view, it makes no difference whether a partial or a full handshake takes place in the hardware. The differences between the two types of handshake relate to the timing details of the signal exchange between the interface circuit and the I/O device.

5.5 MULTIFUNCTION PARALLEL INTERFACE

A computer interface is a circuit that translates and arranges the signals and information on a computer bus into a format that matches the characteristics of an input or an output device. We have discussed several of the features needed in an interface circuit for various applications. A considerable saving in cost and design effort can

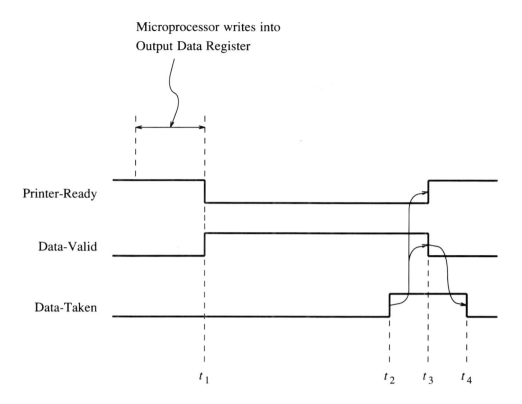

Figure 5.24 Full handshake between a parallel interface circuit and a printer.

be achieved if a general circuit is defined that can serve a range of devices. We will call an interface circuit that offers the desired flexibility a multifunction interface circuit. It is particularly useful if such a circuit is manufactured in the form of a single integrated circuit chip.

A multifunction interface circuit has several modes of operation. For example, it may provide both input and output functions, or may operate with or without input latching. At any given time, the way in which the interface behaves is determined by the contents of one or more internal control registers, which can be changed under program control. The ACIA described in Section 5.3.3 exemplifies a flexible chip for use in asynchronous serial interface circuits. Many of its operational parameters, such as the transmission speed and the use of parity checking, are determined by the information deposited in the Control and Command registers.

In this section, we will introduce examples of commercial integrated circuit chips that can serve as parallel interfaces in a variety of applications. The Parallel Interface Adapter (PIA), which is one of the support chips in Motorola's 8-bit microprocessor family, will be described first. We will then discuss a similar, but more flexible chip called a Versatile Interface Adapter (VIA). Both of these chips can also be used with 16-bit microprocessors such as the 68000.

Before discussing specific parallel interface chips, we will present one feature that is common to most of them. The individual data lines of each parallel port may be programmed independently to function either as input or as output lines, as explained below.

5.5.1 Combined Input/Output

The flexibility of an interface circuit can be enhanced by making it possible for a given port to be used for either input or output transfers, depending upon the application at hand. This flexibility may be achieved by providing circuitry for both input and output functions in each port, together with a mechanism to activate one or the other under program control. In this way, one program may use the port to read input data from a device. In a different program, or even later in the same program, the same port may be used to send data to an output device.

A typical arrangement is shown in Figure 5.25. The interface circuit incorporates three 8-bit registers: an input data register, an output data register, and a data direction register. The data direction register has one bit associated with each data line of the port, to specify the direction of transfers on that line. The line functions as an input line when the corresponding data direction bit is equal to 0, and as an output line when this bit is equal to 1. Hence, not all lines need to behave in the same way. For example, if the pattern $F0_{16}$ is deposited in the data direction register, the four high-order lines will serve as outputs, and the four low-order lines as inputs. Any data sent by the microprocessor will be loaded into the output data register. However, the data will appear only on those port lines programmed as outputs. A bit position defined as an input will have the output data register electrically disconnected from the port line, to enable an external device to transmit its input data. Since the contents of the data-direction register can be changed at any time, the configuration of the port for input and output operations is fully under program control.

In general, different data and control registers in a chip should be assigned distinct addresses, to enable program instructions to refer to them individually. However, the input and output data registers of a given port need not have separate addresses. The microprocessor uses the output data register to send data to an external device. It never needs to read the contents of this register. On the other hand, the microprocessor only needs to read the input data register. Therefore, these two registers may be assigned the same address. A read operation at that address results in transferring the contents of the input data register to the microprocessor, while a write operation places new data in the output data register.

We will now discuss in some detail the two commercial examples of parallel interface chips mentioned earlier: the PIA and the VIA.

5.5.2 Peripheral Interface Adapter (PIA)

The PIA is a 40-pin IC chip, manufactured by Motorola as part number 6821 [1, 2]. Its internal organization is depicted in Figure 5.26. Two parallel ports, A and B, are provided, each having eight data lines, PA_{7-0} and PB_{7-0}. Each port also has two

Microprocessor Bus

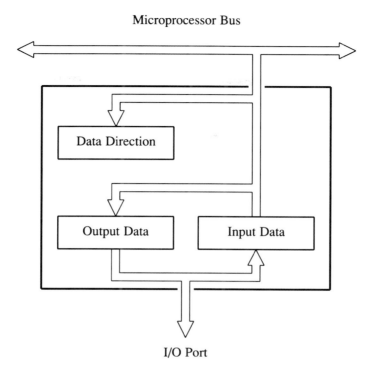

I/O Port

Figure 5.25 Combined input and output in a parallel interface circuit.

control lines, CA1 and CA2 for port A, and CB1 and CB2 for port B. Three addressable locations are associated with each port—a Data Direction Register (DDRA and DDRB), a Peripheral Register (PRA and PRB), and a Control Register (CRA and CRB). The two peripheral registers are used to transfer data to and from an I/O device. A read operation from PRA (PRB) transfers an input data byte to the microprocessor bus. A write operation to the same location loads the corresponding output register with a data byte from the bus. There is no input latching at the peripheral ports; that is, when either PRA or PRB is selected during a read operation, the data that are on the port data lines at that time are transferred to the microprocessor bus. However, a buffer register is provided for output operations. The Data Direction Registers enable individual port lines to be programmed for either input or output. Line PA_i (PB_i) operates as an output line if bit i in DDRA (DDRB) is set to 1. Otherwise it operates as an input line.

The PIA is connected to the microprocessor bus using the lines that appear on the left side of the figure. This connection will be discussed in detail in Chapter 6. However, for the purposes of the present discussion, we should note that the PIA has two register-select lines, RS0 and RS1, and thus it occupies only four locations in the address space of the microprocessor. Since there are six internal registers that need to be accessed, the register-select lines are not sufficient to completely specify the

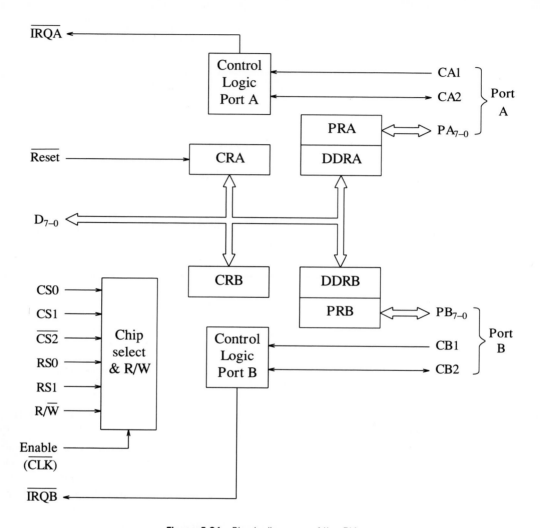

Figure 5.26 Block diagram of the PIA.

desired register during a read or a write operation. To overcome this difficulty, the Peripheral Register and the Data Direction Register for each port are assigned the same address, as shown in Table 5.2. Bit 2 of the port's control register, CRA or CRB, determines which of the two is selected. The organization of CRA is shown in Figure 5.27. Bits 0 to 5 of this register are control bits, which may be set by the software to select a particular mode of operation. Bits 6 and 7 are status bits used in association with the port control lines, CA1 and CA2, as will be explained shortly. When bit 2 of this register is equal to 1, RS1=RS0=0 refers to PRA. Otherwise, the same address refers to DDRA. A similar arrangement is used for port B.

Table 5.2   REGISTER SELECTION IN PIA

RS1	RS0	Location selected
0	0	Peripheral Register PRA, if $CRA_2 = 1$ Data Direction Register DDRA, if $CRA_2 = 0$
0	1	Control Register CRA
1	0	Peripheral Register PRB, if $CRB_2 = 1$ Data Direction Register DDRB, if $CRB_2 = 0$
1	1	Control Register CRB

The way in which the control lines are used is determined by the information in the corresponding control register. Let us examine port A in detail. Control line CA1 is an input line, which has an associated status bit, SA1, in register CRA. This bit is set to 1 when CA1 is activated. It is often convenient to call one of the two logic states of a control line the active state and the other the inactive or idle state. The voltage transition from the idle to the active state is called the *active edge* or *active transition* of the control signal. Depending on the details of the hardware, the active state of a line may be either the high-voltage or the low-voltage state. Hence, the active edge may be a low-to-high (also called positive) or a high-to-low (negative) transition. For CA1, the active edge is selected by bit 1 of the control register, as described in the figure. Thus, if bit 1 is equal to 0, an I/O device can cause SA1 to be set to 1 by changing the voltage on the CA1 line from high to low. Bit 0 of the control register determines whether an interrupt request is to be sent to the microprocessor at that time, as will be explained in Chapter 7. Bit SA1 is cleared when the peripheral register, PRA, is read by the microprocessor.

The mode of operation of the CA2 control line is determined by bits 3 to 5 of register CRA. This line may be used either as an input or as an output, as determined by bit 5 of CRA. When this bit is equal to 0, CA2 acts as an input line that controls the state of status bit SA2. An active transition, as defined by bit 4, sets SA2 to 1. Once set, SA2 remains equal to 1 until the microprocessor performs a read operation involving the port's data register, PRA. If it is desired that the PIA generate an interrupt request when CA2 is activated, bit 3 should be set to 1, as will be discussed in Section 7.2.

As an output line, CA2 may be used in conjunction with CA1 to implement the handshake scheme described in Section 5.4.3, to control input transfers on port A. This mode of operation is selected by setting bits 5 to 3 of CRA to 100. CA2 is normally in the high-voltage state, which is its inactive state. A typical sequence of events is illustrated in Figure 5.28, where CA1 is also assumed to be active when low. An input device places data on the data lines of port A and activates CA1, which serves as the Data-Valid line, at time t_1. As a result, status bit SA1 is set.

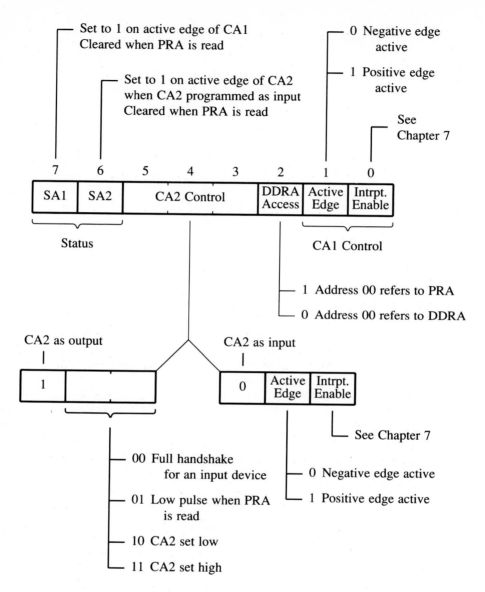

Figure 5.27 Organization of CRA, control register for port A of the PIA.

After the microprocessor has read the input data, the PIA activates CA2 by setting it in the low-voltage state (t_2). Thus, CA2 performs the function of Data-Taken. The PIA maintains CA2 in the active state until it receives another transition on CA1 (t_1'), indicating that new data is available. The sequence of events depicted in Figure 5.28 is called a "full handshake" in the manufacturer's literature. However, it is not truly a full handshake as defined in Figure 5.21, because CA1 returns to the inactive

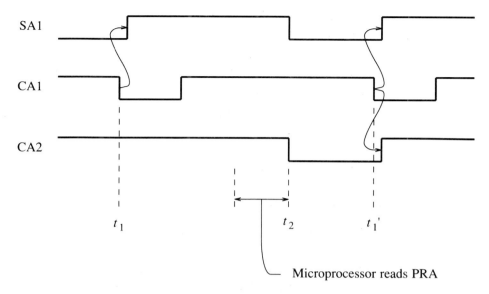

Microprocessor reads PRA

Figure 5.28 A full handshake using the PIA control lines.

state before CA2 becomes active. These and other signal timing details must be taken into account in the design of the hardware. They will be discussed in Section 6.6.3.

The CA2 line may be used to send a pulse to an I/O device. When bits 5 to 3 of CRA are set to 101, a low pulse is transmitted on CA2 every time the PRA register is read by the microprocessor. An important application of this mode is to implement the partial handshake of Figure 5.20, again using CA1 as the Data-Valid input from the I/O device and CA2 as the Data-Taken response. This mode is illustrated in Figure 5.29. The pulse on line CA2 has a fixed width, and is independent of the state of CA1. This contrasts with the handshake of Figure 5.28, where CA2 is maintained in the active (low) state until CA1 becomes active, indicating the start of a new transfer.

In the remaining two modes, CA2 acts as an output line whose state is independent of all other port lines. It is set high or low when bits 5 to 3 of register CRA are set to 110 or 111, respectively.

The organization of the control register of port B and the modes of operation available for control lines CB1 and CB2 are identical to those of port A, with one exception. The full handshake mode and the pulse mode apply to output instead of input operations. In the case of port B, the CB2 control line carries the Data-Valid signal, which is activated when the microprocessor writes new information into register PRB. The output device responds with Data-Taken on CB1. We should emphasize that ports A and B are restricted to input and output operations, respectively, only when one of these two modes is used. Otherwise, either port may be connected to an input or an output device.

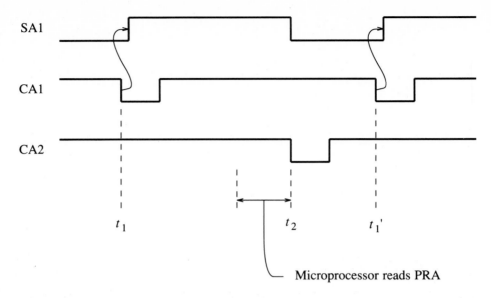

Figure 5.29 A pulse-mode handshake as implemented by the port control lines of the PIA.

RESET

All devices in a microprocessor system must start operation in a known initial state. As mentioned in Section 5.3.1, a reset line is usually provided on the microprocessor bus. It is activated whenever it is desired to put the system in its initial state, usually at the time power is turned on or after a software crash. The Reset input of the PIA should be connected to this line. Activating this input causes the control registers and the two data-direction registers to be filled with 0s. As a result, all bidirectional lines of the PIA ports will be configured as inputs. This is a safe starting state. If a port line is initially configured as an output, but it happens to be connected to an input device, then both the PIA and the input device will attempt to drive this line. The resulting conflict may lead to electrical damage. After the PIA is reset, those lines that are to be used as outputs should be configured as such, by depositing the appropriate information in the data-direction registers.

EXAMPLE

As an example, consider a printer connected to a 6809 microprocessor via a PIA. Since the printer requires handshake signaling during output operations, it should be connected to port B, as shown in Figure 5.30. The two control lines, CB1 and CB2, perform the Data-Taken and Data-Valid functions, respectively. As mentioned earlier, CB2 is always active when low. Hence, we have called the signal on the CB2 line Data-Valid. The other control line, CB1, may be active low or high. We have called this line Data-Taken, assuming that the printer responds with an active-low signal. In this case, bit 1 of CRB should be cleared to 0 to select the negative-edge option for CB1.

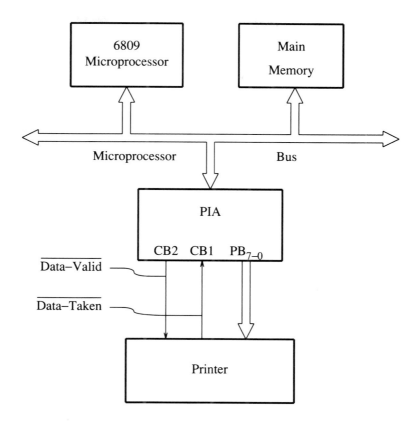

Figure 5.30 Connection of a printer to a 6809 microprocessor via a PIA chip.

Figure 5.31 gives a 6809 program that prints the results of a computation on the printer. The PIA is assumed to occupy the address range CF50 to CF53. At the beginning of the program, CRB is cleared to 0 to provide access to the data-direction register, DDRB. After setting DDRB to an all-1 pattern, to cause port B to function as an output port, CRB is loaded with the binary pattern 00100100, which selects the desired mode of operation for CB1 and CB2. Bit 1 is set to 0 to select negative-edge operation for CB1, and bits 5 to 3 are set to 100, assuming that the printer uses a full handshake. The information to be printed consists of a string of ASCII characters stored in successive byte locations in the main memory, starting at location RESLTS and followed by a null character, that is, an all-0 byte.

The program of Figure 5.31 is repeated in Figure 5.32 for a 68000 microprocessor. It was pointed out earlier that the internal registers of 8-bit devices appear at successive even or odd addresses, depending upon whether the device is connected to the high- or low-order byte of the 68000 bus (see the examples in Section 5.3.1). In the program of Figure 5.32, the PIA is assumed to occupy the four odd-address

```
          DDRB    EQU     $CF52       Declare register addresses.
          PRB     EQU     $CF52
          CRB     EQU     $CF53
                  CLRA                Clear CRB to 0 to select DDRB.
                  STA     CRB
                  LDA     #$FF        Set port B for output.
                  STA     DDRB
                  LDA     #$24        Select PRB, negative edge on CB1
                  STA     CRB            and full handshake.
                  LDX     #RESLTS     Initialize pointer.
          LOOP    LDB     ,X+         Get character.
                  BEQ     FINISH      Exit if null.
                  BRS     SEND        Otherwise print character.
                  BRA     LOOP        Repeat.
          FINISH  .
                  .
                  .
          *
```

Subroutine to print the character in accumulator B.
```
          *
          SEND    LDA     CRB         Test SB1 (CRB7).
                  ANDA    #$80
                  BEQ     SEND        Wait if ≠ 1.
                  STB     PRB         Send character.
                  RTS
```

Figure 5.31 A 6809 program that sends a message to a printer via port B of the PIA.

locations in the range EA0001 to EA0007. Otherwise, the program uses the 68000 instructions to perform the same operations as the program of Figure 5.31.

5.5.3 Versatile Interface Adapter (VIA)

Our second example of a flexible parallel interface is the VIA. This is a 40-pin IC manufactured by Synertek as part number 6522 [2]. It is similar to the PIA described in the previous section, but it provides more features and flexibility. Its internal organization is shown in Figure 5.33. The VIA provides two 8-bit parallel ports, PA and PB, and four associated control lines, CA1, CA2, CB1, and CB2. The facilities for data transfer over these ports are essentially the same as in the PIA, though they differ in some details as will be described shortly. In addition, the VIA contains two timer circuits and a shift register. The timers may be used in applications involving precise time intervals, as in the frequency counter example in Section 5.4.3. Since the timers are used often in conjunction with interrupts, they will be described in Chapter 7. The shift register provides a limited capability for serial I/O using control lines CB1 and CB2. Its operation will also be described in Chapter 7.

```
DDRB      EQU        $EA0005         Declare register addresses.
PRB       EQU        $EA0005
CRB       EQU        $EA0007
          CLR.B      CRB             Clear CRB to 0 to select DDRB.
          MOVE.B     #$FF,DDRB       Set port B for output.
          MOVE.B     #$24,CRB        Select PRB, negative edge on CA1,
*                                      and full handshake.
          MOVEA      #RESLTS,A0      Initialize pointer.
LOOP      MOVE.B     (A0)+,D0        Get character.
          BEQ        FINISH          Exit if null.
          BSR        SEND            Otherwise print character.
          BRA        LOOP            Repeat.
FINISH    .
          .
          .
*
Subroutine to print the character in accumulator B.
*
SEND      BTST.B     #$80,CRB        Test SB1 (CRB7).
          BEQ        SEND            Wait if ≠ 1.
          MOVE.B     D0,PRB          Send character.
          RTS
```

Figure 5.32   A 68000 program that sends a message to a printer via port B of the PIA.

The VIA has a total of 18 internal registers. Of these, 11 are shown explicitly in Figure 5.33. There is an input register, an output register, and a data-direction register associated with each port. The shift register mentioned above is associated with port B, and four registers containing status and control information are common to both ports. The remaining seven registers are associated with the two timer circuits. The VIA has four register-select lines, RS0 to RS3, which are used to address the registers in Figure 5.33 according to Table 5.3. Note that the input and output registers for each port have the same address. The input registers, IRA and IRB, are selected during read operations, and the output registers, ORA and ORB, during write operations. Individual port lines can be programmed as input or output lines by setting the corresponding bit in the data-direction register, DDRA or DDRB, to 0 or 1, respectively.

Three of the four control and status registers of the VIA are described below. The fourth, IER, will be discussed with interrupt-controlled I/O in Chapter 7. The control registers used for setting the mode of operation of the port lines are called the Peripheral Control Register (PCR), and the Auxiliary Control Register (ACR). The status register is called the Interrupt Flag Register (IFR). Although its name suggests that it is used in conjunction with interrupts, it in fact contains the status information needed for any I/O operation. The formats of the three control and status registers are given in Figure 5.34, and the functions of various fields within them are described in Tables 5.4 to 5.6. Table 5.4 describes the modes of operation of control

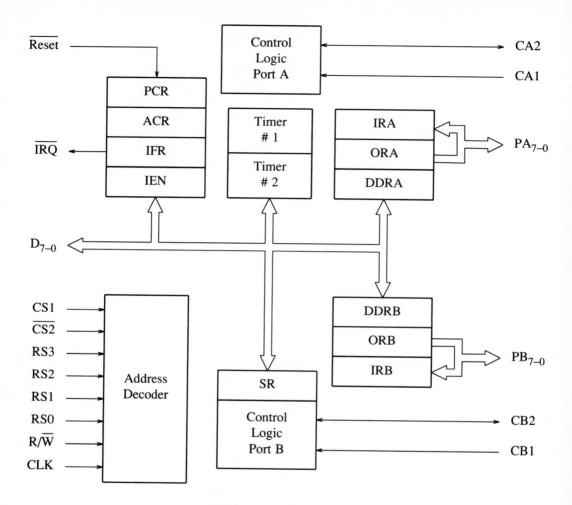

Figure 5.33   Internal organization of the VIA.

lines CA1 and CA2 of port A. A particular mode of operation is selected by fields MA1 and MA2 of the PCR register. Fields MB1 and MB2 determine the mode of operation of the control lines of port B. The two ports behave in essentially the same manner, except for the differences mentioned in footnotes 2 and 3 in the table, which will be discussed shortly. Table 5.5 describes the functions controlled by register ACR. As mentioned earlier, the timer facility and serial transmission using the shift register in the VIA will be discussed in Chapter 7. Finally, Table 5.6 lists the status bits in register IFR. Four of these bits, IFR_{1-0} and IFR_{4-3} are used during data transfers through port A and port B, in conjunction with the port control lines. Their behavior will be described in detail below.

For each setting of the MA1 and MA2 control fields, Table 5.4 gives the mode of operation of the control line affected and the conditions under which the

Table 5.3 REGISTER SELECTION IN VIA

RS3-0	Operation*	Selected register
0000	Read	IRB — Input register, port B
0000	Write	ORB — Output register, port B
0001	Read	IRA — Input register, port A
0001	Write	ORA — Output register, port A
0010	x	DDRB — Data direction register, port B
0011	x	DDRA — Data direction register, port A
0100 - 1001	x	Timer registers — see Table 7.1
1010	x	SR — Shift register
1011	x	ACR — Auxiliary Control register
1100	x	PCR — Peripheral Control register
1101	x	IFR — Interrupt Flag Register (status register)
1110	x	IEN — Interrupt Enable register

* The symbol "x" means that the register indicated is selected during both read and write operations.

corresponding bit in the status register is set and cleared. The first two rows describe the effect of the MA1 field, consisting of bit PCR_0, on the mode of operation of the CA1 control line and the SA1 status flag in register IFR. Setting MA1 to 0 or 1 causes CA1 to be considered active when low or high, respectively. The reception of an active transition on CA1 will cause the SA1 flag to be set to 1. Also, if input latching is enabled (see Table 5.5), the data on lines PA_{7-0} will be loaded into the input data register, IRA. Hence, bit LA (ACR_0) should be set to 1 if it is desired to capture the data on the port lines at the time CA1 is activated. When LA is equal to 0, the only effect of an active transition on CA1 is to set the corresponding status flag, SA1. In either case, the SA1 flag remains set until it is cleared by a program instruction that reads IRA or writes into ORA. This flag may also be cleared directly by writing a 1 into bit position 0 of the IFR register.

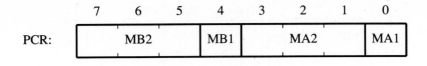

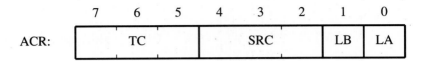

(a) Control registers PCR and ACR

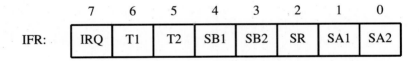

(b) Status register IFR

Figure 5.34 Control and status registers of the VIA.

The remaining entries in Table 5.4 describe the available modes of operation for the CA2 control line and its associated status bit, SA2. Any of the first four entries, MA2 = 000, 001, 010, or 011, causes the Port Control Logic to treat CA2 as an input line. The active edge, which sets the SA2 flag, can be selected as either a positive or a negative transition. Two options are also available for clearing the status bit. If MA2 is set to 000 or 010, the SA2 flag is cleared as a result of a read or a write operation involving the data registers of port A or by depositing a 1 in bit IFR_0. Alternatively, MA2 may be set to 001 or 011, in which case the status flag can be cleared only by writing a 1 into IFR_0. Reading the contents of register IRA or writing into ORA will have no effect on the status bit when either of these two modes is selected.

In the remaining four modes, MA2 = 100 to 111, CA2 becomes an output line. The available options, which are summarized in Table 5.4, are exactly the same as their counterparts in the PIA. The reader is referred to Figure 5.27 and the discussion in Subsection 5.5.2 for details.

Table 5.4   REGISTER PCR: MODES OF OPERATION OF PORT A[1] CONTROL LINES OF THE VIA CHIP

Control Field		Control Line		Status Flag		
Name	Value	Name	Mode of Operation	Name	Set	Cleared
MA1	0	CA1	Input; negative edge active.	SA1	High-to-low transition on CA1.	Reading IRA, writing into ORA, or writing a 1 in bit IFR_1.
	1		Input; positive edge active.		Low-to-high transition on CA1.	
MA2	000	CA2	Input; negative edge active.	SA2	High-to-low transition on CA2.	Reading IRA, writing into ORA, or writing a 1 in bit IFR_0.
	001		Input; negative edge active.		High-to-low transition on CA2.	Writing a 1 in bit IFR_0.
	010		Input; positive edge active.		Low-to-high transition on CA2.	Reading IRA, writing into ORA, or writing a 1 in bit IFR_0.
	011		Input; positive edge active.		Low-to-high transition on CA2.	Writing a 1 in bit IFR_0.
	100		Output; active low; full handshake[2].			
	101		Output; pulsed low[3]; partial handshake possible[4].			
	110		Output; set to low.			
	111		Output; set to high.			

[1] Port B behaves the same way, except as noted.

[2] CA2 goes low following a read IRA operation, and high at the next active edge on CA1. CB2 behaves the same way, but for a write ORB operation.

[3] A low pulse is transmitted on CA2 for one clock period following a read IRA operation. CB2 behaves the same way, but for a write ORB operation.

[4] CA1 (CB1) and CA2 (CB2) operate independently in this mode. But, together they may be used to implement a partial handshake.

Table 5.5   REGISTER ACR: INPUT LATCHING, SHIFT-REGISTER AND TIMER CONTROL IN THE VIA

Control Field		Description
LA	0	Select unlatched inputs during a read operation at port A.
	1	Select latched inputs during a read operation at port A.
LB	0	Select unlatched inputs during a read operation at port B.
	1	Select latched inputs during a read operation at port B.
SRC		Shift register control (see Chapter 7)
TC		Timer Control (see Chapter 7)

Table 5.6   STATUS AND INTERRUPT FLAGS OF THE VIA (REGISTER IFR)

Bit	Name	Description
0	SA2	See Table 5.4.
1	SA1	See Table 5.4.
2	SR	Status flag for the shift register.
3	SB2	See Table 5.4.
4	SB1	See Table 5.4.
5	T2	Status flag for timer T2.
6	T1	Status flag for timer T1.
7	IRQ	Interrupt flag.

The mode of operation of the control lines of port B are determined by fields MB1 and MB2 of register PCR, and the corresponding status bits are SB1 and SB2 of register IFR. Control lines CB1 and CB2 behave in the same manner as CA1 and

CA2, respectively, except in the two handshake modes. When field MB2 is set to 100 or 101, line CB2 is activated following a write instead of a read operation. Hence, when it is desired to use either of these two modes for controlling the transfer of data between the VIA and an I/O device, an input device should be connected to port A, while an output device should be connected to port B.

As in the case of the PIA, the VIA has a reset input called $\overline{\text{Reset}}$. Activating this line causes all status, control and data-direction registers to be filled with 0s. As a result, all bidirectional lines are configured as inputs.

As seen from the above description, the VIA offers many options to the user. Each of its two ports may be configured for input, output or both. Input may be either latched or direct. The control signals associated with each port may operate in any of a number of modes, including the highly reliable handshake mode. A particular mode of operation is selected via the control registers, according to Tables 5.4 and 5.5.

5.5.4 Application Examples

An example involving the connection of a printer to a microprocessor through a PIA chip was given in Subsection 5.5.2. The reader should rewrite the programs of Figures 5.31 and 5.32 for the VIA. We will now discuss two more examples. The first is a program that reads the frequency meter introduced in Subsection 5.4.3, and displays the result on a single-line display. The second example uses control line CB2 as an output line.

The frequency meter of Figure 5.19 is an input device that uses handshake signaling to transfer data to the interface circuit. Hence, it should be connected to port A of the VIA. We will rename the two handshake signals $\overline{\text{Data-Valid}}$ and $\overline{\text{Data-Taken}}$, assuming that they are both active when low. They should be connected to CA1 and CA2, respectively, as shown in Figure 5.35. The single-line display is connected to port B. It consists of several digits, similar to the displays found in hand-held calculators. The control circuit associated with the display consists of an 8-bit-wide shift register. The ASCII code for a digit to be displayed is entered into the least-significant digit position when a low pulse is received on the $\overline{\text{New-Digit}}$ line. If a second pulse is received, the first digit is shifted one place to the left, and the new digit is entered into the least-significant position, and so on. The $\overline{\text{New-Digit}}$ line is connected to CB2, which should be programmed to operate in the pulse mode. Each pulse on this line will cause the display to be shifted one digit to the left. We will assume that the display is cleared at the time power is turned on. It may also be cleared at any time by the software by shifting 0s into it.

A 6809 program that reads the counter and displays the result is given in Figure 5.36. At the beginning of the program, port B is configured as an output port by setting its data-direction register to all 1s. Then, the appropriate patterns are deposited into the control registers, PCR and ACR, to select the desired mode of operation for the port control lines. Full handshake and latched input are selected for port A, and pulse-mode operation for port B. All port-control lines are set to be active when low. The remainder of the main program consists of an infinite loop that repeatedly

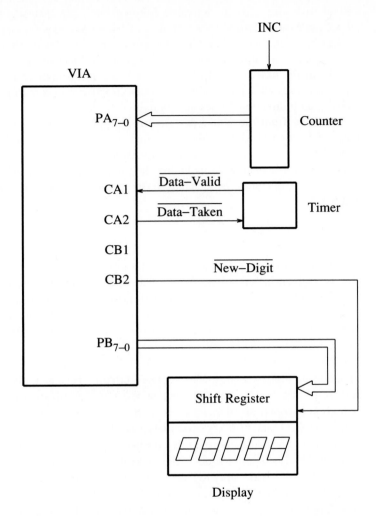

Figure 5.35 A frequency meter and a single-line display connected to a microprocessor by means of a VIA.

reads the counter, converts the count to a decimal representation and sends it to the display. Two subroutines are used in the process, DIVIDE and DISPLY.

At the beginning of the loop, the program waits for the status bit, SA1, to become equal to 1, then reads the pulse count from the input register and puts it in accumulator A. The value read from the counter is an 8-bit number in the range 0 to 255. Hence, its decimal representation has three digits, for the hundreds, tens, and units positions. To obtain these digits, we have used a simple approach based on the following steps:

```
* Initialize the VIA.
                LDA     #$FF        Set port B for output.
                STA     DDRB
                LDA     #$A8        Full handshake on A and pulse
                STA     PCR            mode on B.
                LDA     #1          Latched input on port A.
                STA     ACR
*
* Main program loop.
                ORG     $1000
                LDU     #$1000      Initialize user stack pointer.
WAIT            LDA     IFR
                ANDA    #2          Check bit SA1 (IFR₁).
                BEQ     WAIT
                LDA     PRA         Get count and put it in Accum A.
                LDB     #100        Push digit weight on user stack.
                PSHU    B
                BRS     DIVIDE      Divide by 100.
                BRS     DISPLY      Display 100s' digit (which is in Accum B).
                LDB     #10         Divide by 10.
                STB     ,U          Update the value at the top of the stack.
                BSR     DIVIDE
                BSR     DISPLY      Display 10s' digit.
                TFR     A,B         Put remainder in B.
                BSR     DISPLY      Display units' digit.
                LEAU    1,U         Remove top element from the stack.
                BRA     WAIT        Get next measurement.
*
* Division subroutine — counts the number of times the divisor, which is at the top of the user
* stack, goes into the dividend (in Accum A).  It leaves remainder in A and quotient in B.
DIVIDE          CLRB                Initialize quotient to 0.
LOOP            CMPA    ,U
                BLO     RETURN
                SUBA    ,U          Subtract divisor from dividend.
                INCB                Increment quotient.
                BRA     LOOP
RETURN          RTS
*
* Subroutine to display value in Accum B, after converting it to ASCII.
DISPLY          ORB     #$30        Convert to ASCII.
                STB     PRB         Send to port B.
                RTS
                END
```

Figure 5.36 A 6809 program to operate the frequency meter in Figure 5.35.

1. Get the quotient and remainder resulting from dividing the count by 100.

2. The quotient, which will be in the range 0 to 2, is the most-significant digit of the decimal representation. Convert it to ASCII and send it to the display.

3. Divide the remainder by 10.

4. The new quotient is the 10s' digit. Convert it to ASCII and send it to the display.

5. The remainder will now be in the range 0 to 9. It is the desired digit for the unit's position. Convert it to ASCII and send it to the display.

Division is a difficult operation to perform when using the 6809 microprocessor. Because we are dealing with small integers, the program in Figure 5.36 takes a very simple approach. To divide a number N by another number K, we simply count the number of times K goes into N, by repeatedly subtracting the divisor, K, from the dividend, N, until the remainder is smaller than K. In the program of Figure 5.36, the count (the dividend) is in accumulator A. The value by which it is to be divided is pushed on the top of the user stack, U. Subroutine DIVIDE repeatedly subtracts the value at the top of the stack from accumulator A, and uses accumulator B to keep track of the number of times the subtraction operation is performed. Thus, upon return from the subroutine, accumulators A and B contain the remainder and the quotient, respectively. The main program calls the DISPLY routine, which sends the digit in accumulator B to port B. Before doing so, it converts it to the equivalent ASCII code by setting bits 4 and 5 to 1.

Note that the digits are sent to the display starting with the most-significant digit. Because the pulse mode has been selected, each write operation into port B causes a pulse to be sent to the display on line CB2. This pulse loads a new digit into the right-most position, and causes higher-order digits to be shifted to the left in the display's shift register. Thus, the digits will be displayed in the correct order. Also, no wait loop is included in subroutine DISPLY, because the display hardware does not generate any response signal.

The reader will recall that in the pulse mode the two control lines, CB1 and CB2, are independent. In this example we have used CB2 alone to send pulses to the display, while in other applications the pulse mode may be used to implement a partial handshake with an output device. In the latter case, the software should be organized to wait for a response on CB1, i.e., for status bit SA1 to be set, before writing new information into register PRB.

Consider now an application that is quite different from the familiar input and output examples. Suppose that we want to generate a square wave signal whose frequency can be changed under program control. Such a signal may be used to record binary data on an audio cassette recorder. In a popular scheme, known as the *Kansas City interface*, two tones, having frequencies of 1200 Hz and 2400 Hz, are used to represent 0s and 1s, respectively. Since both tones are within the audible range, they can be easily recorded on an audio magnetic tape. It is possible to use one of the output lines of the VIA to generate the required tones. Let us use CB2 for this purpose. The last two entries in Table 5.4 indicate that CB2 can be set either high or low by depositing the appropriate patterns in the MB2 field of register PCR. Thus, in order to produce a signal of frequency f, the following sequence of steps should be performed repeatedly:

1. Set CB2 low.
2. Wait for a period of $1/2f$.
3. Set CB2 high.
4. Wait for a period of $1/2f$.

The waiting periods may be implemented in the program by executing a delay loop an appropriate number of times. To change the frequency of the output signal, we simply change the wait period by changing the number of times the delay loop is executed.

A simple program that implements this scheme using a 68000 microprocessor is given in Figure 5.37. The VIA is assumed to occupy the 16 even addresses in the range DA2000 to DA201E. When referring to Table 5.3 to determine the address of a particular register, it should be remembered that the values applied to the register-select lines, RS0 to RS3, are the same as address bits A_1 to A_4. Bit A_0 is equal to 0 for all registers to select the high-order (even) byte on the 68000 bus. The program in Figure 5.37 consists of a loop starting at LOOP. The wait period is determined by the value stored in location PERIOD. The contents of this location are loaded into register D1; then the required delay is implemented using the DBRA instruction. This instruction will be executed n times, where n is the number specified as the count in register D1. Each pass through the loop sets CB2 low then high, producing a complete cycle. The entire loop is repeated to produce the desired number of cycles, as specified in location CYCLES.

For a 68000 microprocessor driven by a 10 MHz clock, the DBRA instruction takes 1 microsecond to execute (see Section 4.10). Hence, the output frequency will be $10^6 / 2n$. When the value stored in PERIOD is $n = 417$, the frequency is approximately 1200 Hz. Changing this value to 208 increases the frequency to 2400 Hz.

```
PCR       EQU      $DA2018            Declare address of PCR.
CYCLES    DC       100                Desired number of cycles.
PERIOD    DC       417                16-bit delay period.
          MOVE.B   CYCLES,D0          Initialize cycle counter.
LOOP      MOVE.B   #%11000000,PCR     Set CB2 low.
          MOVE     PERIOD,D1          Initialize delay counter.
DELAY1    DBRA     D1,DELAY1
          MOVE.B   #%11100000,PCR     Set CB2 high.
          MOVE     PERIOD,D1          Initialize delay counter.
DELAY2    DBRA     D1,DELAY2
          DBRA     D0,LOOP
```

Figure 5.37 A program to produce a square wave on the CB2 control line of a VIA chip.

5.6 CONCLUDING REMARKS

This chapter expanded the basic ideas of input/output programming introduced in earlier chapters. Typical serial and parallel interfaces were examined from the programmer's point of view. The reader should now be able to prepare the software needed to deal with such devices as the ACIA, PIA, or VIA to transfer data to or from I/O devices.

Input and output operations involve close interaction between the software and the hardware. Chapter 6 will discuss the hardware details of interface circuits, followed in Chapter 7 by a presentation of interrupt-driven I/O and the use of direct memory access techniques.

5.7 PROBLEMS AND REVIEW EXERCISES

All programming exercises may be done using either the 6809 or the 68000 microprocessor.

REVIEW QUESTIONS

1. What is an input buffer, and what functions does it perform?

2. What is the difference between a microprocessor that uses memory-mapped I/O and one that has separate address spaces for I/O devices and for the main memory? Why must the latter type of microprocessor have special instructions for performing input and output operations?

3. A serial link between a video terminal and a microprocessor normally uses the same transmission speed for sending characters in either direction. In this case, it is not essential to check the output status flag when echoing back characters to the terminal. Why? What is the danger in not checking the output status?

4. A status flag in an I/O interface circuit is a facility that enables the microprocessor to synchronize its operation with the outside world. Explain.

5. Give two examples of input operations and two examples of output operations that do not require the use of status flags.

6. Is double buffering more important in the input section or in the output section of the ACIA? Why?

7. What are the bit patterns that should be deposited in the Control Register and the Command Register of the ACIA for operation at 2400 baud, using 8 bits per character, one stop bit and no parity?

8. The input status bit in an interface circuit is cleared as soon as the input data buffer is read. Why is this important?

9. The status flags in an interface circuit are intended to synchronize the activities of a microprocessor with external events. What does this statement mean?

PRIMARY PROBLEMS

10. Write a program that reads a 2-digit unsigned hexadecimal number from a video terminal and stores it in the main memory as a single 8-bit number.

11. Repeat Problem 10 for a 4-digit number, which should be stored as a 16-bit quantity.

12. Write a program that reads an unsigned hexadecimal number from a video terminal, entered as any number of hex characters, followed by a space or a Carriage Return. The program should display a 16-bit binary number of equal value. If more than 4 characters are typed, only the last 4 should be used.

13. Write a program for displaying a message on the screen of a video terminal connected to a serial port. The interface circuit uses an ACIA at locations $DB50_{16}$ to $DB53_{16}$. The message is stored in the main memory as a string of ASCII characters starting at address MSG. The end of each line in the message is marked with a Carriage Return character. This must be sent to the terminal as a Carriage Return followed by a Line Feed, in order to display the lines one below another. The last character of the message is followed by a Null character.

14. Write a program that displays the contents of 10 bytes of the main memory in hexadecimal format on the video terminal of Problem 13. Start at location LOC, and use two hex characters per byte. Contents of successive bytes should be separated by a space.

15. Write a program to accept an input line from the keyboard of a video terminal. The terminal is connected to a microprocessor via an ACIA occupying addresses $CB00_{16}$ to $CB03_{16}$. The input characters should be echoed back to the screen and stored in the main memory, starting at address LINE and ending with a Carriage Return. The operator should be allowed to correct typing mistakes using the Backspace character. (Make sure that you erase the displayed characters.)

16. Eight switches are used to monitor eight doors in a building, where a switch is closed when the corresponding door is closed. Eight LEDs display the state of the doors on an operator's panel. When a door is opened, it is required to light up the corresponding LED and display the message "Door number n is open" on a video screen, where n is a number between 1 and 8. The message "Door number n is closed" should be displayed and the LED turned off when the door is closed. Propose a suitable arrangement for connecting these devices to a microprocessor, and write the required program. (Ignore switch bouncing.)

17. Consider a printer that does not generate the Data-Taken response signal in Figure 5.22. What would have to be done in software to ensure that no output characters are lost? Write a suitable program to print a one-line message.

18. An analog-to-digital (A/D) converter is a device that accepts an analog input signal and converts it into a digital representation. Consider the A/D converter shown in Figure 5.38. After receiving a low pulse on the $\overline{\text{Start}}$ input, it measures the input voltage, v, produces an 8-bit representation of this voltage at its

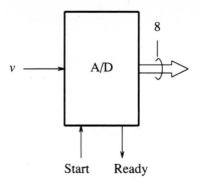

Figure 5.38 An A/D converter.

output, then sends a low pulse on the $\overline{\text{Ready}}$ line to indicate that its output lines carry valid data. The output data remains valid until another start pulse is received.

How would you connect this A/D converter to a PIA or a VIA, and which mode of operation would you choose? Write a program to record 10 voltage readings, one every 2 seconds.

ADVANCED PROBLEMS

19. Repeat Problem 18 assuming that the A/D converter produces a 12-bit representation of the input voltage.

20. A 10-digit decimal number, $D_9D_8 \cdots D_0$, is stored in the main memory starting at location DIGITS, as shown in Figure 5.39. Individual digits are represented using the packed binary-coded-decimal (BCD) format. Write a program to display this number on the screen, assuming an interface similar to that used in the program of Figure 5.12 or Figure 5.13.

21. Repeat Problem 20 for an n-digit number, where the number of digits, n, may be odd or even. Assume that the value of n is stored in a memory location called N. Do not display leading 0s.

22. It is required to build a simpler frequency counter than the one used in Figure 5.19. Instead of using external components, the INC line is connected directly to one of the input lines of a PIA, and counting and time measurement are done in software. Which of the PIA lines is best suited for this use? Write a program to count the number of pulses per second and store the count in location COUNT in the main memory. What is the range of error in a reading obtained using this frequency meter?

23. It is required to read a three-digit decimal number from the keyboard of a video terminal via an ACIA, and convert it into a single 8-bit binary number. If the value entered is larger than 255, an error message should be displayed. Write a suitable program.

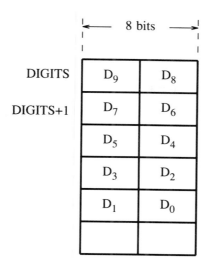

Figure 5.39 Organization of BCD digits in the main memory.

24. A liquid-crystal display panel consists of an array of dots organized in 16 rows of 80 dots each. Information is loaded into the display via 16 data input lines, one for every row of dots, as shown in Figure 5.40. A pulse on the Strobe line causes the information on the data inputs to be loaded into the right-most column. The dots, which normally have the same brightness as the background, become dark where the input signal is equal to 1. Each Strobe pulse causes the contents of the entire display to be shifted one column to the left.

Suggest a way for connecting this display to a microprocessor via an interface circuit that uses a PIA or a VIA chip. Write a program that configures the chip as needed, and displays a "+" sign that is 10 dots high, 10 dots wide, and vertically centered.

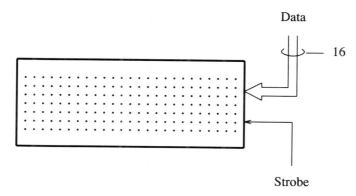

Figure 5.40 A dot-matrix display.

25. The results of an experiment to record temperature as a function of time are stored in the main memory of a microprocessor. Each reading is represented by a 4-bit number, which is stored in the low-order half of a byte. The high-order bits of the byte are all 0s. Readings are stored in 50 bytes, starting at location RECORD.

Write a program to display the temperature as a function of time, using the display described in Problem 24. The display should begin with a solid dark column on the left side representing the temperature axis. The bottom row of dots should represent the time axis. Then, each of 50 columns should have one dark dot representing the corresponding temperature reading.

26. A digital-to-analog (D/A) converter is a device that accepts a digital number and produces an analog output having the same value. Figure 5.41 shows a D/A converter that produces an output voltage proportional to a 10-bit binary number loaded into its input register. The input data are loaded in the converter's register by a pulse on the Strobe line.

Assume that the converter is connected to a microprocessor via a PIA or a VIA. Write a program that produces a triangular waveform at the output, v. Your program should include the instructions needed to select the appropriate mode of operation of the interface chip.

LABORATORY EXERCISES

27. The display unit of a video terminal accepts the commands given in Table 5.7. These commands may be sent from the microprocessor to the terminal to cause the actions specified. For example, the sequences Esc[4A and Esc[6D move the cursor four lines up and six character positions to the left.

Write a program to draw a large square on the screen, using the characters "_" and "I." Assuming that this square represents a railroad track, display a train consisting of five "*" characters travelling slowly around this closed loop. It should be possible to stop and start the motion of the train by pressing the space bar, and to quit the program by typing "q."

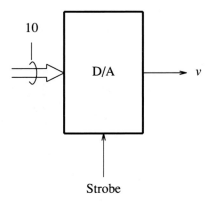

Figure 5.41 A D/A Converter.

Table 5.7   SOME DISPLAY COMMANDS FOR THE VT-100 TERMINAL

Code*	Action
ESC[2J	Clear screen, leaving the cursor at its original location.
ESC[0K	Clear all characters from the cursor position to the end of line inclusive, leaving the cursor in its original location.
ESC[1K	Clear all characters from the beginning of the line to the cursor position inclusive, leaving the cursor in its original location.
ESC[n ;m H	Move cursor to character position m on line n. The number 1 refers to the top line on the screen or the left-most character position.
ESC[n A	Move cursor n lines up in the same column position.
ESC[n B	Move cursor n lines down in the same column position.
ESC[n C	Move cursor n character positions to the right.
ESC[n D	Move cursor n character positions to the left.

* ESC stands for the Escape character.

Note: The commands given in Table 5.7 are applicable to Digital Equipment Corp.'s VT-100 video terminals. You should use the corresponding commands for the terminal available to you.

28. As a debugging aid, a program is required to enable an operator to use a video terminal to check the contents of any location in the main memory. The program displays a "%" character as a prompt. To examine the contents of a given location, the operator types the letter M followed by a four-digit (six-digit in the case of 68000) hexadecimal address. After echoing these digits back, the program adds one space followed by a hexadecimal number representing the contents of the memory location specified. Then, it displays another prompt character on a new line. Write the required program and test it.

After testing your program, modify it to enable the operator to test successive locations in memory and change their contents. After displaying the contents of a memory location, the program should send a ":" followed by a space, then wait for new input. At this point, the operator may do one of three things

1. Type a hexadecimal number, which should be stored in the memory to replace the value displayed.

2. Type a carriage return, to which the program should respond by displaying the contents of the next higher address location on a new line, or

3. Type an escape character, which should end the current examination sequence. This should cause the program to display a prompt on a new line and wait for further input.

After entering a new value, the operator may type a carriage return or an escape to which the program should respond as in items 2 or 3 above. If at any time an illegal entry is detected, the program should end the examination sequence and display a prompt on a new line.

29. Write a program to display a 3 × 3-square tic-tac-toe board, on a terminal of the type described in Problem 27. Each of the nine squares should be 8 characters wide and four lines high. It should be possible for two players to play the game by writing **X**'s and **O**'s in the squares of their choice. Define and implement suitable single-character commands that enable the players to choose a square then write an X or an O. The players should be able to restart the game at any time.

30. A full-duplex communication channel is formed between two microcomputers by connecting the PIA (or VIA) ports of one microcomputer to the other. (A full-duplex channel can carry data in both directions simultaneously.) How would you connect two PIAs (VIAs) for this purpose, and which mode of operation would you recommend for each port?

Assume that the messages exchanged by the two microprocessors have the format shown in Figure 5.42. The message text is preceded by an STX (Start of text) character and followed by an ETX (End of text) character and a one-byte checksum for error checking. The checksum byte is the arithmetic sum modulo 256 of the entire message, including the STX and ETX characters. The communications software in each microcomputer consists of a main program that accepts commands from a video terminal, and two subroutines to receive and transmit messages, as follows:

Transmitter: This subroutine sends a message stored in the main memory over the communication link. After sending the message, it waits for an acknowledgment. The acknowledgment consists of the single character ACK. If the transmitter subroutine receives an acknowledgment from the other microcomputer, it returns a parameter to the main program informing it that transmission

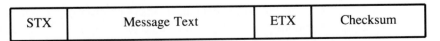

| STX | Message Text | ETX | Checksum |

Figure 5.42 Message format for Problem 30.

has been completed successfully. If it receives anything other than an ACK character, or if it receives no reply at all within 3 seconds, it retransmits the message. It retries sending the message at most five times, after which it assumes that the link is broken and it returns an appropriate indication to the calling program.

Receiver: The receiver subroutine waits for a message to be received. If the error checksum is correct, it displays the message on the terminal and sends an acknowledgment message over the communication link. Otherwise, it discards the message, displays an error indication on the terminal, and waits for another message. Once the receiver subroutine is called by the main program, it continues to receive and display messages until a character is entered at the keyboard of this microcomputer's terminal. When any key is activated, the receiver subroutine returns immediately to the calling program, even if it is in the middle of receiving a message.

Main program: The main program accepts three commands from the terminal, R, T, or E. A command is followed by a Carriage Return. The R command causes the main program to call the receiver subroutine. When the T command is typed, the program accepts a message of up to 256 characters from the terminal, where the message is terminated by the Escape character, ESC. Then, the program calls the transmitter subroutine to send the message over the communication link. Upon return from the transmitter subroutine, the main program displays an appropriate information message and waits for a new command. The E command causes the execution of the program to be terminated.

Prepare and test the communications software described above. Test the error recovery capability by intentionally disconnecting one of the wires that interconnect the two PIAs (VIAs).

5.8 REFERENCES

1. *Microprocessor Interface Chips.* Austin, TX: Motorola, Inc., 1986.

2. *Rockwell 1985 Data Book.* Newport Beach, CA: Rockwell International, 1985.

Chapter
6

Input/Output Hardware

Chapter 5 introduced some basic concepts used in the transfer of data between a microprocessor and the outside world. The discussion was given primarily from the programmer's point of view. In this chapter, we will consider the hardware details involved in connecting I/O devices to the bus of a microprocessor. There is a wide variety of I/O devices, but the required interconnections can be implemented using only a few basic hardware schemes. The information given in this chapter should enable the reader to design interface circuits for simple devices. The buses of our example microprocessors, the 6809 and 68000, will be presented, together with information about some commercial chips that are commonly used in serial and parallel interface circuits. The connection of devices requiring the use of interrupts or direct memory access (DMA) will be discussed in Chapter 7.

The study of interface circuits requires a basic understanding of logic components and circuits. Readers who do not have this background should consult a book on logic design [1–3], before studying the circuit details given in this chapter.

The medium through which data transfer between a microprocessor and all other devices takes place is the microprocessor bus. The concept of a bus was introduced briefly in Chapter 1. Since this concept and its implications are central to the discussion of I/O structures, we begin by examining some fundamental issues relating to the operation of a bus.

6.1 BUS STRUCTURE

A bus provides the basic means for interconnecting functional units in a computer system. A simple microcomputer system is shown in Figure 6.1, in which a bus interconnects a microprocessor, the main memory, a video terminal, a printer, and a magnetic disk. The bus acts as a shared data highway, over which these devices exchange data. Since the bus can only carry data from one device at any given time, a bus *protocol* is needed to coordinate data transfers. That is, all devices must observe a common set of rules that determine when they may transmit or receive information. The responsibility for enforcing these rules resides in the interface circuits that connect various devices to the bus.

A typical interface circuit contains three functional units: a data buffer, an address decoder, and control logic. An example involving a keyboard, which is a part of any video terminal, being used as an input device is shown in Figure 6.2. The data buffer isolates the I/O device from the microprocessor bus and provides the electrical drive capabilities needed by the latter. It may also act as a temporary

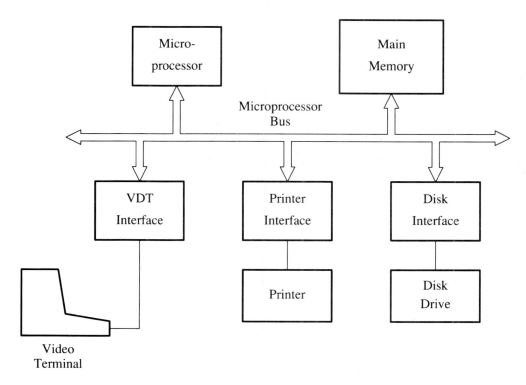

Figure 6.1   A single-bus microcomputer system. VDT=video display terminal.

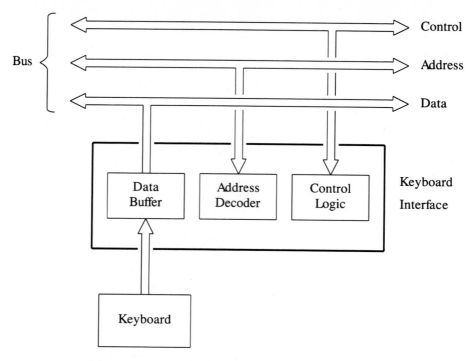

Figure 6.2 Functional components of an interface circuit.

storage location for the data being sent from the device to the microprocessor bus. In order to identify the device participating in a given data transfer, each device is assigned a unique address. The address decoder circuit decodes the information on the address lines to determine whether the interface circuit is being asked to transmit or receive data. Finally, the bus carries control signals that various device interface circuits exchange to coordinate their operation. The control block is responsible for sending and receiving these signals, and controlling the operation of the remainder of the interface accordingly. Thus, in general, the signals on a bus can be divided into three functional groups: address, data, and control. We will assume that there is a separate group of bus lines dedicated to each function. However, such is not always the case. For example, a microprocessor may use a set of lines to carry address information during one phase of an operation and data in a subsequent phase. In such a case, the microprocessor is said to have a *multiplexed* bus.

At any given time data transfer over a bus is controlled by one device, which we will refer to as the *bus master,* normally the microprocessor. When the bus master is ready to transfer a unit of data, it uses the address lines to specify the device to or from which the transfer is to take place. Only the device that recognizes its address on the bus participates in the operation. This device is referred to as a *slave.* The bus master uses one of the control signals to specify the direction of the transfer, that is, to indicate whether it wishes to read data from the slave device or write data into it. A word of data is then transmitted over the data lines. Throughout this process the control signals provide timing information to indicate when a new address has been placed on the address lines, when the receiving device should accept the data, and so forth.

The discussion above illustrates two important functions of the control signals — to specify the direction and timing of data transfers. The control signals are also needed to schedule the use of the bus. Consider, for example, the situation when a mass-storage device is ready to transfer data to the main memory. It needs to become the bus master. The interface circuit of the mass-storage device uses one of the control signals to request bus mastership from the microprocessor. It waits until it receives a response indicating that this has been granted. Then it proceeds to transfer data to the main memory. This mode of operation, which is known as *Direct Memory Access* (DMA), will be discussed in Chapter 7.

Device interface circuits are responsible for interpreting all signals on the bus and generating appropriate responses. They differ in complexity depending upon the nature of the device involved and the modes of operation supported. Most popular microprocessors have their own families of interface chips, which are easily interconnected to construct an inexpensive, yet powerful, microcomputer system. In this chapter we will examine I/O interface circuits in detail. We will show typical circuits and features that may be incorporated in a general-purpose interface chip, using some commercially available products as examples.

There is an important consideration that arises when several devices are to be connected to a common bus. Such connections usually use a special type of logic gates known as *tri-state drivers,* in which the bus is called a *tri-state bus.*

6.1.1 A Tri-State Bus

The output of one logic gate may be connected to the inputs of several logic gates. From the point of view of the electrical circuits used to implement logic gates, this arrangement presents no particular difficulty. However, the bus arrangement of Figure 6.1 gives rise to a requirement that normal logic gates cannot meet. Let A and B be two of the devices connected to the bus. Since data may be transferred from device A to device B as well as from device B to device A, each device requires an input gate and an output gate connected to the bus, as shown in Figure 6.3. This means that the outputs of two gates, 1 and 3 in the figure, are connected to each other. Normal logic gates cannot operate in this manner. The desired bus operation requires a special driver gate that can be disconnected from the bus when it is not being used to transmit data. The required driver is equivalent to a normal gate followed by a switch, as shown in Figure 6.4. When the gate is enabled, that is, when its Enable input is equal to 1, the switch is closed, and the logic state at the data input is transferred to the output. When the Enable input is 0 the switch is open, and the gate is said to be in the *high-impedance* state; it does not affect the voltage level of the bus line. This means that the output of the gate can be in one of three states: 1, 0, or high impedance, hence the name "tri-state gate."

Figure 6.5(a) shows two tri-state gates connected to a bus line, D. The timing diagram given in part (b) of the figure indicates that the logic state at D follows

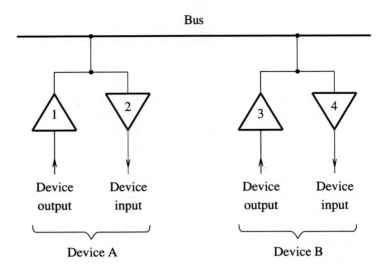

Figure 6.3 A tri-state bus.

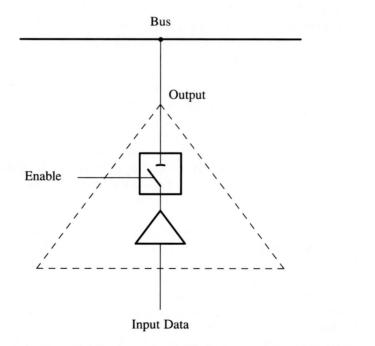

Figure 6.4 Equivalent circuit for a tri-state gate.

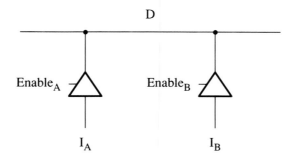

(a) Two tri-state gates connected to a common bus line

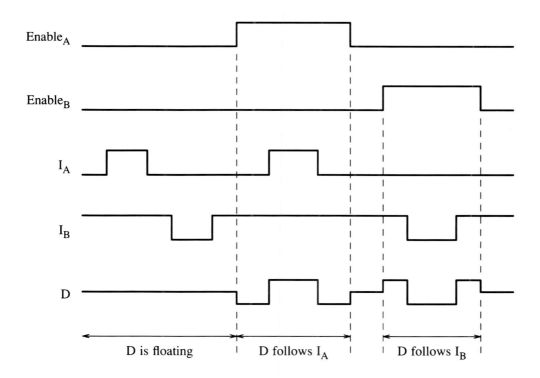

(b) Timing diagram

Figure 6.5 Tri-state bus.

either input I_A or input I_B, depending upon which of the two gates is enabled. When neither gate is enabled, that is when both $Enable_A$ and $Enable_B$ are 0, the state of the bus is unpredictable, because the bus is not being driven to any particular logic state. During such periods, the bus is said to be *floating* — a condition depicted in the timing diagram by an intermediate signal level, halfway between the levels corresponding to the 0 and 1 states. However, we should note that if the signal on the bus is monitored on an oscilloscope, the waveform during the high-impedance state, i.e., when the bus is floating, will not necessarily appear as shown in the figure. One is more likely to observe random noise, or incomplete logic transitions caused by pick-up from adjacent wires.

One of the functions of the control lines of a bus is to provide information to all interface circuits so that they do not mistakenly react to some bus lines while these lines are in the high-impedance state. Also, by interpreting the information on the address and control lines, the circuitry that generates the Enable signals for the tri-state gates connected to a bus must ensure that only one gate is enabled at any given time. Having two or more tri-state gates enabled simultaneously creates a short-circuit situation, which, if it persists for a long time, may lead to permanent damage of the gates involved. In any case, the state of the bus during such a period would be meaningless.

6.2 A SIMPLE INPUT INTERFACE

A simple input device consisting of eight switches was discussed in Section 5.4. The combined state of the eight switches represents a byte of data that can be read through a parallel interface. We will discuss here the design of the interface circuit needed to connect these switches to a microprocessor bus. A suitable design is shown in Figure 6.6, assuming that the address assigned to the input device in question is $FFF1_{16}$. Let us examine this circuit in some detail.

The logic states at points P_0 to P_7, are transferred to the data bus of the computer, lines D_0 to D_7, via tri-state drivers. Eight tri-state gates are used, with a common enable input, E, connected to a signal called Enable-Data. Whenever Enable-Data is equal to 1, the input data byte, P_{7-0}, is transmitted on the data bus of the microprocessor.

Let us now consider the generation of the Enable-Data signal. The interface should transmit data only when all of the following conditions are met:

1. The device is being addressed by the microprocessor.
2. The operation requested by the microprocessor is a read operation.
3. Timing constraints are satisfied.

The first two conditions are straightforward. Since the device has been assigned the address FFF1, a decoder circuit can be used to recognize this particular bit pattern on the address lines, A_0 to A_{15}. A 16-input AND gate, together with three inverters, performs this function in the interface of Figure 6.6. The output, Device-Selected, of

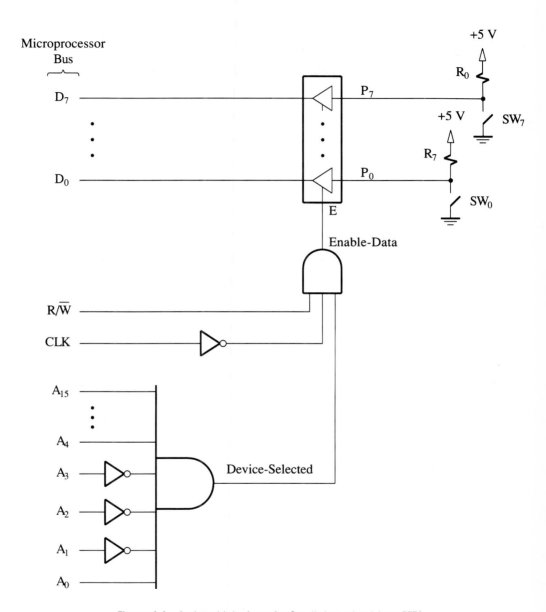

Figure 6.6 An input interface for 8 switches at address FFF1.

the address decoder, is equal to 1 whenever the address lines carry the address FFF1. The operation requested by the microprocessor is specified on one of the bus control lines, called Read/Write (R/$\overline{\text{W}}$) in the figure. The microprocessor transmits a 1 on this line to indicate a read operation, and a 0 for a write operation. Thus, the interface should be enabled when both Device-Selected and R/$\overline{W}$ are equal to 1. The third condition, that timing constraints should be met, is discussed below.

6.2.1 Timing Considerations

Timing constraints arise from the need to consider the delays encountered by logic signals while propagating from one point in the system to another. Setup and hold time requirements for various circuit components must also be satisfied. One technique for dealing with timing constraints is based on providing a reference signal known as a *clock,* relative to which the timing of all bus operations is defined. A bus that uses this approach, such as the 6809 bus, is called a *synchronous* bus. We will use synchronous buses in most of our discussion, because of their simplicity. Asynchronous buses will be introduced in Section 6.8.

The nature and use of a reference clock signal are illustrated in Figure 6.7. The clock signal, CLK, consists of a continuous stream of pulses of known width and frequency. One bus transfer, either an input or an output operation, takes place during one clock period, that is, between the rising edges of two consecutive clock pulses. Figure 6.7 gives the sequence of events on the bus during an input operation. At the beginning of a clock period, the microprocessor places a device address on the address lines. The address lines are shown as if they were moving simultaneously to

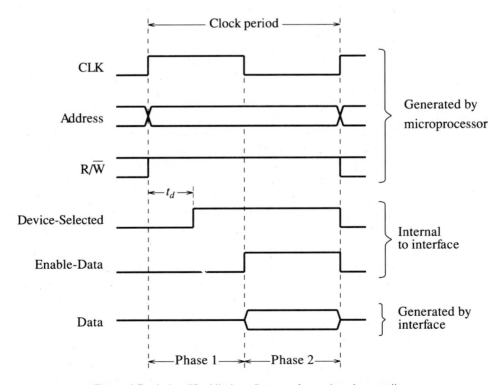

Figure 6.7 A simplified timing diagram for an input operation.

both 0 and 1, because for an arbitrary address pattern, individual address lines may be carrying either a 0 or a 1. The microprocessor also sets the $R/\overline{W}$ line to specify the desired operation. In the figure this line is set to 1, indicating a read operation. Outside the clock period of interest, it is shown to be both high and low. This is meant to indicate that the preceding and succeeding clock cycles may involve either read or write operations. The information transmitted by the microprocessor on the address and $R/\overline{W}$ lines reaches the device interface of Figure 6.6, and after some delay, t_d, the output of the address decoder, labeled Device-Selected, becomes equal to 1.

An important characteristic of buses should be pointed out. The address bus consists of a number of lines. When a new address is placed on the bus by the microprocessor, many lines change their state. The propagation delays over these lines and through different input and output gates cannot be guaranteed to be identical, so the address seen by a given device interface may change several times before settling to the correct value. It is prudent to allow sufficient time for all signals to settle before initiating any action in the interface circuit. The clock signal on the bus provides a simple and reliable mechanism for implementing the required delay. The width of the clock pulse may be chosen such that all signals are guaranteed to have settled before the falling edge of the clock. Thus, device interfaces should respond to the operation requested by the microprocessor following the falling edge of CLK, i.e., while CLK is equal to 0.

Combining all of the conditions discussed above, the Enable-Data signal should be generated only when a valid address is received, the $R/\overline{W}$ line is equal to 1 and the clock signal is equal to 0. That is, it should be derived as

$$\text{Enable-Data} = \text{Device-Selected} \cdot R/\overline{W} \cdot \overline{\text{CLK}}$$

The circuit in Figure 6.6 conforms to this requirement. The tri-state drivers are enabled, hence data are transmitted, while the clock signal is low, as shown in Figure 6.7.

It should now be apparent that the input operation proceeds in two stages, defined by the high and low states of the clock. While the clock is equal to 1, the microprocessor transmits new address and control information, and this information is received and decoded by all device interfaces. The addressed device responds by transmitting the input data when CLK is equal to 0. In general, bus transfers may involve several stages. Therefore, it is convenient to think of a clock period as being subdivided into *phases* during which different suboperations take place. The clock in Figure 6.7 has two phases, designated Phase 1 and Phase 2, as shown.

So far, we have assumed idealized conditions. The rise and fall times, i.e., the time taken by signals to change from 0 to 1 and from 1 to 0, respectively, were assumed to be 0. Moreover, only the propagation delay, t_d, through the address decoder has been accounted for explicitly in Figure 6.7. Many other sources of delay exist in practical circuits. For example, the data lines were assumed to change at exactly the same time as the falling edge of CLK. Inspection of Figure 6.6 indicates that the falling edge of CLK has to propagate through one inverter and one AND gate before Enable-Data becomes equal to 1. Then, the tri-state drivers require some time to become enabled and place the input information on the data bus. For similar reasons, some delay will be encountered following the rising edge of the clock before the address information transmitted by the microprocessor actually appears on the bus. A more realistic picture that takes these factors into account is given in Figure 6.8. All signals have nonzero rise and fall times. The major delay components are identified as t_{PA} and t_{PD} — the propagation delays associated with address and data, respectively — and the address decoder delay, t_d. At the end of the clock period, the Enable-Data signal returns to 0, and the data are removed from the data lines, again after some delay introduced by the gates involved. The address and R/$\overline{\text{W}}$ lines may start to change any time after the rising edge of CLK, as the microprocessor begins a new data transfer operation.

Before concluding the discussion of timing constraints we should consider one more point. In order for data to be stored correctly in a flip-flop, the minimum setup and hold times for the flip-flop must be met. This means that the data at the input of the flip-flop must be stable for some time before and after the edge of the clock that loads the data in the flip-flop. At the end of the input transfer depicted in Figure 6.8, the microprocessor stores the data on the data bus into an internal register. Let us assume that the exact moment at which data are locked in is at the end of Phase 2,

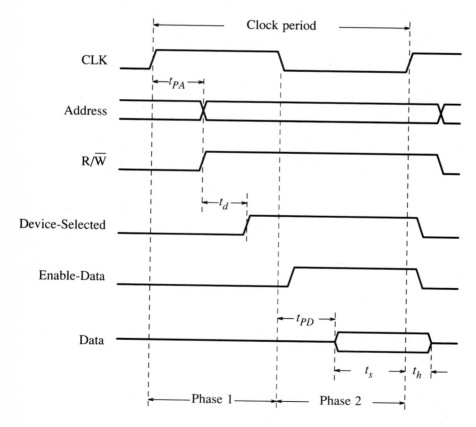

Figure 6.8 A detailed timing diagram for an input operation.

coincident with the rising edge of CLK. Hence, the setup and hold times, t_s and t_h, for loading that buffer are as indicated in the figure. The worst case (minimum) values for these two parameters can be estimated from the width of Phase 2 of the clock and a knowledge of the propagation delays in the interface circuitry. It is the responsibility of the designer of a computer system to ensure that the setup and hold times exceed the minimum requirements of the microprocessor as specified by the manufacturer.

The quantitative aspects of the concepts presented above are illustrated by a numerical example given in the following section. The numbers quoted are intended to provide the reader with an idea about the magnitudes of the delays involved in practical systems.

6.2.2 An Example of Timing Requirements

Consider a microprocessor that has the following specifications:

$$
\begin{aligned}
\text{Maximum address delay} \quad &= t_{PA\,(max)} &= 200 \text{ ns} \\
\text{Minimum set-up time} \quad &= t_{s\,(min)} &= 80 \text{ ns} \\
\text{Minimum hold time} \quad &= t_{h\,(min)} &= 10 \text{ ns}
\end{aligned}
$$

Assume that the logic gates used in the interface have a maximum propagation delay of 15 ns per gate, and that a tri-state driver has a maximum delay of 35 ns from the time its Enable input is set to 1 until its output becomes valid. These numbers correspond to the Low-Power Schottky TTL (Transistor-Transistor-Logic) family.

The delay through the address decoder in Figure 6.6 consists of two gate delays. Therefore, its maximum value is

$$t_{d\,(max)} = 2 \times 15 = 30 \text{ ns}$$

This means that the duration of Phase 1 must be at least equal to $t_{PA\,(max)} + t_{d\,(max)} = 230$ ns. Similarly, the maximum data delay $t_{PD\,(max)}$ may be estimated as follows:

$$t_{PD\,(max)} = 2 \times 15 + 35 = 65 \text{ ns}$$

Hence, to provide a setup time of 80 ns, Phase 2 must be at least 145 ns long.

In order to check that the hold specification is met, we need to know the minimum delay from the time CLK goes to 1 until the interface data is removed from the bus. The minimum delay in logic components is seldom specified in the data sheets supplied by IC manufacturers. As an empirical rule of thumb, one may take it to be one third of the maximum value. Hence, at the end of the read operation of Figure 6.8, the data on the bus will change no earlier than about 22 ns after the rising edge of CLK. This value is well outside the minimum hold interval of 10 ns required by the microprocessor.

With the above parameters, the minimum clock period is $230 + 145 = 375$ ns, which corresponds to a frequency of about 2.6 MHz. It is often convenient to make the two phases of the clock equal in length, in which case the maximum clock frequency would be 2.1 MHz.

6.3 A SIMPLE OUTPUT INTERFACE

Let us now consider an interface that enables the seven-segment display described in Section 5.4 to be used as an output device. A suitable circuit is given in Figure 6.9. Information on the data bus is loaded into a storage buffer, whose clock input is connected to a signal called Load-Buffer. Hence, Load-Buffer should be set to 1 when the display is being addressed, and a write operation is specified. The address

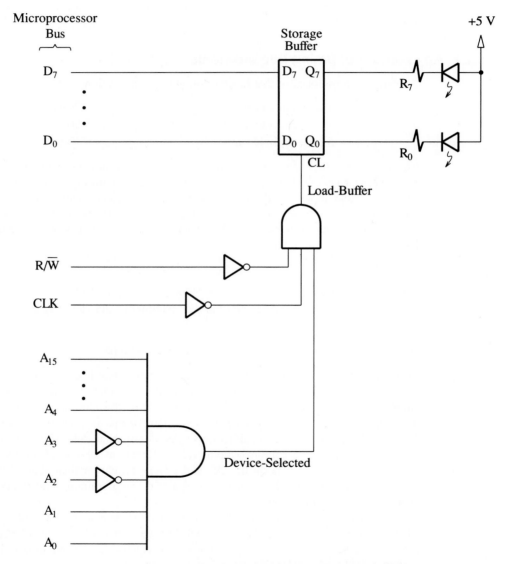

Figure 6.9 A simple output interface for a 7-segment display.

decoder in the figure is based on the assumption that the address assigned to the display is FFF3. When this address appears on the address bus, the Device-Selected signal becomes equal to 1. Thus, the Load-Buffer signal should be generated whenever Device-Selected is equal to 1 and $R/\overline{W} = 0$. The last and most critical aspect in the design of the interface is to determine exactly when within the clock period this should take place.

The time at which Load-Buffer should be activated depends on the characteristics of the flip-flops used in the storage buffer. Let us assume that the buffer is a register consisting of flip-flops having the structure shown in Figure 6.10. Output Q_0 follows input D_0 while the clock input, CL, of the buffer is equal to 1. When CL goes to 0, the buffer is isolated from input D_0, and Q_0 maintains the value that was present just before the transition took place. Hence, to store the information sent by the microprocessor into the data buffer, the Load-Buffer signal in Figure 6.9 should be set to 1 then back to 0 while data on the data lines are valid, and only after ensuring that the Device-Selected signal has reached a stable value. These conditions are guaranteed to be satisfied while the clock signal, CLK, of the microprocessor bus is equal to 0.

The timing diagram for the required output operation is given in Figure 6.11. At the beginning of a clock period, the microprocessor transmits the address and data information, and sets the $R/\overline{W}$ line to 0 to indicate a write operation. These signals reach the device interface after some delay, t_{PA}. Address decoding takes place, and after a delay t_d the Device-Selected signal becomes equal to 1. All signals are guaranteed to be stable at the start of Phase 2 of the clock. Hence, at this time, the Load-Buffer signal is generated to store the data on the data lines into the data buffer. Load-Buffer goes back to 0 immediately following the 0-to-1 transition on the clock line. The reader should verify that the Load-Buffer signal generated by the circuit of Figure 6.9 corresponds to the timing diagram of Figure 6.11.

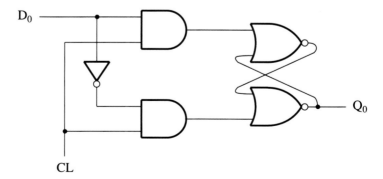

Figure 6.10 One bit of the data buffer of Figure 6.9.

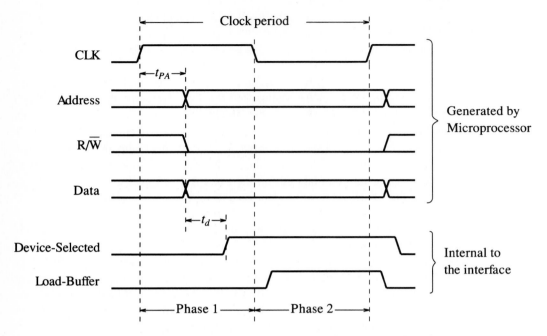

Figure 6.11   A timing diagram for an output transfer.

6.4 STATUS FLAGS

The previous sections introduced examples of interface circuits for connecting very simple devices to the bus of a microprocessor. In this and the following section, we will discuss some additional functions that a device interface should provide, in order to support somewhat more complex devices such as keyboards and printers. Let us start by considering a keyboard connected as an input device. Physically, most keyboards consist of mechanical switches that are normally open. A switch is closed whenever the corresponding key is depressed, and returns to the open position when the key is released. A typical keyboard has about 60 keys. Associating a separate data line with each key is obviously impractical. Instead, the number of wires required is reduced by incorporating an encoder circuit into the keyboard. The function of this circuit is to generate a unique binary code, using a small number of bits, to represent each key. The 7-bit ASCII code described in Appendix A represents 128 (=2^7) different keys and key combinations.

When a parallel interface is used, individual keys of the keyboard may be connected directly to the interface buffer, as was done in Figure 6.6. In this case, the actions of observing that a key has been activated, of accounting for switch bouncing, and of observing that the key has been released must all be performed in software, as was done in the program given in Figure 5.17. These functions can also be implemented in hardware, in which case the hardware reports the result to the microprocessor in the form of an input *status bit*, or *status flag*, as it is often called. The status flag is set to 1 whenever input data are available to be read by the microprocessor and is cleared as soon as the data are read.

To provide the status flag, the keyboard should generate a Data-Valid signal as shown in Figure 6.12. Data-Valid should be set to 1 whenever a key is pressed and switch bouncing has subsided. A suitable interface for use with the modified keyboard is shown in Figure 6.13. It incorporates a status flag circuit whose purpose is to generate a signal that enables the microprocessor to check the status of the input data. The output of this circuit, Keyboard-Ready, is connected to bus line D_0 via a tri-state gate. It should be set to 1 whenever Data-Valid is activated, and returned to 0 as soon as the microprocessor reads the keyboard data. The address decoder produces two outputs, Data-Selected and Status-Selected, which are set to 1 in response to the addresses FFF1 and FFF0. After gating with CLK and R/$\overline{\text{W}}$, these two signals produce Enable-D and Enable-S, which enable the tri-state gates for the data and status lines, respectively. As a result, the microprocessor can read the input data at address FFF1 and the status flag at the least-significant bit of location FFF0. Note that during a read operation at this address, no data are transmitted on lines D_1 to D_7. The microprocessor should ignore the data on these lines, because they will be in an undefined state.

The status flag circuit is shown in Figure 6.14. The rising edge of the Data-Valid signal sets the Q output of flip-flop FF1 to 1. The Keyboard-Ready flag is generated by the NOR latch. It is set to 1 when Q goes to 1, but only during phase 1 of the clock. If Q goes to 1 while CLK = 0, the setting of the latch will be delayed until CLK changes to 1 at the beginning of the next clock period. The use of CLK in this manner ensures that the state of the Keyboard-Ready bit will always be stable during Phase 2 of CLK, just in case it is being read by the microprocessor. As was explained in conjunction with Figure 6.8, it is during Phase 2 that data are transmitted from the interface circuit to the microprocessor.

Once set, the Keyboard-Ready latch is independent of the Data-Valid signal. It remains set until the input data are read by the microprocessor. At that time, both the latch and flip-flop FF1 will be cleared by Enable-D. They will not be set again until another key is pressed, generating a positive-going edge on the Data-Valid line.

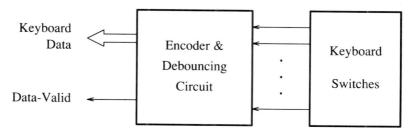

Figure 6.12 A keyboard that incorporates an encoder and a debouncing circuit.

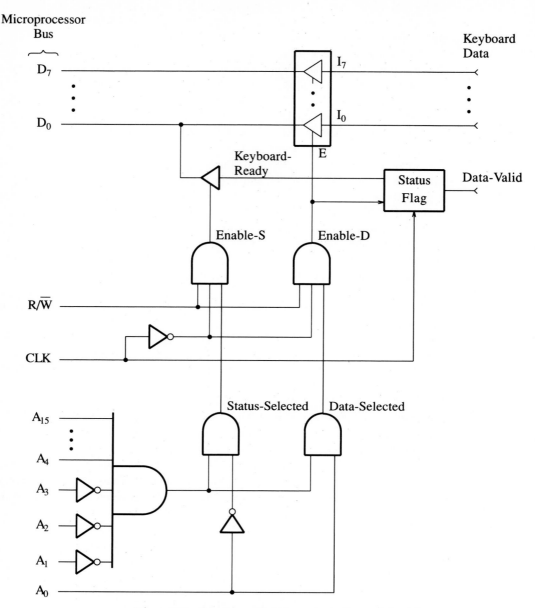

Figure 6.13   A keyboard interface, including a status bit.

It was pointed out in Section 5.4.2 that it is sometimes desirable, or necessary, to store the input data into a register in the interface. This is accomplished simply by inserting a storage buffer at the input port, as shown in Figure 6.15. Positive edge-triggered flip-flops are used in this buffer, so that the input data is loaded by the

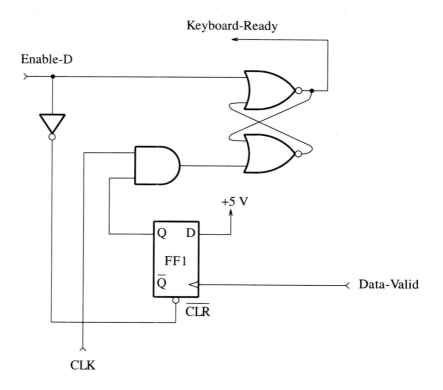

Figure 6.14 Details of status flag circuit in Figure 6.13.

rising edge of Data-Valid. Of course, the same transition on Data-Valid sets the status flag, as before. With the availability of the Data-Valid line and the storage buffer for the input data, the frequency meter circuit of Figure 5.18 may be connected directly to the interface circuit of Figure 6.15.

6.4.1 Handshake Signaling

The exchange of signals known as a *handshake* is needed whenever the communication between two devices involves a request being generated by one device and a response being generated by the other. The concept of a handshake was introduced in Subsection 5.4.3. An example of its use for communicating with a printer was given in Figure 5.22, which is reproduced in Figure 6.16. A timing diagram describing the required signal exchange is given in Figure 6.17, assuming that a full handshake is used. The desired interface circuit may be derived from that of Figure

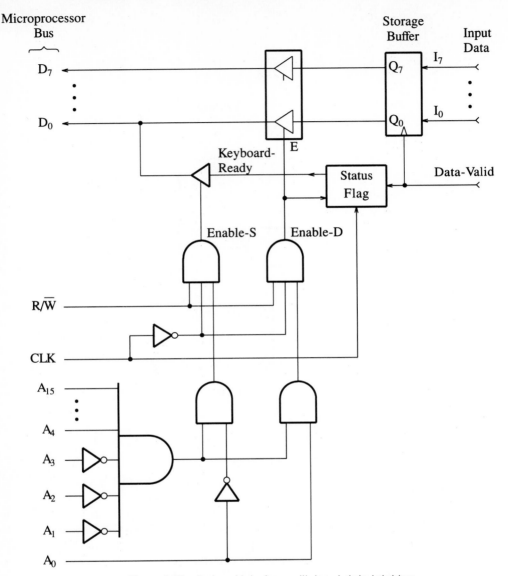

Figure 6.15   An input interface with input data latching.

6.9 by adding the handshake control block shown in Figure 6.18. The function of this block is to handle the Data-Valid and Data-Taken control signals, and to produce a status bit, Printer-Ready. The status bit is connected to the data lines in the same way as the Keyboard-Ready signal in Figure 6.15, so that it may be read by the microprocessor. We have modified the address decoder to produce an enable-status signal, Enable-S, which is set to 1 in response to a read operation at address FFF2. The Load-Buffer signal is activated when the microprocessor writes into location FFF3, as before.

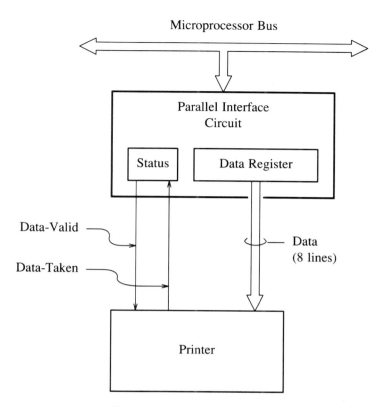

Figure 6.16 A printer connected via a parallel interface.

The design of handshake signaling circuits requires great care. The difficulty arises because we have two systems communicating, each having its own timing

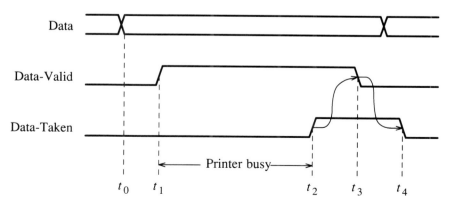

Figure 6.17 Handshake signaling between the printer and the computer interface.

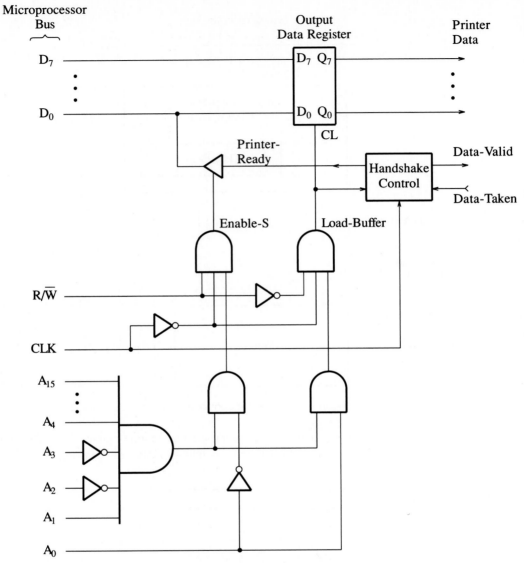

Figure 6.18 Printer interface.

constraints. The interface in Figure 6.18 uses the bus clock signal, CLK, to control its operation, while the I/O device operates independently of that clock. The Data-Taken input, which is generated by the I/O device, may change at any time relative to the clock. Moreover, the device may respond to changes in Data-Valid quickly, or after a long delay. In general, it is best to make as few assumptions as possible about characteristics of the I/O device when designing the interface circuit, to make the interface suitable for connection to a variety of devices. We will examine the design of the handshake control circuit in detail, because it illustrates typical design problems encountered in microprocessor systems. Let us begin by examining the desired behavior of the circuit. Several requirements can be identified:

1. The Printer-Ready bit should be cleared to 0 when new information is loaded in the output data register, i.e., when the Load-Buffer signal becomes active.

2. Data-Valid should not be asserted before the output data being sent to the printer has become stable. We will choose to set Data-Valid to 1 at the beginning of Phase 2 of the clock period following that in which new data is deposited in the output register.

3. Data-Valid should return to 0 soon after an active signal is received at the Data-Taken input. This action is independent of the clock signal.

4. Printer-Ready should be set to 1 after Data-Valid has returned to 0. However, Printer-Ready should not change its state during Phase 2 of the clock, where it may be read by the microprocessor. So, the change in Printer-Ready should wait for the following Phase 1 of the clock.

5. Once Printer-Ready is set to 1, the microprocessor may start another output operation. However, Data-Valid should not be activated again until Data-Taken has returned to 0. Recall that Data-Taken may return to 0 soon after Data-Valid is deactivated, or it may stay active for a long time.

A first attempt at designing the handshake circuit is shown in Figure 6.19. A positive edge-triggered JK flip-flop is used to generate the Printer-Ready signal. This signal goes to 0 at the positive edge of CLK at the end of a clock period in which Load-Buffer is equal to 1, thus satisfying condition 1 above. Data-Valid is generated by a latch, which is set to 1 by the Set-Valid signal. In turn, Set-Valid is

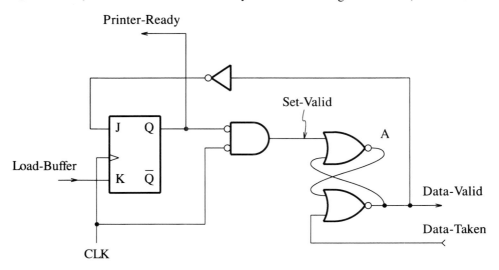

Figure 6.19 A first attempt at the design of the handshake control circuit of Figure 6.18.

activated while both CLK and Printer-Ready are low, thus satisfying condition 2. The Data-Valid latch is cleared by the Data-Taken signal, as required by condition 3. Also, since this latch cannot be set unless Data-Taken is equal to 0, condition 5 is met. After Data-Valid returns to 0, the next positive clock edge sets Printer-Ready to 1, as required by condition 4.

Although the circuit of Figure 6.19 appears to satisfy all the requirements, closer inspection reveals a serious flaw, which is illustrated in the timing diagram in Figure 6.20. As long as Printer-Ready is equal to 0, pulses will be produced on the Set-Valid line during Phase 2 of every clock period. When Data-Taken becomes equal to 1, Data-Valid will go to 0. But, in order to set Printer-Ready the J input of the flip-flop must be high, which means that Data-Valid must remain equal to 0 until the next positive edge of CLK. Unfortunately, this is not guaranteed. Once Data-Valid goes to 0, the I/O device may set Data-Taken to 0 at any time. If this happens before the end of the clock period, that is while Printer-Ready is still equal to 0, Data-Valid will be set to 1 again by the pulse on Set-Valid, as shown in the diagram. The I/O device will take this to be the beginning of a new output transfer, which of course is not correct.

There are many ways to fix the problem. One solution is given in Figure 6.21. An active state on the Set-Valid line is needed only to set the Data-Valid latch. Hence, Set-Valid may be returned to 0 as soon as the latch is set. To do so, a Disable-Set-Valid latch has been added, which when set to 1 forces Set-Valid to 0. This latch is set to 1 as soon as Data-Valid becomes equal to 1, and remains set until Printer-Ready returns to 1. As before, Printer-Ready goes to 1 only after the device's response, Data-Taken, causes Data-Valid to go to 0. A timing diagram illustrating the operation of the modified circuit is given in Figure 6.22.

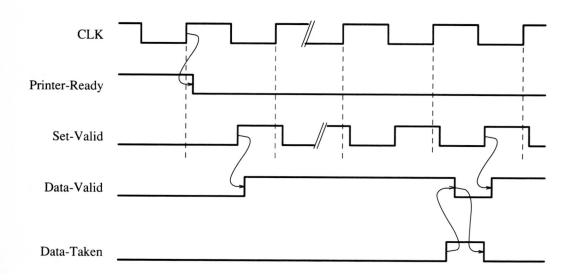

Figure 6.20 A timing diagram for the circuit of Figure 6.19.

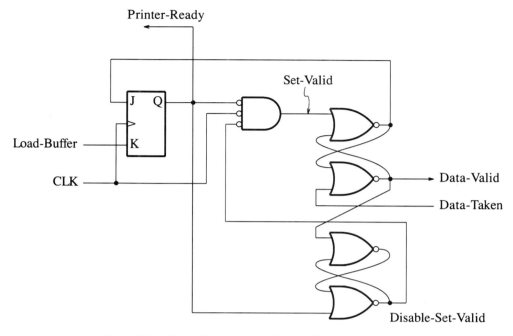

Figure 6.21   A modified handshake circuit, replacing that of Figure 6.19.

6.5 PRACTICAL CONSIDERATIONS

Several interface circuits were presented in the previous sections. The circuits were given in the simplest possible form, in order to focus the reader's attention on the principles involved. In practice these circuits might appear in a slightly different form. Consider, for example, the address decoder section in Figure 6.13, which uses the 15-input AND gate reproduced in Figure 6.23(a). When designing such a circuit, the designer is constrained to use components that are commercially available. There are many logic families to choose from, each having some limitations imposed by the technology used. One important restriction relates to gate fan-in, which is the number of inputs that a gate has. In TTL circuits, gate fan-in is usually limited to 8. Thus, the address decoder circuit cannot be implemented as a single AND gate. Instead, one may use the circuit shown in Figure 6.23(b).

We have tended to use AND and OR gates in most circuits, for the sake of clarity. It is often possible to reduce the number of gates needed, or improve the speed of the circuit, by making use of other types of gates. In fact, designers tend to use NAND and NOR gates wherever possible, because they have a simpler internal structure and introduce slightly less propagation delay than other types. With this in mind, the address decoder circuit may be implemented as shown in Figure 6.23(c).

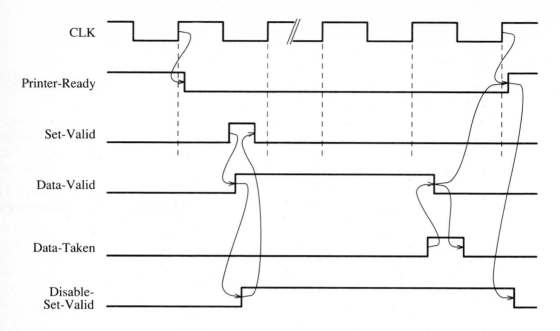

Figure 6.22   A timing diagram for the circuit of Figure 6.21.

The circuits of Figure 6.23 are often drawn as shown in Figure 6.24. While the two sets of diagrams describe the same hardware circuits, the choice of symbols in Figure 6.24 gives a better indication of the intended function of each gate. For example, the inverters in Figure 6.24(a) are shown with the circles on the input side rather than the output side as in Figure 6.23(a). The two symbols are equivalent. Both represent the same electrical circuit. In a given diagram, we choose one symbol or the other to focus attention on the logic state that is of interest. For example, one of the conditions the address decoder is intended to detect is whether A_3 is equal to 0. By placing the circle on the input side of the inverter we emphasize the fact that it is the 0 state of A_3 that the address decoder responds to, thus making the circuit a little easier to understand by a person examining the diagram. Similarly, the NOR gate in Figure 6.23(b) is better represented by the symbol in Figure 6.24(b), which shows an AND gate with inverted inputs. According to De Morgan's rule, the two functions are equivalent. However, it is preferable to use the symbol that more clearly describes the function being performed in the situation at hand. Since the address decoder is intended to recognize the condition $\overline{A_3} \cdot \overline{A_2} \cdot \overline{A_1} = 1$, the AND symbol with inverted inputs is more appropriate. Similar arguments apply to the OR and NOR gates in Figure 6.23(c), which have been replaced by equivalent AND gates in Figure 6.24(c). Inversion circles have been inserted as appropriate to maintain the functional equivalence.

Note that the resulting diagrams in Figure 6.24 contain only AND gates, with or without inversions. This is to be expected from our original simplified view of the

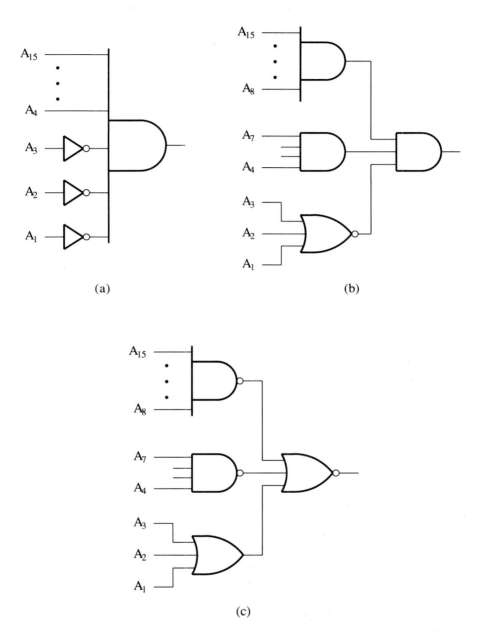

Figure 6.23 Different implementations for the address decoder of Figure 6.13.

address decoder. It performs an AND function on a large number of inputs, to detect the condition where certain inputs are equal to 1 and the others are equal to 0.

6.6 GENERAL-PURPOSE PARALLEL INTERFACE

The idea of a general-purpose interface — one that has several modes of operation and can be programmed to suit the needs of a particular I/O device — was introduced in Section 5.5. Motorola's PIA and Synertek's VIA were described, from the

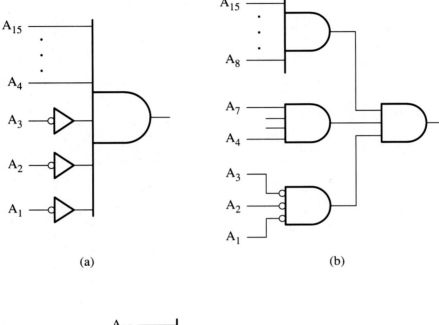

(a)   (b)

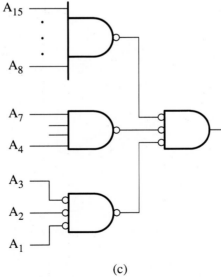

(c)

Figure 6.24   Preferred symbols for the gates in Figure 6.23.

programmer's point of view, as examples of integrated circuit chips that can be used to construct a general-purpose parallel interface circuit. We will now examine the hardware features needed to design a programmable interface circuit of this type. We will do this by introducing these features one by one, until we obtain an interface circuit that has the functional capabilities of a PIA or a VIA.

The idea of a general-purpose interface circuit is particularly useful when such a circuit is manufactured in integrated-circuit form. The large investment needed in the design and manufacture of an IC can be justified when the IC has a wide variety of applications and a large potential market. In the following discussion, we will assume that the general-purpose interface circuit being designed is intended for manufacture in the form of an IC. However, it should be emphasized that the ideas discussed are quite general, and are encountered in many hardware design tasks.

6.6.1 A Simple Single-Function Configuration

Consider the two interface circuits of Figures 6.15 and 6.18. We can combine them into one, to obtain an interface that may be used with both input and output devices. Components common to both circuits, such as parts of the address decoder, may be shared. Also, some changes to the original circuits can be made to make the combined circuit useful in a variety of applications.

The most restrictive feature in the interfaces of Figures 6.15 and 6.18 is that they are designed to respond to predefined addresses. A general-purpose interface chip should be usable with many different devices and in a variety of computer systems. Its flexibility is enhanced if address decoding is carried out, at least partially, in external circuits, thus making it possible for a system designer to choose addresses that are consistent with the requirements of a particular system.

Eliminating or reducing the amount of address decoding on the chip has another important benefit. One of the major objectives when designing circuits for production in integrated form is to keep the number of outside connections, and hence the number of pins on the IC package, to a minimum. A full address decoder requires a large number of external connections, because it has to be connected to all address lines — 16 in the cases of Figures 6.15 and 6.18. This number can be reduced if only a part of the address decoder is implemented on the chip. For example, we may choose the output NOR gate in Figure 6.23b to be incorporated on the chip, because it requires only three input connections. The remainder of the address decoder circuit can be implemented externally.

A possible combined circuit is shown in Figure 6.25. This interface has an input port and an output port. An input device may be connected to the data lines of port A, labeled PA_{7-0}. Its status line, e.g., Data-Valid, can be connected to CA, which serves as the control input for port A. Similarly, an output device may be connected to the output port, B, using data lines PB_{7-0}. The two control lines, CB1 and CB2, provide handshake capability for this port. Details of the control circuitry for the input and output control lines were given earlier in Figures 6.14 and 6.21.

The circuit enclosed by dashed lines in Figure 6.25 is the portion of the address decoder that has been incorporated in this interface circuit. There are four addressable locations: input data, input status, output data, and output status, which are

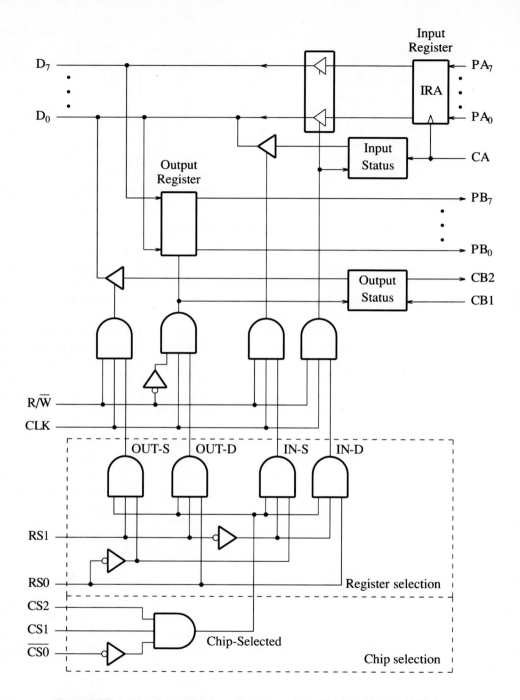

Figure 6.25   Internal organization of a general-purpose input/output interface.

selected by the signals IN-D, IN-S, OUT-D, and OUT-S, respectively. As mentioned in Chapter 5, such locations are often referred to as *interface registers*. The inputs to the address decoder in Figure 6.25 are divided into two groups. The first group consists of three *chip-select* lines called $\overline{CS0}$, CS1, and CS2. All three signals must be activated whenever any of the four interface registers is being addressed. The second group comprises the *register-select* inputs, RS1 and RS0. They determine which of the four registers will be accessed during a given read or write operation. Note that a complement sign — an overbar — is used to indicate that the $\overline{CS0}$ line is active when in the 0 state, while the other two chip-select inputs, CS1 and CS2, are active when equal to 1. The chip is selected when all chip-select inputs are active, that is, when $\overline{CS0} = 0$ and CS1 = CS2 = 1.

When several chip-select inputs are provided on an interface chip, it is useful to have some of them active when high and some active when low. This provides the system designer with added flexibility when choosing addresses and designing the address decoder. It is of course possible to have only one chip-select input on the chip, in which case all address decoding must be done externally. From the chip designer's point of view, such choices are heavily influenced by the desire to keep the number of external connections, hence the size and cost of the chip package, to a minimum.

Let us represent the interface of Figure 6.25 by the symbol in Figure 6.26, which shows all external connections. We will call this interface circuit a *General-Purpose Parallel Interface* (GPPI). Figure 6.27 shows how it may be used to connect a keyboard and a printer to the bus of a microprocessor. The keyboard and the printer are connected to Ports A and B, respectively. With the exception of the chip-select inputs, the data and control lines on the left side of the interface are connected directly to the corresponding lines of the microprocessor bus. The chip-select inputs are connected to the address lines of the bus via an appropriate decoder. Finally, the register-select lines are connected to the two least-significant address bits. Thus, the four registers in the interface will have adjacent addresses. The address decoding circuitry given in the figure corresponds to addresses FFF0 to FFF3, which were used earlier. The reader should verify that the circuit in Figure 6.27 performs the combined functions of the two interfaces in Figures 6.15 and 6.18.

6.6.2 A Flexible Multifunction Configuration

The GPPI of Figure 6.25 has two ports, A and B, each dedicated to a single function — either input or output. This is rather restrictive, particularly in view of our goal of defining as flexible a circuit as possible. We would like to increase the circuit's flexibility while keeping the number of its external connections small. In this section we will discuss three modifications aimed at achieving this objective, namely,

1. Combined input/output.
2. Multiple mode control.
3. Input latching.

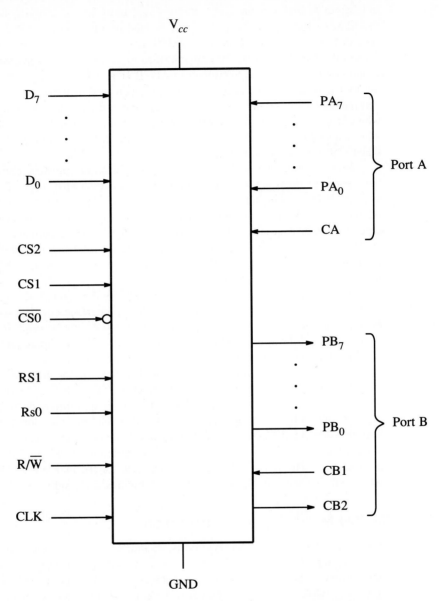

Figure 6.26   A single-chip general-purpose parallel interface.

COMBINED INPUT/OUTPUT

The flexibility of an interface circuit can be enhanced by making it possible for a given port to be used for either input or output transfers, depending upon the application at hand. In order to provide this flexibility, each port should contain circuitry for both input and output functions, together with a mechanism to activate one or the

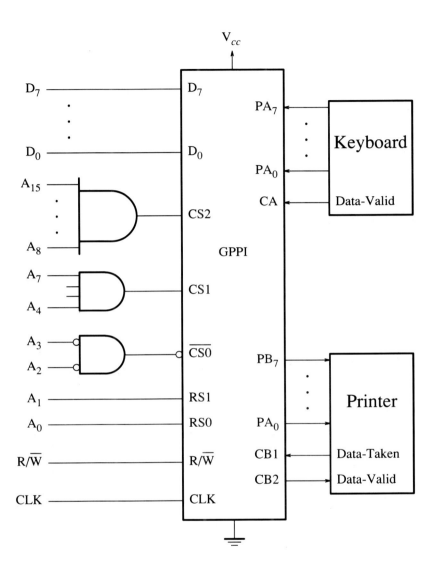

Figure 6.27 A keyboard and a printer connected to a microprocessor bus via a general-purpose parallel interface chip.

other under program control. In this way, a program may use a GPPI port to read input data from a device; then, later in the same program, the configuration of the GPPI can be changed to make the same port serve an output device.

The problem of using a given physical line to carry both input and output information was encountered earlier in Subsection 6.1.1, where it was pointed out that the key element needed is a tri-state driver. The output of a tri-state gate may be disconnected from the line when no output function is being performed. Using this approach an arrangement that enables Port A to be used in the desired manner is shown in Figure 6.28. An input register, IRA, and the tri-state drivers that connect it to the data lines of the microprocessor bus are provided, as before. An output register, ORA, similar to that of port B in Figure 6.25, is also provided. This register is connected to Port A data lines (PA_{7-0}) via another set of tri-state drivers, whose individual Enable inputs are controlled by an 8-bit *data direction register,* DDRA. Note that because the Enable inputs of different drivers are connected individually to the corresponding bits of DDRA, they need not all be enabled or disabled at the same time. When all 8 bits of DDRA contain 0s, the output drivers are disabled, and Port A functions as an input port in exactly the same manner as before. When DDRA contains 1s, the drivers are enabled, causing the data in ORA to appear on lines PA_{7-0}; that is, port A becomes an output port.

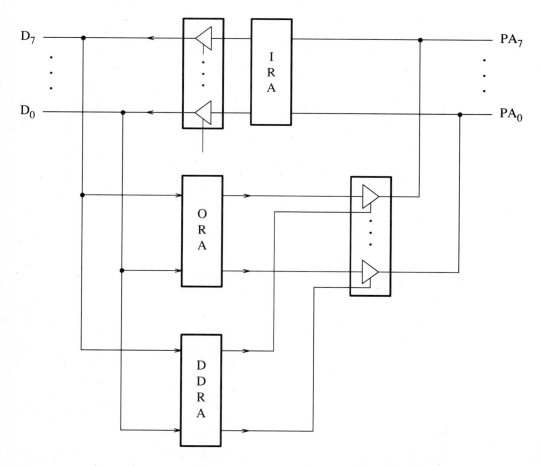

Figure 6.28 Combining input and output functions into one port.

In general, it is not necessary to configure all port lines in the same way. By appropriately setting the corresponding bits in DDRA, some PA lines may be used for input and some for output. The inputs to register DDRA are connected to the data bus, lines D_{7-0}, making it possible for the register contents to be updated at any time by the microprocessor. Thus, the goal of placing port configuration under program control has been achieved.

Different data and control registers in a chip should be assigned distinct addresses, to enable program instructions to refer to them individually. However, as mentioned in Subsection 5.5.1, the input and output data registers of a given port, such as IRA and ORA, need not have separate addresses. They can be treated as a single location. A read operation at this location results in transferring the input data in IRA to the microprocessor, while a write operation places new output data into ORA. An implementation of this idea is illustrated in Figure 6.29. The address decoder recognizes the common address when it appears on the register-select inputs of the chip, and it asserts PA-Data. Then, either IRA or ORA is selected, depending upon the state of $R/\overline{W}$.

MULTIPLE-MODE CONTROL

When the status flag circuits in Figures 6.14 and 6.21 are used, the port control lines behave in a certain way. We have also encountered examples of other modes of operation for the control lines, such as the pulse-mode handshake. In order to enhance the flexibility of the GPPI in Figure 6.25, we may replace each of the status and handshake control circuits by a more flexible block, which we will call the Port Control Logic, and replace the single control line of port A, CA, by two control lines, CA1 and CA2. The Port Control Logic should be designed to operate in a variety of modes, selectable under program control. As additional flexibility, it may also allow the user to define the active edge on the control lines to be either positive or negative. To make the choice of the mode of operation programmable, a control register may be introduced as shown in Figure 6.30. The contents of the Control Register define the functions performed by the Port Control Logic. A particular bit pattern loaded in this register may cause the Port Control Logic to behave in the same manner as the Status Flag circuit in Figure 6.14. Another pattern may cause it to perform the function of the Handshake Control circuit in Figure 6.21. The Port Control Logic generates a status bit and, if appropriate, sends a pulse to store data in register IRA, depending upon the contents of the Control Register.

A full development of the design of the Port Control Logic block with all its modes of operation would be quite involved, and would not necessarily contribute to the understanding of the idea of a programmable circuit. For the purposes of the present discussion, it suffices to show a simple example. Assume that in one mode of operation, the input data is latched into register IRA by the active edge of CA1. When CA1 is connected directly to the clock input of register IRA, as in Figure 6.25, the active edge is determined by the type of flip-flops that make up IRA. Alternatively, the circuit shown in Figure 6.31 may be used to enable the programmer to define the active edge to be either positive or negative. The choice is made by depositing a 1 in bit 0 of the Control Register to select the positive edge or a 0 to select the negative edge. Register IRA is assumed to be positive-edge triggered. When

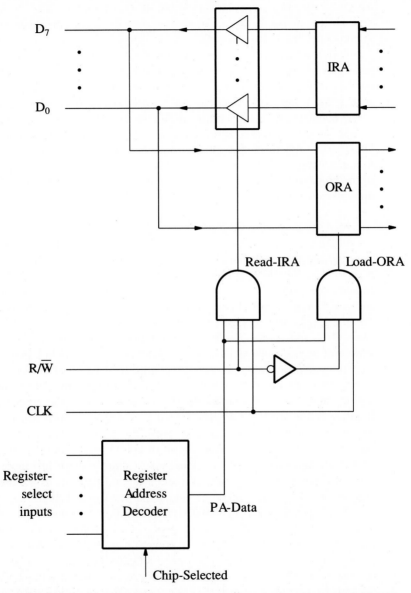

Figure 6.29 Assignment of the same address to the input and output registers of port A.

FF0 contains a 1, i.e., $\overline{Q} = 0$, then

$$\text{Load-IRA} = 0 \oplus \text{CA1} = \text{CA1}$$

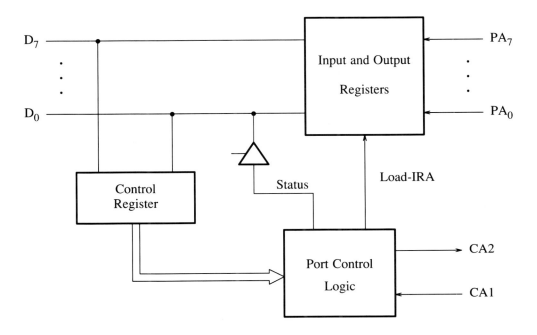

Figure 6.30 Mode selection for port A control lines.

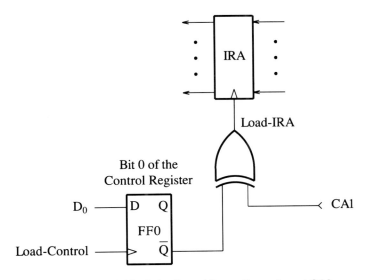

Figure 6.31 Selection of the active edge of CA1.

Hence, a low-to-high transition on CA1 will result in a low-to-high transition on Load-IRA and will load the input data into IRA. If FF0 contains a 0, Load-IRA = CA1, making a high-to-low transition on CA1 the active edge.

INPUT LATCHING

The input interface circuit of Figure 6.15 uses an edge-triggered storage buffer to store 8 bits of input data. Register IRA in the GPPI performs a similar function. However, the simpler interface of Figure 6.13 does not use a storage buffer between the data bus and the input device. It would be beneficial to be able to use the same GPPI in either situation, and configure it to operate in the desired mode. To this end, the GPPI should incorporate a facility to disable input latching in register IRA when not required. When input latching is disabled, the microprocessor reads the data on lines PA_{7-0} directly. The flexibility to choose either latched or direct input is found in the VIA chip. Figure 6.32 gives an example of the circuitry needed. To keep the figure simple, only one input line is shown. A control flip-flop, LA (latching control for port A), which is assumed to be one of the bits of the Control Register in Figure 6.30, selects the input latching mode. During a read operation, Enable-D goes to 1 to place the input data on the data bus. This signal enables one of the two tri-state drivers labeled Latched and Direct, depending upon the state of the control flip-flop, LA. Recall that input data is loaded into IRA under control of the Load-IRA signal, which is generated by the Port Control logic in Figure 6.30. This is the latched data, which is read when LA contains a 1. When LA contains a 0, the port data lines are read directly.

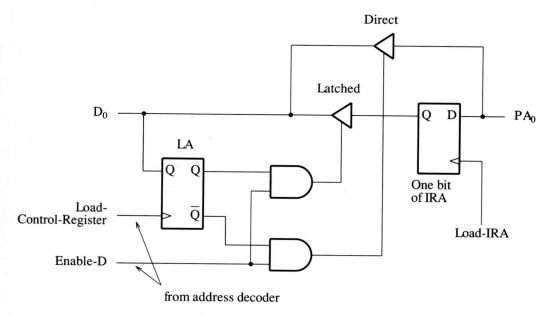

Figure 6.32 Selection of direct or latched inputs.

6.6.3 PIA Example

We will now return to the two general-purpose parallel interface chips that were used as examples in Chapter 5. The PIA will be discussed in this section and the VIA in the next. Because of the similarity of the two chips, most of the discussion given here applies equally to the VIA, and it will not be repeated in the next section.

The internal organization of the PIA is shown in Figure 6.33. The programmer's view of the operation of the PIA was discussed in detail in Subsection 5.5.2. Here, we will discuss two hardware design issues. The first is the connection of the PIA to a microprocessor bus, and the second involves the timing details relating to the operation of the port-control lines.

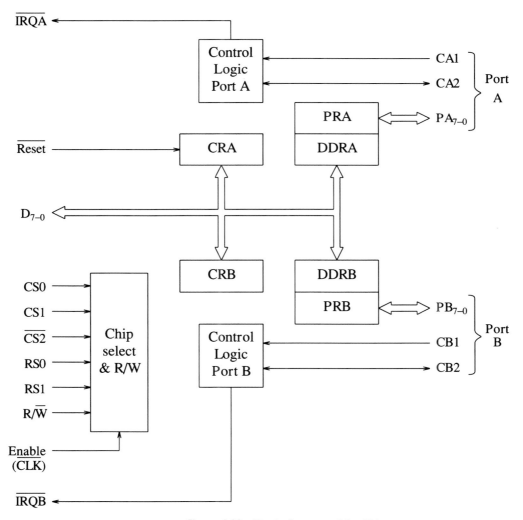

Figure 6.33 Block diagram of the PIA.

The PIA is connected to a microprocessor bus using the signals on the left side of Figure 6.33. There are three chip-select inputs, CS0, CS1, and $\overline{\text{CS2}}$, which should be connected to the address lines via an address decoder, depending upon the address assigned to the PIA. The two least-significant address lines should be connected to the register-select lines RS0 and RS1, as was discussed in conjunction with Figure 6.27. The data lines, D_{7-0}, the R/$\overline{\text{W}}$, and the $\overline{\text{Reset}}$ lines should be connected to the corresponding lines of the microprocessor bus.

The clock input to the PIA is called Enable. This signal is the complement of the clock signal used in our earlier examples. Figures 6.8 and 6.11 show CLK equal to 1 during Phase 1. Data are read from the interface registers or written into them while CLK = 0. For the PIA, data transfer takes place while Enable = 1. This choice is compatible with the signal conventions on the 6809 bus. The clock signal on the 6809 bus is called E, for enable, where E = 0 during Phase 1 and E = 1 during Phase 2 of a clock period. Hence, the Enable input of the PIA may be connected directly to the E signal on the 6809 bus.

The last two bus signals in Figure 6.33 are $\overline{\text{IRQA}}$ and $\overline{\text{IRQB}}$. These are interrupt-request signals used in conjunction with interrupt driven I/O. They will be discussed in Subsection 7.2.1.

In summary, connecting the PIA to a synchronous bus of the type discussed earlier in this chapter is a straightforward task. In the case of the 6809 bus, all PIA signals, except the three chip-select inputs, are connected directly to their counterparts on the bus. The chip-select inputs are connected via an address decoder. The simplicity of this connection is a result of the idea of a family of compatible chips that are easily interconnected to construct a complete microprocessor system. When using chips from different families, the designer must examine the functional and timing specifications of each chip carefully to determine their compatibility. Intervening circuitry may have to be provided to allow chips from different families to work together in a single system.

We now turn our attention to the port control lines, CA1, CA2, CB1, and CB2. The functional behavior of these lines was described in Subsection 5.5.2. The reader is advised to review that discussion, particularly Figures 5.27 to 5.29, before studying the timing details presented here. When handshake signaling is used, an input device is always connected to port A and an output device to port B, as shown in Figure 6.34. Lines CA1 and CB1 may be programmed to be active either when high or when low, by depositing the appropriate pattern in the control registers, CRA and CRB. The polarity used must be chosen to suit the requirements of the I/O devices involved. Assume that the device connected to port A generates a status signal called $\overline{\text{Data-Valid}}$, which is active low. This signal should be connected to input CA1 of the PIA. Therefore, CA1 should be programmed to be active low, by setting CRA_1 to 0. The response of the PIA is generated on line CA2. The port circuitry in the PIA restricts this line to be active low in any of the handshake modes. The CA2 line should be connected to the response input of the device, which must also be active low. We have called this input $\overline{\text{Data-Taken}}$.

For the output device, the PIA activates CB2 to indicate that valid data is available, and waits for the device's response on line CB1. Hence, these two lines are

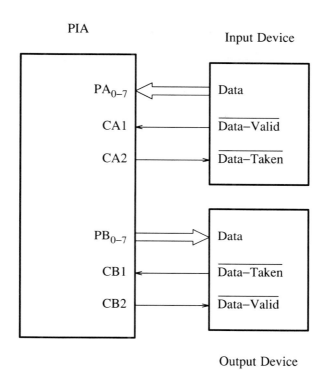

Figure 6.34 Connection of input and output devices to the PIA using the handshake signals.

connected to the $\overline{\text{Data-Valid}}$ control input and $\overline{\text{Data-Taken}}$ control output of the device, respectively. We have again assumed that both signals are active when low.

Let us consider first an input operation controlled by a pulse-mode handshake. The program that controls the input operation reads status bit SA1 repeatedly, until it becomes equal to 1. Then, the program reads the input data from peripheral register PRA. The times at which the control signals and the status bit change state during this process are illustrated in Figure 6.35. The input device sends a pulse on CA1 whenever it has data ready for transfer to the microprocessor. Since the input device is not controlled by the clock signal on the microprocessor bus, the pulse on CA1 is asynchronous relative to the clock signal on the Enable line. That is, the time at which the leading edge of the pulse is received, t_1, bears no relation to the rising and falling edges of the clock. For this reason, the PIA requires the pulse on CA1 to be wider than one clock period to guarantee that it will be recognized. A narrow pulse may be missed completely. The activation of CA1 causes the microprocessor to set its input status bit, SA1, to 1. This happens at time t_2, which is the end of the clock period during which the active state on CA1 is recognized. Some time later, the microprocessor reads register CRA (Control Register of Port A) and checks the state

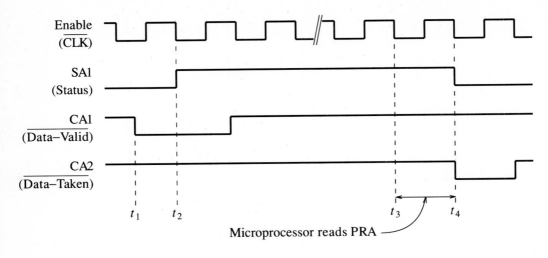

(a) Input operation

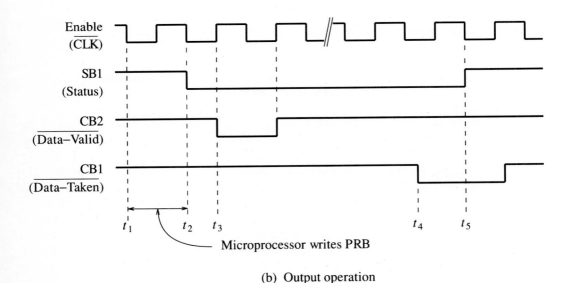

(b) Output operation

Figure 6.35 Pulse mode handshake between the PIA and an I/O device.

of bit SA1. To keep the figure simple, this event is not shown. Observing that SA1 is now equal to 1, the microprocessor proceeds to read the data from register PRA (Peripheral Register of Port A) at time t_3. Upon completion of this read operation, at t_4, the PIA clears the status bit. It also sends a pulse on CA2, indicating to the input device that the microprocessor has read the data. The width of the pulse on CA2 is equal to one clock period.

Part (b) of Figure 6.35 illustrates an output handshake on port B. The microprocessor checks status bit SB1, then writes new data into register PRB (Peripheral Register of Port B). The write operation starts at time t_1 in the figure and ends at t_2, at which point the PIA clears the status bit. The new data appear on the port data lines at t_2. In order to provide a setup period for the output device, the PIA waits until t_3, then sends a pulse on its output control line, CB2, indicating that the data are valid. In this way, the output data are guaranteed to be valid for half a clock period before CB2 is activated. Sometime later, at t_4, the device responds by sending a pulse on CB1, which causes the PIA to return the status bit to 1, indicating that the device is now ready for another output operation. As in the case of CA1, the pulse on CB1 is not synchronized with the clock input of the PIA. Hence, the status bit is actually set at t_5, the end of the clock period during which CB1 is activated. The pulse sent by the device on the CB1 control line must be wider than one clock period to guarantee that it will be recognized by the PIA.

In the pulse-mode timing diagrams in Figure 6.35, the pulses sent by the PIA on the CA2 and CB2 lines have a fixed width of one clock period. The main difference encountered in the full handshake mode is that the width of these pulses is not fixed. Instead, the PIA keeps the line active until the I/O device activates CA1 or CB1, as appropriate. A timing diagram for a full handshake on port A is shown in Figure 6.36. The operation begins at t_1 with a request from the I/O device on line CA1, which causes the status bit to be set at time t_2. At the same time, the PIA deactivates CA2, regardless of the previous state of that signal. The microprocessor checks the status register and then reads the input data. The figure shows the input data being read during the clock period between t_3 and t_4. As a result, the PIA clears the status bit and activates CA2 at t_4. The latter action informs the input device that the microprocessor has read the data. The input device may return CA1 to the inactive state at any time, but it must do so before CA2 becomes active, that is, before t_4. This condition is a departure from the fully interlocked scheme described in Figure 5.21 in Chapter 5, where each signal transition on one control line is a response to a transition on the other. However, the scheme in Figure 6.36 is still referred to as a "full handshake" in the manufacturer's literature. After activating CA2, the PIA maintains it in the active state until the input device signals the beginning of a new input transfer by activating CA1, as shown at time t'_1 in the figure. The PIA responds at time t'_2 by setting SA1 to 1 and deactivating CA2, and so on.

The above discussion is based on the assumption that an input transfer is initiated by the input device, which activates CA1 each time it places new data on the port data lines. The names Data-Valid and Data-Taken for the signals on CA1 and CA2 are consistent with this view. We may also interpret the CA1/CA2 handshake in a slightly different way. Activation of CA2 by the PIA may be regarded as a request for data, or an indication that the PIA is ready to receive new data from the

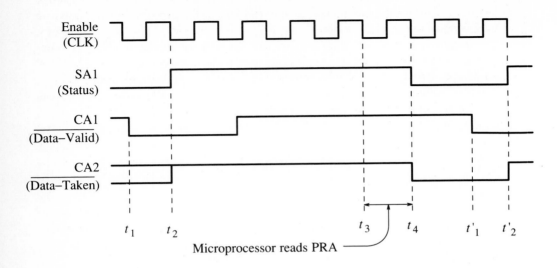

Figure 6.36   Timing of a full handshake between port A of the PIA and an input device.

input device. The device responds by placing data on the port data lines and activating CA1. With this interpretation, the signals on the CA2 and CA1 lines may be called Ready-for-Data and Data-Available, respectively, again assuming that they are both active when in the low-voltage state. Since, in this case, the device waits for a request from the PIA before sending data, CA2 must be activated first, as shown at time t_2 in Figure 6.37. We can arrange for that to happen by issuing a dummy "read PRA" operation at the beginning of the program that uses the PIA. At

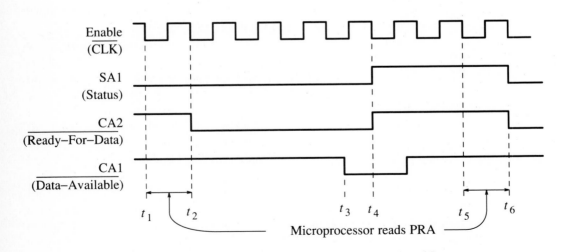

Figure 6.37   An alternative interpretation for the handshake sequence of Figure 6.36.

the end of that operation the PIA activates CA2. Of course, the data read in such an operation are meaningless and should be discarded, because nothing has yet been sent by the input device. Valid input data will be available after the input device activates CA1, which happens at t_3. The status bit is set to 1 at t_4. After checking the status bit, the microprocessor reads the input data at t_5. At t_6, after the data have been read, the PIA clears the status bit and activates CA2 a second time, to request a second byte of data. Hence, the dummy read operation is only needed at the very beginning, as part of the initialization procedure.

Full handshake signaling during an output operation on port B is illustrated in Figure 6.38. After checking the output status flag, SB1, in register CRB, the microprocessor writes the output data into register PRB at time t_1. At the end of the write operation, at t_2, the PIA clears SB1. As in the pulse mode, the PIA activates the CB2 line on the first rising edge of Enable following the write cycle, thus allowing a setup time of half a clock period for the output data. At t_4 an active transition is received on CB1, indicating that the output device has accepted the data. In response, the PIA returns CB2 to the high state and sets the output status flag to 1 at t_5. The time at which the input device should return CB1 to the inactive state is not constrained by the PIA. But, this transition should take place before the PIA activates CB2 to signal the beginning of a new transfer. Also, the resulting pulse on CB1 must be wider than one clock period, to guarantee that it will be recognized by the PIA.

6.6.4 VIA Example

The VIA parallel interface chip is similar to the PIA in its basic functions, but offers several additional features. Some of these features were described in Subsection 5.5.3. Other features, in particular the timer and shift register facilities, will be introduced in Chapter 7.

The hardware organization of the VIA is shown in Figure 6.39, where the signals on the left side of the figure are those used to connect the VIA to the bus of a microprocessor. There are two chip-select lines, CS1 and $\overline{CS2}$, which should be connected to an address decoder, as explained earlier. The four register-select lines, RS0 to RS3, are usually connected to the four least-significant address bits. Thus, the internal registers of the VIA will appear as 16 contiguous locations in the addressable space of the microprocessor.

The handshake modes used to control data transfer through the ports of the VIA are essentially identical to those of the PIA. The timing details for these operations were discussed in conjunction with Figures 6.35 to 6.38. One difference between the two chips is encountered in input operations. The VIA incorporates a facility for latching input data in a storage buffer, which may be enabled or disabled under program control. With reference to Figure 6.35a, the microprocessor reads port A during the clock period between t_3 and t_4. When input latching is disabled, the microprocessor receives the data that are actually on the port lines at that time. On the other hand, if input latching is enabled, the data are latched at the time an active transition is received on CA1, that is at time t_1. Thus, when the microprocessor

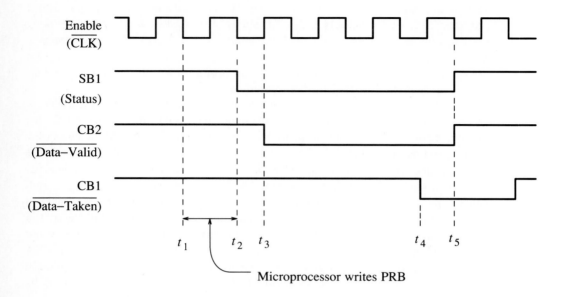

Figure 6.38 Timing of a full handshake during an output operation on port B.

reads port A, it receives the data that existed on the port data lines at time t_1. The input device is not required to maintain valid data at the port lines from t_1 to t_4. However, it must guarantee that the input data meet the setup and hold times required by the input latch of port A, relative to the active edge on CA1.

6.7 SERIAL INPUT/OUTPUT

A parallel connection requires a large number of wires. It is expensive to implement when I/O devices are physically far away from the computer. Serial transmission, where bits are transmitted one at a time, is more suitable in such circumstances. For this reason, serial transmission is widely used for connecting video terminals to computers.

The idea of a serial interface was introduced in Figure 5.3, which is reproduced here as Figure 6.40. Parallel data from an input device such as a keyboard are transmitted one bit at a time using a parallel-to-serial converter. At the receiving end, the serial interface converts the data back to the parallel format, which is the format used on the microprocessor bus. Similar conversions take place during transmission from the microprocessor, via the serial interface, to an output device.

Chapter 5 described a typical serial interface that uses asynchronous transmission from the programmer's point of view. Serial transmission and serial interfaces pose a number of interesting problems, which we will now discuss in some detail.

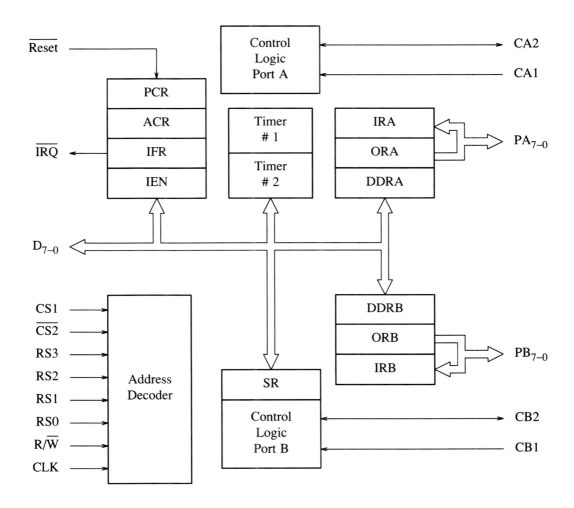

Figure 6.39 Internal organization of the VIA.

6.7.1 Serial Transmission

The basic idea for serial transmission of data is illustrated in Figure 6.41(a). Assume that parallel data are to be transferred from point A to point B. The data at A are loaded into a shift register. Then, under control of the transmitter clock signal, TCLK, the data are shifted out one bit at a time, and transmitted on the serial link. At the receiving end, the process is reversed. The data received on the data link are shifted into a shift register using the receiver clock, RCLK, then delivered in parallel at point B. Let the data pattern loaded into the shift register at A be 1000 1011. The transmitted data and their relationship to the transmission clock are shown in part (b)

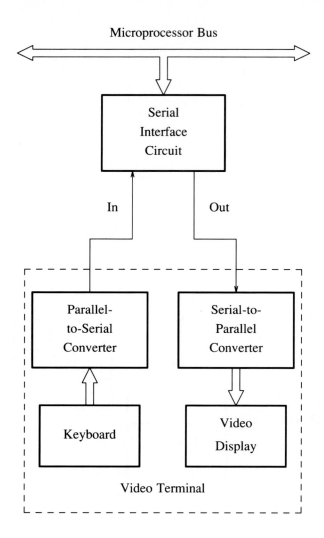

Figure 6.40 Serial connection of a video terminal.

of the figure. It has been assumed that the shift register is arranged so that the least-significant bit is shifted out first.

For data transfer to take place correctly, the transmitter and receiver clocks must not only have the same frequency, but they must also have the same *phase relation-ship* to the serial data stream. That is, the relative position of the rising edges of RCLK and the state transitions of the received data must be the same as for TCLK and the transmitted data. Thus, in its simplest form, serial transmission requires two separate links, one for the data and one for the clock signal, as shown in Figure 6.41(a).

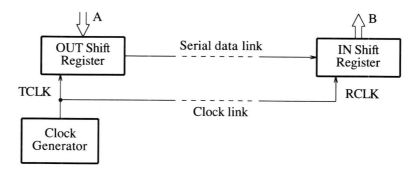

(a) Two-link scheme

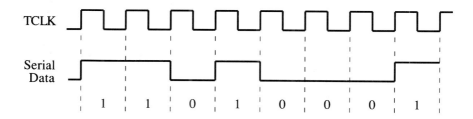

(b) Time relationship between clock and data

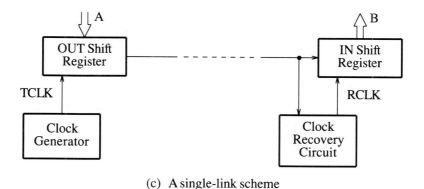

(c) A single-link scheme

Figure 6.41 Synchronous transmission.

The use of a separate transmission link to carry the clock signal has two draw-backs. First, it increases the number of wires required for transmission. The second and more serious problem is that, due to possible differences between the propagation delays on the two links, this arrangement does not guarantee proper phase relationship between the clock and data at the receiving end. Several schemes have been developed that enable clock information to be transmitted along with the data on one transmission link. Serial links that use such schemes may be represented schematically as in Figure 6.41(c). A clock recovery circuit is used to extract clock information from the serial data stream and generate the receiver clock signal, RCLK.

Serial transmission schemes are divided into two broad categories — synchronous and asynchronous. They differ in the way in which clock information is incorporated into the data stream.

SYNCHRONOUS TRANSMISSION

The term "synchronous transmission" refers to schemes in which the receiver and transmitter clocks are kept in synchronism; that is, transitions of the two clock signals are kept in step. Ideally, this could be achieved by using two identical oscillators to generate the two clock signals TCLK and RCLK. However, the frequency of any oscillator circuit varies with operating conditions, such as temperature, humidity, aging, etc. Therefore, no two oscillators can be guaranteed to have exactly the same frequency over an extended period, even though they may have the same design. In synchronous transmission, the oscillator at the receiving end is designed so that its frequency of oscillation is adjustable within a small range around its nominal value. The function of the clock recovery circuit is to continuously monitor the phase relationship between the received signal and RCLK, and, whenever necessary, to adjust the frequency of the local oscillator to maintain the required phase relationship. A circuit that implements this idea is known as a *phase-lock loop*.

Correct operation of the clock recovery circuit is dependent upon the presence of frequent state transitions in the data stream. Absence of such transitions for a long time, e.g., in the middle of a long stream of 0s or 1s, may result in the oscillator at the receiving end drifting away from the correct frequency and causing errors. Synchronous transmission systems use a variety of encoding techniques to ensure that this does not happen. The details of encoding and synchronization techniques are beyond the scope of this book. Interested readers may refer to books on digital communications [4, 5].

ASYNCHRONOUS TRANSMISSION

In asynchronous transmission, no attempt is made to keep the receiving clock in synchronism with the clock at the sending end. Instead, RCLK is derived from a clock generator that has approximately the same frequency as the oscillator at the transmitting end. In order to illustrate this transmission scheme, let us examine the relationship between the data and clock signals at the receiving end. Four bits of data, b_k to b_{k+3}, are shown in Figure 6.42. Assume that they are loaded into the IN shift register of Figure 6.41 at consecutive rising edges of RCLK. Hence, the data will be received correctly if one rising edge of RCLK occurs within each bit period.

Assume that for bit b_k, the rising edge of RCLK occurs exactly in the middle of the bit period, and that RCLK is slower than the transmitting clock. In this case, the rising edge of the receiver clock will occur after the mid-point of bit b_{k+1} by some

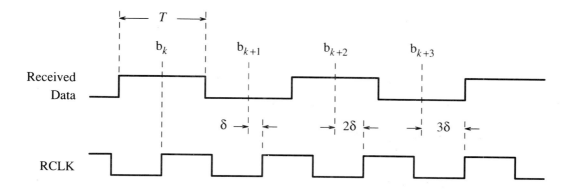

Figure 6.42 Relative timing of data and receiver clock in an asynchronous system.

amount δ. Bit b_{k+1} will still be received correctly, provided that δ is smaller than $T/2$, where T is the bit period of the received data. The relative shift between RCLK and the received data increases in subsequent bit periods, until some bit b_{k+n} is reached where the accumulated shift, $n\delta$, exceeds $T/2$. At that point erroneous reception will occur. A similar situation develops if RCLK is faster than TCLK, except that in this case δ will have a negative value.

In asynchronous transmission, the possibility of errors being caused by accumulation of clock shift is avoided by transmitting data bits in small groups of fixed size. The two clocks TCLK and RCLK are designed to have frequencies as close as practically possible, usually through the use of crystal-controlled oscillators. Moreover, a control circuit in the receiver adjusts the position of the rising edge of RCLK to be near the mid-point of the first bit in a group. With a small number of bits in a group and a small value for δ relative to the bit period, all bits are guaranteed to be received correctly.

A standard scheme that implements the above idea is known as the *Start-Stop* transmission protocol, illustrated in Figure 6.43. When no data are being transmitted, the transmission line is maintained in the 1 state. Data are transmitted in groups of five to eight bits, where each group is preceded by a 0 bit, called the *Start bit*, and followed by a 1 bit, called the *Stop bit*. Another bit, called the *parity bit*, may be transmitted before the Stop bit as shown. Its use will be explained shortly. The data bits, which may consist of the ASCII code for one character, for example, are transmitted starting with the least significant bit. The receiver circuit uses the falling edge of the Start bit to estimate the position of the mid-point of the first bit of data, and it generates RCLK pulses accordingly. After receiving all data bits in the group, the receiver checks for the presence of the Stop bit. If a 0 is encountered during the period in which the receiver expects the Stop bit, a *framing error* is said to have occurred. This may result from an erroneous Start bit, possibly caused by electrical noise on the transmission line, or from the difference in frequency between TCLK and RCLK being too large.

In a typical connection between a video terminal and a computer, the asynchronous Start-Stop scheme is used, where eight data bits are transmitted for each

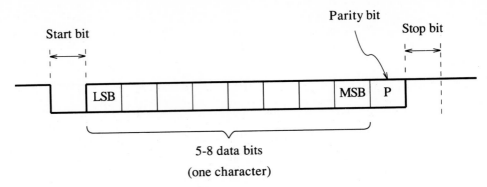

Figure 6.43 Start-stop format.

character. The first 7 bits represent the ASCII code of the character, starting with the least-significant bit. The eighth bit is usually set to 0. As mentioned before, a parity bit may or may not be added at the end of the character. When included, this bit is used to make the overall parity either even or odd. For odd parity, for example, the parity bit is set to either 0 or 1, depending upon the state of the eight data bits, to make the total number of 1s transmitted odd. The receiver computes the parity of the received character, to verify that it is odd. If a transmission error occurs causing a change in one bit the error will be detected, because the parity of the received data will be changed from odd to even. In general, there is a 50 percent chance that a transmission error affecting any number of bits will result in a reversal of the parity of the received character, and hence will be detected.

In many keyboards, a special key called the *Break* key is provided. Pressing this key causes a character to be transmitted with an intentional framing error, i.e., with the Stop bit equal to 0. In fact, pressing this key results in transmitting a very wide 0 bit, usually five times or more the width of a normal character. Hence, the receiving end will detect a 0 character and a framing error. Pressing the *Shift* key at the same time as the break key causes an even longer break to be transmitted. The Break key is used in many computer systems to force a particular action, such as causing a program to be terminated. Such a facility provides a means for recovering from situations where a computational process has gone astray.

6.7.2 A Serial Interface

As in the case of a parallel interface, a flexible serial interface chip can be defined, and several are commercially available. Typically, a single chip contains all the circuits needed for the transmitter and the receiver sections of the interface. For reasons similar to those discussed in Section 6.6, full address decoding is not included on the chip. Instead, a few chip-select and register-select inputs are provided. A general-purpose serial interface chip is often called a *Universal Asynchronous Receiver Transmitter,* or UART, and many such chips are available in the families associated with particular microprocessors.

Motorola's Asynchronous Communications Interface Adapter (ACIA), was described in detail in Chapter 5 as an example of a UART chip. Its internal organization is shown in Figure 6.44. The ACIA may be connected to the microprocessor bus in the same way as the parallel interface chips described earlier. The data, R/$\overline{\text{W}}$ and clock (called ϕ_2) lines may be connected directly to the corresponding lines on the 6809 bus. The clock input of the ACIA is compatible with the clock line, E, on the 6809 bus, which will be described in the next section. The E signal is the complement of CLK as defined in Section 6.2. The chip-select lines, $\overline{\text{CS0}}$ and CS1, should be connected to the microprocessor bus through a suitable address decoder, depending upon the address range assigned to the ACIA. The two register-select lines, RS0 and RS1, are usually connected to the least-significant address bits of the bus. Thus, the internal registers of the ACIA appear as four adjacent locations in the address space of the microprocessor.

It was pointed out in Chapter 5 that the ACIA may be reset by a write operation to address 01 (RS1 = 0 and RS0 = 1), which causes the Command register to be reset to the pattern XXX00000. As a result, both the transmitter and the receiver are disabled, but the parity mode remains unchanged. The ACIA may also be reset by the hardware, by activating its $\overline{\text{Reset}}$ input line. A hardware reset clears both the Command register and the Control register to an all-0 pattern. The $\overline{\text{Reset}}$ input is usually connected to the reset line of the microprocessor bus.

The interrupt request output, $\overline{\text{IRQ}}$, is used in conjunction with interrupt-driven I/O, as will be discussed in Chapter 7. We will describe the RS-232-C signals and the way the serial input and output lines are connected to a transmission link in Chapter 11.

6.8 BUS TIMING ALTERNATIVES

The discussion of bus timing and input/output interfaces presented in Sections 6.2 to 6.4 was based on the assumption that all data transfer operations can be completed within one clock period. Hence, the clock period must be chosen to accommodate the slowest device connected to the bus. For example, if the delay involved in transmitting data during the input operation of Figure 6.8 varies from one device to another, the duration of Phase 2 of the clock must be longer than the largest possible value of this delay.

It is often desirable to provide a mechanism whereby a slow interface can be connected to a bus without having to slow down data transfers for other devices. In this section we discuss two alternatives for dealing with devices that have widely differing speeds of operation. The first involves the introduction of a wait state during data transfer operations, and the second uses asynchronous handshake signaling.

6.8.1 Wait States

Slow devices can be accommodated on a synchronous bus by a simple modification to the timing diagrams of Figures 6.8 and 6.11. The device interface may be allowed

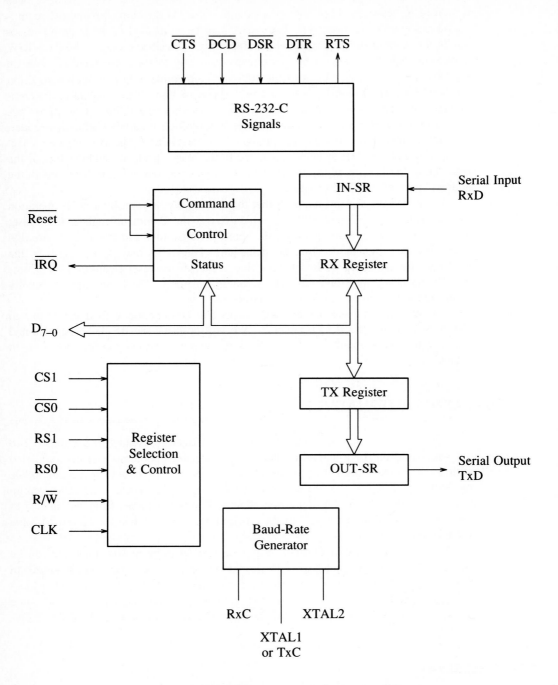

Figure 6.44 Internal organization of the ACIA.

to skip one or more clock cycles before completing its part of the transfer operation. During this period the microprocessor must be informed, so that it may suspend its internal operation until the device is ready. Let us assume that a signal called $\overline{\text{WAIT}}$ is provided on the bus for this purpose. This signal is normally in the inactive state, which we have assumed to be the high-voltage state for reasons that will become apparent shortly. Data transfers involving fast devices that can respond within one clock period proceed as described earlier. During these transfers, the $\overline{\text{WAIT}}$ signal remains inactive. For a slow device, the input operation of Figure 6.8 is modified as shown in Figure 6.45. The device interface, having recognized its address on the bus and knowing that it cannot transmit the requested data immediately, activates the $\overline{\text{WAIT}}$ signal. As a result, the microprocessor enters a wait state, maintaining the address and control information on the bus. In a subsequent clock cycle, when the device interface is ready to transmit the data, it returns $\overline{\text{WAIT}}$ to the inactive state and places the data on the data lines. It does so at the same point within the clock period where input data should be transmitted, which is at the beginning of Phase 2 in our example. At this point, the microprocessor resumes normal operation. The duration of an output transfer may be extended in a similar manner.

To guarantee correct operation, the wait signal must meet certain timing constraints. It may change state only at certain times relative to the bus clock. For example, the addressed device may be required to activate $\overline{\text{WAIT}}$ only during the first phase of a clock period, to provide sufficient time for the microprocessor to take

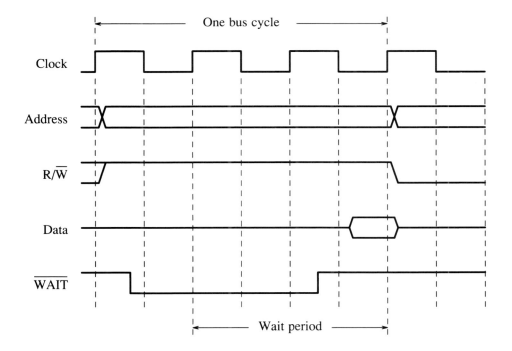

Figure 6.45 Use of a wait signal to extend the duration of a bus transfer operation.

appropriate action. During the wait period, the clock signal may continue to run, as is the case in Figure 6.45. Alternatively, the second phase of the clock may be extended until the wait signal becomes inactive. The latter approach is used on the 6809 bus, which will be introduced in Section 6.9.

In some microprocessors, device interfaces are allowed to activate the wait signal for an indefinite period of time. Other microprocessors cannot wait beyond a maximum period, after which some of the information stored in the microprocessor circuitry may be lost. This is a characteristic of a type of logic circuits known as *dynamic logic,* commonly used in microprocessors. A dynamic logic circuit stores information in the form of a charge on a capacitor. During normal operation, this charge is refreshed periodically. However, if the operation of the microprocessor is suspended for a long time, the stored charge may be lost. For this reason, the duration of the wait signal may not be allowed to exceed a specified maximum value, usually on the order of several microseconds.

The wait signal gives rise to a special requirement. In order to ensure that the microprocessor will not enter a wait state erroneously, the wait signal must be in the inactive state at all times, except when driven into the active state by some device interface. The tri-state arrangement described in Subsection 6.1.1 is not suitable for this purpose, because when all tri-state gates connected to a bus line are turned off, the state of the line is indeterminate. An alternative arrangement that is suitable for the $\overline{\text{WAIT}}$ line is the *open-collector* connection, which uses special open-collector gates, as explained below.

OPEN-COLLECTOR BUS

Figure 6.46 illustrates the principle of operation of open-collector gates. The output of each gate is equivalent to a switch to ground. A number of such gates may be connected to a single bus line, as shown in the figure. One resistor, known as a *pull-up resistor,* connects the bus line to the positive supply voltage, V_{cc}. When all gate inputs are inactive, all switches are open, and the voltage on the bus line is equal to V_{cc}. If one or more gates are active, the corresponding switches are closed, and the voltage on the line drops to 0. Thus, an open-collector line is regarded as inactive when in the high-voltage state. It is in the active state when it is "pulled down" toward ground voltage by one of the gates connected to it.

With this interpretation of the signals on an open-collector bus, the bus is active when one or more of the signals that drive it is active. Hence, the bus performs a logical OR function. The origin of the name "open-collector" relates to the details of the circuit that implements the switch inside each gate. Such details are not essential to our discussion.

The operation of an open-collector bus bears an interesting similarity to the operation of signaling bells in public transportation vehicles. A bell is operated by a wire strung along the ceiling of the vehicle. Any passenger who wishes to get off the bus can alert the driver by pulling down on this wire, at any location in the vehicle.

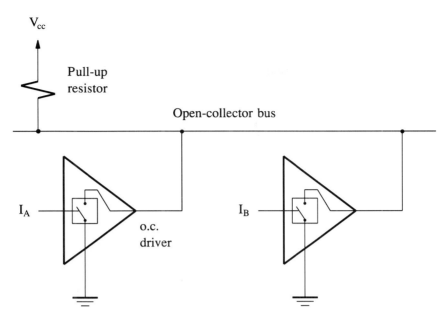

Figure 6.46 A schematic diagram representing the operation of an open-collector bus.

6.8.2 Asynchronous Transfers

The bus timing schemes discussed so far are synchronous schemes. They depend on a clock signal to provide a common timing reference relative to which all other signals are specified. The transfer of one unit of data may be completed in one or several clock periods, and the number of clock periods may be fixed or variable. But, in all cases, the devices connected to the bus must synchronize their actions with the bus clock. A device may transmit new information on bus lines, such as address, data, $\overline{WAIT}$, and so on, only at certain times within a clock period.

When long wires are involved, synchronous buses become difficult to design, particularly when a high speed of operation is desired. The difficulty arises because of the delays encountered by various signals as they travel from one end of the bus to the other. As a result of these delays, the clock signal will reach different devices at different times, and hence it no longer constitutes an absolute time reference for all devices. The clock period may be increased to take into account the worst case delays on the bus. Alternatively, we may abandon altogether the idea of having all devices synchronized to one reference.

An asynchronous bus differs from a synchronous bus in one fundamental respect. It does not depend on a common clock signal to synchronize various stages of a data transfer operation. Instead, handshake signaling is used to control the timing of the operation. The idea of handshake signaling was introduced in Subsection 5.4.3, and some of its hardware implications were discussed in Subsection 6.4.1.

When applied to a microprocessor bus, handshake signaling may be used to control a read operation as shown in Figure 6.47. Instead of a clock line, two lines are provided, Request and Acknowledge. We have assumed the signals on these lines to be active when low, because they are often implemented using the open-collector scheme described in the previous section. The microprocessor places an address on the address lines, sets the R/$\overline{W}$ line to 1, and after some delay, d_1, it activates the Request line. When the Request signal is received by the addressed device, the device places its data on the data lines. Then, after waiting for a short period, d_2, it activates the Acknowledge line, indicating to the microprocessor that the requested data are available. The microprocessor loads the data into its input buffer and deactivates the Request line. After some delay, d_3, it also removes the address and control information from the bus. Finally, the device deactivates Acknowledge and then removes the data from the bus, in response to the Request signal becoming inactive.

The transfer mechanism of Figure 6.47 can be used with devices of any speed, because the microprocessor activates the request signal and waits for the arrival of an acknowledgment. A fast device will respond quickly, while a slow device will delay the acknowledge signal until it has completed the requested input or output operation. Thus, each data transfer operation proceeds at a speed compatible with the nature of the device involved.

Delays d_1, d_2, and d_3 in Figure 6.47 are introduced by the microprocessor and the device interfaces to ensure that the information on the bus is interpreted correctly. Having placed an address on the address lines, the microprocessor waits for a short

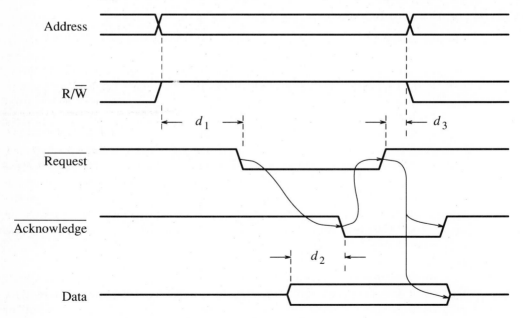

Figure 6.47 Timing of an input operation on an asynchronous bus.

period, d_1, before activating the request signal. The main purpose of this delay is to compensate for a delay parameter called *bus skew*. Propagation delays on different bus lines cannot be guaranteed to be identical. Two signals transmitted by a device at the same instant may reach another device at different times. The term "bus skew" refers to the difference in propagation delay between different lines of a bus. In order to ensure that all devices on the bus interpret the address information properly, delay d_1 should be at least equal to the maximum possible value of the bus skew. In this way, when a device detects an active state on the request line, it can safely assume that the information on the address and R/W lines is correct. We should emphasize that the bus skew is not the total delay from the sender to the receiver. Rather, it is a measure of the variability in this delay among different lines of the bus.

Another delay component that may be covered by d_1 is the address decoding delay. That is, in addition to compensating for the bus skew, the microprocessor may delay the request signal further, to allow all devices sufficient time to decode the information on the address lines. In this case, when a device receives the request signal, it can assume that not only the address but also the output of its internal decoder is valid. When the addressed device transmits the requested data, it delays its acknowledgment in a similar manner. That is, delay d_2 is intended to compensate for the bus skew while the data are propagating from the device to the microprocessor. The specifications of a given bus may also require this delay to be long enough to allow sufficient setup time for the input buffer of the microprocessor.

Finally, delay d_3 should be at least equal to the maximum bus skew. This delay guarantees that the information on the address and control lines as seen by any device on the bus remains valid as long as the Request line is in the active state. If d_3 is less than the bus skew, it is possible for a change on one of the address lines to reach some device before the rising edge of the Request signal. That device would then see a new address and, at the same time, an active request signal. Hence, if the new address happens to match its own, the device would erroneously start a read or a write operation.

The asynchronous scheme of Figure 6.47 has an important feature that may not be immediately apparent. While a synchronous bus requires all devices to monitor a common clock line to derive timing information, the asynchronous bus allows each device to transmit its own timing signal. The microprocessor generates the request signal. This is the reference relative to which the timing of other information transmitted by the microprocessor, namely, the information on the address and R/W lines, is specified. The limits on delays d_1 and d_3 constitute the timing constraints that must be observed by the microprocessor. Similarly, the addressed device generates the acknowledge signal, which is the timing reference for the data it transmits. The device is responsible for ensuring that d_2 exceeds some minimum value. Thus, in general, each device connected to the bus is only responsible for the relative delays among the signals it transmits. The handshake between the request and acknowledge signals accounts automatically for the delays introduced by the bus wires while the signals are traveling from one device to the other.

We should caution that although the use of handshake signaling takes the propagation delays on the bus into account, this does not mean that the length of an asynchronous bus can be increased indefinitely. As the length of the bus increases, so does the bus skew. Hence, a longer bus requires larger minimum values for the delays d_1, d_2, and d_3.

There is another important consideration related to the use of asynchronous buses. The microprocessor itself is almost always a synchronous device, which requires a clock signal to control its operation. Each clock transition causes the microprocessor to move from one state to another, to perform the next step in the sequence of actions needed to carry out its function. In many microprocessors, the clock frequency is sufficiently high that a single bus transfer occupies several clock periods. All output signal transitions produced by the microprocessor are always synchronized to the clock edges. Similarly, input signals are sampled by the microprocessor at one or the other edge of the clock. Examples of the relationship between the microprocessor clock and the signals on the bus will be given in Section 6.10, as part of the discussion of the 68000 family of buses.

The acknowledge signal received by the microprocessor during data transfer on an asynchronous bus is not constrained to have any particular relationship to the microprocessor clock. That is, when it is sampled by the microprocessor it is not guaranteed to meet any setup or hold requirements. Under such circumstances, the microprocessor should synchronize the incoming signal with its own clock before using this signal internally. An example of a synchronizer circuit consisting of an edge-triggered flip-flop is shown in Figure 6.48(a). While the input to the flip-flop may change at any time, its output changes only following the active edge of the clock, as indicated by the solid lines in the timing diagram.

The dotted lines in Figure 6.48(a) represent a particular situation that may arise, albeit with a small probability. Since the input is allowed to change state at any time, it may change very close to the clock edge. In this case, the propagation delay through the flip-flop can be considerably larger than the delays associated with signals that meet the setup and hold requirements of the flip-flop. The microprocessor designer must make allowance for this abnormal delay. For example, a second synchronization stage may be added, as shown in part (b) of the figure. The output of the second flip-flop is much less likely to suffer an abnormal delay. With this approach, state transitions at the asynchronous input are delayed by a minimum of one clock period before they are recognized internally. We will see the effect of this delay in the timing diagrams of the 68000 family of microprocessors in Section 6.10.

Let us now consider the design of an interface circuit for an asynchronous bus. The Request signal can be used in the interface in much the same way as the clock signal in the circuits discussed in previous sections. For example, the keyboard interface of Figure 6.13 may be modified as shown in Figure 6.49 for connection to an asynchronous bus. When the interface is addressed, an active state on the Request line causes the requested data or status information to be placed on the bus and the Acknowledge signal to be activated. According to Figure 6.47, the Acknowledge signal should be delayed by an amount d_2. The required delay is realized in Figure 6.49 by inserting an additional gate in the Acknowledge line, as shown. The reader should verify that this circuit implements the handshake sequence of Figure 6.47.

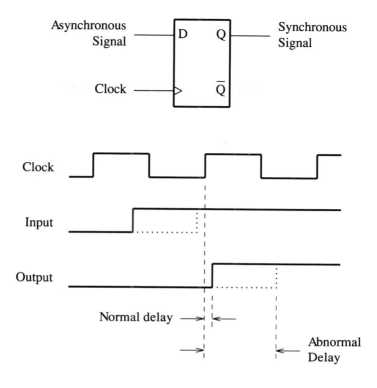

(a) A simple synchronizer

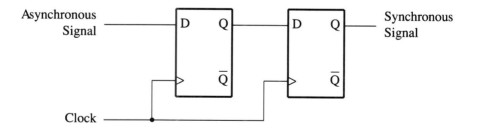

(b) A more reliable synchronizer

Figure 6.48   Synchronizer circuits.

The acknowledge signal represents a positive response from the addressed device, indicating to the microprocessor that the transfer operation has been completed properly. This is a useful feature that enhances the reliability of data transfer

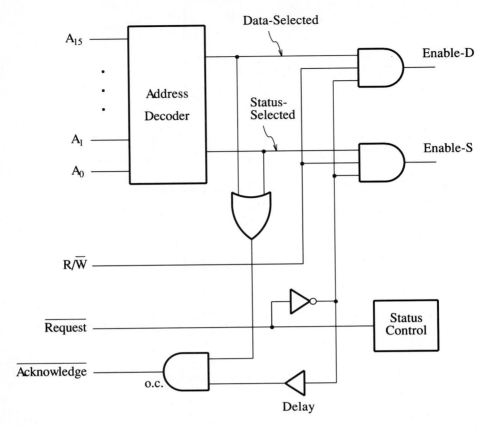

Figure 6.49 Part of the keyboard interface of Figure 6.13 modified for an asynchronous bus.

over an asynchronous bus. However, a microprocessor that uses asynchronous transfers must guard against the possibility of a device not responding at all. If no response is received, and after waiting for a sufficiently long time, the microprocessor should abort the transfer operation and take some corrective action. Such a situation is called a *bus error exception*. It will be discussed in Subsection 7.3.4.

6.9 THE 6809 BUS

In this section and Section 6.10, we describe the buses of our example microprocessors, the 6809 and the 68000. Only those aspects that are used in data transfers between the microprocessor and the main memory or I/O devices are discussed in detail. Other features, related to interrupts and direct memory access, will be examined in Chapter 7.

The 6809 microprocessor is an integrated circuit, housed in a 40-pin package. It is available in several versions, which have different speeds and different electrical characteristics. However, all versions have identical architecture — that is, they have the same instruction set and the same functional behavior. The external signals of a 6809 chip are shown in Figure 6.50. Of the 40 pins, 36 define the bus lines to which other chips, such as the main memory and I/O interfaces, are connected. The remaining four comprise the power supply lines, V_{cc} and V_{ss}, and the two lines labeled XTAL and EXTAL. The latter two are intended for connecting an external crystal that defines the frequency of the clock oscillator inside the microprocessor chip. This is the oscillator that generates the clock signals needed internally by the microprocessor, as well as externally on the microprocessor bus. It is also possible to drive the microprocessor directly by an external clock signal. In this case, a clock signal, which should have four times the desired bus-clock frequency, should be connected to EXTAL, and XTAL should be connected to ground.

The 6809 microprocessor has a synchronous bus. Its bus signals comprise several functional groups, as shown in Table 6.1. The first two groups, data transfer and timing signals, are used in normal memory access and input/output transfers. They are discussed further below. Interrupt, bus scheduling, and control operations constitute the main subject of Chapter 7.

The address, data, and R/$\overline{\text{W}}$ lines of the 6809 bus serve the usual functions of identifying a particular location in the main memory or in an I/0 interface, carrying the data to be transferred, and specifying the direction of the transfer, respectively. Timing information is carried on three lines, called E, Q, and MRDY. The main bus clock is the Enable signal, E. Unlike the clock signal used in our earlier discussions, a bus cycle begins with E = 0, as shown in Figure 6.51. A second clock signal, called the Quadrature clock signal, Q, is also provided on the bus. It has transitions at the 25 percent and 75 percent points of the cycle, and may be used in interface circuits where certain actions are to be initiated at those points. Only the E signal is needed in the majority of device interfaces. The third timing signal, MRDY (memory ready), can be used to extend the duration of a clock cycle when slow devices are involved. It performs the function of the wait signal described in Subsection 6.8.1.

The timing of normal read and write operations is illustrated in Figure 6.52. The maximum and minimum values for the delays shown are given in Table 6.2 for one of the 6809 models known as 68B09, which can operate with a bus clock frequency of up to 2 MHz. The microprocessor places a new address on the address lines and sets the R/$\overline{\text{W}}$ line as appropriate at the beginning of a cycle. These signals are guaranteed to reach a stable state at the corresponding output pins of the chip after some delay, t_{AD} (address delay), from the time E reaches its low-voltage state. The maximum value of t_{AD} is given in Table 6.2. In a read cycle, the microprocessor loads the received data into its input buffer at the end of the cycle. The timing requirements for this buffer are the Data Setup time (t_{DS}) and Data Hold During Read time (t_{DHR}). Table 6.2 gives the minimum values of t_{DS} and t_{DHR}, which are the parameters of interest in this case. Hence, in a system operated with a 2 MHz

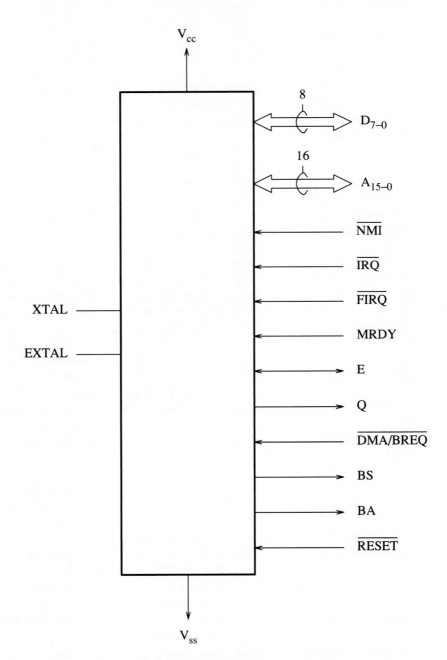

Figure 6.50 The 6809 microprocessor.

Table 6.1   6809 BUS SIGNALS

Function	Name	Input/Output	Description	Number
Data transfer	A0 – A15	O	Memory and I/O address	16
	D0 – D7	I/O	Data	8
	R/$\overline{\text{W}}$	O	Read or write indication	1
Timing	E	O	Enable (Clock signal)	1
	Q	O	Quadrature clock	1
	MRDY	I	Memory ready	1
Interrupt	$\overline{\text{IRQ}},\overline{\text{FIRQ}},\overline{\text{NMI}}$	I	Interrupt request lines	3
Bus scheduling	BS,BA	O	Status information	2
	$\overline{\text{DMA/BREQ}}$	I	Bus request	1
Control	$\overline{\text{RESET}}$	I		1
	$\overline{\text{HALT}}$	I		1
			Total	36

clock, the time available for the main memory to decode the address and place the requested data on the bus is

$$T_c - t_{AD\,(max)} - t_{DS\,(min)} = 500 - 120 - 50 = 330 \; ns$$

where T_c = clock period.

A write operation has similar timing requirements. Data, in this case supplied by the microprocessor, are guaranteed to be valid no later than the Data Delay time (t_{DD}) from the beginning of the cycle. The data will remain valid for a period equal to Data Hold During Write time (t_{DHW}) after the end of the cycle. Hence, the data setup time available for the main memory or I/O device receiving the data is

$$T_c - t_{DD\,(max)} = 500 - 280 = 220 \; ns$$

These requirements are easily met by presently available main memory components.

The need for specifying the address hold time, t_{AH}, may not be immediately apparent. Consider a data transfer involving the input interface circuit of Figure 6.6. A premature change on the address lines at the end of a read operation changes the state of the Enable-Data signal. The latter change removes the data from the data

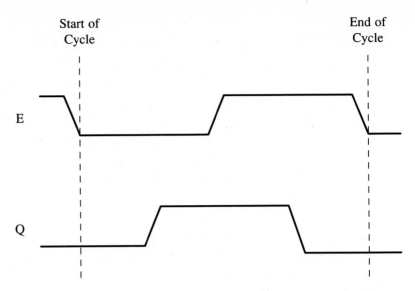

Figure 6.51   Main (E) and Quadrature (Q) clock signals on the 6809 bus.

Table 6.2   TIMING SPECIFICATIONS FOR 68B09

Symbol	Description	Min ns	Max ns
T_c	Clock period	500	–
t_{AD}	Address Delay — A_{15-0} and R/$\overline{W}$ guaranteed valid after this time.	–	120
t_{AH}	Address Hold — A_{15-0} and R/$\overline{W}$ will not change before this time.	30	–
t_{DS}	Data Setup — input data must be valid before this time.	50	–
t_{DHR}	Data Hold during Read — data must not change before this time.	50	–
t_{DD}	Data Delay — Data output guaranteed valid after this time.	–	280
t_{DHW}	Data Hold during Write — Data output guaranteed not to change before this time.	40	–

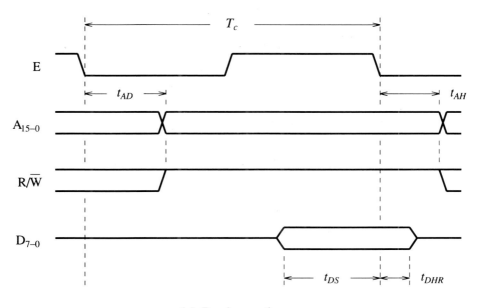

(a) Read operation

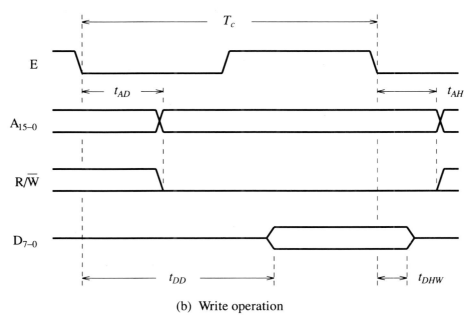

(b) Write operation

Figure 6.52 Timing of read and write operations on the 6809 bus.

lines, thus potentially affecting the data hold time t_{DHR} seen by the microprocessor. The designer of an interface circuit should check the minimum value of t_{AH}, to ensure that the interface will remain selected until all functions have been properly completed.

The 6809 bus accommodates slow devices by extending the clock period, using the MRDY signal. The MRDY line is an open-collector bus line that is regarded as active when in the high-voltage state. It is normally maintained in the active state by a pull-up resistor. An active state on MRDY indicates that the addressed device (usually the main memory) is ready, that is, it can complete a data transfer operation within one clock period. Hence, normal bus cycles, as described in Figure 6.52, take place. An I/O device may pull the MRDY line down to the low-voltage state if it wishes to extend the duration of a transfer cycle. However, it must do so before the last quarter of the clock cycle, that is, before the falling edge of the quadrature clock, Q, of Figure 6.51. In response, the microprocessor extends the second phase of the clock, i.e., keeps E equal to 1, until MRDY returns to the active state. The MRDY signal should not be held in the low state for more than 10 microseconds.

6.10 BUSES OF THE 68000 FAMILY

The basic architecture of the 68000 microprocessor family was presented in Chapter 4. All members of the family—the 68000, 68020, and 68030— have the same basic bus structure. The main differences among them relate to the width of the data bus. The 68000 bus has 16 data lines, while the 68020 and 68030 buses have 32. Also, while the basic bus of all three microprocessors is asynchronous, the 68030 bus offers an additional synchronous transfer mode and a burst mode for transferring blocks of data at high speed. We will describe first the 16-bit bus of the 68000 microprocessor. Then, we will discuss briefly the additional features found on the buses of the 68020 and 68030.

6.10.1 16-Bit Bus (68000)

The 68000 chip comes in a 64-pin package, thus having enough pins to provide separate data and address lines and a number of control lines. The signals on the 68000 bus are summarized in Table 6.3. Those used in data transfer operations between the microprocessor and the main memory are described in this section. The remaining signals, which are needed for interrupts and direct memory access, are discussed in Chapter 7.

The 68000 microprocessor handles data in bytes, words, or long words. Its external data bus is only 16 bits wide. Hence, external data transfers between the microprocessor and the main memory or I/O devices take place using, at most, one word at a time. Long words are transferred as two consecutive words. Transfers that involve only one byte of information take place on the low-order data lines, D_{7-0}, for odd address locations, and on the high-order lines, D_{15-8}, for even address locations.

Table 6.3 68000 BUS SIGNALS

Function	Name	Input/Output	Description	Number
Data transfer	A1 – A23	O	Memory and I/O address	23
	D0 – D15	I/O	Data	16
	R/$\overline{\text{W}}$	O	Read or write indication	1
Timing	$\overline{\text{AS}}$	O	Address strobe	1
	$\overline{\text{UDS}}$	O	Upper data strobe (for D8 – D15)	1
	$\overline{\text{LDS}}$	O	Lower data strobe (for D0 – D7)	1
	$\overline{\text{DTACK}}$	I	Data acknowledge	1
	CLK	I	Processor clock	1
Compatibility	E	O	Enable	1
	$\overline{\text{VPA}}$	I	Valid Peripheral Address — indicates that address corresponds to a 6800 peripheral.	1
	$\overline{\text{VMA}}$	O	Valid memory address	1
Scheduling	$\overline{\text{BR}}$	I	Bus Request	1
	$\overline{\text{BG}}$	O	Bus Grant	1
	BGACK	I	Bus Grant Acknowledge	1
Interrupts	$\overline{\text{IPL0}}$ – $\overline{\text{IPL2}}$	I	Interrupt Priority Line	3
Bus Function	FC0 – FC2	O	Bus function — type of data being transferred.	3
System Control	$\overline{\text{BERR}}$	I	Bus Error	1
	$\overline{\text{RESET}}$	I	Reset	1
	$\overline{\text{HALT}}$	I	Halt instruction execution	1
			Total	60

The 68000 bus uses the asynchronous handshake scheme of Figure 6.47, as shown in Figure 6.53. Two request signals are provided, one for each byte of the data bus. The Lower Data Strobe ($\overline{\text{LDS}}$) and the Upper Data Strobe ($\overline{\text{UDS}}$) are used to request transfers of the low-order and high-order data bytes, respectively. Hence, the microprocessor activates $\overline{\text{UDS}}$ for a byte transfer at an even address and $\overline{\text{LDS}}$ for an odd address. Both $\overline{\text{LDS}}$ and $\overline{\text{UDS}}$ are activated during word transfers. The acknowledge signal in Figure 6.47 is called Data Acknowledge ($\overline{\text{DTACK}}$) in 68000 nomenclature. Note that all of these signals are active when low. The direction of the transfer is determined by the R/$\overline{\text{W}}$ line.

Although the 68000 microprocessor generates 24-bit addresses, there are only 23 address lines on the bus. They carry the 23 most-significant bits of the address, A_{23-1}. For word transfers, the least-significant bit, A_0, is always equal to 0; hence,

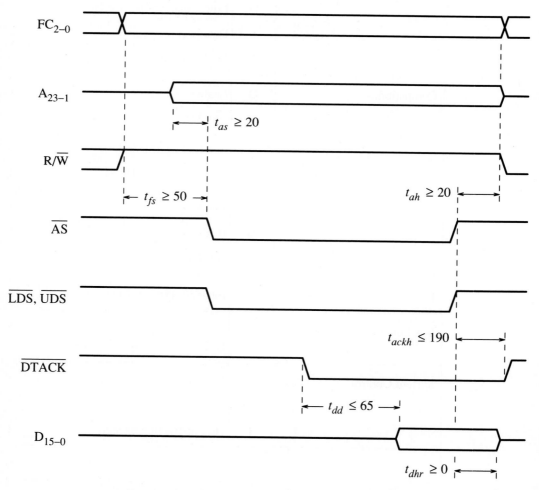

Figure 6.53 Timing of a read operation on the 68000 bus.

it is not needed. During byte transfers, however, A_0 determines whether the high-order (even address) or low-order (odd address) byte of a word is involved. Instead of transmitting A_0, the microprocessor activates only one of the two data-strobe signals, as described above.

In addition to the data strobe and acknowledge signals, there is a fourth timing signal called Address Strobe, $\overline{AS}$. It is used to indicate that a valid address is available on the address lines. The microprocessor activates $\overline{AS}$ after it places a new address on the address lines and after allowing sufficient time for the new address to become stable. $\overline{AS}$ remains active for the duration of the bus transfer. It is deactivated, together with the data strobe signals, after the response of the addressed device has been received on $\overline{DTACK}$. The address lines should be ignored by all devices on the bus while the $\overline{AS}$ line is inactive. The availability of separate strobe signals for the address and data lines offers some flexibility in controlling data transfers, as will become apparent from the remainder of this discussion.

The timing diagram of Figure 6.53 shows some of the delay parameters that are necessary to guarantee correct transfer of data, where the limits on delay values are given in nanoseconds. The delay values shown are applicable to one model of the 68000 microprocessor known as MC68000G10, which is capable of operating at a clock frequency of 10 MHz.

At the beginning of a bus cycle, the microprocessor sets the function lines, FC_{2-0}, to indicate the processor's mode of operation and the type of information being transferred. The purpose of these lines will be discussed later. If the $R/\overline{W}$ line was in the low-voltage state during the previous bus transfer, it is set high at the same time a new code is transmitted on the function lines. Next, the microprocessor places the address of the desired memory location on the address lines, and after a short delay it activates the Address Strobe and the Upper and Lower Data Strobe lines. The information on the function and address lines is guaranteed to be valid for a minimum of $t_{fs} \geq 50$ ns and $t_{as} \geq 20$ ns, respectively, before the address strobe is activated.

The addressed device responds by placing the requested data on the data lines and activating $\overline{DTACK}$. The device is allowed to delay the transmission of data by $t_{dd} \leq 65$ ns after it has activated $\overline{DTACK}$. After completing the handshake between the strobe and acknowledge signals, all lines return to their inactive state. The microprocessor guarantees a hold time of 20 ns on the function and address lines after it returns the strobe lines to the inactive state. The addressed device must release the data lines and deactivate $\overline{DTACK}$ within $t_{ackh} \leq 190$ ns after the strobe signals become inactive. This condition is necessary to guard against the addressed device interfering with the next data transfer operation on the bus.

Figure 6.54 shows a timing diagram for an output operation. Note that in this case, the data strobe signals are delayed slightly relative to the address strobe, to allow sufficient time for device interfaces to decode the address information. The data, which are transmitted by the microprocessor, are guaranteed to be valid at least 20 ns before the data strobe signals are activated.

There is no limit on the amount of delay that may be introduced by a device interface before activating the $\overline{DTACK}$ line. In fact, activation of $\overline{DTACK}$ may be

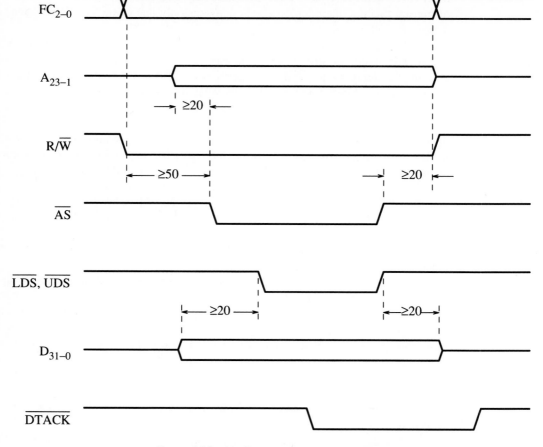

Figure 6.54 A write operation on the 68000 bus.

postponed indefinitely, thus completely suspending the operation of the microprocessor. This is a useful feature during circuit debugging.

ROLE OF THE MICROPROCESSOR CLOCK

The 68000 microprocessor is a synchronous device, whose operation is controlled by a clock signal, CLK. It is illustrative to examine the relationship between
the signals on the bus and the microprocessor clock. Consider, for example, the read
operation shown in Figure 6.53. In order to carry out this bus transfer, the microprocessor goes through eight internal states. The sequence of events is illustrated in
Figure 6.55, where the eight states are labeled S0 to S7. The actions taken in each
state are summarized in Table 6.4.

The first step in the operation takes places while the microprocessor is in state
S0, where it transmits a new code on the function lines and sets the $R/\overline{W}$ line to 1.
At this time, the strobe outputs of the microprocessor are inactive and the address

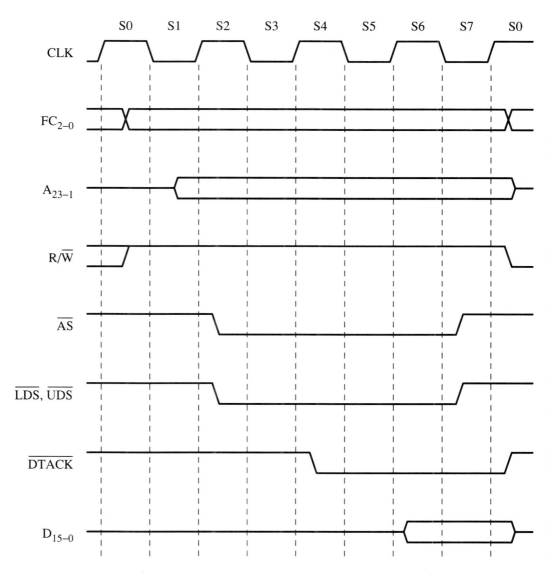

Figure 6.55 Sequence of microprocessor states during a read operation.

and data lines are in the high-impedance state. The falling edge of CLK causes the microprocessor to leave state S0 and enter state S1. While in S1, the microprocessor enables the tri-state drivers connecting its internal address buffer to the address lines. As a result, the address information appears on the address lines, after some delay. The following rising edge of CLK causes the microprocessor to enter state S2, where it activates the strobe signals.

Table 6.4   SEQUENCE OF STATES OF THE 68000 MICROPROCESSOR DURING A READ OPERATION

State	Action
S0	Transmit the function code on lines FC_{2-0}, and set $R/\overline{W}$ to 1.
S1	Place address on the address lines.
S2	Activate $\overline{AS}$, $\overline{LDS}$ and $\overline{UDS}$.
S3	No action.
S4	Sample $\overline{DTACK}$ — remain in state S4 until $\overline{DTACK}$ becomes active.
S5	No action.
S6	Load input data into input buffer on the next falling edge of CLK.
S7	Remove address and strobe signals.

In state S3, the microprocessor does not perform any bus function. Thus, this state represents a delay period during which device interfaces can decode the information on the address lines and prepare the data for transmission to the microprocessor. A fast device may complete the requested operation and assert its response, DTACK, by the time the microprocessor enters state S4. A slow device may not. The microprocessor begins inspecting its $\overline{DTACK}$ input in state S4. If $\overline{DTACK}$ is active, the microprocessor proceeds to state S5. Otherwise, it inserts a wait state and samples $\overline{DTACK}$ at the next falling edge of the clock. Eventually, $\overline{DTACK}$ becomes active, and the microprocessor enters state S5, which is a delay state, then S6. The input data received from the addressed device are loaded into the input buffer of the microprocessor by the falling clock edge at the end of state S6.

After loading the input data, the microprocessor terminates the read operation by deactivating the strobe signals at the beginning of state S7. Then, it places the address lines in the high-impedance state after it enters state S0, which is the beginning of the next bus cycle.

The microprocessor samples $\overline{DTACK}$ at the end of state S4 and inserts wait states if necessary until $\overline{DTACK}$ is activated by the device. This portion of the input cycle is shown enlarged in Figure 6.56. An active state on $\overline{DTACK}$ will be recognized at the end of state S4 if it occurs sufficiently early so that its setup time, t_q, exceeds the minimum requirement. For the 10 MHz chip, $t_{q(min)} = 20$ ns. The microprocessor allows one clock period for the $\overline{DTACK}$ synchronizer circuit to settle (see the discussion in Subsection 6.8.2). Then, it loads the input data into its input buffer at the end of state S6. For correct operation, the input data must meet the minimum setup requirement shown in the figure, $t_{dsr} \leq 10$ ns. We should note that the relative delay constraint between $\overline{DTACK}$ and the input data, t_{dd}, in Figure 6.53 is intended to guarantee that once the $\overline{DTACK}$ signal is received by the micro-

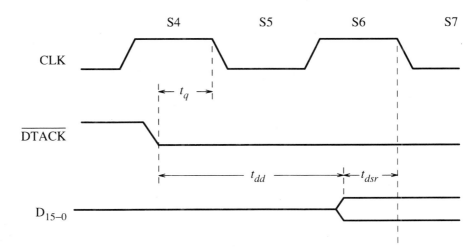

Figure 6.56 Acknowledgment and data timing.

processor, the data will be available sufficiently early to meet the minimum requirement for t_{dsr}.

The timing diagram given in Figure 6.56 has one important implication for interface circuit design. No matter how quickly the addressed device activates $\overline{\text{DTACK}}$, the microprocessor will not load the input data until the end of state S6. Hence, a fast device that can transmit data in time to meet the t_{dsr} setup constraint need not worry about the relative timing of the data and $\overline{\text{DTACK}}$. This fact may be useful in simplifying the design of interface circuits, as we will see in the main memory interface example given in Chapter 8.

READ-MODIFY-WRITE

In addition to the read and write operations of Figures 6.53 and 6.54, the 68000 bus has a combined operation called a read-modify-write cycle. It consists of a sequence of two data transfers, in which the microprocessor reads the contents of a memory location, modifies the data read, and then writes the data back into the same location. Of course, this task can be accomplished in two separate bus transactions. In the first, the microprocessor reads the contents of the desired location. Then, in a second transfer, it writes the modified information using the same address. However, the main purpose of introducing the read-modify-write operation is to have the two transfers appear as a single, indivisible bus transaction. Such an operation is needed in multiprocessor systems, where the main memory is shared among several processors. After one processor completes a data transfer operation, another processor may begin accessing the memory. In the discussion in Chapter 10, it will be pointed out that it is sometimes necessary to guarantee that one processor can read a main memory location and then write into it without any chance of another processor gaining access to that location between the read and write operations. The read-modify-write cycle is intended to provide the guarantee needed in such situations.

In the 68000 microprocessor, the read-modify-write cycle is used in conjunction with the test and set (TAS) instruction.

The read and write portions of the cycle proceed according to the timing diagrams in Figures 6.53 and 6.54, with one exception. The microprocessor does not deactivate the address strobe line, $\overline{AS}$, at the end of the read operation. By maintaining $\overline{AS}$ active, it indicates to other devices that it has not yet completed all the transfers that it needs. No device on the 68000 bus can initiate a new data transfer operation until the device that is currently using the bus has deactivated $\overline{AS}$. Hence, no device can use the bus between the read and write parts of a read-modify-write cycle.

FUNCTION LINES

The microprocessor uses the function lines to transmit additional information about the data being transferred and about its mode of operation. It was mentioned in Chapter 4 that the 68000 has two modes of operation, User mode and Supervisor mode. For each of these, the data lines may carry program instructions or data. This information is encoded into 3 bits, according to Table 6.5. The function lines are also used for acknowledging interrupts, as will be described in Chapter 7.

The information on the function lines is used primarily by a companion chip in the 68000 family, known as the Memory Management Unit (MMU). The MMU enables the system software to allocate parts of the available physical memory to different types of information, as well as to different users of the microcomputer in the case of multiuser systems. Details of the operation of MMU chips will be discussed in Chapter 8.

Table 6.5   FUNCTION LINES OF THE 68000 BUS

FC2–FC0	Cycle type
000	Reserved for future use
001	Data of a User-mode program
010	Instructions of a User-mode program
011	Reserved for future use
100	Reserved for future use
101	Data of a Supervisor-mode program
110	Instructions of a Supervisor-mode program
111	Interrupt Acknowledge (see Chapter 7)

The function code may be used to increase the size of the main memory beyond the 16-megabyte limit imposed by the number of address bits. The memory may be regarded as consisting of four separate modules, one each for the supervisor program, the supervisor data, the user program, and the user data. By using the function lines to choose the appropriate module during each bus access cycle, up to 16 Mbytes of storage may be included in each module. That is, the main memory may be expanded up to 64 Mbytes. While such an arrangement is possible, at least in theory, it is not as convenient as increasing the size of the addressable space by providing more address lines. Newer members of the 68000 family, namely, the 68020 and 68030, have a 32-bit address bus.

BUS ERROR HANDLING

In Section 6.8.2, it was pointed out that a computer with an asynchronous bus must incorporate some means for recovery when the addressed device fails to respond. A timeout mechanism is usually provided, such that if the processor does not receive a response within a predefined time-out period, an exception occurs that directs the processor to execute an error recovery program.

The 68000 does not have an internal time-out mechanism. Hence, the response signal, $\overline{\text{DTACK}}$, may be delayed indefinitely without an error indication. The Bus Error line, $\overline{\text{BERR}}$, is provided instead. It may be activated by any device connected to the bus, when a malfunction of any kind is detected. Activation of $\overline{\text{BERR}}$ causes an exception situation to be recognized by the microprocessor. The $\overline{\text{BERR}}$ line may be, and should be, used to implement a time-out facility on the 68000 bus. A time-out device can monitor the bus activity, for example, by testing the $\overline{\text{AS}}$ line. If $\overline{\text{AS}}$ stays in the active (low) state and $\overline{\text{DTACK}}$ in the inactive (high) state for more than the desired time-out period, the device should activate the $\overline{\text{BERR}}$ line.

INTERFACE CIRCUITS

Let us now consider the design of the interface circuits needed to connect I/O devices to the 68000 bus. The only difference between these circuits and those discussed in Sections 6.2 to 6.4 is that the clock signal used on synchronous buses is replaced by the handshake signals on the 68000 bus. So, the 68000 interfaces follow the general organization of Figure 6.49. We should note that the timing constraint for input data on the 68000 bus is less stringent than in the general case depicted in Figure 6.47. The 68000 microprocessor will read the input data correctly if it arrives after $\overline{\text{DTACK}}$. That is, delay d_2 is allowed to have a negative value. Hence, the delay gate used in Figure 6.49 may not be needed. We will examine this point in more detail later.

An example of a simple output interface circuit for the 68000 bus is shown in Figure 6.57. The Load-Data signal becomes equal to 1 when the device is addressed, the $\text{R}/\overline{\text{W}}$ line indicates a write operation and all three strobe signals are low. As a result, the information on the data lines is loaded into the output buffer, which is assumed to be a 16-bit wide positive edge-triggered register. The interface's response, $\overline{\text{DTACK}}$, is derived from the Load-Data signal as shown, using an open-collector gate. When the microprocessor receives $\overline{\text{DTACK}} = 0$, it sets the strobe

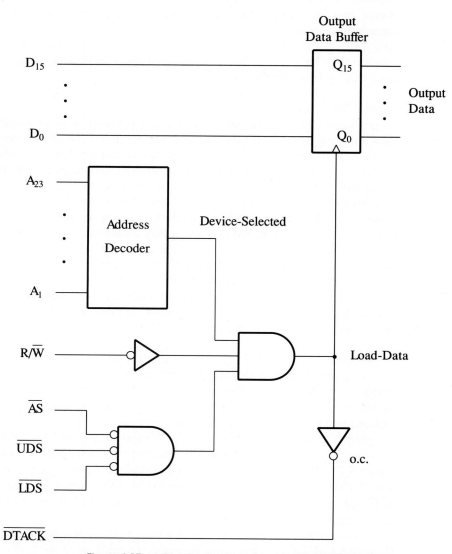

Figure 6.57   A 16-bit output interface for the 68000 bus.

signals back to the high-voltage state, causing both Load-Data and $\overline{\text{DTACK}}$ to be deactivated.

According to Figure 6.54, the microprocessor guarantees that the information on the data lines is valid at least 20 ns before the data strobe signals are activated.

Allowing a minimum of 5 ns for the propagation delay in each gate, valid data will be available at the input of the data buffer in Figure 6.57 at least 30 ns before Load-Data becomes active. The designer of the interface circuit should ensure that this is sufficient setup time for the particular register used to implement the data buffer. Should more than 30 ns of setup time be needed, the activation of Load-Data and DTACK must be delayed, by inserting a delay circuit at the output of the AND gate that generates Load-Data. For a small delay, one or two logic gates would suffice, as was used in Figure 6.49.

Longer delays may be realized by means of a shift-register arrangement, as shown in Figure 6.58. The clock signal, CLK, that drives the 68000 microprocessor may be used by the interface circuit whenever a continuously running clock signal is needed. It is used in Figure 6.58 to drive a shift register consisting of two flip-flops. A 0-to-1 transition at point P will cause Load-Data and DTACK to become active after a delay of between one and two clock periods. The amount of this delay depends upon the instant at which P changes from 0 to 1 relative to the rising edge of CLK. When the microprocessor returns AS and LDS to the inactive state, the two flip-flops will be cleared immediately, because of the connection to their reset (R) inputs. As a result, DTACK will become inactive a few gate delays after AS and LDS become inactive. Without the connection to the R inputs of the flip-flops, a further delay of one to two clock periods would be introduced unnecessarily before deactivating DTACK.

COMPATIBILITY MODE

A number of I/O chips are available to provide parallel and serial interfaces in a 68000 system. Those developed specifically for the 68000 microprocessor can be

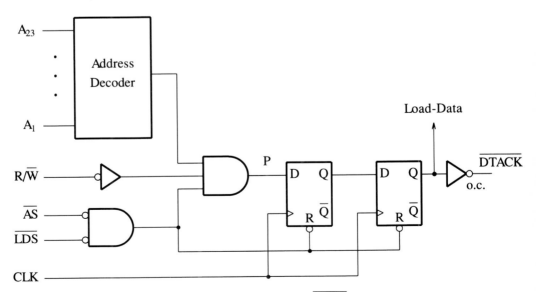

Figure 6.58 Generation of the $\overline{\text{DTACK}}$ signal.

connected directly to the bus, requiring only an external address decoder to generate the chip-select signals. However, the concept of a microprocessor family of chips requires that newer members of the family be compatible with existing chips. To maintain this compatibility, the 68000 bus includes a number of signals that enable I/O chips introduced for the 6800 and 6809 microprocessors, such as the PIA and the ACIA, to be easily used in 68000 systems. The main difference between the 68000 bus and that of its predecessors is that while the 6800 and 6809 are synchronous, the 68000 bus is asynchronous. I/O chips developed for synchronous buses are not capable of generating the acknowledgment signal, $\overline{\text{DTACK}}$, needed on the 68000 bus. Also, these chips are not able to make use of the address and data strobe signals; they require a clock signal instead.

Three signals — two outputs, E and $\overline{\text{VMA}}$ (Valid Memory Address), and one input, $\overline{\text{VPA}}$ (Valid Peripheral Address) — are provided on the 68000 bus to maintain compatibility with the 6800 family of I/O chips. The 68000 transmits a continuously running clock signal on its E output, which should be used as the clock input (also called Enable) of 6809-compatible devices. Figure 6.59 shows a typical arrangement in which an ACIA is connected to a 68000 microprocessor. The address decoder recognizes all addresses assigned to I/O devices that require compatibility mode operation. Whenever one of these addresses is placed on the bus by the microprocessor and $\overline{\text{AS}}$ is activated, the address decoder activates the Chip-Select input of the corresponding device. At the same time, it activates $\overline{\text{VPA}}$ (Valid Peripheral Address), indicating to the microprocessor that compatibility-mode operation is required for this address. In response, the 68000 activates the $\overline{\text{VMA}}$ (Valid Memory Address) line, maintaining correct timing relationship to the clock signal, E. Unlike normal 68000 bus cycles, a compatibility-mode transfer is completed without waiting for a response on the $\overline{\text{DTACK}}$ line.

6.10.2 32-Bit Bus (68020 and 68030)

The buses of the 68020 and 68030 microprocessors offer several features not found on the 68000 bus. The main difference is that the 68020 and 68030 buses carry up to 32 bits of data in parallel. Since a given transfer operation may involve a byte, a word, or a long word, sufficient information must be provided to indicate the desired size of the operand. A related problem is that of "address alignment," which results from the fact that the microprocessor may request a long word transfer starting at any address. When the microprocessor requests access to a long word that does not start at a long word address boundary, that is whose address is not divisible by 4, two bus cycles are needed to transfer all bytes of the long word. Also, some of the devices connected to the bus may only be able to transfer 8 or 16 bits at a time. Such devices are said to have a *port width* of 8 or 16, respectively. The bus used in the 68020 and 68030 microprocessors handles the operand size and alignment as well as the port size automatically, in a manner that is completely transparent to the programmer.

The bus signals used for data transfer are summarized in Table 6.6. In comparing these signals to the 68000 bus signals given in Table 6.3, we note a few differences. The 68000 bus does not carry the least-significant address bit, A_0.

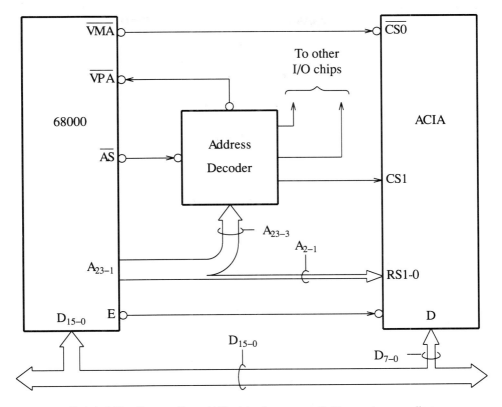

Figure 6.59 Connection of I/O chips for compatibility-mode operation.

Instead, it uses two data strobe signals, $\overline{\text{LDS}}$ and $\overline{\text{UDS}}$, to specify the desired byte location. In contrast, the 32-bit bus carries the full 32-bit address and only one data strobe signal, $\overline{\text{DS}}$. Several control signals have also been added. Two lines, SIZ0 and SIZ1, are used by the microprocessor to indicate the size of the data it wishes to read or write. The addressed device uses two acknowledgment signals, $\overline{\text{DSACK0}}$ and $\overline{\text{DSACK1}}$, to respond to a data transfer request. These two signals enable the device to communicate its port width to the microprocessor at the same time it acknowledges that it has completed a read or a write operation. They replace $\overline{\text{DTACK}}$ of the 68000 bus.

We will first illustrate how the operand size and alignment are handled on the 32-bit bus. Figure 6.60 shows a portion of the main memory, which is assumed to be 32 bits wide. The 68020 and 68030 microprocessors allow operands to start at any address, irrespective of their size. However, when the main memory reads or writes data, it can only deal with one of its 32-bit words at a time. The figure shows the information transmitted on the address, size, and acknowledgment lines when four operands having different sizes and alignments are read. In the first three cases the entire operand is transferred in a single bus cycle. For example, to read a long

Table 6.6   DATA TRANSFER AND TIMING SIGNALS ON THE 68020 AND 68030 BUSES

Function	Name	Input/Output	Description	Number (20/30)
Data transfer	A0 – A31	O	Memory and I/O address	32
	D0 – D31	I/O	Data	32
	R/$\overline{\text{W}}$	O	Read or write indication	1
	SIZ0,1	O	Operand size	2
	$\overline{\text{ECS}}$	O	External Cycle Start — asserted at the beginning of every bus cycle.	1
	$\overline{\text{OCS}}$	O	Operand Cycle Start — marks the first bus cycle of an operand or an instruction.	1
	$\overline{\text{RMC}}$	O	Read Modify Write cycle	1
	$\overline{\text{DBEN}}$	O	Data Buffer Enable — can be used to enable external bus drivers	1
Timing	$\overline{\text{AS}}$	O	Address strobe	1
	$\overline{\text{DS}}$	O	Data strobe	1
	$\overline{\text{DSACK0,1}}$	I	Acknowledge and port width indication	2
	$\overline{\text{STERM}}$	I	Synchronous termination (68030 only)	–/1
	CLK	I	Processor clock	1
Scheduling	$\overline{\text{BR}}$	I	Bus Request	1
	$\overline{\text{BG}}$	O	Bus Grant	1
	BGACK	I	Bus Grant Acknowledge	1
Interrupts	$\overline{\text{IPL0}}$ – $\overline{\text{IPL2}}$	I	Interrupt Priority Line	3
	$\overline{\text{IPEND}}$	O	Interrupt Pending — asserted when the microprocessor accepts an interrupt request	1
	$\overline{\text{AVEC}}$	I	Autovector — asserted for autovectoring	1
Bus Function	FC0 – FC2	O	Processor mode and type of data.	3
System Control	$\overline{\text{BERR}}$	I	Bus Error	1
	$\overline{\text{RESET}}$	I	Reset	1
	$\overline{\text{HALT}}$	I	Halt instruction execution	1
Other				1/6
			Total	91/97

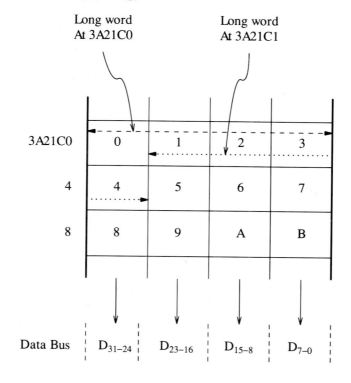

Figure 6.60 Handling of operand size and alignment on the 32-bit bus.

Operand	Bus Cycle	Address	Operand Size	Acknowledgment (Port Width)
LW@3A21C0	1	3A21C0	4	4
W@3A21C1	1	3A21C1	2	4
B@3A21C2	1	3A21C2	1	4
LW@3A21C1	1	3A21C1	4	4
	2	3A21C4	1	4

word starting at address 3A21C0, the microprocessor places this address on the address lines, and indicates the size of the operand, 4 bytes, on the size lines. When the main memory sends its acknowledgement, by means of the $\overline{\text{DSACK0}}$ and $\overline{\text{DSACK1}}$ signals, it indicates that it has a port width of 4 bytes. Based on this information the microprocessor concludes that the transfer will be completed in one cycle. The requested word is transmitted on lines D_{31-0}.

In the second example, the microprocessor reads a word from location 3A21C1. It sends this address on the bus, and sets the size information to 2. Again, the memory uses the acknowledgment signals to indicate a port width of 4 bytes. The requested word is transmitted on lines D_{23-8}. Hence, the microprocessor uses internal multiplexers to shift it to the proper place on its internal data paths, namely, to bit positions 15 to 0.

In the last example shown in the figure, the microprocessor requests to read a long word at address 3A21C1. The four bytes that make up this long word are stored in two different locations in the memory, and hence cannot be accessed at the same time. Knowing that, the microprocessor accepts only the three most-significant bytes of the operand, which are on lines D_{23-0}, in the first bus cycle. Then, it automatically starts a second bus cycle during which it receives the remaining byte from location 3A21C4. Note that the value transmitted on the size lines represents the number of bytes remaining to be transferred, not the size of the entire operand. This is necessary to prevent the addressed device from altering the contents of byte locations not involved in the transfer, in the case of a write operation. With this arrangement, both the microprocessor and the addressed device know at all times how many bytes are being transferred and where they are to be found on the data bus.

Consider now the question of port size. A 16-bit device is connected to lines D_{31-16}, and an 8-bit device is connected to lines D_{31-24}. An example of a transfer involving a 16-bit device is illustrated in Figure 6.61. The microprocessor does not know the port size of the addressed device until it receives the acknowledgment signals. To read a long word at address 3A21C1, it places the address and size information on the bus in the usual manner. The device's acknowledgment indicates that it is only capable of transferring two bytes at a time. This means that only the most-significant byte of the long word requested is actually available on the bus, on lines D_{23-16}. The microprocessor loads this byte into its input buffer and, using internal multiplexers, shifts it to the most-significant position. Then, it initiates a second bus cycle requesting the remaining three bytes. Again, the device indicates it can only transfer two bytes. The microprocessor accepts the two bytes, shifts them to the proper place, then requests the last byte in the third cycle. A similar sequence of events takes place to read a long word from a device that has a port width of 8-bits, except, of course, four bus cycles would be needed. In this case, all bytes would be transferred on lines D_{31-24}.

During a read operation, the microprocessor is able to determine the active portion of its data bus based on the port width indicated by the two acknowledgment signals. In a write operation, the microprocessor has to place the data on the bus before it gets the width information. For this reason, when it is transmitting fewer than four bytes, it duplicates some or all the bytes as many times as needed on the four lanes (bytes) of the data bus to cover all possibilities. For example, to write a word at location 3A21C2, it places the same data on lines D_{15-0} and D_{31-16}. A 32-bit device accepts the two bytes on lines D_{15-0}, and a 16-bit device accepts the two bytes on lines D_{31-16}. An 8-bit device accepts only the byte on lines D_{31-24}. In this case, the microprocessor initiates a second cycle to send the second byte of the 2-

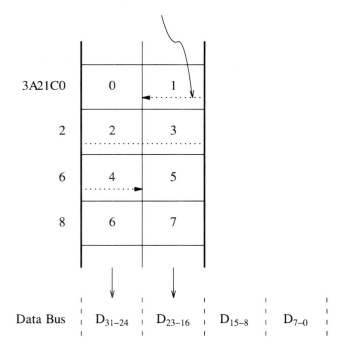

Figure 6.61 Handling a 16-bit device on the 32-bit bus.

Operand	Bus Cycle	Address	Operand Size	Acknowledgment (Port Width)
LW@3A21C1	1	3A21C1	4	2
	2	3A21C2	3	2
	3	3A21C4	1	2

byte operand. Since it is only sending one byte, it duplicates it on all four lanes of the bus.

The flexibility in handling the operand size and alignment, as well as the port size of the addressed device simplifies the design of I/O interfaces, and enables I/O devices or main memory circuit boards of any width to coexist within the same system. Because this flexibility is provided by the microprocessor hardware, the programmer need not worry about operand alignment or port size. However, it should be remembered that there is a potential for performance degradation when operand alignment is ignored, because more bus cycles are needed to access a long word if it does not begin at a long word address boundary.

The timing of data transfer over the 32-bit bus is illustrated in Figure 6.62, which shows a read operation. The sequence of events is essentially the same as in Figure 6.55, with the exception that the transfer is completed in six instead of eight internal states. The microprocessor sends the function code, address, and read/write information in state S0, and activates the strobe signals in state S1. It begins sampling the two $\overline{\text{DSACK}i}$ lines at the end of state S2, and it loads the input data at the

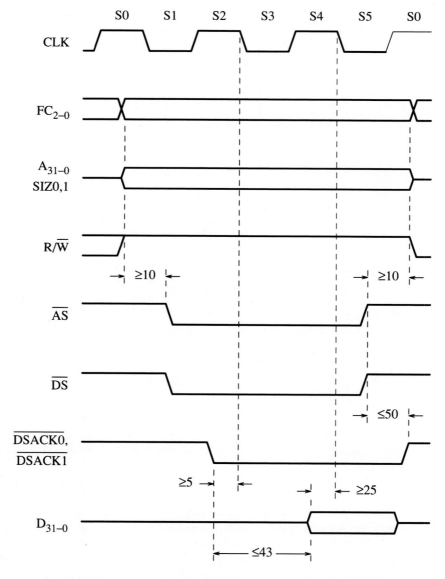

Figure 6.62 An asynchronous read operation on the 32-bit bus.

end of state S4. The delay values given in the figure are applicable to a 20-MHz microprocessor chip.

6.10.3 Synchronous Operation (68030)

The 68030 microprocessor bus can operate as a synchronous bus to achieve higher transfer speeds than those possible during asynchronous operation. It is up to the addressed device in any given bus transaction to determine which mode it wishes to use, synchronous or asynchronous. A device that uses asynchronous control responds to a transfer request by activating the $\overline{\text{DSACK}i}$ signals. In this case the transfer operation proceeds as described in the previous section, and it occupies a minimum of three clock periods. A device that is capable of synchronous operation does not activate $\overline{\text{DSACK}i}$. Instead it activates a response line called Synchronous Termination, $\overline{\text{STERM}}$. This line serves a similar function to $\overline{\text{DSACK}i}$. However, while the device may assert $\overline{\text{DSACK}i}$ at any time, it must guarantee certain setup and hold times for $\overline{\text{STERM}}$ relative to the clock. The synchronous mode can be used only during 32-bit transfers.

The timing of a synchronous read operation is shown in Figure 6.63. When the addressed device is ready to transmit the data, it sends a low pulse on the $\overline{\text{STERM}}$ line to meet the setup and hold requirements shown. It need not wait for the strobe signals to become active before doing so. The microprocessor samples $\overline{\text{STERM}}$ at successive rising edges of the clock until it becomes active. Then, it loads the input data at the following edge of the clock. Similar constraints apply during output operations.

Figure 6.63 shows an input operation being completed in two clock periods. This represents a substantial improvement relative to the three clock periods needed during asynchronous transfers. The key difference is in the way the response signals, $\overline{\text{DSACK}i}$ and $\overline{\text{STERM}}$, are treated. The microprocessor begins sampling $\overline{\text{STERM}}$ at the end of state S1, which is half a clock period earlier than it samples $\overline{\text{DSACK}i}$. Also, it was pointed out in Section 6.8.2 that the microprocessor must use a synchronizer circuit to synchronize an asynchronous input with its clock. One clock period is allowed for the synchronization of $\overline{\text{DSACK0}}$ and $\overline{\text{DSACK1}}$ in Figure 6.62. On the other hand, because $\overline{\text{STERM}}$ is a synchronous input, less time is required to test its state. Hence, the microprocessor is able to load the input data at the end of state S2.

BURST MODE
The highest data transfer rate attainable on the 68030 bus is realized during *burst mode* operations. This mode can only be used to read a contiguous block of long words from the main memory to fill a small internal memory called a *cache*. Data may be read at a rate of up to one long word every clock period. The purpose and use of the cache will be discussed in Chapter 8. Here, we will only describe the data transfer mechanism used.

In order to support a burst read, the main memory must be capable of operating in the synchronous mode described above. A burst read begins as a normal read

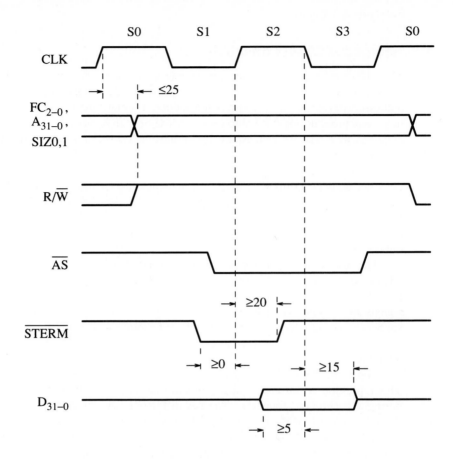

Figure 6.63 A synchronous read operation on the 68030 bus.

operation, during which the microprocessor also activates a control signal called Cache Burst Request ($\overline{\text{CBREQ}}$). The read request must be for a long word at a long-word address boundary. The main memory responds using the synchronous termination signal, and the requested long word is transferred in the usual way. The memory also activates a control signal called Cache Burst Acknowledge ($\overline{\text{CBACK}}$), indicating its readiness to participate in a burst read. Instead of terminating the transfer, the microprocessor maintains all its output signals — the function code, address, size, R/$\overline{\text{W}}$, address strobe, and data strobe — on the bus, and simply waits for subsequent transfers. The main memory access circuit should then repeatedly increment the address by 4, send a long word, and activate $\overline{\text{STERM}}$. If the memory is sufficiently fast, transfers can take place in successive clock periods. The process continues until the microprocessor deactivates the strobe signals.

6.11 CONCLUDING REMARKS

This chapter has introduced the basic facilities needed on a microprocessor bus and the ways in which they can be used to transfer data between the microprocessor and the main memory or I/O devices. Example interface circuits have been given, together with a discussion of some commercially available chips. At this point, the reader should be able to design and build a simple interface for the exchange of data with an I/O device, either serially or in parallel. Details of main memory circuits will be given in Chapter 8.

An important aspect of input and output operations is the use of interrupts and direct memory access, which is the subject of Chapter 7.

6.12 REVIEW QUESTIONS AND EXERCISES

REVIEW QUESTIONS

1. Both a tri-state bus and an open-collector bus allow several devices to drive a common line. To what extent are these two types interchangeable? Describe at least one situation in which one type can be used, but not the other.

2. Why is it necessary to use a storage buffer in the interface of Figure 6.9, but not in Figure 6.6?

3. The Enable-Data signal in the interface of Figure 6.6 is asserted only during the second phase of the clock. In fact, this condition is not necessary, but it is a recommended practice. Why? What problems, if any, might arise if the CLK signal is not used in this interface? (*Hint:* Consider what is happening in other interfaces.)

4. Repeat Question 3 for the interface of Figure 6.9, and comment on the difference.

5. It was suggested in Question 3 that the CLK signal is not essential in the interface of Figure 6.6. The interface of Figure 6.13 is identical, except for the addition of a status flag circuit. Is it still possible to do without the CLK signal? Why? (*Hint:* There are two reasons.)

6. What is the advantage of assigning the input and output registers in Figure 6.29 the same address?

7. Many interface chips, such as the one in Figure 6.26, have more than one chip-select input. Is one chip-select input sufficient? If so, why is more than one often provided?

8. Why is input latching a useful feature to provide in a GPPI chip?

9. Consider the timing diagram for the asynchronous bus transfer in Figure 6.47. How is this diagram affected as the distance between the two devices increases?

10. The microprocessor and the I/O device shown in Figure 6.64 are interconnected by a long asynchronous bus. Draw a timing diagram similar to Figure 6.47 in which each signal appears twice, once as seen by the microprocessor and once as seen by the device, that is at points A and B, respectively.

PRIMARY PROBLEMS

11. Design an address decoder circuit to recognize address $CB36_{16}$, using AND, OR, and NOT gates. Give an alternative circuit using NAND and NOR gates.

12. Design an address decoder circuit to recognize the block of addresses $D9E0_{16}$ to $D9E7_{16}$.

13. A peripheral device is assigned the address $EF74_{16}$. However, the address decoder in the device's interface circuit ignores address lines A_{13} and A_{12}. What are all the addresses that this device will respond to?

14. Design an address decoder to generate the Status-Selected and Data-Selected signals at addresses $D3A0_{16}$ and $D3A1_{16}$, respectively. Use NAND and NOR gates wherever possible.

15. The interface of Figure 6.9 uses a data register of the type shown in Figure 6.10. What changes would you make to the circuit in Figure 6.9 if this register is replaced by

    (a) A positive edge-triggered register.

    (b) A negative edge-triggered register.

    Give a timing diagram showing the Load-Buffer signal in each case.

16. Redesign the handshake circuit of Figure 6.21 assuming that Data-Valid and Data-Taken are active low signals.

17. An alternative way of defining the operation of the handshake signals of Figure 6.17 is shown in Figure 6.65. The printer accepts the data and asserts Data-Taken immediately. Then, it keeps Data-Taken active until it completes the printing operation. Redesign the handshake circuit of Figure 6.21 for this case.

18. Consider the start-stop asynchronous transmission scheme depicted in Figures 6.42 and 6.43. Assume that the receiver clock, RCLK, is derived from a local

Figure 6.64 Microprocessor system in Problem 10.

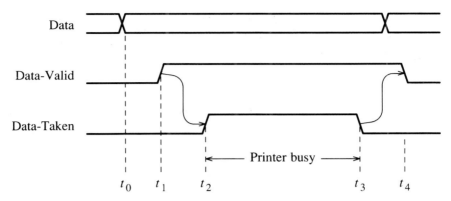

Figure 6.65 Handshake timing for Problem 17.

oscillator in the receiver whose frequency is four times the bit rate. Figure 6.42 suggests that the rising edge of RCLK can be adjusted to coincide with the center of the first data bit. Is that always possible? What is the maximum error?

19. Assume that the frequency of the receiver oscillator mentioned in Problem 18 is 16 times the bit rate. What is the maximum frequency error that can be tolerated to receive the character shown in Figure 6.43?

20. The delays for the 68B09 microprocessor chip are given in Table 6.2. Consider a combined input and output interface, similar to Figure 6.25. The minimum setup time needed for the output buffer is 80 ns. The input buffer requires 200 ns from the time its enable input, E, is activated until data on the microprocessor bus can be assumed valid. Assume that all logic gates have a maximum delay of 15 ns. What are the minimum allowable durations for the first and second phases of the clock?

21. Let d_{1M}, d_{2M}, and d_{3M} be the delays seen by the microprocessor in the system of Problem 10, corresponding to delays d_1, d_2, and d_3 in Figure 6.47. The delays seen by the device are d_{1D}, d_{2D}, and d_{3D}, respectively. Give an expression for each of these delays as a function of the following parameters (add the suffixes max or min where appropriate):

    1. Propagation delay on the bus, T

    2. Bus skew, S

    3. Address decoding delay in the slave, t_{dc}

    4. Setup and hold times, t_{sM} and t_{hM}, of the microprocessor input buffer

22. Repeat Problem 21 for a write operation. Assume that the setup and hold times of the input buffer of the device interface are t_{sD} and t_{hD}, respectively.

23. The pen of a plotter is driven by a stepper motor that has four control coils: A, B, C, and D. In order to cause the pen to move in the forward direction,

electrical pulses should be applied to these coils in the sequence A, B, C, D, A, B, and so on. Each pulse to a coil causes the pen to move 0.1 mm. The pulses must be 25 ms wide and 50 ms apart. The four coils of the motor are connected to lines PB_{3-0} of the GPPI described in Section 8.7, via appropriate amplifiers.

(a) Write a 6809 (or 68000) subroutine that moves the pen forward a given distance. The distance traveled is specified by an 8-bit integer representing the number of millimeters. It is passed to the subroutine as a parameter on the stack. Assume the clock frequency of the microprocessor to be 2 MHz (10 MHz for the 68000).

(b) Write another subroutine to move the pen backward. This requires a reverse sequence of pulses, i.e., D, C, B, A, D, etc.

24. Combine the two subroutines of Problem 23, so that the pen is moved either forward or backward depending on the distance parameter being positive or negative.

ADVANCED PROBLEMS

25. Consider the timing diagram in Figure 6.8. Assume that the microprocessor and the I/O device are physically far apart, and that any signal transmitted by one reaches the other exactly T seconds later. Which of the delay components in the figure are affected by the transmission delay, and by how much? Assume that the clock signal is generated at the microprocessor end of the bus. (*Hint:* To examine this situation carefully, draw two timing diagrams, one showing the signals at each end of the bus.)

Let P be the minimum clock period for correct operation for the case of a short bus ($T = 0$). What is the minimum clock period for the long bus?

26. Repeat Problem 25 for the case where the clock is generated at the device end.

27. Repeat Problem 25 for the output transfer of Figure 6.11. Does the delay T have the same effect on the clock period requirements for read and write operations? Why?

28. An important factor that has been ignored in Problem 25 is the bus skew. In reality, the value of T varies from one bus line to another. Derive a new expression for the minimum clock period needed by a read operation, assuming that T varies from a minimum of T_{min} to a maximum of T_{max}. Repeat for a write operation and comment.

29. The asynchronous scheme of Figure 6.47 automatically adapts to the speed of a given interface circuit and to the transmission delays on the bus. It is possible to adopt timing specifications for the bus such that each interface circuit is required to take into account only the delays introduced within it. The design of a given interface should not be affected by the timing requirements, such as setup time, of other devices on the bus. This goal may be accomplished by properly choosing the way in which delays d_1 to d_3 in the figure are specified. Examine the timing of a bus transfer carefully, and suggest what each delay should be to account for the parameters mentioned in Problems 21 and 22.

LABORATORY EXERCISES

30. A hex keypad is a simple keyboard having 16 keys organized in a 4 × 4 array. The keys are interconnected by row and column wires, as shown in Figure 6.66. It is desired to use this keypad as an input device in a 6809 (or 68000) microprocessor system. The eight wires of the keypad are connected to port A of a PIA (or VIA) chip. The PIA port has built-in pull-up resistors, which will keep all lines in the high-voltage state, when they are programmed as inputs.

In order to detemine whether a key is pressed, the microprocessor pulls the row wires to the low-voltage state, by writing into the appropriate bits of the data direction and output registers of the PIA. Thus, when a key is pressed, one of the column wires will go to the low-voltage state. By reading the state of the column wires, the microprocessor can determine the column number of the key that is being pressed. Then, by repeating the procedure with the roles of the columns and rows interchanged, it can determine the row number.

Write a program that waits for a key to be pressed, determines the row and column numbers, then prints them on the screen of a video terminal. The video terminal is connected via an ACIA chip. The program should allow about 10 ms for key bouncing.

31. Design a circuit that can receive data transmitted in the start-stop format. The circuit is to be driven by a clock signal whose frequency is 16 times the baud rate. Show how you may use a counter and appropriate logic circuits to generate pulses for shifting the input data into a shift register. (*Hint:* The counter should be initialized when the start bit is detected. Then, it should be used to generate active-edge transitions at the mid-points of subsequent bits.)

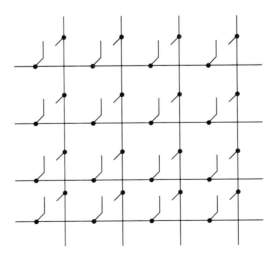

Figure 6.66 Hex keypad for Problem 30.

6.13 REFERENCES

1. Mano, M.M., *Digital Design.* Englewood Cliffs, NJ: Prentice-Hall, 1984.

2. Comer, D.J., *Digital Logic and State Machine Design.* New York: Holt, Rinehart and Winston, 1984.

3. Fletcher, W.I., *An Engineering Approach to Digital Design.* Englewood Cliffs, NJ: Prentice-Hall, 1980.

4. Roden, M.S., *Digital and Data Communication Systems.* Englewood Cliffs, NJ: Prentice-Hall, 1982.

5. McNamara, J.E., *Technical Aspects of Data Communication.* Maynard, MA: Digital Equipment Corp., 1978.

Chapter 7

Interrupts and Direct Memory Access

In all input and output schemes discussed in Chapter 5, the program enters a wait loop in which it repeatedly tests a status flag in the I/O device interface. When the flag indicates that the device is ready, data are transferred to or from the device. In practical applications, it is often necessary to perform many tasks. Some of these require input or output operations to be performed, whereas others may involve lengthy computations. A considerable amount of time may be wasted waiting for an I/O device to become ready. In this chapter we will discuss an alternative approach, the use of interrupts, in which other tasks can be performed during this wait time. We will also introduce an I/O mechanism known as direct memory access. This is a particularly useful scheme when dealing with fast devices, such as magnetic disks and graphic displays.

7.1 INTERRUPTS

An *interrupt* is a request for service generated by a device interface whenever the corresponding device is ready to send or receive data. The device interface sends an interrupt-request signal to the microprocessor as one of the control signals on the bus. In response, the microprocessor may interrupt the program it is executing at the time, and switch to another program that provides the requested service.

Before examining the software and hardware details associated with interrupts, we will present a simple example to illustrate how such a mechanism can be used in a computer system. Consider an application in which some computations are performed, then the results are printed, followed by more computations, and so on. Let us assume that the printer accepts one character at a time. That is, the microprocessor sends one character to the printer, waits for it to be printed, then sends the next character.

Performance of this system can be improved substantially if, instead of doing nothing while the character is being printed, the microprocessor can proceed with other computations. When the printer becomes ready, it signals the microprocessor by means of an interrupt request, which can be regarded as a request for the next character to be sent for printing. The microprocessor should then temporarily stop the computational task it is performing, and execute a program that sends the next character to the printer. This scheme makes it possible for the two tasks of computing and printing to proceed in parallel, where computations are carried out during the waiting periods in the printing process.

The use of interrupts is conceptually very simple. However, its implementation requires careful attention to the design of both software and hardware, and an awareness of the ways in which they interact. We will first examine interrupts as they appear to the programmer. This will be followed by a presentation of the hardware details.

7.1.1 The Interrupt Mechanism

The sequence of events associated with the use of interrupts is illustrated in Figure 7.1. A computer is executing a program that contains instructions $i-1$, i, $i+1$, and so on. An I/O device connected to this computer sends an interrupt-request signal whenever it is ready to receive or send data. Assume that this signal arrives at the microprocessor while instruction i is being executed. Before responding, the microprocessor first completes execution of instruction i. Then, instead of starting to fetch and execute instruction $i+1$, it jumps to the first instruction of a special routine, which we will refer to as the *interrupt-service routine*. This routine performs the data transfer requested by the device. Following its completion, the microprocessor resumes execution of the interrupted program. That is, it fetches and executes instruction $i+1$.

The reader will undoubtedly have observed a similarity between the notion of an interrupt-service routine as depicted in Figure 7.1 and that of a subroutine. However, two major differences should be noted. First, execution of the interrupt-service routine is initiated by an external event. It can take place at any point within the

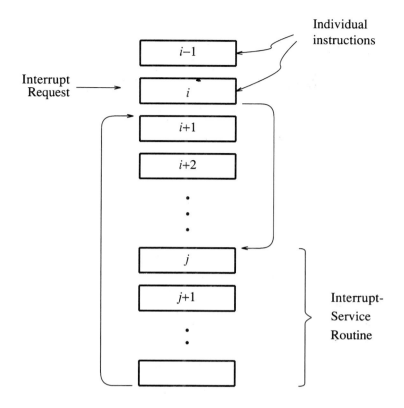

Figure 7.1 Basic idea of an interrupt.

interrupted program. In contrast, execution of a subroutine is initiated by a specific instruction within the calling program.

The second difference is that the function performed by a subroutine constitutes an essential part of the task being carried out by the calling program. The interrupt-service routine, on the other hand, does not necessarily have anything in common with the interrupted program. Therefore, the mechanism that the microprocessor uses to effect a jump to the interrupt-service routine, followed by a return to the interrupted program, must ensure that the two programs do not interfere in any way with each other. It is the responsibility of the programmer to make sure that neither program accidentally changes data items used by the other.

Figure 7.2 illustrates the actions taken by the microprocessor when it receives an interrupt-request signal. As pointed out earlier, the microprocessor first completes the instruction it is executing at the time the interrupt request arrives. Suppose that instruction i consists of 3 words and that the interrupt request arrives after the first word has been read from the main memory. The microprocessor proceeds to read the remaining two words and perform the operation specified. At this point, the program counter (PC) contains the address of the first word of instruction $i+1$. This is the return address, i.e., the address at which the microprocessor must continue

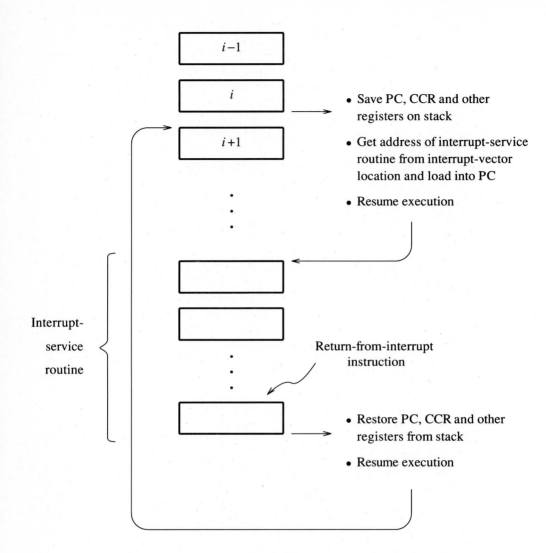

Figure 7.2 Functions performed by CPU hardware at the beginning and end of an interrupt-service routine.

processing when it completes execution of the interrupt-service routine. It must be saved in a suitable location in the microprocessor or in the main memory.

It was pointed out in Chapter 2 that most modern computers use a stack structure in the main memory to save return addresses of subroutines. The stack structure offers considerable flexibility because it is well suited to nested calls. For similar reasons, the same stack is also used for saving the return address when a program is interrupted.

It is not sufficient to save only the return address before starting to execute the interrupt-service routine. Consider, for example, the case where instructions i and $i+1$ are

| Decrement | POINTER |
| Branch_if_>_0 | LOOP |

A branch to LOOP will take place if the result of the Decrement instruction is greater than 0. The required information is stored in the condition code register, CCR, whose contents are updated after every instruction. Therefore, the contents of CCR must be saved along with the return address, then restored before resuming execution of the interrupted program. Otherwise, the condition to be tested by the Branch instruction may be destroyed as a result of the execution of the interrupt-service routine.

In many microprocessors, the contents of all registers are saved before starting the execution of an interrupt-service routine. This is desirable, because the interrupt-service routine is likely to use these registers for its own purposes, and the original register contents must be preserved for the interrupted program. The delay involved in saving and later restoring the registers can be considerable, particularly in a microprocessor that has a large number of registers. For this reason, some microprocessors save only the contents of the PC and CCR registers. Then, it is up to the programmer to save the contents of any other registers that may be needed in the interrupt-service routine. Where necessary, registers may be saved by pushing them on the stack at the beginning of the interrupt-service routine, then pulled off the stack at the end of the routine. A microprocessor that can accommodate different types of interrupts may provide both facilities. That is, some interrupt requests result in saving all registers, while for others, only the PC and the CCR are saved.

The reader will recall that no registers are saved automatically as a part of the subroutine linkage process. Because a subroutine performs a function needed by the calling program, the registers are often used for passing parameters between the two.

Having saved all vital information, the microprocessor begins preparations for execution of the interrupt-service routine. First, it has to obtain the starting address for that routine. There are several techniques that can be used for this purpose. Let us assume that the required starting address is stored at a fixed location in the main memory, whose address is known to the microprocessor. In this case, the starting address is referred to as the *interrupt vector*. In order to cause a jump to the interrupt-service routine, the microprocessor reads the contents of the interrupt-vector location and loads them into the program counter. This simple approach is used in many microprocessors. Other schemes based on variations of this idea are also used, as will be explained later.

The interrupt-service routine ends with a Return_from_interrupt instruction. Execution of this instruction results in the reversal of all the actions taken at the time the routine was entered. That is, the contents of all registers that have been saved, including PC and CCR, are restored from the stack. As a result of the restoration of the contents of the PC, the next instruction to be executed will be instruction $i+1$.

7.1.2 Enabling and Disabling Interrupts

The sequence of steps given in Figure 7.2 and discussed in the previous section represents the response of the microprocessor to an interrupt request. The interrupt request is a signal received over the computer bus, sent by an input or an output device when it is ready for transferring data. Obviously, the microprocessor should respond to such a request only when the program being executed can be interrupted safely. Thus, a mechanism for enabling and disabling interrupts must be provided.

Let the microprocessor operate in one of two modes. In the first mode interrupts are disabled; that is, interrupt requests are ignored. The microprocessor responds to interrupts only when operating in the second mode, in which interrupts are enabled. An internal flag in the microprocessor, called an "interrupt mask," determines its mode of operation. Interrupts are disabled, or "masked," when this flag is equal to 1 and enabled when it is equal to 0. The state of the flag may be changed under program control, thus providing the programmer with the ability to alter the mode of operation of the microprocessor.

The details of the instructions that enable and disable interrupts vary considerably from one microprocessor to another. For the purposes of this general discussion, we will assume that two instructions are available. The first, Enable_interrupts, clears the interrupt mask, thus causing the microprocessor to operate with interrupts enabled. Any interrupt request received after this instruction has been executed triggers the sequence of actions described in Figure 7.2. The second instruction, Disable_interrupts, changes the mode of operation to one in which interrupt requests are ignored.

Following is a very simple example that illustrates the use of interrupts and shows when interrupts should be enabled or disabled. Consider a program that, among other things, transfers one character to a printer. Assume that the character to be printed is stored in location CHAR in the main memory. The required operation may be performed as shown in Figure 7.3, using subroutine PRINT. In this program, PRS and PRD refer to the printer status and data buffers, respectively.

In order to use interrupts to accomplish the same task, the program of Figure 7.3 may be reorganized as shown in Figure 7.4. Instead of calling a subroutine, the main program loads the address of the interrupt-service routine—PRINT in this case—in the interrupt vector location INTVECTOR. Then it enables interrupts. The microprocessor is assumed to have been operating with interrupts disabled up to that point. After enabling interrupts, the main program starts performing other tasks.

Whenever the printer interface is ready to receive data from the microprocessor, it activates the interrupt-request signal. Since interrupts are now enabled, the microprocessor interrupts the execution of the main program and starts executing the interrupt-service routine. The starting address for this routine is fetched from location INTVECTOR, which now contains the address PRINT.

The first instruction of routine PRINT is Disable_interrupts. It returns the microprocessor to the state in which interrupt requests are ignored, to ensure that the interrupt-service routine cannot be interrupted by another interrupt request. The remainder of the interrupt-service routine is straightforward. It simply transfers the character to be printed to the printer data buffer, PRD, then returns control to the

MAIN   _____

               _____

               Call   PRINT

               _____

               _____

PRINT   Move   PRS, ACCUM   Wait for printer.
               AND   # 1, ACCUM
               Branch_if=0   PRINT
               Move   CHAR, PRD   Send character.
               Return_from_subroutine

Figure 7.3 A program to print one character using status checking.

interrupted program. There is no need to check the Printer-Ready bit in PRS, because an interrupt request is generated by the printer interface only when Printer-Ready is equal to 1. When the printer completes printing, it will send another interrupt request. However, this time the request will be ignored, because an Interrupt_disable instruction has been executed. Interrupts should be enabled again when a second character is to be printed.

The need for disabling interrupts at the beginning of an interrupt-service routine may be explained as follows. The interrupt-request signal is generated by the device interface, and is maintained active as long as the device is ready for an input or an output operation. In the example of Figure 7.4, the interrupt request will remain active until the Move instruction in routine PRINT is executed. At this point, the

MAIN   _____

               _____

               Move   #PRINT, INTVECTOR
               Enable_interrupts

               _____

               _____

PRINT   Disable_interrupts
               Move   CHAR, PRD   Send character to printer
                                                        This action causes the printer to
                                                        turn off its interrupt request.

               Return_from_interrupt

Figure 7.4 A program to print one character using interrupts.

printer drops its interrupt request, and starts printing the character. In general, the service requested by an I/O device may take several instructions to complete. During this period, the interrupt-request line remains in the active state. Hence, if interrupts are still enabled after executing the first instruction in the interrupt-service routine, a second interruption could occur. The microprocessor would begin executing the same interrupt-service routine one more time, and the process would be repeated indefinitely. In other words, if interrupts are not disabled at the beginning of an interrupt-service routine, the microprocessor may never get past the first instruction.

Because of the importance of disabling interrupts at the beginning of interrupt-service routines, many microprocessors perform this function automatically. When an interrupt is accepted, further interrupts are disabled before starting execution of the interrupt-service routine. If the program of Figure 7.4 is to be executed on a microprocessor that has this feature, the Disable_interrupts instruction would not be needed.

We should note that a special interrupt-request line is provided in some microprocessors to be used for interrupts that cannot be disabled by the programmer. Interrupt requests on such a line are usually referred to as *nonmaskable* interrupts.

7.1.3 Interrupt Software

The program of Figure 7.4 uses interrupts to transfer one character to the printer. Let us consider a more realistic example. A computer is required to perform some computation, print a one-line message containing the results, then proceed with other computations. Figure 7.5 shows a possible program, in which subroutine PRINT

MAIN	_____		Compute and store results
	_____		at MSG.
	_____		
	Move	#MSG,POINTER	
	Call_subroutine	PRINT	
	_____		Continue computing.
	_____		
PRINT	Move	PRS,ACCUM	Wait for printer.
	AND	#1,ACCUM	
	Branch_if_=0	PRINT	
	Move	[POINTER],PRD	Send character.
	Compare	#CR,[POINTER]$_{autoinc}$	Check for end of message.
	Branch_if_=0	MSGEND	
	Branch	PRINT	
MSGEND	Return_from_subroutine		

Figure 7.5 Printing using a subroutine which waits for the printer to become ready.

prints a one-line message using the status-checking approach of Chapter 5. It is assumed that the message to be printed consists of characters stored in successive memory locations, starting at location MSG. The address MSG is passed to the subroutine as a parameter in microprocessor register POINTER. The end of the message is indicated by the Carriage-Return character, CR. In this case, no computations are performed while the results are being printed.

If we wish to carry out some other computations while waiting for a character to be printed, the program of Figure 7.5 may be reorganized to use interrupts, as shown in Figure 7.6. Instead of calling the PRINT routine directly, the main program puts the address of that routine in the interrupt-vector location, INTVECTOR. The address parameter MSG should not be passed to the interrupt-service routine in a register, because most microprocessor registers are likely to be used during the computations that will proceed in parallel with the printing process. The required parameter is passed through a memory location, which we have called MEMPOINTER. After storing the address MSG in this location, the main program enables interrupts and continues with its computational tasks. Obviously, these computations should not alter the contents of either location MEMPOINTER or the area in the main memory where the message is stored.

Register POINTER is used in the PRINT routine. Hence, its contents are saved on the stack at the beginning of the routine, to be restored just before returning to the interrupted program. The PRINT routine fetches the character to be printed using the address in the pointer location, MEMPOINTER, and sends it to the printer. Then, it increments the pointer to point at the next character in the message.

MAIN	_____		Compute and store results at MSG.
	_____		
	Move	#MSG,MEMPOINTER	
	Move	#PRINT,INTVECTOR	
	Enable_interrupts		
	_____		Continue computing, but do not use locations MSG or MEMPOINT.
	_____		
	_____		
PRINT	Disable_interrupts		
	Push	POINTER	Save register to be used.
	Move	MEMPOINT,POINTER	
	Move	[POINTER],PRD	Send character. Printer will turn off its interrupt request.
	Increment	MEMPOINT	Increment pointer in memory.
	Compare	#CR,[POINTER]$_{autoinc}$	Check for end of message.
	Branch_if_=0	MSGEND	
	Enable_interrupts		
MSGEND	Pull	POINTER	Restore microprocessor register.
	Return_from_interrupt		

Figure 7.6 Printing a one-line message using interrupts.

After sending a character to the printer, the PRINT routine checks for the end of the message. If the carriage-return character has not been reached, interrupts are enabled, and a Return_from_interrupt instruction is executed, causing the execution of the interrupted program, MAIN, to be resumed. When the printer completes printing a character, it sends a request for interrupt to the microprocessor, which causes the PRINT routine to be executed again. As a result, the next character is sent to the printer, and so on. After sending the last character of the message, which is Carriage Return, the PRINT routine leaves interrupts disabled. Thus, once the main program is resumed, it will continue uninterrupted until it enables interrupts again in order to print another message.

The printer removes its interrupt request after it receives the service it is requesting, that is after a character is sent for printing by the microprocessor. As pointed out earlier, it is important to ensure that interrupts remain disabled until that happens. In the interrupt-service routine of Figure 7.6, interrupts are enabled after the Move instruction that sends a character to the printer. Hence, there is no danger of a second interruption occuring as a result of the interrupt request that caused this routine to be entered.

7.1.4 Interrupt Hardware

In order to use interrupts for input and output operations, the device interface must include a circuit that generates an interrupt-request signal. One or more lines of the computer bus are usually provided to carry the interrupt requests, where each line may have several I/O devices connected to it.

The device interface should activate the interrupt-request signal when it is ready to receive or transmit data, that is, when the Ready bit in its status register is equal to 1. Hence, the interrupt-request output of the interface may simply follow the state of the Ready bit. However, since several I/O devices may be involved, it should be possible to selectively enable and disable interrupts from individual devices under program control. Therefore, the interrupt-request output of a device interface should follow the state of the Ready bit only when interrupts are enabled for that device. Otherwise, the interrupt-request output of the interface should be inactive. In Subsection 7.1.2, the concept of enabling and disabling interrupts in the microprocessor was introduced. It relates to whether the microprocessor should accept an interrupt request when one is received. On the other hand, the idea of enabling and disabling interrupts in the interface of a given device relates to whether that interface is allowed to generate an interrupt request when it requires service. At a given moment, interrupts may be enabled in some device interfaces and disabled in others. The mechanism by which interrupts from a given interface circuit can be enabled or disabled will be discussed shortly.

In order to allow several devices to send their interrupt requests to the microprocessor via a common bus line, the line is usually implemented as an open-collector bus, which was described in Subsection 6.8.1. Figure 7.7 shows the interrupt-request outputs from several devices connected in this manner. The common line, $\overline{\text{IRQ}}$, is active when in the low voltage state. Hence, the microprocessor will see an active

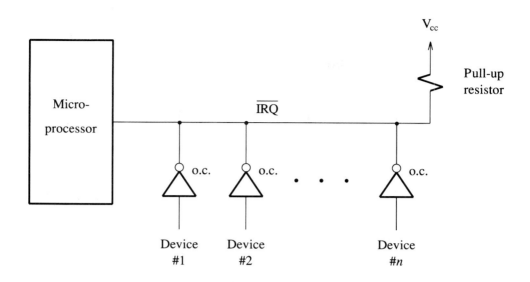

Figure 7.7 An open-collector interrupt-request line.

interrupt-request signal whenever at least one of the interrupt-request outputs is active.

Consider the keyboard interface discussed in conjunction with Figure 6.15. An interrupt-request output for this interface may be derived from the Keyboard-Ready status bit. In order to provide a facility for enabling and disabling interrupt requests, a control register may be added and given a unique address, so that its contents may be updated by the microprocessor at any time. Only one control bit, Interrupt-Enable, is needed in this simple interface, as shown in Figure 7.8. The Interrupt-Enable flip-flop is set to 1 whenever an instruction is executed that writes a 1 into the least significant bit of the control register. When a key is pressed on the keyboard, Keyboard-Ready becomes equal to 1, and the interrupt-request output, $\overline{\text{IRQ}}$, is activated. The $\overline{\text{IRQ}}$ signal remains active, that is, in the low state, until an instruction is executed that reads the contents of the input data buffer. The read operation causes the Enable-D signal to become active, thus resetting the Keyboard-Ready signal and removing the interrupt request.

7.2 EXAMPLES OF THE USE OF INTERRUPTS

The previous section has presented the basic principles of interrupts and the software and hardware needed for their implementation. Before introducing some of the more advanced aspects of interrupt systems, we will examine how interrupt capabilities are incorporated in the general-purpose serial and parallel interfaces introduced in Chapters 5 and 6. We will also describe timer circuits, which are important

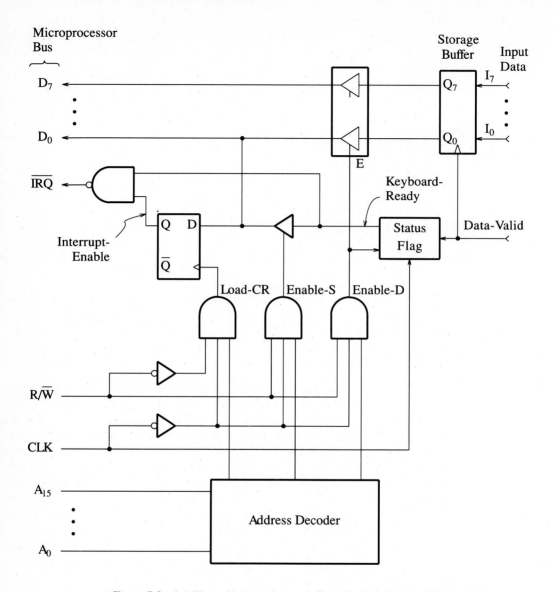

Figure 7.8   Addition of interrupt capability to the interface of Figure 6.15.

components that are almost always used in conjunction with interrupts to provide some real-time capabilities in a microprocessor system.

7.2.1 Serial Interfaces—the ACIA

The ACIA was introduced in Subsection 5.3.1, as an example of a general-purpose serial interface chip. Some of its hardware characteristics were described in Section

6.7.2. An interrupt capability is incorporated in this chip in the form of an interrupt request bit in the Status register and an interrupt-enable facility in the Command register, as shown in Figure 7.9. For convenience, separate control fields, TIE and RIE, are provided for the transmitter and the receiver sections, respectively. The ACIA has one interrupt-request output, $\overline{\text{IRQ}}$, which should be connected to the interrupt-request line on the microprocessor bus. The interrupt-request bit, IRQ, in the status register is equal to 1 whenever $\overline{\text{IRQ}}$ is active.

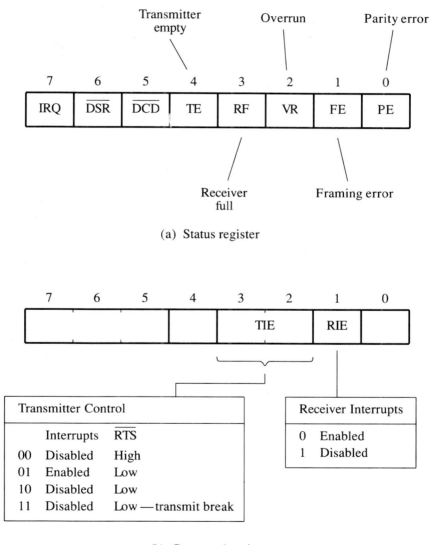

(a) Status register

(b) Command register

Figure 7.9 Status and Command registers of the ACIA.

When interrupts are to be used for input and output operations, a program instruction should enable the transmitter and the receiver interrupts by setting TIE to 01 and RIE to 0 in the Command register. Then, whenever the ACIA receives a character, it activates its interrupt-request output, $\overline{\text{IRQ}}$, and sets bits RF and IRQ in the status register to 1. When the microprocessor reads the input character, the ACIA clears the two status bits and deactivates $\overline{\text{IRQ}}$. Similarly, when the transmitter buffer is empty, the ACIA activates $\overline{\text{IRQ}}$ and sets bits TE and IRQ to 1. As soon as the microprocessor sends a character for transmission, the interrupt request is removed and the status bits are cleared. They remain in this state until the character is transferred from the transmitter buffer, TX, to the OUT shift register for transmission (see Fig. 5.7). Note that the IRQ bit is set for both receiver and transmitter interrupts. This simplifies polling operations, as will be described in Subsection 7.3.1.

7.2.2 Timers

Many microprocessor applications require certain tasks to be performed at specific times. The generation of a square wave signal, which was discussed in Subsection 5.5.4, is a simple example of such an application. Many other examples are encountered in instrumentation and industrial control systems, where strict timing constraints must be met. Such systems are known as *real-time* systems.

It is often convenient to incorporate timing circuits in a microprocessor system, as hardware aids for real-time applications. A timer may be used to measure the elapsed time between two events, or to perform a given task periodically. For example, a microcomputer controlling a furnace may have to obtain a temperature reading once every second. This can be accomplished by arranging for the timer to interrupt the microprocessor at one-second intervals. The interrupt-service routine reads the furnace temperature, then returns to the interrupted program, which performs other control functions.

In yet another application a timer may be used in a *one-shot* mode. Consider, for example, the case of one microcomputer sending a message to another over a telephone link. Typically, after receiving a message, the destination computer sends back an acknowledgment, confirming that it has received the message correctly. If the sending computer does not receive the acknowledgment within a predefined period, it retransmits the message, assuming that the first copy has been lost or corrupted by noise on the line. To implement this scheme, a timer is initialized to the required time-out period. As soon as the transmission of a message is completed, the timer is allowed to run, until it is either stopped by the microprocessor, or the time-out period expires. The microprocessor stops the timer when it receives the awaited acknowledgment. Otherwise, after the preset period has elapsed, the timer sends an interrupt request to the microprocessor, indicating that retransmission should take place.

The discussion above suggests that there are two useful modes of operation for a timer, a free-running mode, and a one-shot mode. In the free-running mode, interrupt requests are generated at regular intervals. In the one-shot mode, only one interrupt request is sent to the microprocessor at the end of a predefined period.

A timer circuit may be implemented in a variety of ways. In its simplest form, it consists of a counter driven by an appropriate clock signal and some control logic. Additional hardware may be used to provide some flexibility and place the operation of the timer under program control. The block diagram in Figure 7.10 gives an example of the main hardware parts and functional capabilities that may be incorporated in a timer circuit. A counter is connected to the microprocessor bus via a latch. The output of the counter is also connected to the bus via a buffer, which enables the contents of the counter to be read by the microprocessor. The latch is used to hold the count value respresenting the desired timing period. When the timer is started, the contents of the latch are transferred to the counter, and a timing period begins. The counter is driven by pulses from a suitable source, usually the clock line on the microprocessor bus. It is decremented by each clock pulse until it reaches zero. At this point, the Zero Detector circuit signals the interrupt-request logic to send an interrupt request to the microprocessor, provided that the timer interrupts are enabled. If the timer is operating in the free-running mode, the contents of the timer latch are automatically reloaded into the counter, and another timing period is started immediately. In the one-shot mode, a second timing period is started only when requested by a program instruction.

Once a timer interrupt request is activated, it remains active until removed by an instruction in the interrupt-service routine. For example, the interrupt request may be removed as a result of loading a new value into the timer latch or reading the

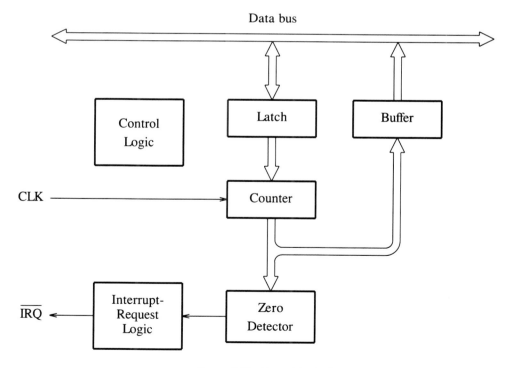

Figure 7.10 Timer block diagram.

contents of the counter. We should note that in the one-shot mode, the counter may continue to be decremented through the values $11\cdots11$, $11\cdots10$, and so on after reaching zero. Thus, by reading the contents of the counter, the microprocessor can determine the time that has elapsed since the interrupt request was generated. The counter will eventually reach zero again. However, a second interrupt will not be raised. In the free-running mode, an interrupt request is generated every time the counter reaches zero.

A useful feature is for the timer to be *retriggerable* when operating in the one-shot mode. That is, while a one-shot timing period is in progress, the microprocessor may restart a new timing period before the first one expires. This may be done by sending a command to the timer that would cause it to reload the counter from the latch. The second timing period need not have the same duration as the first one, because the microprocessor can change the contents of the latch at any time. A particularly useful application of the retriggering feature is in the implementation of "watchdog" timers (see Problem 23).

Many commercially available general-purpose interface chips have timers that incorporate variations and additions to the ideas introduced above. For example, the timer latch and the counter in Figure 7.10 may be 8 or 16 bits wide, providing a maximum count of 256 or 65,536, respectively. A useful feature found in some timers is the ability to use an external clock. In this case, one of the control bits determines whether the clock signal on the microprocessor bus or the external clock signal is to be used for decrementing the timer counter. Also, the timer may have an output signal whose state changes at the end of a timing period. With this feature it becomes possible to produce pulses of variable width or signals having complex waveforms, all under program control.

The timer function may be implemented in a separate IC chip, such as Motorola's 6840. Alternatively, it may be incorporated in a general-purpose chip, as in the case of the VIA discussed in Section 7.2.4.

7.2.3 Parallel Interface—the PIA

Two examples of a general-purpose parallel interface chip were introduced in Chapter 5, the PIA and the VIA. We study their interrupt facilities in this and the following section. Let us consider first Motorola's PIA.

The PIA has two 8-bit bidirectional ports, PA and PB. It has two interrupt-request outputs, $\overline{IRQA}$ and $\overline{IRQB}$, one for each port. Let us consider port A as an example. The corresponding control register, CRA, is shown in Figure 7.11. Bits CRA_{1-0} and CRA_{5-3} determine the mode of operation of the CA1 and CA2 control lines, respectively. They also determine the conditions under which the two status bits, SA1 and SA2, are set and cleared, as described in Subsection 5.5.2. For interrupt operation, two interrupt-enable bits are provided. When CRA_0 is set to 1, an active transition on line CA1 will activate $\overline{IRQA}$ at the same time it sets SA1 to 1. The interrupt request will remain active until the microprocessor reads the data register of port A. The second source of interrupt is an active transition on line CA2 when this line is programmed as input, that is when $CRA_5 = 0$. Interrupts from CA2 are enabled by setting CRA_3 to 1. In this case, an active transition on CA2 will set

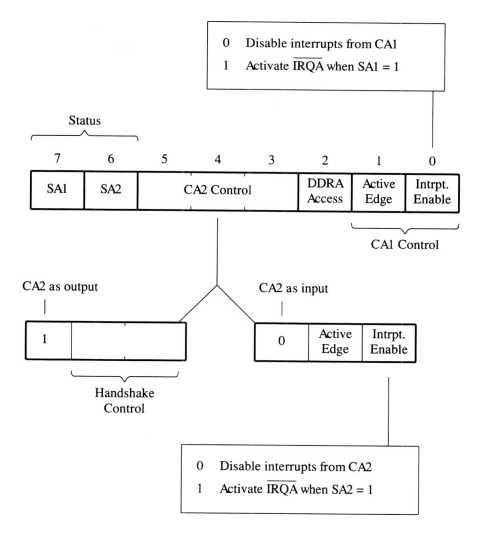

Figure 7.11 Control register CRA of the PIA.

SA2 to 1, and at the same time will activate $\overline{\text{IRQA}}$. Again, SA2 will remain set and $\overline{\text{IRQA}}$ active until the microprocessor reads the data register. When CA2 is programmed as an output line, its associated status bit, SA2, carries no information; it is always equal to 0. Interrupts can only be generated by transitions on CA1 in this case.

Interrupts are generated by port B in exactly the same manner. Output $\overline{\text{IRQB}}$ of the PIA is activated whenever one of the two status bits, SB1 and SB2, as well as its associated interrupt-enable bit, is equal to 1.

Because the PIA has two interrupt-request outputs, its two ports appear as two separate devices. The two outputs may be connected to the same interrupt-request line, or to separate lines if the microprocessor has more than one interrupt-request input. Some of the options available for handling multiple devices will be explored in Section 7.3.

7.2.4 Parallel Interface—the VIA

The internal organization of the VIA is shown in Figure 7.12. Its use for transferring data via the two 8-bit ports was described in detail in Subsections 5.5.3 and 6.6.4. Here, we will describe the facilities provided for interrupt-controlled data transfer, as well as the timer and shift register features of the VIA.

The mode of operation of each VIA port is controlled by the information deposited in the peripheral control register (PCR) and the auxiliary control register (ACR). The status register, which is also called the Interrupt Flag Register, IFR, has

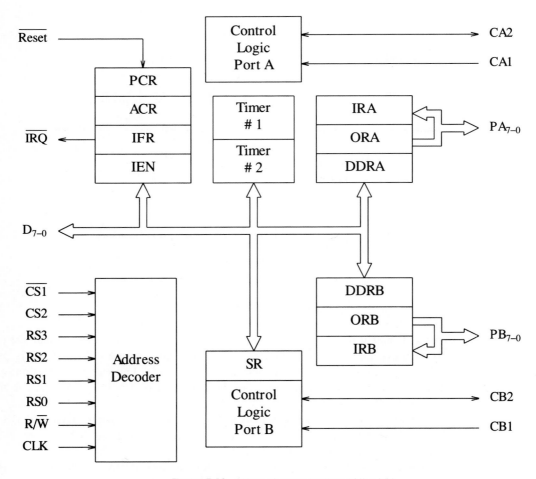

Figure 7.12 Internal organization of the VIA.

7 status bits, IFR_{6-0}, that describe the state of various functional entities within the VIA, as shown in Figure 7.13(a). The eighth bit, IRQ, reflects the state of the interrupt-request output, $\overline{IRQ}$. This bit is equal to 1 whenever $\overline{IRQ}$ is active.

The seven status bits in register IFR represent seven independent sources of interrupt. To maximize the VIA's flexibility, a separate interrupt-enable flag is provided for each source. These flags constitute the seven least-significant bits of the Interrupt Enable Register, IER, which is shown in part (b) of the figure. Bit IER_1, for example, is the interrupt-enable flag that can be used to enable interrupts from control line CA1. When set to 1, any event that causes the associated status flag, SA1, to be set will also result in the interrupt-request flag, IRQ, being set and the interrupt-request output, $\overline{IRQ}$, being activated. The interrupt request will remain active until SA1 is cleared. The conditions under which the four status flags associated with the port control lines are set and cleared were given in detail in Table 5.4. The timer and shift-register flags will be discussed shortly.

The way in which the IER register is used to enable and disable interrupts differs slightly from our earlier examples. In the interface of Figure 7.8, interrupts are enabled by writing a 1 directly into the interrupt-enable flag, bit 0 of the control register. According to this approach, writing the binary pattern 00001010 into

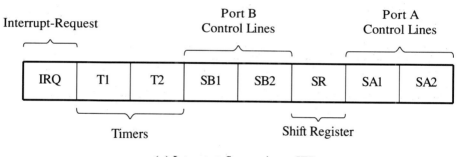

(a) Interrupt flag register, IFR

(b) Interrupt enable register, IER

Figure 7.13 Status and interrupt-enable registers of the VIA.

register IER would result in enabling interrupts from control lines CA1 and CB2 and disabling interrupts from all other sources. This, in fact, is not the case. The VIA allows interrupts from individual sources to be enabled or disabled, without affecting interrupts from other sources. This flexibility is provided by interpreting the information written into the IER register in a slightly different way. Bit 7 of this register is called S/C for Set/Clear. When the pattern written into IER has S/C = 1, the interrupt-enable flags will be set to 1 for every bit location where the pattern contains a 1. Other interrupt sources will not be affected. That is, they remain either enabled or disabled, according to their state before the write operation took place. Similarly, a write operation that has S/C = 0 will clear the interrupt-enable flags for every bit location of IER that has a 1 written into it. Other locations will not be affected. Thus, to enable interrupts from CA1 and CB2, the microprocessor should write the pattern 10001010 into register IER. To disable these interrupt sources, the pattern 00001010 should be written into IER. Neither of these write operations will affect the interrupt enable flags for CA2, CB1, or any other interrupt source.

The microprocessor can check the state of the interrupt-enable flags at any time by reading the contents of register IER. Each bit location will contain a 1 where interrupts are enabled, and 0 otherwise. Bit S/C has no significance in a read operation, and will always appear equal to 1.

VIA TIMERS

The VIA chip incorporates two timers, T1 and T2, which have several modes of operation. They can be used to produce programmable delays and to control the operation of the shift register, SR.

Timer T1 consists of a 16-bit latch, T1L, and a 16-bit counter, T1C, which are connected to the bus data lines of the VIA as shown in Figure 7.14. The high- and low-order bytes of the latch and the counter are accessible at the addresses given in Table 7.1. As can be seen in the figure, there is no direct path for transferring data from the microprocessor bus to the counter. In order to load a starting value into the counter, the microprocessor writes first into the low-order byte, T1C-L, at address 0100. However, the data being written do not get loaded directly into the counter; they are stored in the low-order byte of the latch. Similarly, when the microprocessor writes into the high-order byte of the counter, T1C-H, the data are loaded into the high-order byte of the latch. But, as soon as this last transfer is completed, the 16 bits in the latch are automatically copied into the counter, and the timer operation is started. Thus, by writing into T1C-H, the microprocessor loads an initial value into the counter, and at the same time provides the timer with the trigger signal needed to start a timing period. Since the initial value is stored in the latch, the timer can automatically reload the counter and start a new timing period when the current one expires, as was explained in Subsection 7.2.2.

Although the microprocessor cannot write directly into the counter, it can read its contents at any time, because the counter output is connected directly to the bus. The microprocessor can also access the low- and high-order bytes of the latch directly during read and write operations, at the register addresses given in Table 7.1. Writing into the low-order byte, T1L-L, has exactly the same effect as writing into

Microprocessor Bus

D_{7-0}

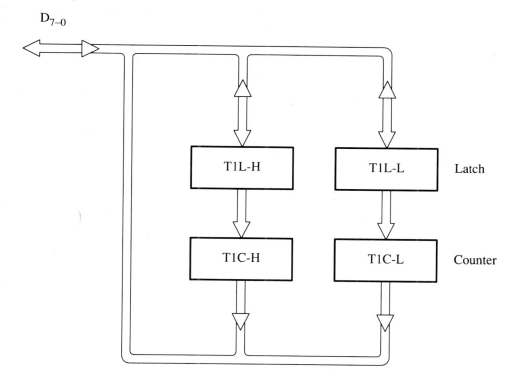

Figure 7.14 Timer T1 of the VIA.

T1C-L. However, writing into the high-order byte, T1L-H, loads new information into the latch without triggering a transfer to the counter or starting timer operation.

The T1 timer has two modes of operation: one shot or continuous. In either mode, the timer may be programmed to produce an output signal on line PB_7 of port B. These options are selected by bits 6 and 7 of the Auxiliary Control Register, ACR, as shown in Figure 7.15. Bit 6 selects either the one shot or the continuous mode when set to 0 or 1, respectively. If bit 7 is set to 1, an output signal is produced on line PB_7. In this case, line PB_7 must be programmed as an output line, by setting bit 7 of the data direction register of port B to 1. When this option is selected, and before the timer is triggered, line PB_7 goes to the high-voltage state. Figure 7.16(a) shows the output signal when the timer is operated in the one-shot mode. Line PB_7 goes low at the end of the clock period in which the microprocessor writes into T1C-H to start the timer. It returns to the high-voltage state at the end of the clock period in which the timer counter reaches 0. Thus, a single low pulse is produced on PB_7, with a width of $N+1$ clock periods, where N is the 16-bit number that was loaded into the timer counter. If the timer is operated in the continuous mode, the state of PB_7 changes each time the counter reaches 0, thus producing the

Table 7.1   TIMER AND INTERRUPT CONTROL REGISTERS IN THE VIA

RS3-0	Designation	Description
0100	T1C-L	Low-order byte of T1 counter. Read — transfers the contents of T1C-L to the microprocessor bus. Write — loads data from the microprocessor bus into T1L-L.
0101	T1C-H	High-order byte of T1 counter. Read — transfers the contents of T1C-H to the microprocessor bus. Write — loads data from the microprocessor bus into T1L-H, then copies the contents of T1L into T1C and starts the timer.
0110	T1L-L	Low-order byte of T1 latch — read and write.
0111	T1L-H	High-order byte of T1 latch — read and write.
1000	T2C-L	Low-order byte of T2 counter. Read — transfers the contents of T2C-L to the microprocessor bus. Write — loads data from the microprocessor bus into T2L-L.
1001	T2C-H	High-order byte of T2 counter. Read — transfers the contents of T2C-H to the microprocessor bus. Write — loads data from the microprocessor bus into T2C-H, transfers the contents of T2L-L to T2C-L and starts the timer.
1010	SR	Shift register.
1011	ACR	Auxiliary control register.
1101	IFR	Interrupt flag register—contains status flags.
1110	IER	Interrupt enable register.

square wave shown in Figure 7.16(b). We should note that a waveform is produced only when $ACR_7 = DDRB_7 = 1$. If either ACR_7 or $DDRB_7$ is equal to 0, line PB_7 is independent of the timer, and it behaves in the same manner as other data lines of port B.

The state of the timer may be checked by the microprocessor at any time by inspecting bit 6 of the Interrupt Flag Register, IFR (see Fig. 7.13). This bit is set to 1 when the timer counter reaches 0. It is cleared when the microprocessor reads the timer counter, or when it starts another timing period by depositing a new value in T1C-H. It may also be cleared directly by writing into IFR. When the timer is

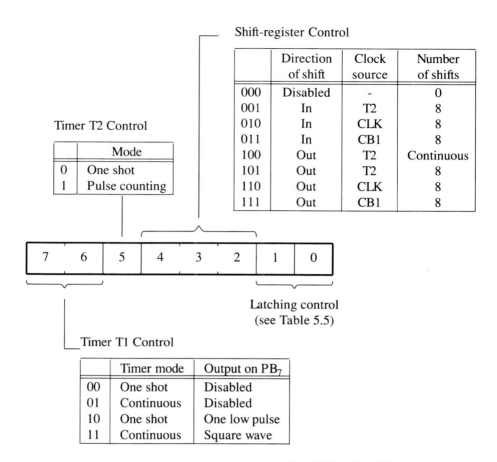

Figure 7.15 Auxiliary control register (ACR) of the VIA.

operated in the one-shot mode, the timer flag is set only once, at the end of the timing period. The counter continues to be decremented after the flag is set. However, the flag, once cleared by the microprocessor, will not be set again. On the other hand, when the timer is operated in the continuous mode, the timer flag is set every time the counter reaches 0. Thus, by operating the timer in this mode and repeatedly inspecting IFR the microprocessor can perform a given task at regular intervals.

Instead of inspecting IFR in a wait loop, the timer is more likely to be used in conjunction with interrupts. Interrupts from timer T1 are enabled by setting bit 6 of the Interrupt Enable Register (IER) to 1. In this case, the interrupt request output, $\overline{\text{IRQ}}$, of the VIA is activated at the end of each timing period, as shown in Figure 7.16. The interrupt request is cleared at the same time the timer flag in IFR returns to 0, that is, when the microprocessor reads the timer counter or triggers a new timing period.

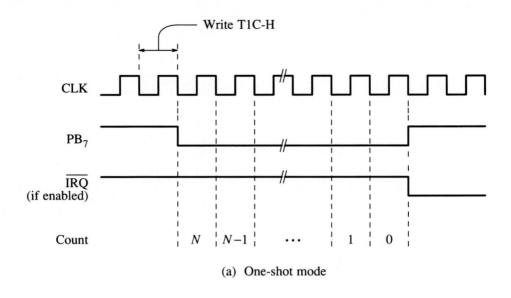

(a) One-shot mode

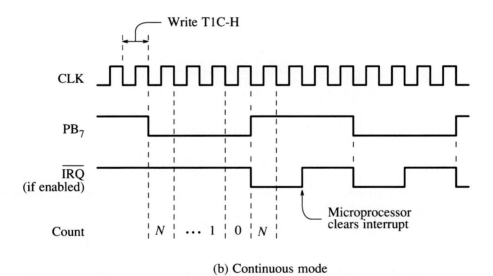

(b) Continuous mode

Figure 7.16   Output of timer T1 on line PB$_7$ of the VIA.

An example of a program that uses timer T1 to produce an output pulse on line PB$_7$ and then send an interrupt request to the microprocessor is given in Figure 7.17. This program generates the output waveform shown in Figure 7.16(a). The number loaded into the timer counter is $2C88_{16}$ (= 11,400). Therefore, assuming a microprocessor clock frequency of 2 MHz, the width of the output pulse will be 5.7 ms.

Move	#%10000000,DDRB	Set PB_7 as output.
Move	#%10000000,ACR	Set timer T1 to one-shot mode and to produce output on PB_7.
Move	#%11000000,IFR	Enable timer T1 interrupts.
Move	#$88,T1C-L	Load low-order byte of counter.
Move	#$2C,T1C-H	Load high-order byte of counter, and start timer.

Figure 7.17 Generation of an output pulse using timer T1.

Note that the low-order byte of the counter must be loaded first, because the timer starts running immediately following the write operation that loads the high-order byte of the counter. It is instructive to consider using a single 16-bit move instruction to load both bytes of the counter. For example, when the VIA is connected to a 6809 microprocessor, the instruction

$$\text{STX} \qquad \text{T1C}$$

appears to do the job. The instructions MOVEP and MOVE can be used in the 68000 and 68020 microprocessors to transfer multiple bytes to 8-bit devices. The reader is encouraged to check register address assignments and the order of byte transfers in each case to determine whether this approach is, in fact, possible. (See Problems 24 to 26.)

Timer T2 is similar to T1, but offers slightly different options. It consists of a 16-bit counter and an 8-bit latch, as shown in Figure 7.18. Only the low- and high-order bytes of the counter are assigned register addresses, as given in Table 7.1. Timer T2 may be operated in the one-shot mode, in the same way as timer T1. This mode is selected by writing a 0 into bit 5 of the ACR register (see Fig. 7.15). The low-order byte of the desired count is deposited into the latch by writing into location T2C-L. Then, the high-order byte is deposited into T2C-H, which automatically starts the timer operation. At the end of the specified timing period, the T2 flag in register IFR will be set, and if T2 interrupts are enabled, an interrupt request will be sent to the microprocessor.

In the second mode, that is, when ACR_5 is equal to 1, the operation of timer T2 is controlled by port line PB_6, which should be programmed as an input by writing a 0 into bit $DDRB_6$. In this mode, T2 does not operate as a timer at all. It simply counts input pulses on line PB_6 and sets the status flag when a specified number of pulses has been received. The microprocessor loads the low- and high-order bytes of the desired count in the T2 counter as usual. After loading T2C-H, each negative edge on PB_6 decrements the contents of the counter by 1. When the counter reaches 0, the T2 flag in IFR is set and an interrupt request is generated. The counter continues to be decremented by input pulses on PB_6. However, once the status flag and

Microprocessor Bus

D_{7-0}

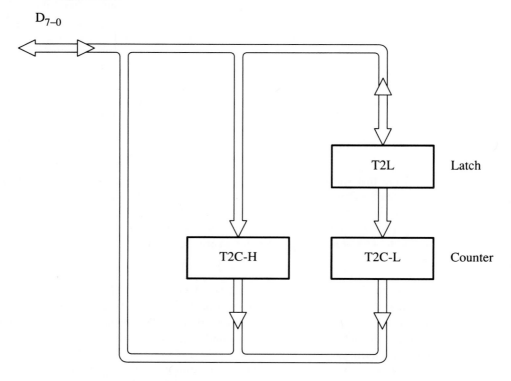

Figure 7.18   Main components of timer T2 of the VIA.

the interrupt are cleared by the microprocessor, they will not be set again until another pulse counting operation is initiated by the microprocessor.

The T2 timer has a third mode of operation similar to the continuous mode of T1. This mode can only be used in conjunction with the shift register, as explained below.

VIA SHIFT-REGISTER

The shift register in the VIA can be used to receive or transmit serial data on control line CB2. The datum on CB2 is loaded into the least-significant bit of the shift register during input operations, and the state of the least significant bit appears on CB2 during output operations, as illustrated in Figure 7.19. The transmission clock used to shift data in or out can be derived from one of three sources: the clock signal on the microprocessor bus, the T2 timer, or an external signal applied to control line CB1. In the first two cases, where the transmission clock is generated within the VIA, the clock signal appears on CB1 for use by the external device that is transmitting or receiving the serial data.

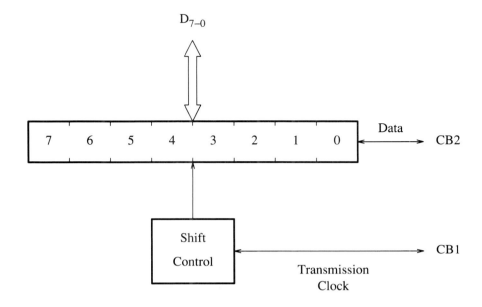

Figure 7.19 Shift register in the VIA.

To use the shift register, bits 2 to 4 of the ACR register must be set to select the desired mode according to Figure 7.15. The shifting operation is initiated by reading the shift register for shifting in or by writing into it for shifting out. In all modes except mode 4, the shift register status flag in register IFR (see Fig. 7.13) is set after eight shifts are completed, and the shifting operation stops. At the same time, an interrupt request is generated if the shift-register interrupt is enabled. The status flag and the interrupt request are cleared when the microprocessor starts another shift operation. They may also be cleared directly by writing into register IFR.

In mode 4, data are shifted out using timer T2 as the source of the transmission clock. Once started, the shifting operation continues indefinitely in a rotating manner; that is, the bit shifted out is automatically loaded into the most-significant bit position of the shift register. Thus, the 8-bit pattern that was initially loaded into the shift register will be transmitted repeatedly on line CB2. The process continues until it is stopped by the microprocessor when it changes the mode selection in ACR. In this mode, only the low-order byte of the T2 counter is used. The width of each phase of the transmission clock is $N + 2$ times the period of the microprocessor bus clock, where N is the value written into T2C-L before the shifting operation is started.

The inclusion of the timer and shift register capabilities makes the VIA a very versatile chip. It can be used to connect a variety of I/O devices to a 6809 or a 68000 microprocessor and to provide the basic timing facilities needed in a real-time system. It also offers the hardware designer many options where elaborate signaling arrangements are involved, as in instrumentation applications.

7.3 HANDLING MULTIPLE DEVICES

In many computer systems, it is desirable to use interrupts to perform input and output operations involving several devices. In this section, we discuss some of the techniques for dealing with multiple interrupt requests. We start by presenting a simple and widely used technique known as polling.

7.3.1 Polling

When several devices are connected to one interrupt-request line and an interrupt occurs, the interrupt-service routine needs to determine which of the devices is requesting service. It may do so by polling the devices, that is, by checking their status registers one by one, until it reaches a device that is requesting an interrupt.

In order to illustrate the polling technique, consider the 6809 microcomputer system shown in Figure 7.20. Four video terminals are connected to the microprocessor bus, each via an ACIA chip. Assume that the ACIA for terminal i is assigned four adjacent addresses, starting at address Terminal_i. The addresses of the four internal registers of this ACIA will be Terminal_i to Terminal_i+3, according to Table 5.1. The polling process is illustrated in the program of Figure 7.21, which is an example of an interrupt-service routine that may be used to receive data from the terminals. The POLL routine is entered as a result of an interrupt caused by typing a character at one of the terminals. It checks the status register in each serial interface until it finds one with the IRQ status bit set, indicating that the interface is requesting an interrupt and that a character is available in its input data register. The POLL routine reads the input character and calls a subroutine called STORE (not shown) to store it in an appropriate location in the main memory.

During the initialization process, which is not shown in the figure, the address of the POLL routine should be loaded into the interrupt-vector location and interrupts should be enabled, both in the device interfaces and in the microprocessor. The way

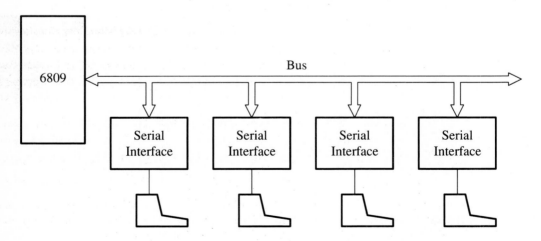

Figure 7.20   Four video terminals connected to a microprocessor bus.

ADRSLIST	FDB	Terminal_1	Address list for status registers of
	FDB	Terminal_2	4 terminals.
	FDB	Terminal_3	
	FDB	Terminal_4	

* Following is the interrupt-service routine.

POLL	LDX	#ADRSLIST	Initialize X as a pointer to address list.
	LDB	#4	Initialize terminal count.
LOOP	LDY	,X	Get address of ACIA.
	TST	1,Y	Check if IRQ=1.
	BMI	FOUND	
	LEAX	2,X	Increment X to point to the address of the next terminal.
	DECB		Check if list is exhausted.
	BHI	LOOP	Poll next terminal.
	BRA	ERROR	Error if none found.
FOUND	LDA	,Y	Read data register.
	BSR	STORE	
	RTI		

Figure 7.21 A 6809 interrupt-service routine for polling four terminals.

in which these actions are done in the 6809 will be discussed in Section 7.4. In order to facilitate device addressing, the addresses of the ACIAs are stored in a list, starting at location ADRSLIST. The interrupt-service routine uses index register X to read successive entries in that list, and load them into register Y. For each ACIA, the TST instruction checks the sign bit of the status register, which is at location [Y]+1. A negative value indicates that the IRQ bit is set, meaning that an input character is available in the corresponding data register. The contents of that register, whose address is in Y, are read by the LDA instruction at FOUND. Then, subroutine STORE is called. At this time, the input character and the terminal number are in accumulators A and B, respectively. These are the parameters passed to the subroutine.

If the status register contains a positive value, index register X is incremented by 2, because addresses occupy 2 bytes, and the terminal count in accumulator B is decremented by 1. The process is repeated until the list of status addresses is exhausted. If all four terminals are polled without finding one requesting service, a branch is made to a routine called ERROR, because this situation can only arise as a result of either a hardware or a software error.

7.3.2 Priority

The example above illustrates a commonly used technique for handling a number of input/output devices. Since interrupts are disabled when the interrupt-service routine is entered, interrupt requests from any device will be ignored until execution of this routine is completed. In some applications, it is desirable to allow certain devices to interrupt the microprocessor during the execution of the interrupt-service routine of another device. This is often the case when the operation of a device is time-critical, for example, when a delay in responding to an interrupt request may lead to loss of data. The desired behavior may be achieved by making it possible for different devices to be assigned different priorities. When an interrupt request from one device is being serviced, interrupts from devices with equal or lower priority should be disabled. However, higher-priority interrupts should continue to be accepted. An interrupt accepted during the execution of an interrupt-service routine for an earlier interrupt is called a *nested interrupt.*

Many microprocessors have an interrupt priority feature, where each interrupt received by the microprocessor is accompanied by an indication of its priority. The microprocessor decides whether to accept or to ignore a given interrupt request depending upon the request's priority. The way in which the priority of an interrupt request is indicated varies from one microprocessor to another. For example, several interrupt-request lines may be provided, with each line assigned a priority level. The priority of service for a given device is determined by the priority of the line to which that device is connected. In another scheme, the priority of a given request may be encoded on a few bus lines provided for this purpose. The 6809 and 68000 microprocessors provide examples of both these approaches, as will be described in Sections 7.4 and 7.5.

When several priority levels are supported by a microprocessor, a separate interrupt-vector location is usually provided for each level. Thus, if only one device is connected on a given level, it is possible to have a separate interrupt-service routine for each device. The routine is entered automatically when an interrupt request is received from the corresponding device, obviating the need for polling. However, if several devices are connected to the same interrupt-request line, the polling scheme would still be needed to determine which of them is requesting service.

7.3.3 Vector-Address Transfer

It is sometimes desirable to eliminate, or reduce, the time spent polling I/O devices. Instead of polling, a hardware mechanism may be provided to enable a device whose request for interrupt has been accepted to identify itself to the microprocessor. The identifying information is typically in the form of an address that points to an appropriate interrupt vector in the main memory.

The transfer of the interrupt-vector address takes place in a manner similar to normal data transfers. The address of the interrupt-vector is transmitted by the device, usually over the bus data lines, when requested by the microprocessor. The microprocessor indicates its readiness to receive this information by activating a special bus signal, which we will call Vector-Address-Enable. A vector-address transfer for a microprocessor that has a synchronous bus is illustrated by the timing diagram in Figure 7.22. (For details of data transfer over a synchronous bus refer to Section 6.2, particularly Figure 6.8.) Let us assume that there is only one device activating the interrupt-request line, IRQ. When the microprocessor accepts the interrupt request, it activates the Vector-Address-Enable signal as an indication that it has recognized the interrupt request, and that it is ready to receive the interrupt-vector

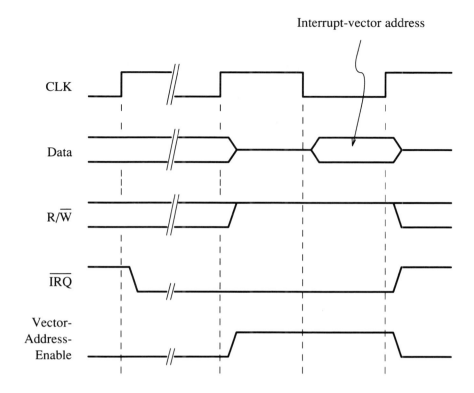

Figure 7.22 Transfer of interrupt-vector address over a synchronous bus.

address. The device that requested the interrupt responds by sending its interrupt-vector address to the microprocessor, in the same way as in normal read operations.

After receiving the interrupt-vector address, the microprocessor proceeds with interrupt processing in the normal manner. It saves the program counter, condition-code register, and other registers that may have to be saved. Then, it reads the starting address of the interrupt-service routine from the memory location whose address was supplied by the device. We should note that some of these transfers may require more than one bus cycle, if memory addresses are longer than the word length of the microprocessor bus. The vector address supplied by the device may be a full-length address, or it may be only a part of the address of the interrupt vector. In the latter case, the remainder of the address is supplied by the microprocessor and is always the same, such as being all 0s or all 1s.

In a practical situation, several I/O devices may request an interrupt at the same time. In order to ensure that only one is allowed to transmit its interrupt-vector address, the Vector-Address-Enable signal is usually transferred sequentially from one device interface to the next. The first device requesting service that receives this signal transmits its interrupt-vector address and blocks the propagation of the signal to other devices. This arrangement, which is illustrated in Figure 7.23, is known as a

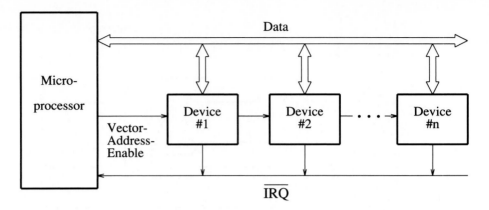

Figure 7.23 Daisy-chain connection.

daisy-chain. Note that if two or more devices request service simultaneously, the one electrically closest to the microprocessor will be serviced first. Other devices will continue to activate the interrupt-request line. However, their requests will be ignored while the first device is being serviced, because interrupts are disabled during execution of the interrupt-service routine. At the end of the routine, interrupts are enabled, and since there is an outstanding interrupt request, the microprocessor is immediately interrupted a second time. Once more, the microprocessor generates the Vector-Address-Enable signal, which will now be passed to one of the devices waiting for service.

7.3.4 Software Interrupts

So far, the discussion of interrupts has concentrated on interrupt requests received from I/O devices. Most computers have one or more special machine instructions called *software interrupts,* or *traps.* The execution of a software-interrupt instruction results in exactly the same sequence of events as an interrupt request received from an I/O device. That is, the microprocessor saves the contents of some or all of its registers and starts executing an appropriate interrupt-service routine. The starting address of this routine is obtained from a known interrupt-vector location in the main memory. In many microprocessors, several software interrupt instructions are available, each having a separate interrupt vector.

Superficially, the software-interrupt instruction appears no different from a subroutine call. A software interrupt causes the microprocessor to start executing an interrupt-service routine, at the end of which a return-from-interrupt instruction returns execution to the program that issued the interrupt request. Essentially the same thing happens as a result of a subroutine-call instruction. However, the two instructions differ in two important respects. First, a subroutine-call instruction specifies the address of the subroutine it calls. A program and all the subroutines it calls may be regarded as a single entity that performs some task. On the other hand, a software-interrupt instruction does not specify any address. The starting address of the interrupt-service routine is contained in the interrupt-vector, which is stored at a fixed address in the main memory. It is often the case that the interrupt-service routine is a part of a set of routines that constitute the *operating system* of the computer. This is the software that controls the operation of the system and provides certain services to the user. The services offered by the operating system may include performing input/output operations, reading data from a disk file, communicating with other computer systems, etc, as will be discussed in Chapter 12.

A user preparing a program need not know the details of the operating system and the addresses of various routines within it. It is the responsibility of the operating system to load the appropriate addresses in the interrupt-vector locations before starting the execution of any user programs. In order to use one of the services provided by the operating system, a user's program places the related parameters on the stack, then issues a software interrupt. As a result, control is transferred to the operating system. After performing the requested service, the operating system returns control to the user's program by issuing a return-from-interrupt instruction. Thus, unlike subroutine calls, software interrupts enable communication between two programs that were prepared independently.

The second, and equally important, difference between subroutine calls and software interrupts relates to the concept of priority. It was pointed out earlier that a microprocessor may have several interrupt-request lines, each assigned a different priority level. Software interrupts are integrated in this priority hierarchy by assigning a priority level to each software-interrupt instruction. When a program uses one of these instructions to call an interrupt-service routine, all hardware interrupt-requests of lower priority are automatically disabled during execution of that routine. No equivalent notion exists in relation to subroutine calls.

EXCEPTIONS

Yet another source of interrupts in a microprocessor is the occurrence of abnormal or unexpected events, often called *exceptions.* Such events usually arise as a result of a programming error or a hardware malfunction. The microprocessor responds to the occurrence of an exception in much the same way it responds to interrupt requests. It suspends the execution of the program during which the exception occurred, and after saving the appropriate state information on the stack, it starts executing an interrupt-service routine.

Consider, for example, a microcomputer that has 16K bytes of memory, occupying locations 0 to 3FFF of its addressable space. What happens if, as a result of a programming error, an instruction attempts to perform a read or a write operation outside this range? If the execution of this instruction is allowed to proceed, the error will go undetected. It was pointed out in Subsection 6.8.2 that such an error is easily detected in a microprocessor that uses a handshake mechanism for data transfers over its bus. If the address on the bus is not recognized by any device, the microprocessor will not receive the Acknowledge signal. Thus, after waiting for a reasonable time-out period, it aborts the offending instruction. The program containing this instruction is interrupted, and an appropriate interrupt-service routine is executed.

An attempt to execute an illegal instruction is another example of an event that results in an exception. Since the instruction is not recognized by the microprocessor, it cannot be executed. A number of other exceptions may arise during execution of arithmetic instructions, such as an attempt to divide by zero. Usually, each exception is assigned a unique interrupt-vector location in the main memory, enabling the programmer to prepare different interrupt-service routines that take appropriate actions in each case.

As mentioned earlier, all sources of interruption, hardware, software and exceptions, must be incorporated into an overall priority structure. Consider, for example, a microprocessor that has two interrupt-request lines and two software-interrupt instructions, and assume that this microprocessor is capable of detecting bus errors. A possible priority ordering, starting with the highest priority, is

1. Hardware interrupt #1
2. Bus-error exception
3. Software interrupt #1
4. Hardware interrupt #2
5. Software interrupt #2

This priority structure means that if a request arrives on interrupt-request line #1 during execution of an instruction that causes a bus error, the interrupt-service routine for the hardware interrupt will be executed first. After its completion, the routine that deals with bus errors will be executed. Also, software interrupt #1 will automatically disable hardware interrupt #2, but will have no effect on hardware interrupt #1.

7.4 INTERRUPTS IN 6809

The previous sections have dealt with many features associated with interrupts. A small microprocessor may have only one interrupt line and either limited or no interrupt-vector facilities. A more powerful microprocessor is likely to have several interrupt lines, multiple priority levels, and a flexible vectored-interrupt scheme. In this and the following section, we will examine the interrupt capabilities in our example microprocessors—Motorola's 6809 and 68000.

7.4.1 Hardware and Software Interrupts

The 6809 bus has three hardware-interrupt lines: Interrupt Request ($\overline{\text{IRQ}}$), Fast Interrupt Request ($\overline{\text{FIRQ}}$), and Non-Maskable Interrupt ($\overline{\text{NMI}}$). A fourth line, $\overline{\text{RESET}}$, functions in a manner that is somewhat similar to interrupts, as will be explained later. Signals on these lines are active when in the low-voltage state. The 6809 microprocessor also has three software-interrupt instructions: SWI, SWI2, and SWI3. Each interrupt source has a separate 16-bit interrupt vector location, which contains the starting address of the corresponding interrupt-service routine. The interrupt vectors are stored in the main memory near the top end of the addressable space, as shown in Table 7.2. Thus, when an interrupt is received on the $\overline{\text{IRQ}}$ line, for example, the two-byte address at locations FFF8 (high-order byte) and FFF9 (low-order byte) is used as the starting address of the interrupt-service routine. The priorities assigned to different interrupts are also given in the table, where priority 1 is the highest. The 6809 microprocessor does not provide an interrupt facility for handling exceptions such as bus errors or illegal instructions.

It was pointed out in Subsection 7.1.1 that the contents of the program counter and the processor status, or condition-code register, must be saved at the time of interruption. In the case of an FIRQ interrupt, this is the only information saved by the 6809 microprocessor. The contents of the PC and the CCR registers are pushed onto the stack pointed at by the S stack-pointer register. However, an interrupt request on either the $\overline{\text{IRQ}}$ or $\overline{\text{NMI}}$ lines causes all microprocessor registers to be saved. The order in which registers are pushed onto the stack is shown in Figure 7.24.

At the end of the interrupt-service routine, execution of the interrupted program is resumed by means of the Return-from-interrupt instruction, RTI, which restores the contents of the microprocessor registers from the stack. At this point, the microprocessor needs to determine whether to restore all registers, or only the PC and the CCR. The information required to make that decision is recorded in the copy of the CCR stored on the stack. The format of the CCR is shown in Figure 7.25. When an FIRQ interrupt is accepted, bit 7 of this register, which is called the Entire bit (E), is cleared by the microprocessor before pushing the contents of the CCR on the stack. For other interrupts E is set to 1. Thus, during execution of an RTI instruction, the microprocessor can determine which registers to restore by inspecting the E bit in the copy of the CCR at the top of the stack. If E is equal to 0, the microprocessor pulls the top three bytes off the stack and loads them into CCR and PC. Otherwise, it restores the contents of all registers.

Two bits are provided in the CCR for enabling and disabling interrupts. They are bits F and I in Figure 7.25, which serve as interrupt masks for FIRQ and IRQ,

Table 7.2   MEMORY MAP FOR INTERRUPT VECTORS IN THE 6809 MICROPROCESSOR

Interrupt Source	Address of Interrupt Vector		Priority
	High Byte	Low Byte	
RESET	FFFE	FFFF	1
NMI	FFFC	FFFD	2
SWI	FFFA	FFFB	3
IRQ	FFF8	FFF9	5
FIRQ	FFF6	FFF7	4
SWI2	FFF4	FFF5	6
SWI3	FFF2	FFF3	6

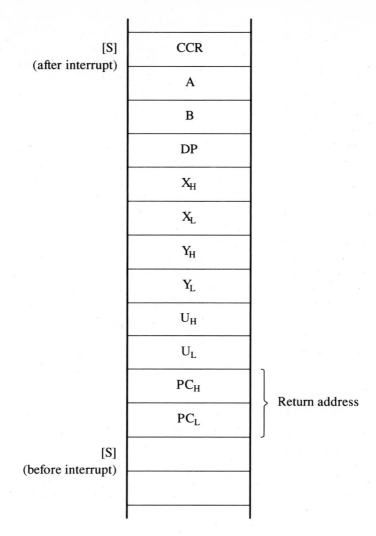

Figure 7.24 Top of the processor stack following an IRQ or an NMI interrupt in the 6809 microprocessor.

respectively. An interrupt is enabled when the corresponding mask bit is equal to 0. It is disabled, or *masked*, otherwise. The state of these bits can be changed by program instructions at any time. Also, they may be automatically set to 1 at the time an interrupt is accepted, as will be explained shortly. The reader will have noticed that no mask bit is provided for the third interrupt line, $\overline{\text{NMI}}$, because interrupts on this line cannot be disabled by program instructions.

Table 7.2 indicates that hardware and software interrupts are assigned different priorities. The $\overline{\text{RESET}}$ line has the highest priority. Activation of this line results in the initialization of all internal circuitry of the microprocessor to a predetermined

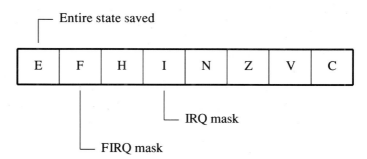

Figure 7.25 Interrupt-control bits in the Condition Code Register of the 6809 microprocessor.

state. Then, the microprocessor starts executing instructions at the address given in the Reset vector, at location FFFE. Typically, the service routine pointed at by the Reset vector initializes various parts of the microcomputer system, so that all components start operation in a well defined state. These actions are needed at power-up time, or whenever it is desired to restart the operation of the microcomputer system, perhaps following a software crash. The RESET line differs from other interrupt lines only in that no information is saved on the stack at the time the line is activated.

Hardware interrupts are enabled or disabled based on their priority. Hence, after activating the RESET line, all interrupts, including the nonmaskable interrupt, NMI, are disabled. The NMI interrupt is enabled as soon as a value is deposited in the S stack-pointer register. It remains enabled until RESET is activated again. The two interrupt mask bits, F and I, are manipulated by various interrupts in order to enforce the priority rules for FIRQ and IRQ. Both F and I are set to 1 when a Reset, an NMI or an FIRQ interrupt is accepted. Both bits are also set as a result of executing an SWI instruction. When an IRQ interrupt is accepted, only the I mask is set to 1. Thus, if an FIRQ interrupt request arrives while the interrupt-service routine for IRQ is being executed, a second interruption will occur. The interrupt-service routine for FIRQ will be entered and executed to completion, then the IRQ service routine will be resumed. The hardware interrupts that can be accepted during execution of the FIRQ service routine are Reset and NMI. The two software interrupts SWI2 and SWI3 have lower priority than all hardware interrupts. They have no effect on the interrupt mask bits. Note that the priority of software interrupts is significant only with respect to their effect on hardware interrupts. Since it is not possible for two instructions to occur simultaneously, it is meaningless to assign different priorities to SWI2 and SWI3.

It is important to point out that following the acceptance of an interrupt, the contents of the CCR are saved on the stack before the interrupt mask is modified. Thus, when the CCR is restored from the stack at the end of the interrupt-service routine, the interrupt masks will be restored to their state before the interrupt occured. Interrupts that were disabled at the beginning of the interrupt-service routine will be enabled automatically. Figure 7.26 illustrates the changes in the E, F, and I bits during a sequence of nested interrupts. An IRQ interrupt request is received while all interrupts are enabled. In response, the 6809 pushes all its registers on the stack, and sets bit E to 1 in the saved copy of the CCR. The F and I bits in the saved copy have their values prior to the arrival of the interrupt request. However, the I bit in the microprocessor is set to 1, to disable further IRQ interrupts. A while later, but before the execution of the IRQ interrupt-service routine is completed, an FIRQ interrupt request is received. Since the F mask in the 6809 is equal to 0, the request is accepted. This time, only the PC and the CCR registers are saved. Hence, the microprocessor sets the E bit to 0, before saving the CCR on the stack. Both the F and the I masks are now set to 1 in the microprocessor. If an NMI interrupt is received during execution of the FIRQ interrupt-service routine, it will be accepted, and the mask bits will be set as shown. The contents of the processor stack at this point are shown in Figure 7.27. As the execution of each interrupt-service routine is completed, an RTI instruction restores the contents of the microprocessor registers from the stack, in the reverse of the order in which they were saved.

7.4.2 Programming Examples

Let us now consider a simple example in which a video terminal is connected to a 6809 microcomputer via the ACIA chip described in Subsections 5.3.1 and 7.2.1. Assume the interrupt-request output of the ACIA to be connected to the $\overline{\text{IRQ}}$ line of the microprocessor bus. The overall organization of the program needed to send a

		CCR (in 6809) E F I	CCR (on stack) E F I
		x 0 0	- - -
IRQ	→		
		x 0 1	1 0 0
FIRQ	→		
		x 1 1	0 0 1
NMI	→		
		x 1 1	1 1 1
RTI	→		
		x 1 1	0 0 1
RTI	→		
		x 0 1	1 0 0
RTI	→		
		x 0 0	- - -

Figure 7.26 Changes in the E, F, and I bits of CCR during a nested interrupt sequence. (x indicates either 0 or 1.)

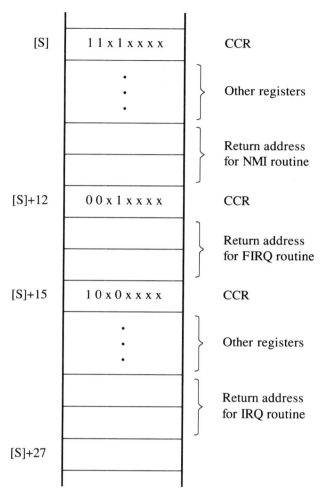

Figure 7.27 Contents of the processor stack immediately
following the NMI interrupt in Figure 7.26.

one-line message to an output device using interrupts was given in Figure 7.6. An
equivalent 6809 program for a video terminal is given in Figure 7.28, assuming that
the ACIA registers occupy four addresses starting at CB50. The section for initialization of parameters stores the address of the interrupt-service routine, DISPLY, in
the interrupt vector location and sets the mode of operation of the ACIA. It also
stores the starting address of the message to be printed in memory location MPNTR.
Then, interrupts are enabled in the 6809 by clearing the I mask in the CCR.

Since the 6809 microprocessor disables interrupts automatically before entering
the interrupt-service routine, no Disable_interrupts instruction is needed at the beginning of the interrupt-service routine. Also, for an IRQ interrupt, all registers are

```
CR        EQU    $0D                Carriage Return character.
INTVEC    EQU    $FFF8              Interrupt-vector address.
OUTA      EQU    $CB50              Output data register in ACIA.
STATS     EQU    $CB51              Status register.
CMNDR     EQU    $CB52              Command register.
CNTLR     EQU    $CB53              Control register.
MPNTR     FDB    0                  Character pointer.
MSG       FCC    'HELLO. THIS IS A 1-LINE MESSAGE'
          FCB    CR
INIT      LDX    #DISPLY            Initialize interrupt vector.
          STX    INTVEC
          LDX    #MSG               Initialize message pointer.
          STX    MPNTR
          LDA    #$1E               Select 8-bit word, one stop bit
*                                      and 9600 baud.
          STA    CNTLR
          LDA    #7                 Enable transmitter interrupts in ACIA
*                                      and activate DTR.
          STA    CMNDR
          ANDCC  #$EF               Clear I flag in CCR
*                                      (Enable IRQ interrupts in 6809).
MAIN      .
          .
          .
DISPLY    LDX    MPNTR              Get message address pointer.
          LDA    ,X+                Get character and
          STX    MPNTR                 increment pointer.
          STA    OUTA               Send character to Output register.
          CMPA   #CR                Check for end of line.
          BNE    RTRN
          CLR    CMNDR              Disable interrupts in ACIA.
          LDA    ,S                 Set I mask in copy of CCR at
          ORA    #$10                  top of stack.
          STA    ,S
RTRN      RTI
```

Figure 7.28   A 6809 program to send a one-line message to a video display via an ACIA chip, using interrupts.

automatically saved on the stack by the microprocessor, allowing processor registers to be used freely within the interrupt-service routine. The interrupt-service routine transmits one character of the message. This action causes the ACIA to turn off its interrupt request and to clear the Transmitter Empty bit. The RTI instruction restores all 6809 registers, including the CCR, from the stack. As a result, interrupts are enabled when execution of the interrupted program is resumed. When the ACIA is ready to accept another character, it sets the Transmitter Empty bit and generates an interrupt request; the interrupt-service routine is once again entered, and another character is sent to the ACIA.

At the end of the message, the ACIA should be prevented from generating further interrupt requests. Hence, the Transmitter Control field in the Command register is set to 00 when the carriage-return character is encountered. Assuming that displaying information on the video screen is the only task requiring interrupts, interrupts should also be disabled in the microprocessor. To do so, the I mask is set to 1 in the copy of the CCR at the top of the stack. When the RTI instruction is executed, this information will be loaded into the CCR register, thus disabling interrupts in the microprocessor.

In order to illustrate the use of interrupts with the PIA described in Subsections 5.5.2 and 7.2.3, let us consider again the case of a printer connected to port B. A one-line message may be printed using interrupts by means of the program in Figure 7.29. This is essentially the same program as that given in Figure 7.28, with additional instructions to select the mode of operation of the PIA. Register addresses have been chosen according to Table 5.2, assuming that the PIA is assigned a block of addresses starting at location CB00. The reader will recall that handshake signaling can be used with output data transfers on port B (see Subsection 6.6.3). Let the CB1 and CB2 lines be used for the $\overline{\text{Data-Taken}}$ and $\overline{\text{Data-Valid}}$ signals, respectively. Hence, the PIA is initialized to select handshake signaling, and interrupts are enabled for a high-to-low transition on CB1. The PIA will generate an interrupt request whenever it receives a response from the printer on the $\overline{\text{Data-Taken}}$ line. The pattern deposited in the Control Register clears bit CRB_2 to 0 to select the Data Direction Register. After setting DDRB as desired, this bit is set to 1, so that data may be sent to the output data register, PRB. The interrupt-service routine, PRINT, has the same organization as the corresponding routine in Figure 7.28.

7.4.3 Interrupt Acknowledgment

The 6809 obtains the starting address for the interrupt-service routine by reading the contents of the interrupt-vector locations described in Table 7.2. These locations are normally in the main memory, in which case no further action is required from the device. However, the microprocessor provides sufficient information on the bus to make it possible for the interrupting device to supply its own interrupt vector, instead of using the interrupt vector stored in the main memory. Two output signals, called Bus Available, BA, and Bus Status, BS, describe the state of the microprocessor during each clock period, according to Table 7.3. During the two cycles in which the microprocessor reads the contents of the interrupt vector, it sets BA = 0 and BS = 1. At the same time, the three least-significant address bits indicate which interrupt request is being acknowledged and which byte of the interrupt vector is being read. Hence, by decoding the information on BA, BS, and the address lines, a Vector-Address-Enable signal similar to that shown in Figure 7.23 may be generated for each interrupt-request line. The highest-priority device requesting an interrupt on a given line may respond by transmitting the appropriate byte of its interrupt vector on the data lines in each of the two read cycles. These two bytes will be taken by the

```
CR        EQU     $0D              Carriage Return character.
INTVEC    EQU     $FFF8            Interrupt-vector address.
OUT/DD    EQU     $CB02            Output data (PRB) or DDRB.
CNTLB     EQU     $CB03            Control register for port B.
*
MPNTR     FDB     0                Character pointer.
MSG       FCC     'HELLO. THIS IS A 1-LINE MESSAGE.'
          FCB     CR
INIT      LDX     #PRINT           Initialize interrupt vector.
          STX     INTVEC
          LDX     #MSG             Initialize character pointer.
          STX     MPNTR
          CLR     CNTLB            Select DDRB.
          LDA     #$FF             Set port B as output.
          STA     OUT/DD
          LDA     #$25             Select full-handshake mode and negative
          STA     CNTLB                edge on CB1 with interrupts enabled,
*                                      select PRB.
          ANDCC   #$EF             Clear I flag in CCR
*                                      (Enable IRQ interrupts in 6809).
MAIN      .
          .
          .
PRINT     LDX     MPNTR            Get message address pointer.
          LDA     ,X+              Get character and
          STX     MPNTR                increment pointer.
          STA     OUT/DD           Send char. to output register,
*                                      (clears interrupt and SA1 bit).
          CMPA    #CR              Check for end of line.
          BNE     RTRN
          CLR     CNTLB            Disable PIA interrupts.
          LDA     ,S               Set I mask in copy of CCR at top of stack.
          ORA     #$10
          STA     ,S
RTRN      RTI
```

Figure 7.29   An interrupt-driven 6809 program to send a one-line message to a printer via a PIA chip.

microprocessor to be the starting address of the interrupt-service routine. Of course, when this approach is used for any of the interrupt-request lines, the main memory should not respond to read requests at the corresponding interrupt-vector address.

The 6809 offers a special interrupt feature in conjunction with an instruction called SYNC. When this instruction is executed, the microprocessor simply suspends its operation, and waits for any of its interrupt-request lines to become active. When this happens, it continues execution at the location following the

Table 7.3 6809 BUS STATUS SIGNALS

BA	BS	State Description
0	0	Normal execution of program instructions.
0	1	Interrupt-vector read cycle.
1	0	Microprocessor waiting for synchronization signal.
1	1	Bus grant or microprocessor halted.

SYNC instruction. It does not push any information on the stack or read an interrupt vector. Hence, the signal on the interrupt-request line does not actually cause any interruption. It simply causes the microprocessor to resume its operation. The microprocessor sets BA = 1 and BS = 0 to indicate that it is waiting for the synchronization signal.

The synchronization feature can be used to synchronize the execution of a particular portion of a program to an external event. For example, consider an application in which the microprocessor reads data from a high-speed device. The SYNC instruction may be placed at the beginning of a short program loop. During each pass through the loop, the microprocessor waits for the device to send a signal on an interrupt line. When it receives this signal, it reads one byte of data, stores it in the main memory, then goes to the beginning of the loop to wait for another byte. The microprocessor can read data in this manner at a faster rate than can be achieved in any other way.

The last entry in Table 7.3 is used in conjunction with DMA operations. It will be discussed in Section 7.7.

7.5 INTERRUPTS IN 68000

Like the 6809, the 68000 microprocessor uses vectored interrupts. However, it supports more priority levels and offers considerably higher flexibility in determining the interrupt vector location. The bus signals and their use in normal data transfers were described in Section 6.10. An understanding of those aspects is essential as background to the discussion of interrupt handling given below.

The 68000 is able to recognize a wide range of hardware interrupts, software interrupts, and exceptions. In Motorola's nomenclature, all such events are called exceptions. To be consistent with Motorola's literature for the 68000 [1], which the reader may consult for further details, we will use the term "exception" in this section to refer to all types of interrupts.

The wide variety of exceptions that can be recognized by the 68000 microprocessor is a reflection of the sophistication of the software systems that it can support. A detailed understanding of exception handling is essential for those involved in the preparation of system software. In this book, we will limit our discussion to an explanation of the way in which external interrupts and bus errors are handled.

The 68000 microprocessor handles all exceptions in the same manner. It interrupts the sequence of instructions being executed and starts executing an exception-handling routine. The location of this routine is defined by a 32-bit address vector in an exception-vector table, which has 256 entries. As a part of exception handling, sufficient information is stored on the system stack to enable the exception-handling routine to determine the cause of the exception. For example, in the case of a bus error, the memory address and the type of bus transfer, typically read or write, at the time the error occurred are stored on the stack, along with other identifying information.

7.5.1 Interrupt Priority

The 68000 supports seven levels of interrupt priority. The priority of an interrupt is indicated to the microprocessor in the form of a 3-bit code on three Interrupt Priority Lines, IPL_{2-0}, with IPL_0 representing the least-significant bit of the code. The lowest-priority code is 001 and the highest is 111. The code 000 is an indication that none of the devices on the bus is requesting an interrupt. Priority level 111 is a non-maskable interrupt, which cannot be disabled by program instructions.

The 68000 does not perform priority arbitration among its peripheral devices. This function is left to external hardware, which should continuously monitor interrupt requests from all peripherals and place the priority code of the highest-priority request on the IPL lines. A simple circuit that can be used for this purpose is illustrated in Figure 7.30. It uses a priority encoder, which is a logic circuit that produces a 3-bit code at its output equal to the number of the highest-numbered input

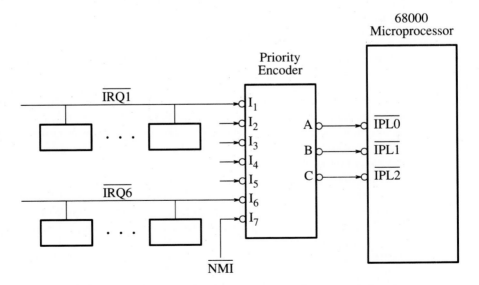

Figure 7.30   Encoding of interrupt-request priorities on the 68000 bus.

that is active. The highest-priority input line to the encoder, labeled $\overline{\text{NMI}}$, is a non-maskable interrupt. The remaining six lines, $\overline{\text{IRQ}i}$, are interrupt-request lines having different priorities, with $\overline{\text{IRQ1}}$ having the lowest priority. Several devices may be connected to each of the $\overline{\text{IRQ}i}$ lines, by means of open-collector gates.

The processor status word, PSW, which was given in Figure 4.2, contains a 3-bit interrupt mask in bit positions 8 to 10. Any program being executed on the 68000 is considered to be running at a certain level of priority indicated by this mask, with a value of 0 representing the lowest and 7 the highest priority. At the end of the execution of an instruction, the microprocessor checks the code on the IPL lines. If it is higher than zero, the processor compares the priority of the incoming request to that of the program currently being executed, as defined in the PSW. An interrupt request with priority level 1 to 6 is accepted only if it has a higher priority level than that indicated in the PSW. Requests at level 7 are always accepted, irrespective of the contents of the PSW. Hence, interrupts at levels 1 to 6 can be masked, or disabled, by setting the priority bits in the PSW to a sufficiently high value. On the other hand, interrupt requests at level 7 are nonmaskable.

When an interrupt is accepted, the general exception-handling procedure of the microprocessor is invoked. The microprocessor is switched to the Supervisor mode (see Chapter 4); the state of the microprocessor prior to the exception is saved; then the starting address of the appropriate exception-handling routine is fetched from the exception-vector table and loaded into the program counter. In the case of interrupts, the saved information consists of the contents of the PSW at the time the interrupt is received and the address of the instruction that would have been executed had the interruption not occurred. This information is pushed on the Supervisor stack. At the same time, the priority code in the status register (the new PSW) is updated to that on the IPL lines. This step disables interrupt requests of equal or lower priority.

7.5.2 Exception Vectors

Let us now turn our attention to the exception-vector table and the way in which a particular entry is selected. The exception vector table occupies the lowest 1024 bytes of the address space of the microprocessor. The table entries are 4 bytes long, and are numbered from 0 to 255 as shown in Table 7.4. The vector number, which can be represented in 8 bits, $v_7 v_6 \cdots v_0$, can be regarded as a short form for the address, where the full address is derived as illustrated in Figure 7.31. Each table entry gives the starting address for the handling routine intended for a particular exception. The Reset exception is assigned two entries in the table, numbers 0 and 1. The first gives an initial value for the Supervisor stack pointer (SSP) and the second gives the starting address of the initialization software.

The exceptions that appear in Table 7.4 are described briefly below.

- *Reset*: A Reset exception occurs either at the time power is turned on or when the Reset line on the microprocessor bus is activated by an external device (often manually). The handling routine for the Reset exception may be a loader program that loads a fresh copy of the operating system from a disk into the main memory (see Chapter 12).

Table 7.4 68000 EXCEPTION VECTOR TABLE

Vector Number	Address (Hex)	Assignment
0	000-007	Reset - initial values for SSP and PC
2	008	Bus error
3	00C	Address error
4	010	Illegal instruction
5	014	Zero divide
6	018	CHK instruction
7	01C	TRAPV instruction
8	020	Privilege violation
9	024	Trace
10,11	028,02C	Software emulation of new instructions
12-14	030-03B	Reserved for future use
15	03C	Uninitialized interrupt vector
16-23	040-05F	Reserved for future use
24	060	Spurious interrupt
25-31	064-07F	Interrupt autovector, levels 1-7
32-47	080-0BF	TRAP instruction vectors
48-63	0C0-0FF	Reserved for future use
64-255	100-3FF	User interrupt vectors

- *Bus error*: A 68000 system should incorporate a watchdog timer that monitors bus activity. If a bus transfer starts, as indicated by the activation of the Address Strobe line, $\overline{AS}$, and no device responds within a certain time-out period, the timer should activate the Bus Error line, $\overline{BERR}$. In response, the processor terminates the bus cycle, saves the state information on the supervisor stack, and starts executing the bus-error exception routine. The address of this routine is found in the four bytes that constitute vector number 2.

- *Address error*: An address error exception results from an attempt to access a word or a long word operand or an instruction at an odd address. The bus transfer is aborted, and exception processing begins.

- *Instruction traps*: These are exceptions that may arise during instruction decoding or execution. An *illegal instruction* is one whose bit pattern does not

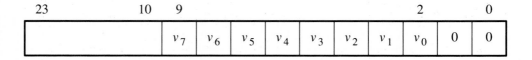

Figure 7.31 Derivation of exception-vector address from its number.

correspond to any 68000 instruction. A *privilege violation* occurs if one of the privileged instructions is encountered while the microprocessor is in the User mode (see Chapter 4).

Some other instructions may cause an exception as a part of their normal execution. The divide instructions DIVS and DIVU result in an exception, using vector 5, if the divisor is equal to zero. Also, the CHK, TRAPV, and TRAP are software-interrupt instructions, which lead to exceptions using their respective exception vectors.

- *Trace traps*: The *trace exception* is a debugging facility that enables the programmer to execute a program one instruction at a time. It is invoked by setting the T bit in the PSW (see Fig. 4.2). If the T bit is set during the execution of any instruction, the microprocessor completes execution of that instruction, then starts exception processing. It saves the current state on the supervisor stack and resets the T bit in the PSW to prevent further trace exceptions. Then, it uses vector 9 to determine the starting address of the exception-handling routine.

  The trace exception-handling routine is usually a part of the debugging facility provided in the system software. When this facility is invoked to debug a given user program, the exception handling routine is performed after each instruction in the program that is being debugged is executed. It offers the programmer a facility to examine, and change if necessary, the contents of any microprocessor register or main memory location. Thus, the programmer can verify that the instruction has performed its intended task, then ask for the next instruction to be executed. The exception-handling routine executes a Return_from_exception instruction (RTE), which restores the contents of the PSW and the PC from the Supervisor stack. Thus, the microprocessor begins execution of the next instruction of the program that was interrupted. Recall that the copy of the PSW saved on the stack had the T bit set. When the PSW contents are restored from the stack as a result of the RTE instruction, the T bit in the microprocessor is set. Thus, after executing one instruction in the user program, another trace exception takes place, returning control to the debugging software.

  Another useful debugging facility that is often provided in conjunction with the ability to execute one instruction at a time is the ability to install *breakpoints*. The TRAP instruction may be used for this purpose. Breakpoints will be discussed in Chapter 12.

- *Interrupts*: In all the exceptions discussed above, the address of the exception vector is generated by internal circuits in the microprocessor, according to the type of the exception. In the case of interrupts, the vector address may be generated internally, or it may be supplied by the device requesting the interrupt. In the latter case, the device supplies only the vector number, which is an 8-bit quantity. The internal generation of the vector address or the transfer of the vector number over the bus takes place during the interrupt-acknowledge cycle, which is discussed below.

7.5.3 Interrupt Acknowledgment

Exception vectors numbered 64 to 255 are intended for use by peripheral devices connected to the 68000 bus. When the microprocessor accepts an interrupt request, it asks the device requesting the interrupt to supply the number of the exception vector that should be used. It does so by starting an interrupt-acknowledge cycle, whose timing is shown in Figure 7.32. An interrupt-acknowledge cycle is a normal read cycle, with the code 111 transmitted over the function lines, FC_{2-0}. Recall that the decision to accept an interrupt request is made based on the priority of that request as indicated on the IPL lines. Since the signals on these lines may change at any time, the processor indicates the priority level it has accepted using address lines A_3, A_2, and A_1. The remaining address lines are set high. The device that is

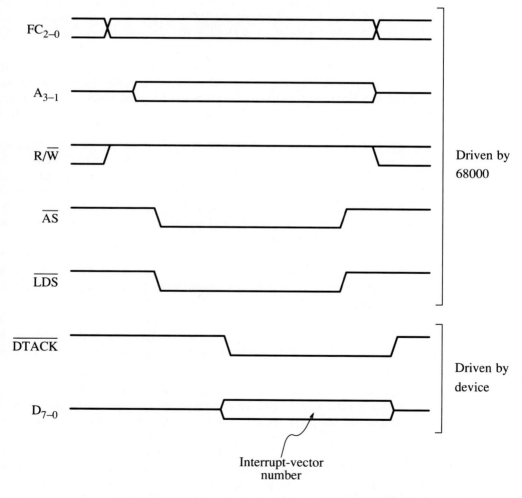

Figure 7.32 Interrupt-acknowledge cycle on the 68000 bus.

requesting an interrupt at the specified priority level responds by placing its interrupt-vector number on the low order data lines, D_{7-0}, and activates the $\overline{\text{DTACK}}$ line (or the $\overline{\text{DSACK}i}$ lines on the 32-bit bus). The microprocessor reads the vector number on the data lines and completes the bus cycle in the normal manner.

Individual interrupt-acknowledge signals that correspond to the interrupt-request signals of Figure 7.30 may be derived as shown in Figure 7.33. The information on address lines A_{3-1} is decoded to activate an interrupt-acknowledge signal, $\overline{\text{INTA}i}$, at the appropriate priority level. Note that the decoder circuit is enabled only when $FC_{2-0} = 111$ and $\overline{\text{AS}}$ is active. The $\overline{\text{INTA}i}$ signal performs the function of the Vector-Address-Enable signal described in Subsection 7.3.3. Hence, the device interface should be designed to place its assigned vector number on the data lines and to activate the $\overline{\text{DTACK}}$ line whenever it receives an active signal on the $\overline{\text{LDS}}$ line while INTAi is active.

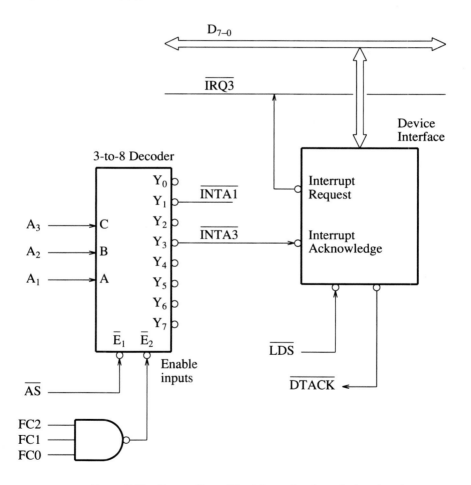

Figure 7.33 Generation of the interrupt-acknowledge signal.

In some 68000 peripherals, the vector number is programmable. It is stored in an addressable register in the interface, which should be initialized before interrupts are enabled for that interface. If, as a result of a software error, the interface generates an interrupt request before this register is initialized and the request is acknowledged, the interface may return 15 as the vector number. The exception-handling routine pointed at by vector number 15 should provide for an orderly recovery from such an error. Another facility intended to enhance a system's ability to recover from error is the spurious interrupt vector, number 24. If an interrupt request is received by the 68000 as a result of some noise or disruption on the bus, there will be no response when the microprocessor attempts to read the interrupt-vector number. Instead, the microprocessor will receive a bus error indication from the watchdog timer on the bus. As a result, it will begin exception processing using vector number 24.

The facility that enables peripheral devices to supply an interrupt-vector number during the interrupt-acknowledge cycle provides considerable flexibility in the design of a microcomputer system. Since individual devices, or groups of devices, can use unique vector numbers, there is no need for a lengthy and time-consuming polling process. A given device may even use different vector numbers, depending upon the cause of the interrupt. For example, a serial interface chip may use one interrupt vector to indicate that an input character is available, and use another vector in the case of a parity error or input buffer overflow.

The flexibility offered by asking the peripheral interface to transmit a vector number comes at the expense of a slight increase in complexity of the interface hardware. The increased complexity may not be warranted in all situations. Moreover, interface chips designed for use with other processor families may not have a built-in provision for storing and transmitting an interrupt-vector number. In order to enable such devices to be easily connected to the 68000 bus, an *autovector* facility is provided. When the microprocessor requests the vector number, the external device may activate the VPA (Valid Peripheral Address) line. This indicates to the microprocessor that the device does not have the capability of transmitting a vector number, and that it wishes to use the autovector facility instead. As a result, the microprocessor terminates the interrupt-acknowledge cycle without waiting for the normal response on the $\overline{\text{DTACK}}$ line. Then, it proceeds with exception processing, using an internally-generated vector number in the range 25-31, depending upon the priority level of the interrupt request.

An example of a 68000 interrupt-service routine for sending data to an ACIA chip is shown in Figure 7.34. We have assumed that the ACIA is connected to the high-order data lines, D_{15-8}, and that its interrupt-request output is connected to $\overline{\text{IRQ1}}$ in Figure 7.30. Hence, the autovector at location 64 (vector number 25 in Table 7.4) is used to hold the starting address of the interrupt-service routine. The ACIA registers will appear at successive even addresses, as shown in the declarations given at the beginning of the program.

The initialization section is assumed to be executed shortly after the microcomputer system has been reset. The reset signal causes the microprocessor to enter the Supervisor mode and sets the interrupt mask to 7. The ACIA registers are initialized

```
CR        EQU        $0D
INTVEC    EQU        $64              Use autovector, level 1.
OUTA      EQU        $3CB50           Address of ACIA registers.
STATS     EQU        $3CB52
CMNDR     EQU        $3CB54
CNTLR     EQU        $3CB56
MPNTR     DS.L       1                Pointer to message.
MSG       DC         'Hello.  THIS IS A 1-LINE MESSAGE'.
          DC.B       CR
*
INIT      MOVE.L     DISPLY,INTVEC
          MOVE.L     #MSG,MPNTR
          MOVE.B     #$1E,CNTLR       Initialize ACIA and
          MOVE.B     #$7,CMNDR             enable transmitter interrupts.
          MOVE.B     #$2000,SR        Enable interrupts in 68000, and leave
                     .                    processor in Supervisor mode.
MAIN               .

                   .
DISPLY    MOVE.L     MPNTR,A0         Get character address.
          MOVE.B     (A0)+,OUTA       Send character.
          MOVE.L     A0,MPNTR         Update pointer.
          CMPI.B     #CR,-1(A0)       Check for Carriage Return.
          BNE        RTRN
          CLR.B      CMNDR            Disable ACIA interrupts.
          RTE
```

Figure 7.34 A 68000 program to display a one-line message on a video terminal, using interrupts.

and its interrupts are enabled in the same way as in Figure 7.29. Then, a pattern is loaded in the status register of the microprocessor to set the interrupt mask, bits SR_{10-8}, to 0. As a result, the microprocessor will accept interrupt requests received from the ACIA, because they have been assigned a priority level of 1. Note that the instruction that changes the PSW is a privileged instruction, which can be executed only when the microprocessor is in the Supervisor mode. In addition to changing the interrupt mask, this instruction sets bit SR_{13} to 1, thus leaving the microprocessor in the Supervisor mode. We have assumed, for simplicity, that the microprocessor will continue to perform other tasks in the Supervisor mode. However, this is not essential. The User mode may be entered either at the same time the processor's priority is lowered, or later in the program. When an interrupt request is received and accepted, the microprocessor automatically switches to the Supervisor mode, changes its priority level, then starts executing the interrupt-service routine.

The interrupt-service routine in Figure 7.34 differs from that in Figure 7.29 in only one respect. When the end of the message is reached, the copy of the PSW at the top of the stack, which will be loaded into the status register by the RTE instruction, is left unchanged. As a result, the processor's priority level will continue to be

0 after the task of displaying the message is completed. However, no interrupts will be received from the ACIA, because its interrupts have been disabled by the instruction that cleared the Command register, CMNDR. We have left the processor's priority at level 0, assuming that there are other interrupt-driven tasks in progress.

Before concluding this discussion, let us consider briefly the case where several devices have the same interrupt-priority level. This situation arises when there are more devices on the bus than there are priority levels. Also, when similar devices are involved, it is often desirable to give them equal priority, so that they may receive equal service. For example, in a microcomputer system that has several video terminals, the speed of response of the system should appear to the user to be the same, regardless of which terminal is being used. If the terminal interfaces are assigned different priority levels, requests for interrupt from one terminal would be accepted while a lower priority terminal is being serviced, thus giving the lower priority terminal less than its fair share of the processing power of the microprocessor. To achieve the desired equality, all terminals should be assigned the same priority by connecting them to the same $\overline{IRQi}$ line.

During the interrupt-acknowledge cycle, the microprocessor resolves conflicts arising from simultaneous requests having different priority levels by transmitting the highest priority code on the address lines. When several devices are connected to the same interrupt request line, the daisy-chain scheme of Figure 7.23 may be used. The interrupt-acknowledge signal, $INTA_i$, at the output of the decoder in Figure 7.33 should be propagated in sequence through the device interfaces.

7.6 DIRECT MEMORY ACCESS

In all the input and output techniques discussed so far, the actual transfer of data requires the intervention of the microprocessor. In an input operation, for example, the microprocessor reads one word of data from the data buffer in the interface, and deposits this word in the main memory. Each word transferred in this manner, whether by means of interrupts or status checking, involves several data transfers over the microprocessor bus. An input operation requires bus cycles to read instructions from the main memory, to check the status register of the input device interface, and to transfer the input data to the main memory. These transfers represent a considerable overhead associated with each input word.

Consider now the case of a high speed device, such as a magnetic disk. The overhead mentioned above may make it impossible for a microprocessor to meet the data transfer rate requirements of the device. A commonly used alternative is to provide specialized hardware that transfers data directly between the device and the main memory. The microprocessor initiates the process, but does not handle individual data items as they are transferred between the memory and the I/O device. This technique is known as *Direct Memory Access* (DMA). We will call an I/O device interface that is capable of performing DMA tasks a *DMA interface*. It is connected between the I/O device and the microprocessor bus, as shown in Figure 7.35.

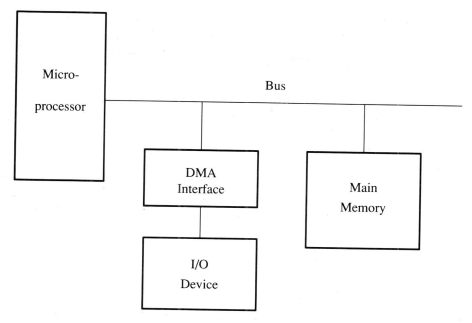

Figure 7.35 Connection of an I/O device via a DMA interface.

In order to initiate a DMA task, it is necessary to specify several parameters. These include the address of the data in the main memory, the amount of data to be transferred and the direction of the transfer, that is, whether data are to be transferred from the device to the memory, or vice versa. Additional information, such as the location of the data on a disk, must also be specified in some cases. The microprocessor deposits this information into registers provided for this purpose in the DMA interface, then issues a GO command. The DMA interface proceeds to transfer the data, independent of the microprocessor. During that time, the microprocessor may simply wait for the transfer to be completed, or it may perform other functions. However, it is important to remember that all data transfers involving the main memory, the microprocessor or the DMA interface take place over a common path—the microprocessor bus. Typically, a complete DMA transfer requires many bus cycles to complete, where one unit of data, such as a byte or a word, is transferred in each cycle. The rate at which bus cycles are used by the DMA interface depends on the speed of the I/O device involved. In practice, it is unlikely that DMA transfers will need a large portion of the available bus cycles. The remaining cycles can be used by the microprocessor to fetch and execute instructions.

Shared usage of the bus means that access to it must be coordinated. During program execution the bus is under control of the microprocessor. All data transfers are initiated and controlled by the microprocessor. Hence, the microprocessor is said to be the *bus master*. If a DMA interface is to transfer data directly to or from the main memory, it must first acquire control of the bus; that is, it must become the bus master. Only the bus master can place an address on the address lines and drive the control lines to indicate a read or a write operation.

7.6.1 Transfer of Bus Mastership

The transfer of bus mastership from the microprocessor to the DMA interface takes place by exchanging control signals between the two devices. Two bus lines, Bus-Request and Bus-Grant, are usually provided for this purpose. When the DMA interface is ready to transfer data, it activates the Bus-Request line. In response, the microprocessor stops executing instructions and disconnects itself from the bus, by putting all its output drivers in the high-impedance state. Then, it activates the Bus-Grant line. This is an indication to the DMA interface that it may start using the bus. When the DMA interface no longer requires the use of the bus, it deactivates Bus-Request, at which point the microprocessor reclaims bus mastership. The required signaling exchange is shown in Figure 7.36.

During the transfer of bus mastership, two precautions have to be taken:

1. The present bus master must turn off its bus drivers before the new bus master activates its drivers.
2. Since the state of the bus address lines is unpredictable during this period, other devices connected to the bus should be told to ignore the bus until the new bus master has placed a valid address on the address lines.

The first precaution is easily assured, because all the signals involved are under the control of the current bus master. The internal logic circuits of the microprocessor simply turn off the bus drivers before activating Bus-Grant. Similarly, the DMA interface turns off its bus drivers before deactivating Bus-Request. The second

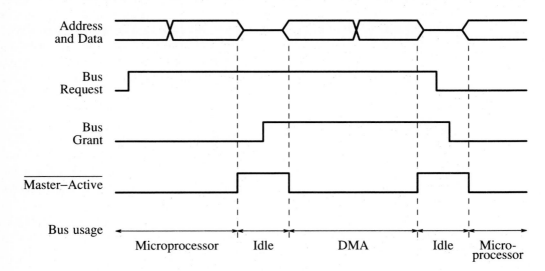

Figure 7.36 Transfer of bus control from the microprocessor to the DMA interface.

precaution is not quite as straightforward to implement, because two devices are involved—the microprocessor and the DMA interface. A possible and often used scheme is to provide an additional open-collector control line, which we will call Master-Active. Device interfaces should ignore the address lines whenever this signal is in the inactive state. The use of the Master-Active line is illustrated in the timing diagram of Figure 7.36. The current bus master maintains Master-Active in the active (low-voltage) state, until it is about to transfer bus mastership to another device. At that time, it releases the Master-Active line. Because of the pull-up resistor used in the open-collector configuration, the Master-Active line returns to its inactive (high-voltage) state, thus preventing other device interfaces from being falsely selected. After completion of the transfer of bus mastership, the new bus master activates the Master-Active line and normal bus operation resumes.

Having become the bus master, the DMA interface proceeds to transfer data in the same manner the microprocessor normally does. As little as one word or as much as an entire file may be transferred before releasing the bus. The DMA interface maintains the Bus-Request line in the active state during this period. When it is ready to relinquish the bus mastership, it deactivates Master-Active and Bus-Request. As a result, the microprocessor once again becomes the bus master. This sequence of events, where bus mastership is passed to the DMA interface and back, is repeated whenever the I/O device is ready for another data transfer.

Let us consider a numerical example to illustrate the time needed for DMA transfers. Assume that a disk drive with a data transfer rate of 100K words per second is connected to a microprocessor bus capable of transferring 2M words/s. In this case, the DMA interface needs to transfer one word every 20 bus cycles. Assuming that one bus cycle is used for transferring bus mastership from one device to another, a total of three bus cycles are needed for every word handled by the DMA interface. Thus, 17 out of every 20 bus cycles are available for the microprocessor to do other things, such as to fetch and execute instructions that perform some other task. Obviously, considerably fewer bus cycles are needed for a given I/O task, compared to the program-controlled transfers described earlier. The DMA interface may be viewed as "stealing" bus cycles from the microprocessor. Hence, a DMA interface that transfers one word of data then returns bus mastership to the microprocessor is said to be operating in the *cycle-stealing mode*.

Because of the idle periods shown in Figure 7.36, it is desirable to keep the number of transfers of bus mastership between a DMA interface and the microprocessor to a minimum. The idle periods in Figure 7.36 are wasted time during which no device is making use of the bus to transfer data. They amount to a significant overhead when the cycle-stealing approach is used. An alternative approach is to arrange for the DMA interface to operate in a *burst* mode, in which it transfers an entire block of data between the main memory and the disk in one DMA operation. Obviously, this approach is beneficial only if the disk is capable of transferring a block of data at a rate comparable to that of the main memory. The burst mode is most useful when the DMA interface contains a storage buffer that can store an entire block of data. In this case, a block may be read from the disk into the buffer in its interface. Then, the interface circuit requests bus mastership and transfers the block to the main memory in a single burst. This approach is often used in practice, where blocks of 256 or 512 bytes are handled at a time. The block size is determined by the way the data are organized on the disk and the size of the storage buffer available in the DMA interface. Data are stored on a disk in blocks, known as *sectors*, as will be discussed in Chapter 9. It is convenient to choose the size of a DMA block to correspond to one disk sector.

7.6.2 Multiple Bus Masters

The need often arises for connecting several devices that require to become the bus master to the same microprocessor bus. For example, the system may contain more than one DMA interface. Also, in Chapter 10 we will discuss multiprocessor systems, in which several microprocessors require access to a common memory.

With several such devices, more than one request for bus mastership may be generated at the same time. The problem of resolving this conflict is essentially the same as that discussed in Subsection 7.3.3, regarding the transfer of interrupt-vector addresses. Hence, the daisy-chain connection of Figure 7.23 can be used, where the Bus-Request and Bus-Grant signals replace the $\overline{IRQ}$ and Vector-Address-Enable signals, respectively, as shown in Figure 7.37.

The question of arbitration among requests from several devices that need to share the use of the bus to transfer data to or from the main memory is an important one. The daisy chain scheme is very simple and effective. For microprocessor buses, it has the advantage of requiring only two bus lines, request and grant, hence only two pins on the microprocessor chip.

When a microprocessor system includes a large number of DMA devices, or when several microprocessors require access to the same main memory module, more elaborate schemes are needed. It is seldom the case that these devices would all be connected directly to the microprocessor bus. Instead, other interconnection networks, such as a *backplane* bus, are used. The design issues that arise in the

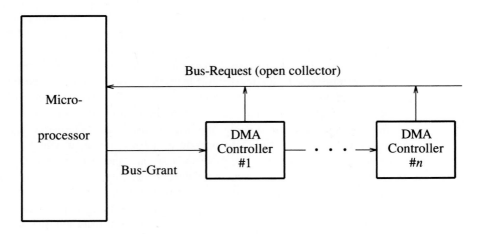

Figure 7.37 Bus arbitration using a daisy chain.

design of backplanes will be discussed in Chapter 10, and some commonly used standard buses will be introduced in Chapter 11.

DMA transfers take place in essentially the same manner for most microprocessors. The 6809 and 68000 microprocessors follow the general approach discussed above for the transfer of bus mastership. A brief description of the specific signals available on the bus of each microprocessor is given in the following two sections.

7.6.3 DMA on the 6809 Bus

The 6809 bus was introduced in Section 6.9 and Table 6.1. The transfer of bus mastership follows the general timing diagram of Figure 7.36. Because of the synchronous nature of the bus, each of the idle periods in the figure is equal to one clock cycle. The Bus-Request signal is called DMA/BREQ. However, no Bus-Grant output is available directly. It can be derived from the Bus Status (BS) and Bus Available (BA) signals, which were described in Subsection 7.4.3 and Table 7.3. The BA output is activated whenever the microprocessor puts its bus drivers in the high-impedance state. When it does so in response to a bus request, it also sets BS to 1. Hence, the bus grant signal may be generated as BA·BS. The microprocessor maintains both BA and BS equal to 1 until the DMA/BREQ line becomes inactive. Transitions on the BS and BA lines are always synchronized with the bus clock, E.

The 6809 microprocessor also does not provide a Master-Active signal directly. However, such a signal can be generated by external hardware for use by various device interfaces on the bus. In Motorola's nomenclature, the Master-Active line is called DMAVMA, for DMA or Valid Memory Address. This signal should normally be in the active state. It should be deactivated during any clock period in which the BA output of the microprocessor changes state. A suitable circuit for generating the Bus-Grant and VMA signals is given in Figure 7.38. The output of the EX-OR gate will be equal to 1 during any clock period in which BA changes from 0 to 1 or from 1 to 0, because the flip-flop remembers the state of BA during the previous clock cycle. A timing diagram illustrating the operation of the circuit is shown in Figure 7.39.

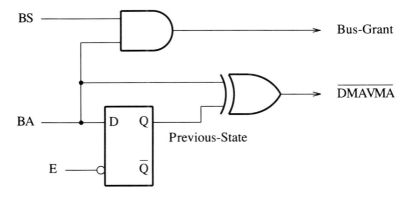

Figure 7.38 A circuit for generating Bus-Grant and DMAVMA on the 6809 bus.

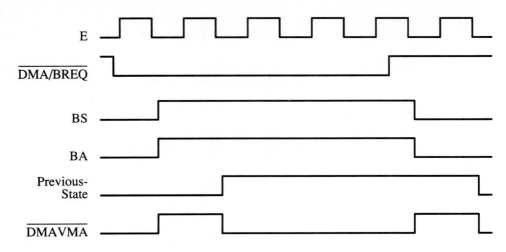

Figure 7.39   A timing diagram for the circuit in Figure 7.38.

7.6.4 DMA on the 68000 Bus

The 68000 bus signals were described in Section 6.10 and Table 6.3. Three signals are used in the transfer of bus mastership, namely, Bus Request, $\overline{\text{BR}}$, Bus Grant, BG, and Bus Grant Acknowledge, $\overline{\text{BGACK}}$. The way in which the $\overline{\text{BR}}$ and BG signals are used is essentially the same as discussed in Subsection 7.6.1, with one important difference. The arbitration operation to select the next bus master takes place in parallel with data transfer on the bus. When a device is notified that it has been selected as the next bus master, it waits until the current bus master releases the bus and until any data transfer that may be in progress is completed; then, it acknowledges receipt of Bus Grant, assumes bus mastership, and starts transferring data.

Figure 7.40 illustrates the sequence of events on the bus during the transfer of bus mastership. A DMA controller requests the use of the bus by activating $\overline{\text{BR}}$. This request may be issued at any time. In the figure, it has been assumed to take place while a data transfer operation is in progress, as indicated by the state of the $\overline{\text{AS}}$ line. In response, the microprocessor issues a bus grant signal on the BG line, thus informing the device that it may start using the bus as soon as the current cycle ends. In order to ensure that the devices involved in the current bus cycle have completed transmission and released the bus, the DMA controller monitors the $\overline{\text{AS}}$ and $\overline{\text{DTACK}}$ lines. When both lines are inactive, it activates $\overline{\text{BGACK}}$ and drops its bus request. It may now proceed to use the bus for as long as it wishes, keeping $\overline{\text{BGACK}}$ in the active state. It should deactivate $\overline{\text{BGACK}}$ only when it is ready to relinquish bus mastership. The microprocessor may start using the bus any time after it sees $\overline{\text{BGACK}}$ in the inactive state.

We should note that the Address Strobe signal used in each data transfer on the 68000 bus obviates the need for the Master-Active signal of Figure 7.36. The latter is intended to ensure that all devices ignore the bus while bus mastership is being

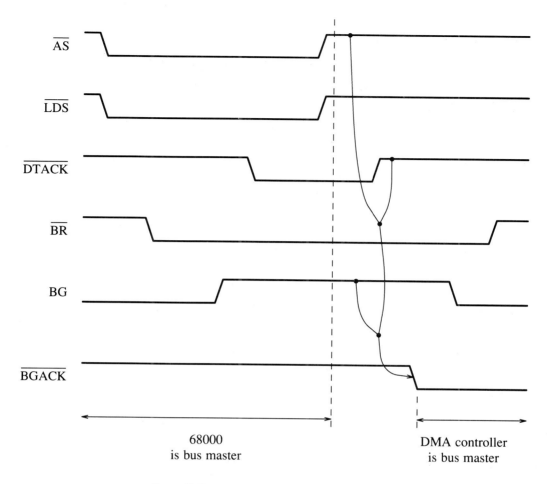

Figure 7.40 Transfer of bus mastership on the 68000 bus.

transferred from the microprocessor to the DMA controller. Since neither the microprocessor nor the DMA controller will activate $\overline{AS}$ during that period, there is no danger of an erroneous data transfer taking place.

A 68000 system may have several devices with DMA capability. In this case, an external priority network should be used to direct the BG signal issued by the microprocessor to the highest priority device requesting the use of the bus. A daisy-chain arrangement is often used for this purpose.

7.7 DMA INTERFACE

The basic functions performed by a DMA interface have been introduced above. In order to simplify the description of such an interface, we will assume that it transfers a single word of data every time it becomes the bus master. The case where a number of words are transferred before the bus mastership is returned to the microprocessor is a straightforward extension.

In order to store the relevant parameters for a given DMA task, the interface contains a register for each parameter. These registers are accessible to the microprocessor in the same manner as the status, control, and data registers of other I/O interfaces. Four typical registers are shown in Figure 7.41. The Memory Address Register stores the starting address of the memory block to or from which data transfer is to take place. The number of data bytes to be transferred is stored in the Byte Count Register. Each of these two registers is shown as 16 bits long. In the case of an 8-bit machine, a 16-bit register may be implemented as two separately addressable byte locations. The size of the Memory Address Register should be the same as the number of address bits of the microprocessor.

The DMA Control Register enables the programmer to define several aspects of the DMA transfer. The control fields in the example of Figure 7.41 are

- **R/$\overline{\text{W}}$**: This bit determines whether the DMA transfer is a read or a write operation. A read operation (R/$\overline{\text{W}}$ = 1) means that the data are transferred from the main memory to the I/O device. A write operation transfers data in the opposite direction.
- **Mode**: DMA transfers may take place in a variety of ways, depending upon the characteristics of the microprocessor. A DMA interface may transfer one

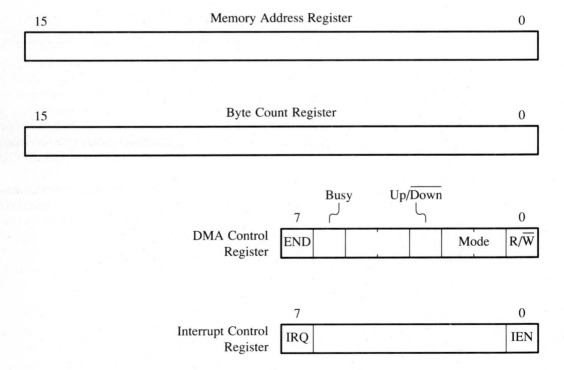

Figure 7.41 Typical registers in a DMA interface.

data word or an entire block, before returning bus mastership to the microprocessor. The word and block modes are two of the options selected by the Mode field. Other options may relate to the way in which bus mastership is acquired.

- **Up/Down**: Many DMA interfaces are capable of transferring data in the order of either increasing or decreasing main memory addresses. The microprocessor loads the starting address in the Memory Address Register. Then, depending upon the state of the Up/Down bit, the contents of this register are either incremented or decremented after each data word is transferred by the interface hardware. This flexibility is useful with devices such as magnetic tape drives, which often allow data transfer to take place while the tape is running forward or backward.

- **Busy**: This bit is set to 1 while a DMA task is in progress. It is cleared when the contents of the Byte Count Register reach zero. Recall that bus mastership may be passed back and forth several times between the microprocessor and the DMA interface during one DMA task. The busy bit enables the microprocessor to check at any time whether the DMA interface is available to initiate a new DMA task.

- **End**: When the End bit is set, it indicates that a previously requested DMA task has been completed.

The fourth register in Figure 7.41 is the Interrupt Control Register, which contains an interrupt-enable bit (IEN) and an interrupt-status bit (IRQ). The interface generates an interrupt request after completing a DMA task, that is when the End bit is set in the control register, provided that IEN is equal to 1. To facilitate the polling process, bit IRQ is set to 1 whenever the interrupt-request output of the interface is active.

The registers in Figure 7.41 contain all the necessary information to control DMA activity for a simple device, such as a printer. However, other devices may require more registers. For example, a disk drive requires the location of the data on the disk to be specified. Also, additional control and status information is needed for such operations as positioning the Read/Write head over a given track on the disk.

7.7.1 General-Purpose DMA Controller

Let us now consider the feasibility of developing a general-purpose DMA interface, in the same manner as we developed the GPPI in Section 6.6. A general-purpose circuit must be independent of the characteristics of a particular I/O device. Hence, we begin by isolating those aspects of the DMA interface that are device dependent. The DMA interface of Figure 7.35 can be thought of as consisting of two distinct parts, as shown in Figure 7.42. The first part is labeled "DMA Controller." This is the portion of the DMA interface that deals with acquiring bus mastership, generating memory addresses, controlling the transfer of individual data words, and keeping track of the amount of data to be transferred. The second part, called the Device Controller, constitutes the data path between the bus and the I/O device.

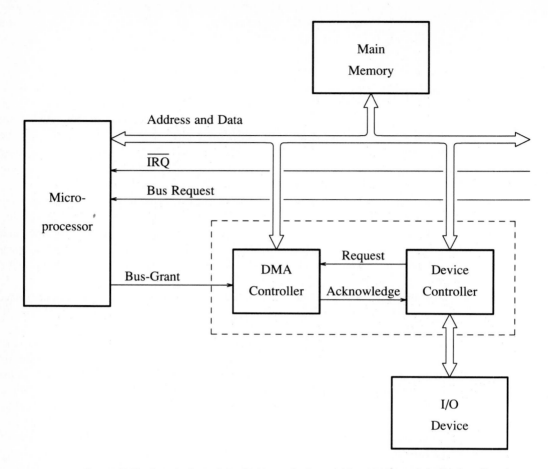

Figure 7.42 Separation of the DMA controller and the device controller functions.

It contains a data buffer through which words of data are transferred. It also enables the microprocessor to send control information directly to the device. For example, in the case of a disk drive, the Device Controller contains registers for storing the disk address of the data to be transferred, as well as control and status information relating to the motion of the read/write heads.

By splitting the DMA interface into two sections, we have isolated device-specific functions from the functions that are common to all interfaces having DMA capability. Thus, it is possible to define a general-purpose DMA controller, which can be used with a wide variety of I/O devices. In fact, several such controllers are commercially available, some in the form of single chips.

The two parts of the DMA interface in Figure 7.42 communicate by means of two signals, Request and Acknowledge. In order to illustrate the function of each of the two parts and the way in which the Request and Acknowledge signals are used, let us consider a DMA transfer from the I/O device to the main memory. Before the

DMA task begins, the microprocessor loads appropriate control parameters in various registers of both the DMA Controller and the Device Controller. Then, it issues a command to the Device Controller to start reading information from the device. When the Device Controller has a word of data ready for transfer to the main memory, it activates the Request signal. In turn, the DMA Controller requests the bus mastership using the Bus-Request control line, BR. Upon receiving a Bus-Grant signal from the microprocessor, it becomes the bus master. Then, it places the main memory address, where the data word is to be deposited, on the address bus and sets the R/W line to indicate a write operation to the main memory. At the same time it activates the Acknowledge signal, instructing the Device Controller to place the data word on the data lines of the bus. The Device Controller transmits a word of data and deactivates its request signal. The transfer of data to the main memory proceeds in the normal manner, with the DMA Controller performing the actions normally carried out by the microprocessor.

After completion of the data transfer, the DMA Controller removes the address and control information from the bus and deactivates the Acknowledge signal. In response, the Device Controller removes its data from the data lines. The DMA Controller updates the contents of its Memory Address Register and Byte Count Register in preparation for the next transfer. Assuming that word mode has been selected, the DMA Controller drops its bus request, to return bus mastership to the microprocessor. The entire process is repeated when the Device Controller obtains the next word of data from the I/O device. When the byte count reaches zero, and assuming that the Interrupt Enable bit is equal to 1, the DMA Controller activates the interrupt-request line, informing the microprocessor that the requested DMA task has been completed.

From the discussion above it can be appreciated that the operation of the DMA Controller is independent of the characteristics of the I/O device. The DMA Controller is only responsible for operations related to the acquisition of bus mastership and generation of memory addresses. Hence, it is possible to design a general-purpose DMA controller that can handle a variety of devices. On the other hand, a Device controller has to be tailored to the requirements of the I/O device. It is responsible for exchanging data with the device and for handling device-dependent commands from the microprocessor.

The DMA Controller of Figure 7.42 need not be dedicated to serve only one I/O device. By providing several Request and Acknowledge signals, a single DMA controller can be used to support several devices. It can communicate with each device on a separate pair of Request and Acknowledge signals. Since data transfers for different devices must be allowed to proceed independently, separate memory address and byte count registers should be provided for each device. The DMA registers and hardware signals dedicated to one device are collectively referred to as a DMA *channel*. A DMA controller may provide several channels, where most of the control logic is shared by all channels. The control and status bits for different channels may also be grouped together in common control and status registers. A typical commercial DMA chip provides four to eight channels.

Whenever multiple devices share the use of a common resource, an arbitration scheme must be provided to deal with situations when there are two or more simultaneous requests for service. A DMA controller that has several channels must incorporate a priority structure to resolve conflicts between simultaneous requests on these channels. Two commonly encountered schemes are fixed priority and ring priority.

In the *fixed priority* scheme, the priority level assigned to a given channel is determined in hardware by the DMA controller chip. This approach is useful when the devices connected to different channels have widely differing demands. For example, a request for a DMA transfer to a high-speed device, such as a disk, should be given high priority, because a delay in responding to this request may lead to loss of data. A lower speed device, such as a printer or a low-speed communications interface, may be given lower priority.

The situation often arises where the devices connected to different DMA channels have similar service requirements. The use of a fixed-priority assignment in this case would give the high-priority device an unfair advantage. In fact, if this device requests service often, it may "hog" the DMA controller, and cause very long delays for other devices. The *ring priority* scheme, also known as *rotating priority*, enables all devices to share the use of the DMA controller equally.

In order to illustrate ring priority, consider a DMA controller that has four channels, numbered 1, 2, 3, and 4. Initially, channel 1 has the highest and channel 4 the lowest priority. After any channel receives service it becomes the lowest-priority channel, and the next channel in numerical sequence becomes the highest-priority channel. For example, after channel 2 requests and receives service, the order of priority, starting with the highest, becomes 3, 4, 1, 2. Thus, on average, all four channels have equal chance of receiving service. If several devices request service repeatedly, the DMA controller will respond to each request once, before it accepts a second request from any of them. A DMA controller that has such facilities must also have control registers to select priority assignment and enable interrupts for each channel.

SEQUENCING AND DATA CHAINING

Each transfer request that a DMA interface receives from the microprocessor describes a block of data consisting of up to several hundred bytes. The data usually correspond to a natural block of data in the I/O device, such as a disk sector or a tape record (see Chapter 9). The interface informs the microprocessor when the transfer of a block is completed, at which point the microprocessor may initiate another transfer. The utility of a DMA interface can be enhanced if we introduce some flexibility in the way data blocks are defined both in the I/O device and in the main memory. Some features that are often incorporated in DMA controllers are discussed in this section.

We have already encountered one feature relating to address sequencing in the simple interface described earlier. A data block may be transferred starting at the main memory location having the lowest address, then successively incrementing the address until the end of the block is reached. Alternatively, the transfer may begin at the end of the block, with the address being decremented after each word transfer. In both cases, one block in the I/O device is mapped onto one contiguous block in the main memory. Additional flexibility can be gained if the main memory is allowed to be divided into sub-blocks, as shown in Figure 7.43. The sub-blocks should be automatically concatenated by the DMA controller, so that they appear as a single block of data to the I/O device. To do so, the information loaded in the memory address and byte count registers of Figure 7.41 should correspond to the first sub-block. The address and count for the second block should also be provided by the microprocessor, and stored in such a way that they can be easily accessed by the DMA controller when it reaches the end of the first sub-block. A convenient arrangement is to provide "continuation" registers in the controller for both the address and the count values. At the end of the first block, the controller transfers the contents of these registers into the memory address and byte count registers and continues its operation without interruption.

To reduce the workload of the microprocessor even further, it may be allowed to specify several block transfers in one request to the DMA controller. Then, the DMA controller interrupts the microprocessor only after all blocks have been transferred. This option is usually called *data chaining*. The microprocessor places the address and count information for all the blocks in suitable locations, either in the DMA controller or in the main memory. In the latter case, an address pointer should be sent to the DMA controller indicating where the block descriptors are in the main memory. The controller reads the information from the main memory for one block at a time, loads it into its internal registers, and performs the specified transfer.

We should note the similarity and complementary relationship between the continuation and data chaining features described above. The first enables a large block in the I/O device to be stored as several smaller and not necessarily contiguous blocks in the main memory. The second allows the microprocessor to deal with data that comprise several blocks in the I/O device. Such flexibility is useful in the implementation of memory management, which is one of the important functions of system software. The system software allocates memory space to individual tasks in

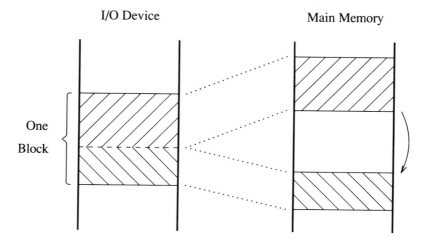

Figure 7.43   Concatenation of sub-blocks during DMA transfers.

blocks, also called *pages*, as will be described in Chapter 8. The features described here make it possible to choose page sizes and other parameters in a manner that is to a large extent independent of the characteristics of any particular I/O device.

IMPLICIT AND EXPLICIT ADDRESSING

The DMA controller in Figure 7.42 communicates with the device controller via the Request and Acknowledge handshake signals. The address lines of the microprocessor bus are used to identify the main memory location to or from which a transfer is to take place. No addressing information appears on the bus to identify the I/O device, or its controller, during the transfer. This arrangement is sometimes referred to as *implicit addressing* of the device. It is possible to operate the DMA controller in a different way, where the I/O device is addressed explicitly. Instead of using Request and Acknowledge control lines to communicate with the device controller, the DMA controller may use the microprocessor bus itself. In this case, each word transferred requires two bus cycles. First, the DMA controller reads a word from the device by placing the device's address on the bus and performing a read operation. Then, it writes the word it received into the main memory in a separate bus cycle. Because two cycles are used, each one with a different address, this approach is called the *dual address* DMA mode.

The Acknowledge signal in Figure 7.42 informs the I/O device controller when to place a word of data on the bus. This signal is not needed when the explicit addressing scheme described above is used. Instead, the device controller waits for a read request on the bus. It is instructive to consider whether we can also do without the Request signal. The main function of this signal is to inform the DMA controller when the I/O device is ready for another transfer. This is essential information in the case of a device such as a disk, where the time at which individual units of data can be transferred is dictated by the mechanical motion of the disk. Hence, the Request signal must be provided. The explicit addressing scheme offers little advantage in this case. There are situations, however, where data may be transferred to or from the device at any time. For example, it was mentioned earlier that the disk controller may contain a buffer that stores a block of data. The timing of data transfer between this buffer and the main memory is not constrained by the mechanical motion of the disk. Hence, the DMA controller can initiate transfer cycles on its own, without requiring any connection to the device controller other than that provided by the microprocessor bus. A similar situation is encountered with a mass storage device that consists of a large semiconductor memory. Again, it can respond to read or write operations generated by the DMA controller at any time.

There are considerable benefits in eliminating the connection between the DMA controller and the device controller. An obvious advantage is the simplification of the physical interconnections. Also, since the DMA controller communicates with the I/O device by placing an address on the bus, it need not be dedicated to a single device. Each DMA channel may now be provided with a Device Address Register, in which the microprocessor writes the address of the I/O device. Thus, the microprocessor can use the DMA controller to transfer data to or from any I/O device on the bus.

Another use for the dual-address mode is in transferring data from one area of the main memory to another. To make such transfers possible, the DMA controller must increment, or decrement, the device address after each transfer, in the same way as the memory address. (See Problems 42 and 43.)

7.8 COMMERCIAL EXAMPLES OF DMA CONTROLLERS

Many DMA controllers are commercially available in the form of a single chip. Three examples found in the Motorola family are designated 6844 [2], 68440 [3], and 68450 [4]. The 6844 is suitable for use with the 6809 microprocessor, while the 68440 and the 68450 are intended for use in 68000 systems. We will describe these chips briefly to illustrate the range of features they offer.

7.8.1 6844 DMA Controller

The DMA registers described in Section 7.7 and shown in Figure 7.41 are equivalent to those found in the 6844 DMA controller. This controller provides four DMA channels. It has one set of address, count, and DMA control registers for each channel. In addition, it has three control and status registers common to all four channels, as shown in Figure 7.44. The first is the Interrupt Control register (ICR), which provides an interrupt-enable bit for each channel. When enabled, an interrupt is generated when the End bit of the corresponding channel control register becomes equal to 1. Bit IRQ is equal to 1 whenever the interrupt-request output of the chip is active. The interrupt is cleared when the microprocessor reads the DMA control register of the channel that generated the interrupt.

The 6844 DMA controller offers a limited data chaining facility, under control of the data-chaining register, DCR. The 2-bit number in bit positions 1 and 2 of this register selects one of channels 0 to 2 for data chaining operation. When data chaining is enabled and the selected channel completes a block transfer, the contents of the address and count registers of channel 3 are automatically copied into the corresponding registers of the data chaining channel, and another block transfer is initiated. Thus, data chaining can only be used with channels 0 to 2.

The third control register is the Priority Control Register, PCR, which is used to enable DMA operations on individual channels and to select the priority scheme used.

The 6844 has four separate Request/Acknowledge handshake signal pairs, one for each channel. It begins accepting requests on a given input only after the corresponding channel-enable bit has been set. If the rotating priority bit is equal to 1, the channels are assigned priorities on a rotating basis, as explained earlier. Otherwise, a fixed priority scheme is adopted.

7.8.2 68440 and 68450 DMA Controllers

The 68440 and 68450 DMA controllers offer a wide variety of options, with the 68440 having a subset of the facilities found in the 68450 chip. The main difference

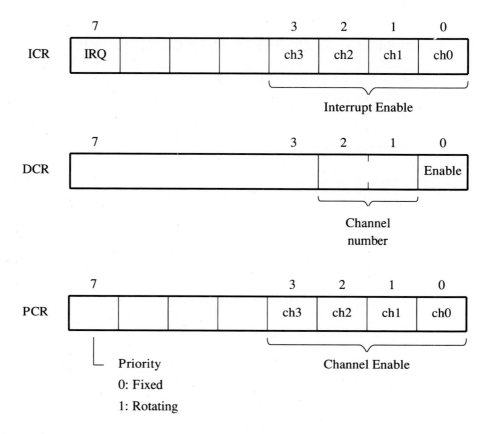

Figure 7.44 Common control registers of the 6844 DMA controller.

is that the 68440 has two channels, while the 68450 has four. The rest of this discussion will refer to the 68450 only. Each of the four channels of this controller has the block continuation and data chaining facilities described in Subsection 7.7.2. The channels may be programmed to use single- or dual-address cycles. In the dual-address mode, requests for individual transfers are generated by the DMA controller itself, rather than by the device. The rate at which these requests are generated is programmable, so that the system software can control the percentage of the available bus cycles that are allocated to a particular DMA transfer.

Figure 7.45 shows the 68450 being used to control DMA transfers for a hard disk, a floppy disk, and a communications controller. The latter is a circuit that enables a microprocessor system to exchange data with other systems over a telephone connection or a communications network, such as a local area network. The DMA controller is connected to the microprocessor bus through a

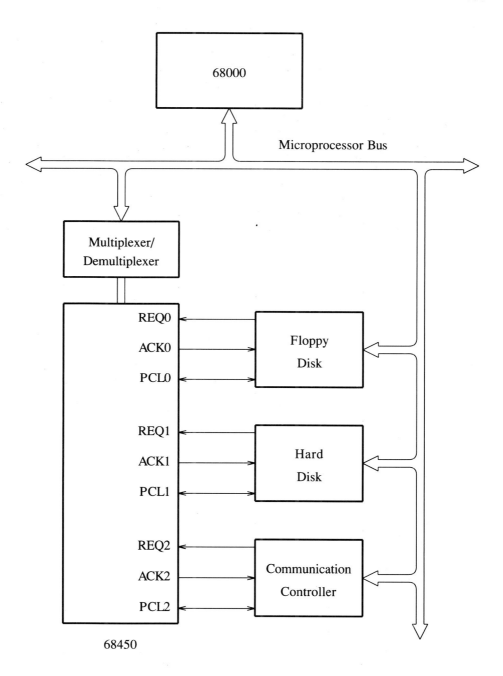

Figure 7.45 Connection of a 68450 DMA controller.

multiplexer/demultiplexer circuit. This circuit is needed because the address and data information are multiplexed on 24 pins of the 68450 chip. There are three signals dedicated to each channel. The first two are the Request and Acknowledge signals described earlier. The third is a bidirectional line called the Peripheral Control Line, PCL. It can be programmed to perform a variety of operations, such as aborting a failed transfer, restarting an operation or providing timing information.

The 68450 has a large number of internal registers. Each channel has eight registers that describe the data block to be transferred, as shown in Figure 7.46. A

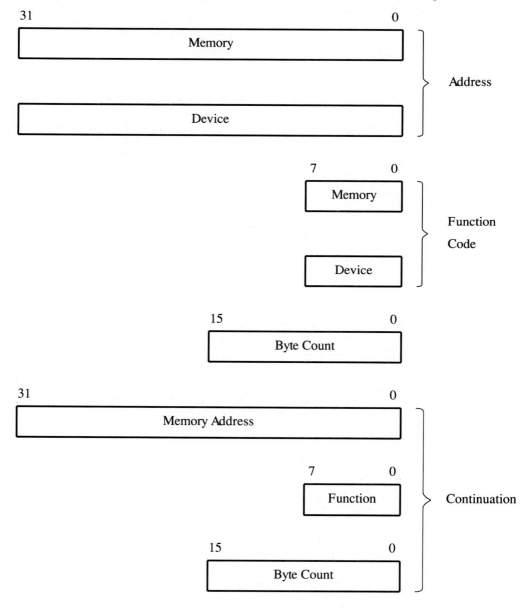

Figure 7.46 Block descriptors in the 68450 DMA controller.

32-bit memory address register, an 8-bit function code register, and a 16-bit byte count register are provided. The function code register holds the 3-bit function code needed on the 68000 bus. Since dual addressing is supported, address and function code registers are provided for device addressing. Also, to enable block continuation, duplicate registers are provided for the memory address, the function code, and the byte count.

When data chaining is enabled for any channel, the continuation registers are used in a slightly different way. The main memory location pointed at by these registers is the beginning of a table that contains the address and byte count information for subsequent blocks to be transferred. This information is read by the DMA controller and stored in the respective channel registers before starting another block transfer.

In addition to the address and byte count registers, there are nine 8-bit registers dedicated to each channel. They are used to store additional information for mode selection, device control, address sequencing, channel status, and so on. They also store the interrupt vectors, which should be sent to the 68000 microprocessor during the interrupt-acknowledge cycle.

The description given above for the DMA controller chips is intended to provide an appreciation of the range of features and the degree of sophistication of currently available microprocessor chip families. For further details the reader should consult references [2-4].

7.9 CONCLUDING REMARKS

This chapter has introduced two powerful I/O techniques that are commonly used in microprocessor systems; namely, interrupts and direct memory access. While details vary from one system to another, the general principles discussed should enable the reader to understand the operation of these schemes with any microprocessor.

Interrupts and DMA transfers are usually handled by system software. As such, their existence may not be seen by users writing applications programs. However, a proper understanding of these concepts is essential, particularly in applications that involve real-time processing.

7.10 REVIEW QUESTIONS AND EXERCISES

REVIEW QUESTIONS

1. What is the difference between a subroutine and an interrupt-service routine?

2. Why is it essential to disable interrupts at the beginning of an interrupt-service routine?

3. Is it important to check the device status in an interrupt-service routine before transferring the data? Why?

4. Interrupts may be enabled or disabled inside the microprocessor, as well as in individual I/O interfaces. Why are both these facilities necessary?

5. Why is the E bit needed in the CCR register of the 6809 microprocessor?

6. Which of the exceptions listed in Subsection 7.5.1 for the 68000 microprocessor take effect in the middle of the execution of an instruction?

7. The timing diagram in Figure 7.40 indicates that a DMA device must wait for both AS and DTACK to return to the inactive state before it assumes bus mastership and asserts BGACK. Why is it essential to check both signals?

8. Give an example of a situation in which each of the following techniques is most suitable for performing I/O operations:
   (a) Checking the device status in a wait loop
   (b) Using interrupts
   (c) Using DMA.

9. What is the difference between DMA transfers in the cycle-stealing mode and in the burst mode? Give examples to illustrate when you would use one or the other.

PRIMARY PROBLEMS

10. A microprocessor does not automatically disable interrupts before starting the execution of the interrupt-service routine. Instead, it guarantees that at least one instruction will be executed without interruption. Is this a workable scheme? Why?

11. There is a short period during which nonmaskable interrupts of the 6809 and 68000 microprocessors are disabled. When is that, and why?

12. Interrupts from a given interrupt-request line must be disabled at the beginning of the corresponding interrupt-service routine. Yet, the nonmaskable interrupt input, NMI, of the 6809 is not disabled during the execution of its interrupt-service routine. Is that likely to cause any problem? Why?

13. The 6809 microprocessor has three interrupt-request lines. Assume that all interrupts are enabled. For each line, what is the maximum number of bus cycles that can elapse from the time the microprocessor recognizes a request until it starts to read the first instruction in the corresponding interrupt-service routine? Assume that only one interrupt-request is received at a time.

14. The 6809 and the 68000 microprocessors each has a nonmaskable interrupt-request line. The microprocessor recognizes an interrupt request on this line only when an active transition (high to low) occurs. After the transition, no further interruptions will occur if the line stays active. Why is this arrangement necessary? (*Hint*: Recall the discussion of the need to disable interrupts at certain times.)

15. In view of the fact that only active transitions on the nonmaskable interrupt line are recognized (see Problem 14), what happens if two or more devices are connected to this line?

16. Rewrite the program for Problem 23 in Chapter 6, using the VIA timer to generate the pulses to the motor.

17. A 6809 program for displaying a message on a video screen using interrupts was given in Figure 7.28. Assume that the program at MAIN will initiate another task to read input characters from the keyboard of the terminal, also using interrupts. Modify the interrupt-service routine to enable these two tasks to proceed in parallel. (*Note*: If the ACIA happens to have both its Receiver Full and Transmitter Empty flags equal to 1, both will be cleared when the status register is read. So, the interrupt-service routine must check both flags every time it is entered.)

18. Repeat Problem 17 for the 68000 program in Figure 7.34.

19. A user program calls the operating system by means of a software interrupt, and the operating system transfers control to a user program by a Return-from-interrupt instruction. Consider a 68000 microprocessor. It is desired to use an RTE instruction to lower the processor's priority in the status register to 0, to change from Supervisor mode to User mode and to start a user program. The starting address of the user program is at location USER. What should the contents of the Supervisor stack be at the time the RTE instruction is executed? Write a suitable program segment to implement this task.

20. Consider the 6809 or 68000 microprocessor. Interrupts from a given interrupt-request line on the microprocessor bus are disabled during the execution of the corresponding interrupt-service routine. How would you enable interrupts from that line in the middle of the routine? Will interrupts remain enabled when execution of the interrupted program is resumed?

21. A microprocessor system uses a VIA to measure the frequency of an input signal. Write a program that uses timer T2 to count the input pulses and T1 to measure the time. The program should determine the number of pulses received in a fixed period of time. Do not use interrupts.

22. Repeat Problem 21 using interrupts with timer T1.

23. A microprocessor is used to monitor the activities of some device. One of the microprocessor tasks is to act as a "watchdog." The device sends a message indicating that it is in good operating order once every 10 ms. If a 15 ms period expires without this message being received, the microprocessor sounds an alarm. Describe how you would implement this scheme using the retriggering feature of the VIA timer.

24. Register X of the 6809 microprocessor contains the hexadecimal value C58A. Assume that a VIA chip connected to this microprocessor is assigned addresses starting at location DA00. What will be the effect of executing the instruction

STX $DA04

on the T1 timer of the VIA?

25. Can the MOVEP instruction of the 68000 microprocessor be used to load the counter of timer T1 of the VIA? If the answer is yes, give an example.

26. The 68020 bus enables the microprocessor to communicate with devices having a port width of 8, 16, or 32 bits. Can you load the timer counter of a VIA connected to the bus with a single MOVE instruction? Give an example.

27. Consider four devices connected to four DMA channels of a DMA controller. Each data transfer operation requested by a device takes a fixed period T to complete. When a device requests service, it may have to wait while higher priority devices are being serviced. Let D be the delay from the time a device requests a transfer operation until the transfer is completed. What are the minimum and maximum values of D for each device in the following cases:

(a) Fixed priority assignment

(b) Rotating priority assignment.

Assume that once a device is serviced, it will not generate a second request until all outstanding requests have been serviced.

28. Consider the case where two DMA devices are connected to the 6809 bus. The Bus-Grant signal of Figure 7.38 propagates from one DMA interface to the other through the daisy chain-scheme of Figure 7.37. The two devices request service simultaneously. Describe what happens on the BR and Bus-Grant lines until both devices receive service.

29. Consider the daisy chain scheme of Figure 7.37. Let the bus grant input signal to interface #i be BG_i, and the internal bus request signal (the input signal to the open-collector gate that drives the bus) be BR_i. Design a suitable circuit to generate BG_{i+1}.

30. Consider the case where two DMA interfaces on the 68000 bus request the bus simultaneously, and assume that they are connected using the daisy chain scheme of Figure 7.37. When the first device receiver becomes the bus master, it activates $\overline{BGACK}$, as shown in Figure 7.40, and drops its bus request, $\overline{BR}$. But, $\overline{BR}$ continues to be activated by the second device. How will the second device eventually receive the bus mastership?

31. The circuit in Figure 6.59 enables 6809 peripherals to be used on the 68000 bus. Show what additions are needed to enable the devices to use interrupts. (Remember that 6809 peripherals cannot generate an interrupt vector number.)

32. What is the order of priority in which terminals will be serviced by the routine in Figure 7.21? Modify this program so that, in the case of simultaneous requests, service will be according to a rotating (round-robin) priority structure.

33. Rewrite the interrupt-service routine in Figure 7.28 for the 68000 microprocessor.

34. Consider two DMA controllers, C1 and C2, connected using the daisy-chain arrangement of Figure 7.37, where C1 appears first on the chain. Controller C1 serves devices A, B, and C, using rotating priority assignment, and controller C2 serves devices D, E, and F on a fixed priority basis, with D having the highest and F the lowest priority. Starting from an idle condition, devices B and E request service simultaneously. Then, during the first DMA transfer, all the remaining devices request service. Give the order in which the requested transfers will take place.

ADVANCED PROBLEMS

35. A microprocessor keeps track of the time of day using timer T1 of the VIA. The timer sends an interrupt request once every second. An interrupt routine increments the time of day, which is stored in the main memory as six BCD digits representing hours, minutes, and seconds. The main program initializes the timer, then waits for input from a video terminal. To set the time, the operator types S followed by six digits. When any other character is received, the program displays the time of day. Write a suitable program. (*Caution*: Be sure not to display a partially updated value for time.)

36. Consider the following program, which is written in some high-level language

> Name := "your name"
> Out Name

It assigns a person's name to a variable called Name; then it sends the name to be displayed on a video terminal. The compiler for the language used generates assembler instructions that store the name into a suitable location, then calls the operating system to perform the output function. Suggest a scheme for passing the necessary information to the operating system, and give the assembler instructions for either the 6809 or 68000 microprocessor.

37. A microcomputer is used in a device that counts cars at an intersection. Each car that crosses the intersection causes a low pulse to appear at the CA1 input of a PIA (or a VIA) chip. Write a 6809 or a 68000 program that consists of a main program and an interrupt-service routine, as follows:

> **1.** The main program accepts input characters from an operator's console, which is connected through an ACIA, and echoes them back.
> **2.** The interrupt-service routine increments an 8-bit count in the main memory every time a pulse is received by the PIA.

Counting should begin only when the operator types the letter S, and it should stop when the count reaches 100. At this point, the main program should display the letter E, and wait for another command to start counting.

38. Consider the counting device described in Problem 37, and assume that it uses a VIA chip. Modify the program to use timer T1 of the VIA to measure the time during which 100 pulses are received on control line CA1. Assume that 100 pulses will be received in less than the maximum time-out period of T1 (2^{16} clock pulses).

39. A microprocessor uses the shift register in the VIA to send and receive serial data. The CB1 and CB2 control lines are connected to their counter parts on an identical system. Each system has a video terminal connected through an ACIA. Consider the case when one microprocessor always sends and the other always receives. The sender waits for a character from the terminal, echoes it back and sends it for transmission through the shift register. It uses a transmission speed that enables a maximum rate of about 4000 bits per second. The receiver waits for a character, reads it from the shift register and displays it on its video terminal. Write the transmitter and receiver programs, without using interrupts.

40. Modify the program of the transmitter in Problem 39 to use shift register interrupts.

41. It is required to use one of the VIA lines to drive a loudspeaker to produce sounds of different frequencies. Suggest a suitable connection, and write a program to repeatedly produce a chirp (a sound that starts at a low frequency and increases in tone to some maximum frequency).

42. It was suggested in Section 7.7 that the dual-address mode may be used for transferring a block of data from one area of the main memory to another. What facilities are likely to be needed in a general-purpose DMA controller to enable this process?

43. A block of data is to be copied from one area of the main memory to another in a 6809 microprocessor system. Calculate the minimum number of clock cycles needed to perform this task

    (a) Using a program loop
    (b) Using cycle-stealing DMA transfers
    (c) Using a burst DMA transfer.

    In cases b and c, ignore the clock cycles needed to initialize the DMA operation and to interrupt the microprocessor after the transfer is completed.

44. The situation often arises in a computer system where the use of a common resource is shared among several devices or tasks. The following are possible priority schemes for dealing with multiple requests:

    - Fixed priority
    - Rotating priority
    - First-come, first-served
    - Random assignment.

    There is often a waiting period from the time the resource is requested until it becomes available to be assigned. In each priority scheme, when should the selection decision be made—at the time the request is received or after the resource becomes available? Why?

45. In the cycle-stealing mode, a transfer is usually requested by the I/O device, which sends a request to the DMA controller. The time at which these requests are generated is determined by some aspect of the device's operation, such as its mechanical motion. If all the data to be transferred are placed in a storage buffer in the device's inteface before DMA transfers begin, then the transfer of individual words may be initiated at any time. Many DMA controllers incorporate a facility to initiate transfers at a programmable rate, which is usually expressed as a fraction of the bus bandwidth. Why is it important to place this rate under program control? Is it not always better to transfer data as quickly as possible?

LABORATORY EXERCISES

46. Design, build, and test a waveform generator consisting of a digital-to-analog converter (DAC) connected to a 6809 or a 68000 microprocessor. The DAC is connected to the microprocessor bus as shown in Figure 7.47. It appears to the microprocessor as a DMA device controlled by a 6844 DMA controller. The waveform to be displayed is stored in the main memory in the form of a sequence of voltage samples, represented by integers in the range 0 to 255. To display this waveform, the samples should be sent to the DAC one after the other, at regular intervals. Timer T1 of the VIA is used to generate pulses at the desired frequency on line PB_7, which is connected to the request input, REQ0, for channel 0 of the DMA controller. For each pulse, the DMA controller starts a read operation on the bus and generates a pulse on its acknowledgment output, ACK0, thus loading a byte of data into the DAC.

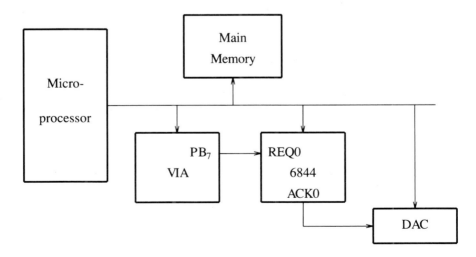

Figure 7.47   Connection of the DMA controller in Problem 46.

Build the system described above using the components available in your laboratory. Then:

(a) Prepare software to test the operation of the system in a step-by-step manner, with interrupts turned off. For example,

1. Initialize the timer and observe the output pulses on PB_7 on an oscilloscope.

2. Initialize the DMA controller to read a block of data repeatedly from the main memory using data chaining. Observe the request/acknowledge handshake.

3. Load a simple pattern, e.g., a square wave or a sawtooth, in the appropriate area in the main memory, and observe the DAC's output.

(b) Prepare a program to enable the user to enter a new value for a point on the waveform being dispayed.

(c) Prepare a program to enable the user to enter a new set of samples. The display should switch to these new values only after they have been completely entered.

7.11 REFERENCES

1. *MC68000 16-bit Microprocessor User's Manual.* Austin, TX: Motorola Inc., 1983.
2. "MC6844 Direct Memory Access Controller (DMAC)." *Motorola Interface Chips.* Motorola Inc., 1986.
3. *MC68440 Dual-Channel Direct Memory Access Controller.* Motorola Inc., 1984.
4. *MC68450 Direct Memory Access Controller (DMAC).* Motorola Inc., 1986.

Chapter
8

Main Memory

In this chapter we consider the structure of the main memory and typical components that are used in its construction. Emphasis is placed on semiconductor memories, in view of their present domination of the microcomputer ambient.

Semiconductor memories differ in size, speed, and the type of operation they can perform. Nevertheless, they have a common underlying structure. Once this basic structure is understood, there is little difficulty in gaining adequate appreciation of the differences found in various memory types. The next section presents the basic structure, and subsequent sections deal with the specifics of the most commonly used semiconductor memories.

The chapter includes a discussion of cache and virtual memory concepts. A cache memory enhances the performance of a microcomputer. A virtual memory provides greater flexibility by simplifying the development of software.

8.1 ORGANIZATION OF MEMORY INTEGRATED CIRCUITS

Main memories in microcomputers contain one or more memory integrated circuit chips (ICs), in a structure that allows reading or writing a unit of data (byte or word) within one "access" cycle. The storage cells that correspond to a particular unit of data are accessed by presenting an appropriate address to the memory, along with a control signal indicating the desired operation, e.g. Read/Write (R/W).

Figure 8.1 shows a conceptual organization of a typical memory IC. The central part is an array of storage cells, each capable of storing one bit of information. The cells are arranged so that a group of k cells corresponds to a given address, where the number k depends upon the organization of a specific IC and is usually equal to 1, 4, or 8. Each cell in a group is associated with a particular bit position. When an address is applied to such an IC, the cells of the selected group are connected to the k data lines. In the technical literature, it is common to refer to a group of k cells as a "word." This usage of the term must not be confused with the word length of a computer, because several memory ICs may be placed side-by-side to provide a main memory of any desired word length, as will be explained in Section 8.2. For purposes of the present discussion, we will assume that a word refers to all cells in the memory that correspond to a particular address.

Address decoding circuitry decodes the incoming address and activates one *word-select* line, thus enabling the cells connected to this line. If the status of the

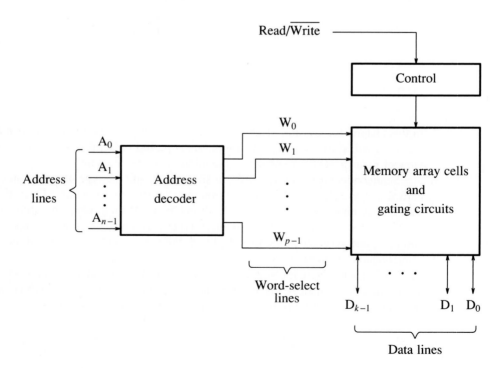

Figure 8.1 Conceptual organization of a memory IC.

$R/\overline{W}$ signal indicates a read operation, then the contents of the enabled cells are placed on the corresponding data lines. In the case of a write operation, new information supplied over the data lines by the device that initiated the memory operation is written into the selected cells.

A more detailed example of a cell array is given in Figure 8.2. The cells are arranged in p rows and k columns. Each cell has two connections: one to a word-select line, which is common to all cells in a row, and the other to a *bit line*, which is common to all cells in a column. All k cells in a given row are selected simultaneously, when the corresponding word-select line is activated. The bit lines constitute the data paths for transfer of data during read and write operations. The address

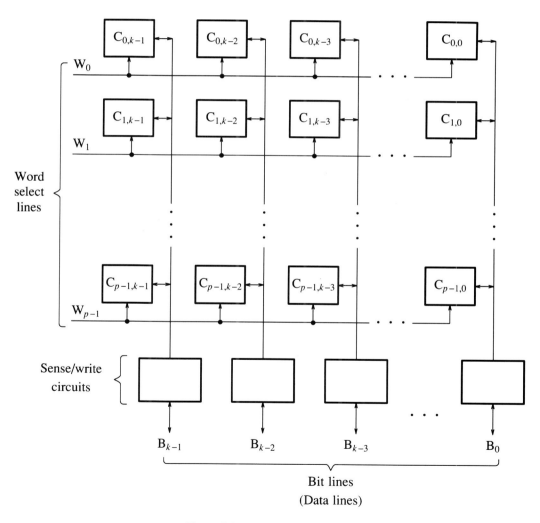

Figure 8.2 A $p \times k$ array of memory cells.

decoder activates only one word-select line at a time. Hence, only one cell is selected on each bit line.

Memory cells can be implemented in a variety of ways, as discussed in the sections that follow. Usually, the storage mechanism is such that the signal levels attained at the cell's connection to the bit line are quite different from the signal levels that the memory IC must present to the outside world. Hence, additional sense/write conversion circuits have to be included in each bit position to achieve the necessary signal level adjustments.

So far we have discussed the structure of memory ICs without considering physical constraints on their size. It would be convenient to be able to use a single IC to provide the main memory in a microcomputer. Suppose that we wish to use a 68000 microprocessor, which transfers data in 16-bit quantities, and that a main memory of 4 Mbytes (2M words) is desired. A single-chip memory, patterned after Figures 8.1 and 8.2, that could be used for this purpose would have to contain $2097,152 \times 16$ storage cells. This is a rather large number of cells, the implementation of which is outside the bounds of present VLSI (very large scale integration) technology. The present limit is 1M bits per IC, but larger chips will be available in the future.

Another factor pertinent to the physical size of ICs is the number of external connections required, that is, the number of pins that an IC must have. Consider, as an example, a relatively small memory chip capable of storing 1K ($= 1024 = 2^{10}$) bits. There are several ways in which this chip can be organized. It can have 1K addressable locations of one bit each. This would require 10 address pins and one data pin. Alternatively, it can be organized as 256 addressable locations of 4 bits each. In this case there must be eight address and four data pins. Going a step farther, an organization of 128 locations of 8 bits each would need 7 address and 8 data pins. This clearly indicates that increasing the number of bits per addressable location results in a larger number of pins that the IC package must have. Of course, we should not forget that in addition to the address and data pins, it is necessary to provide power supply connections and some control signals such as $R/\overline{W}$ and at least one "Enable" signal used to activate the memory.

It is generally useful to keep the number of pins of an IC as low as possible. This leads to smaller packages and hence better utilization of space when memory boards are constructed. In the quest for minimizing the number of pins, it is feasible to time-multiplex the information on some pins. For example, consider the address lines. The m address bits may be divided into two equal groups having $m/2$ bits each. The memory chip could have only $m/2$ address pins and a set of $m/2$ internal address latches. With this arrangement, the first group of $m/2$ address bits may be placed onto the address pins and latched into the address latches. Then, the second group of address bits may be applied. In this way only half as many address pins are needed as there are address bits. But, it may take a little longer to access the memory, due to the longer time that it takes to transfer the complete address to the chip. Such multiplexing of address bits is commonly found in dynamic RAMs, as will be discussed in Section 8.5.

Regardless of the external appearance of a memory chip, its cell array is organized in a way that best utilizes the available chip area. Figure 8.2 shows a conceptual structure of a cell array, which is the picture pertinent to the user. In practice, a 256 × 4 bit memory is likely to be organized as shown in Figure 8.3. A cell array of 32 words having 32 bits each is used, with multiplexing circuits that choose the desired 4 bits out of the 32 bits in the selected word. The 8-bit address is split into two parts: 5 bits select one of the 32 words and the remaining 3 bits serve as select inputs for the 8-to-1 multiplexer.

This section has dealt with the basic organization of a single memory IC. The structure presented above is found in most commercially available products. Later discussion of different types of memories will deal with the particulars of implementation of individual memory cells. It will be shown that considerable differences exist between different types. However, the basic structure remains the same.

Before turning to specific memory types, we will consider the task of constructing larger memory units using ICs as building blocks.

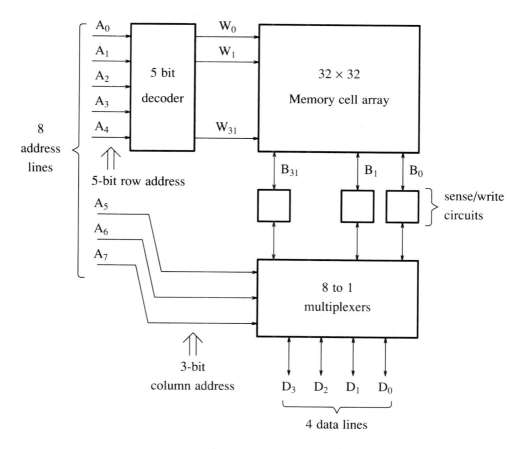

Figure 8.3 Organization of a 256 × 4 memory chip.

8.2 CONSTRUCTION OF LARGER MEMORIES

Technological constraints preclude fabrication of arbitrarily large single-chip memories, forcing us to build large memories using several memory chips. The task is not difficult. Figures 8.4 and 8.5 illustrate two arrangements for constructing a 4 Kbyte memory. We have chosen a small example for simplicity, but the approach is the same for larger memories. In Figure 8.4 the chips used have a 4K × 1 organization. Thus, eight chips are needed to provide the necessary word length of 8 bits. All 8 chips are enabled simultaneously by means of a signal on the Enable line.

Figure 8.5 shows how the same memory can be built with 2K × 4 chips. Since each chip provides 4 bits of data per addressable location, two chips are needed to give an 8-bit word. Four chips suffice for the desired memory. Address bits A0-A10 are used to select one of 2K locations in each of the four chips. Address bit A11, in conjunction with the Enable signal, selects either the upper or the lower pair of chips, thus providing access to all 4K locations.

When building large memories, it is prudent to use chips with as large a capacity as possible. This minimizes the total number of chips and results in lower cost and better utilization of the space on the board on which the components are mounted.

8.3 STATIC RAM CELLS

Having looked at the overall structure of a semiconductor memory, we should next examine the details of individual memory cells. The cells in Figure 8.2 can be implemented in several ways. In this section we consider "read/write" cells, that is, cells that allow both reading and writing of information.

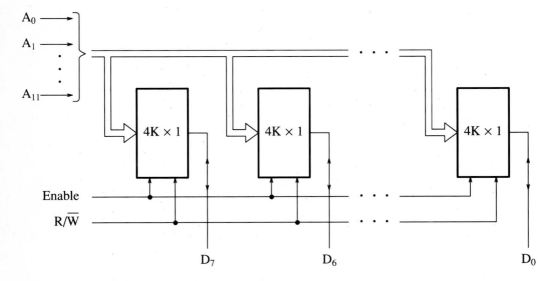

Figure 8.4   A 4K-byte memory using 4K × 1 chips.

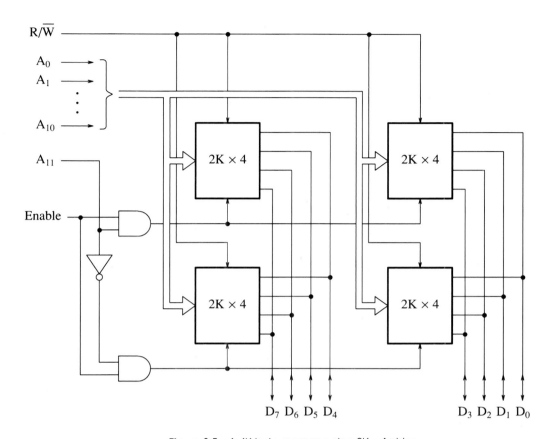

Figure 8.5 A 4K-byte memory using 2K × 4 chips.

It is apparent that the structure of Figures 8.1 and 8.2 is such that any word in the memory can be accessed by applying the corresponding address as an input. This flexibility gave rise to the term "Random Access Memory," or RAM, as a descriptor for this type of memories. Unfortunately, in addition to implying the random access capability, the term RAM has become associated only with those memories that are of the read/write type. Some random access memories can only be read from during normal operation. In keeping with established naming conventions, we will not refer to such memories as RAMs, but will call them "Read Only Memories" (ROM) instead.

A memory cell is said to be *static* if it is capable of retaining the information stored in it indefinitely. Once set in a particular state, 0 or 1, a static cell remains in that state provided that the power supply to the circuit is not interrupted. Such RAM cells are the subject of the following discussion.

There are two basic types of static RAM memories, one using bipolar and the other using MOS (metal oxide semiconductor) technologies. Figure 8.6 depicts a possible bipolar cell. The cross-coupled transistor inverters form a bistable circuit

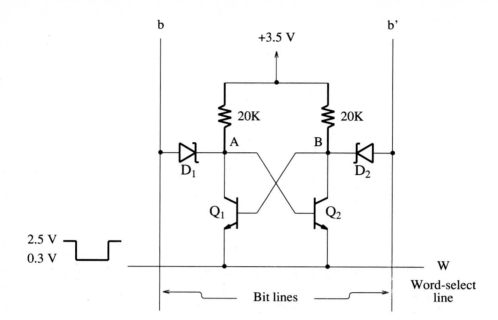

Figure 8.6 A bipolar RAM cell.

(latch). A word-select line connected to the emitters of Q_1 and Q_2 is used to activate the cell during read and write operations. There are two bit lines connected to the opposite sides of the latch. However, the information on each pair of bit lines, as detected by the sensing circuits, is eventually combined into a single bit of data. Hence, the cell fits into the general structure depicted in Figure 8.2.

Under quiescent conditions, the bit lines are kept at a voltage of approximately 1.6 volts, and the word-select line at 2.5 volts. Consider the state of the circuit when transistor Q_2 is conducting and Q_1 is turned off. This is a stable state that can be maintained indefinitely. The current flowing from the 3.5 V power supply through the 20K resistor at point A is sufficient to maintain Q_2 in the conducting state. In turn, the current flowing through Q_2 causes the voltage at B to be at 2.7 V. Hence, Q_1 continues in the off state, because it requires at least 0.4 V across its base-emitter junction to begin conducting. Let us regard this state of the cell as representing the logic value 0. Clearly, another stable state exists, in which Q_1 is conducting and Q_2 is off. We will regard this state as representing a 1. In either state, diodes D_1 and D_2 are reverse biased, thus isolating the cell from the bit lines.

Read and write operations involving the cell of Figure 8.6 are described briefly below:

Read operation: The contents of the cell are read by lowering the voltage on the word-select line to 0.3-volt. As a result, one of the diodes D_1 or D_2 becomes forward biased — D_1 if Q_1 is on and D_2 if Q_2 is on. Thus, a current will flow in bit line b if a 1 is stored in the cell, or in b' if a 0 is stored. The flow of current is detected by sensing circuits connected to the ends of the bit lines.

Write operation: When the cell is selected by a 0.3-volt signal on the word-select line, it can be driven into a desired state by applying a signal of about 3 volts on the appropriate bit line. A signal on line b' will store a 1, and a signal on b will store a 0. For example, starting in the 0 state (Q_1 off), a 3-volt signal on line b' will raise the voltage at point B so that Q_1 will start conducting. When Q_1 turns on, the voltage at point A will drop to the point where Q_2 turns off, thus completing the switching of the cell from state 0 to state 1.

Now, let us consider the second type of static RAM cells, those fabricated with MOS technology. Figure 8.7 shows an example, which closely resembles the bipolar cell described above. The function of transistors Q_1 and Q_2 remains the same. Transistors Q_3 and Q_4 replace the 20K resistors in Figure 8.6, exploiting the fact that MOS transistors can be readily used as resistors. This approach is used in integrated circuits because resistors are more difficult to implement than transistors. Transistors Q_5 and Q_6 replace diodes D_1 and D_2. They serve as switches controlled by the voltage on the word-select line. Closing these switches transfers the contents of the cell to the bit lines during a read operation, and allows overwriting the state of the latch during a write operation.

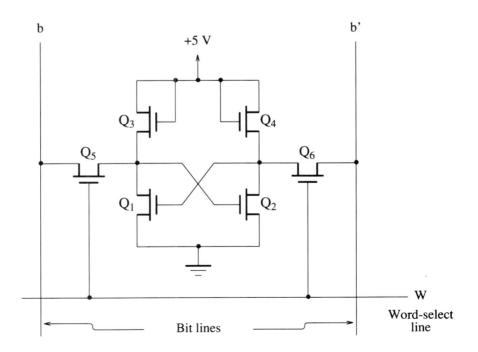

Figure 8.7 A static MOS RAM cell.

MOS technology offers several advantages over its bipolar counterpart. Greater component densities on integrated circuits can be achieved, resulting in larger capacity memory chips. The manufacturing process is simpler and the higher impedance of MOS transistors results in lower power dissipation. They have one drawback; their speed of operation is slower.

8.4 DYNAMIC RAM CELLS

The number of memory cells that can be placed on a single chip depends largely on the complexity of the basic cell, which is a function of the number of transistors and the number of connections needed. Therefore, in order to increase the density of cells it is necessary to reduce the number of transistors per cell. The circuit of Figure 8.7 uses six transistors to implement the basic cell. In this section we discuss simpler circuits that can be used for the same purpose.

MOS transistors have very high input impedance, which makes it possible to use their gate capacitance (produced by the geometry of the device) as a temporary storage element. A charge stored in this capacitance decays through small leakage currents, but it can be usefully interpreted as stored information for a period of a few milliseconds. If a provision is made to refresh the required charge at appropriately short intervals, typically 2 to 4 milliseconds, then the transistor and its gate capacitance can serve as the basis of a memory cell. Such memories are called *dynamic*, because their contents are lost if periodic refreshing is not performed.

Consider the cell shown in Figure 8.8. Transistor Q_2 and its gate capacitance C store the charge corresponding to one bit of data. Transistors Q_1 and Q_3 are switches that enable the cell contents to be read, refreshed, or changed. Two bit lines are used. When Q_1 is conducting, the datum on the write bit line charges capacitor C if a 1 is to be stored, and discharges it if a 0 is to be stored. The presence or absence of charge on C determines whether Q_2 is turned on or off. The state of Q_2 can be sensed on the read bit line, b', when Q_3 is turned on. A two-step process is used to refresh the contents of this cell. First, its state is read on line b'. Then, this information is transferred to line b and written back into the same cell.

It is possible to construct even simpler dynamic RAM (DRAM) cells. One of the simplest structures currently in use is shown in Figure 8.9. A single capacitor is used to store the charge corresponding to the data bit. Transistor Q is a switch that isolates C from the bit line, so that the same bit line can be used for reading and writing cells in other words. The charge representing the data on C is small, which causes some difficulties with the read operation. The problem can be overcome with a sensitive sense amplifier that compares the stored charge level with the charge in a special "dummy" cell. As a part of the process of reading the state of a cell, the charge on capacitor C is restored to its full value. Hence, the cell is automatically refreshed every time its contents are read. The details of this circuitry are outside the scope of this book. They can be found in many books on digital electronics [1–4].

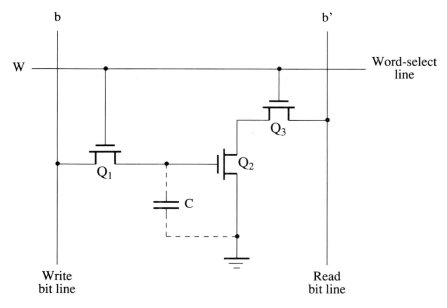

Figure 8.8   A three-transistor dynamic MOS RAM cell.

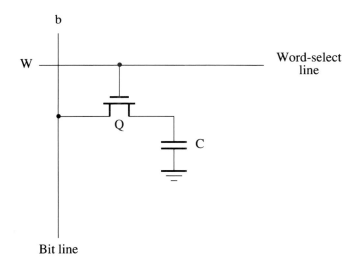

Figure 8.9   A one-transistor dynamic MOS RAM cell.

8.5 CONTROL CIRCUITRY FOR DYNAMIC RAMS

So far we have considered the overall organization of integrated circuit memories and some details of RAM cells. At this point it is worthwhile to examine more closely the possible ways of accessing such memories. We will take one commonly used memory chip as an illustrative example and discuss the access circuitry that is appropriate for it. The choice for our example is a 16K-bit dynamic RAM chip manufactured by several companies, variously known as 2116 (Intel), 4116 (Motorola, Texas Instruments, Mostek), and 6616 (Zilog). This is a rather small memory chip that makes the example easy to present. The approach discussed is applicable to memories of any size.

Figure 8.10 shows the block diagram of the chosen chip. The memory array consists of 128 rows of 128 cells each. Each cell consists of the single-transistor circuit given in Figure 8.9. In order to fit the chip into a 16-pin package, the address lines are multiplexed as discussed in Section 8.1. Seven address bits, used to select one of the 128 rows, are latched under the control of $\overline{RAS}$ (Row Address Strobe) signal. This is followed by latching the next 7 address bits, which select one of the 128 columns, under the control of $\overline{CAS}$ (Column Address Strobe). Input/output gating circuits connect the selected cell with the input and output data buffers, which

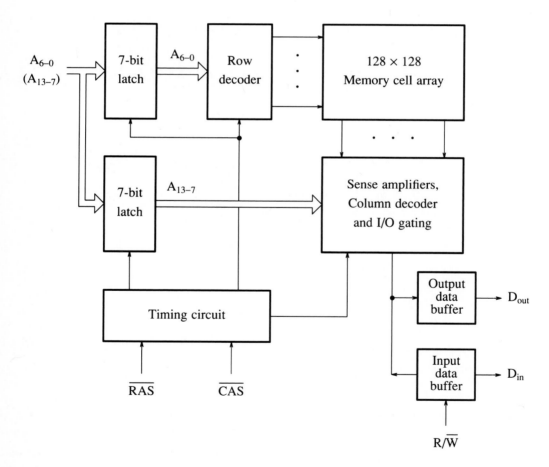

Figure 8.10 Block diagram of a 16K-bit dynamic RAM.

are connected to the input data and the output data pins of the chip. A R/$\overline{W}$ control line determines whether a read or a write cycle will take place.

Access circuitry for the memory must generate the required control signals and provide the row and column addresses in the correct sequence. Consider a typical read cycle, which requires the signaling sequence depicted in Figure 8.11. The process begins by activating the STARTCYC. Once the cycle is started it should proceed uninterrupted to its completion. This may be ensured through a $\overline{\text{BUSY}}$ signal, which prevents initiation of further requests. Row address data must be available at the address pins when $\overline{\text{RAS}}$ is activated. Hence, $\overline{\text{RAS}}$ is delayed by a set-up time t_{rs}. Activation of $\overline{\text{RAS}}$ latches the row address. This is followed by gating the column address data onto the address pins. After a set-up time t_{cs}, the column address is used by activating $\overline{\text{CAS}}$. At this point the R/$\overline{W}$ line indicates that a read operation is required. The output datum bit is available after some delay. The $\overline{\text{RAS}}$ and $\overline{\text{CAS}}$ are deactivated after the datum on D$_{out}$ has been read. Finally, the $\overline{\text{BUSY}}$ signal is returned to high, allowing a new cycle to be started.

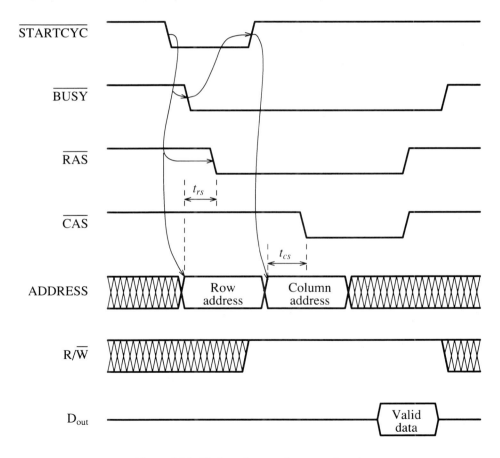

Figure 8.11 Timing diagram for a read cycle.

There are many ways of generating the control signals in Figure 8.11. One possible circuit is shown in Figure 8.12. The $\overline{\text{STARTCYC}}$ signal is activated in response to an external request, $\overline{\text{CYCREQ}}$. The STARTCYC sets the BUSY flip-flop, which remains set for the duration of the access cycle. A shift register provides the timing sequence. Let us assume that the access time of the memory chip is 250 ns. Then, a 5-bit shift register, driven by a 20 MHz clock, may be used. Taking $\overline{\text{RAS}}$ and $\overline{\text{CAS}}$ from the first and second stages of the shift register means that these signals will be 50 ns apart, which is within the timing specifications for the particular IC chosen for this example. After five shifts ($5 \times 50 = 250$ ns) the cycle time has elapsed and the timing circuit is reset through the latch driven by Q_5. A 2 to 1 multiplexer is used to gate first the row and then the column bits from the 14-bit address needed to access 16K locations.

Let us now consider the need for refreshing. As explained in Section 8.4, each cell in this DRAM must be refreshed approximately once every 2 ms. The internal structure of the memory chip is such that reading or writing data in a given row refreshes all cells of that row. This means that if one could guarantee that each row

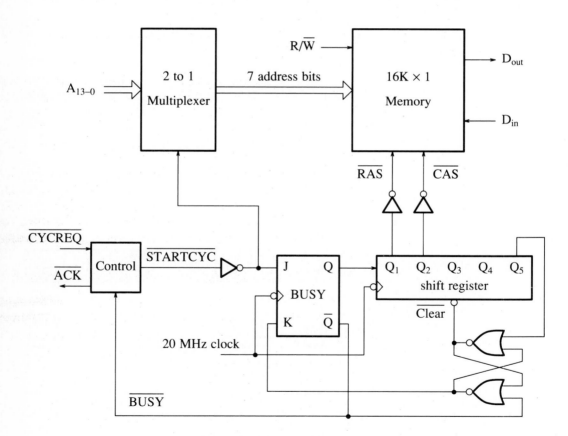

Figure 8.12 Access circuitry for a 16K-bit dynamic RAM.

is accessed at least once during every 2 ms interval as part of the normal operation of the memory, then no special refreshing mechanism would be needed. This is possible only in specialized applications. In general, it is necessary to provide separate means for refreshing dynamic memories. The refresh circuitry should access each row once within each refresh time interval.

Our example memory has 128 rows of cells. Since each row must be refreshed once every 2 ms, it follows that roughly 16 microseconds are available for dealing with one row. Since an access cycle for one row takes only 250 ns, the overhead for refreshing is

$$\text{Refresh overhead} \quad = 250 \times 10^{-9} \, / \, 16 \times 10^{-6}$$
$$= 0.016$$

Therefore, the memory is not available for external access less than two percent of the time.

The refresh circuitry can be combined with the circuit of Figure 8.12, as shown in Figure 8.13. The same control signal timing structure is used. A refresh cycle is just a read cycle, where data read from the memory cells are simply ignored. The refresh circuitry must include a circuit that triggers the refresh cycle. This "refresh request generator" should produce a signal, labeled $\overline{Q_t}$ in Figure 8.13, every 16 microseconds to indicate the need for a new cycle. The specific time interval is determined by the choice of the values of the resistor R_t and the capacitor C_t.

Since one row is refreshed during a refresh cycle, a 7-bit counter is provided to generate the row address. This counter is incremented after each access, so that different rows are accessed in successive refresh cycles. Inclusion of the counter means that a 3 to 1 multiplexer is needed, as indicated in the figure.

A memory access cycle is started when either a normal read/write request is received on the bus or a refresh cycle is requested. The normal read/write cycle is requested by the microprocessor with a $\overline{\text{CYCREQ}}$ signal. A refresh request ($\overline{\text{REFQ}}$) may come from the refresh request generator as shown in the figure, but it can also be provided by some other means. For example, some microprocessors include a feature whereby a refresh request signal is produced at proper time intervals. When the read/write requests and the refresh requests are generated independently, as is the case in Figure 8.13, an arbitration mechanism is needed to deal with simultaneous arrival of two requests. The arbitrator circuit performs this function. It should give higher priority to refresh requests, because delaying them can lead to loss of data stored in the memory. The arbitrator ignores further requests while the $\overline{\text{BUSY}}$ signal is active, so that once an access cycle is started it will proceed to its end.

The circuit of Figure 8.13 is fairly complex and it may seem that quite a few ICs are needed to implement it. However, such circuits are commonly used and special ICs are available to implement the desired functions. For example, the refresh generator and arbitrator come in a single chip known as 3222. The 3-to-1 multiplexer and the 7-bit counter form the 3242 chip. Other chips are also available, which implement most of the control and timing functions needed by DRAMs.

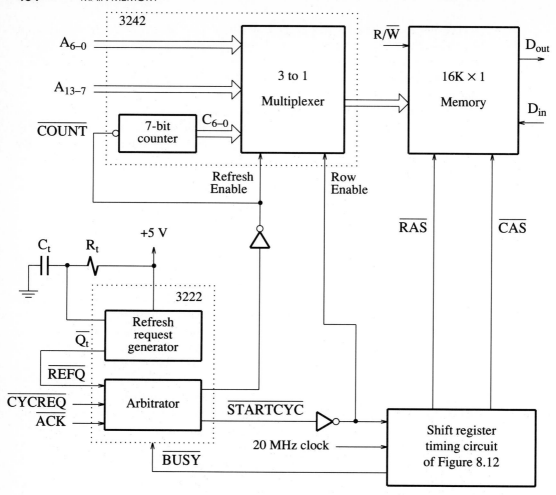

Refresh Enable	Row Enable	Output address
H	–	C_{6-0}
L	H	A_{6-0}
L	L	A_{13-7}

Figure 8.13 Access and refresh circuitry for a 16K-bit dynamic RAM.

8.6 READ ONLY MEMORY (ROM)

There are many applications where fixed programs or data are used. Permanency of such information can be ensured by placing it into read only memories.

A simple ROM cell is shown in Figure 8.14. A bit of information is stored in the cell by either the presence or absence of a connection to ground at point P. A pull-up resistor maintains the bit line in a high-voltage state. A positive signal on the word-select line turns the transistor on and produces a signal near ground level on the bit line, if the indicated connection is made.

Data are stored in the ROM at the time of manufacture by providing the connections at points P as required. ROMs programmed in this manner can be produced economically only in relatively large quantities. However, one often wants to use ROMs in smaller quantities without having to pay an exorbitant price. Special ROM chips are available that can be programmed by the user rather than the manufacturer of the IC. Such ROMs are equipped with a fuse inserted at point P. Then, the user can "program" the memory by burning out the fuses at the desired locations, where a fuse can be blown open by passing a high-current pulse through it. Such memories are called Programmable ROMs (PROM).

8.7 ERASABLE PROM (EPROM)

Programming ROMs and PROMs is an irreversible process. They are suitable for hardware that will not be subject to further alterations. If used in applications where design changes are still being made, they would have to be replaced repeatedly by new ones, at considerable cost. In a development laboratory one can get around this problem by using RAMs in the place of ROMs or PROMs, although considerable inconvenience can be caused by the fact that they may not be pin-compatible and that the contents of a RAM are lost when power is turned off. The problem becomes more difficult when a number of units of the product that is being developed are manufactured and used in field trials. Such units normally use PROMs for storage of

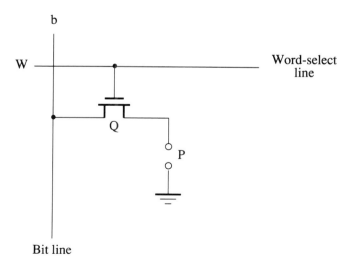

Figure 8.14 A ROM cell.

static software. In this case the only way of making changes in the software is to replace the existing PROMs with updated chips.

The problems encountered during product development may be simplified by using erasable PROMs, called EPROMs. In MOS technology it is possible to construct a transistor that has two gates. One is the regular "select" gate, used to turn the transistor on or off. A second "floating" gate is inserted between the select gate and the substrate. This gate has no electrical connections with the rest of the circuitry, so that any charge placed on it will remain trapped there. The charge can be placed on the floating gate by injection of high energy electrons through the isolating oxide, available when the select gate is energized and a large pulse is applied to the source of the transistor. A charge trapped on the floating gate prevents the transistor from turning on when the select gate is activated. In absence of the trapped charge, the transistor is turned on when its select gate is energized. Thus, the device can be used to store one bit of information.

An example of an EPROM cell is shown in Figure 8.15. A positive signal on the word select line turns the transistor on if the floating gate has no trapped charge. This is interpreted as a 1 by the sense circuit connected to the bit line. On the other hand, a charged floating gate prevents the transistor from turning on, thus resulting in a 0 being read.

The usefulness of EPROMs derives from the ability to erase their contents when desired. This involves the removal of the charge trapped on the floating gate, which is accomplished by exposing the chip to ultraviolet light. EPROM chips are mounted in packages specially equipped with transparent windows for this purpose. Typically, light with a wavelength of 2537 Angstroms is applied for several minutes. The actual time needed depends upon the power rating of the lamp used and the type of IC.

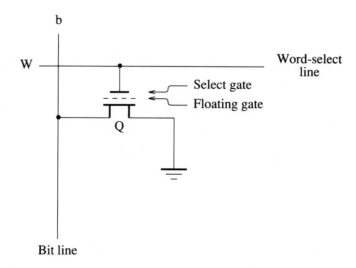

Figure 8.15 An EPROM cell.

Erasure of an EPROM chip results in the removal of trapped charges in the floating gates of all cells. Hence, the erased chip has all bits set to 1. Data are written into the memory by selectively programming the 0 bits, as explained above. Note that selective erasure of single bits is not possible in EPROMs.

There exists another version of erasable PROMS. These memory chips, known as EEPROMs (or E^{2}PROMs), are both programmed and erased electrically. EEPROMs are particularly useful in applications where it is desirable to make occasional changes in the stored data, but it is inconvenient to have to remove the chips from the circuit for reprogramming. For example, many video terminals offer a wide choice of features, where the selection of particular features to be used is made by the user. A possible mechanism for this selection is to provide a set of manual switches that can be set by the user. A more attractive alternative is to use an EEPROM to store the chosen information. In this case, the user makes a selection by typing on the keyboard, as a part of an initialization process, and follows it with a command that results in writing this data into the EEPROM.

A disadvantage of EEPROMs is that they require different voltages for erasing, writing, and reading of stored data, which complicates the power supply requirements.

8.8 EXAMPLE

In Sections 8.4 and 8.5 we discussed DRAMs and the control circuitry needed for these memories. It will now be helpful to consider a specific example of how a DRAM may be connected to a microprocessor. We will choose the 68000 microprocessor, which requires the bus signaling convention described in Section 6.10. Our choice of a DRAM is the 4464 chip, which has 64K addressable locations of four bits each, for a total of 256K bits.

The example presented in this section includes many details that a hardware designer of a microprocessor system should consider. Readers who are not interested in such details may skip this section without loss of continuity.

Table 8.1 lists the pins of the 4464 chip. There are eight address inputs, A0 to A7, which are time multiplexed to accept the 16-bit address needed to select one of the 64K locations. The four data pins are labeled DQ1 to DQ4. They are combined input and output pins that can be connected directly to the data lines of a microprocessor bus. Two pins are provided as inputs for the $\overline{RAS}$ and $\overline{CAS}$ signals. The chip is selected by activating these signals. No separate "chip select" pin is provided. The read/write function is specified by means of the Write Enable pin, $\overline{W}$, indicating a read operation when the signal is high and a write operation when low. There exists one more control pin, called Output Enable, $\overline{G}$, which is used to control the tri-state output buffers of pins DQ1 to DQ4. These buffers are forced into the high-impedance state when either $\overline{CAS}$ or $\overline{G}$ is high. Low signals on both $\overline{CAS}$ and $\overline{G}$ maintain the output buffers in the low-impedance state, in order to place data read from the memory on the bus. Normal read/write operations can be controlled by the $\overline{CAS}$ signal only, providing that $\overline{G}$ is kept low. In our example we will ensure that $\overline{G}$ is low by tying this pin to ground. We should mention that $\overline{G}$ pin is provided to

Table 8.1 PIN DESCRIPTIONS FOR THE 4464 DRAM

Name	Function
A0-A7	Address inputs.
DQ1-DQ4	Data input/output pins.
$\overline{RAS}$	Row address strobe.
$\overline{CAS}$	Column address strobe.
$\overline{W}$	Write enable; denotes a write cycle when low and read cycle when high.
$\overline{G}$	Output enable; enables output buffers when low.
V_{DD}	+5 volt power supply.
V_{SS}	Ground.

facilitate implementation of read-modify-write cycles. This mode of memory access is used in multiprocessor systems, as will be discussed in Section 10.3. Finally, the 4464 chip has two power supply pins that must be connected to +5 V and ground.

As an example of the requirements of the 4464 DRAM, a timing diagram for a read cycle is shown in Figure 8.16. This figure depicts the relationship between various signals in a more detailed format than Figure 8.11, giving most of the information that one finds in manufacturers' data sheets. Specific values for the parameters indicated are given in Table 8.2. Minimum and maximum values are specified as appropriate. Note that for most parameters only the minimum value is of interest, hence it is the only value given. There are several versions of the 4464 DRAM available, differing only in their speed of operation. Table 8.2 shows the data for two such chips—the TMS4464-10 and TMS4464-20 [5], which have row access times of 100 ns and 200 ns, and read or write cycle times of 200 ns and 300 ns, respectively. Their maximum refresh period is 4 ms.

Let us consider a possible implementation of a 256K-byte memory connected to a 68000 microprocessor, using eight 4464 chips. The access circuitry can be provided using controller chips that contain the type of circuits described in Figure 8.13. We will use a TMS4500 chip [1], which performs most of the functions required.

The TMS4500 can be used to control DRAM chips that have 8K, 16K, 32K, or 64K addressable locations. It uses an externally provided clock, that may be the microprocessor clock. Refresh requests can be generated either externally or internally. Internal requests are initiated at a rate determined by dividing the clock frequency by a constant factor. The choice between external and internal refresh requests, as well as the refresh rate, are determined by the state of three strap pins that are connected to logic 0 or 1 as desired.

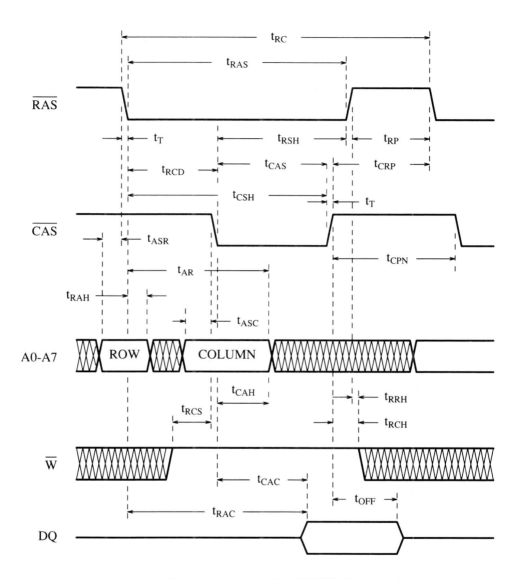

Figure 8.16 Timing diagram for the 4464 DRAM read cycle.

The TMS4500 is implemented as a 40-pin package. The name and function of various pins are shown in Table 8.3. There are 16 input address pins, RA0–RA7 and CA0–CA7, which define 8-bit row and column addresses that are passed on to the DRAM chips in time-multiplexed fashion, via outputs MA0–MA7. The chip can generate two output $\overline{\text{RAS}}$ signals, called $\overline{\text{RAS0}}$ and $\overline{\text{RAS1}}$, allowing two banks of DRAMs to be used. The selection of $\overline{\text{RAS0}}$ or $\overline{\text{RAS1}}$ depends upon the state of the REN1 pin. The output $\overline{\text{CAS}}$ signal is activated only if enabled by an active access control signal, either $\overline{\text{ACR}}$ or $\overline{\text{ACW}}$. The TMS4500 contains latches for storing

Table 8.2 PARAMETERS FOR THE READ CYCLE OF THE TMS4464 DRAM

Parameter		TMS4464-10		TMS4464-20	
Symbol	Description	Min ns	Max ns	Min ns	Max ns
t_{RC}	Read cycle time.	200		300	
t_{CPN}	Pulse duration, $\overline{CAS}$ high.	40		80	
t_{CAS}	Pulse duration, $\overline{CAS}$ low.	60	10,000	120	10,000
t_{RP}	Pulse duration, $\overline{RAS}$ high.	90		120	
t_{RAS}	Pulse duration, $\overline{RAS}$ low.	100	10,000	200	10,000
t_T	Transition time.	3	50	3	50
t_{ASC}	Column address setup time.	0		0	
t_{ASR}	Row address setup time.	0		0	
t_{RCS}	Read command setup time.	0		0	
t_{RAH}	Row address hold time.	15		20	
t_{CAH}	Column address hold time after $\overline{CAS}$ low.	20		45	
t_{AR}	Column address hold time after $\overline{RAS}$ low.	60		125	
t_{RCH}	Read command hold time after $\overline{CAS}$ high.	0		0	
t_{RRH}	Read command hold time after $\overline{RAS}$ high.	10		15	
t_{CSH}	Delay time, $\overline{RAS}$ low to $\overline{CAS}$ high.	100		200	
t_{CRP}	Delay time, $\overline{CAS}$ high to $\overline{RAS}$ low.	0		0	
t_{RSH}	Delay time, $\overline{CAS}$ low to $\overline{RAS}$ high.	60		120	
t_{RCD}	Delay time, $\overline{RAS}$ low to $\overline{CAS}$ low (maximum value is specified only to guarantee access time).	25	40	30	80
t_{CAC}	Access time from $\overline{CAS}$.		60		120
t_{RAC}	Access time from $\overline{RAS}$.		100		200
t_{OFF}	Output disable time after $\overline{CAS}$ high.	0	30	0	35

addresses, $\overline{CS}$ and REN1 inputs. The information is latched at the negative edge of the address latch enable signal, ALE. Another output, RDY, is provided to facilitate

Table 8.3   PIN DESCRIPTIONS FOR TMS4500

Name	Direction	Description
RA0-7	Input	Row address.
CA0-7	Input	Column address.
MA0-7	Output	Memory address outputs; drive address pins of DRAM.
$\overline{CS}$	Input	Chip select.
REN1	Input	RAS enable, enables $\overline{RAS0}$ when low and $\overline{RAS1}$ when high.
ALE	Input	Address latch enable; latches RA0–7, CA0–7, $\overline{CS}$ and REN1 inputs.
$\overline{ACR}, \overline{ACW}$	Input	Access control for read and write. A low on either pin enables generation of $\overline{CAS}$. A low on both pins forces MA0-7, $\overline{RAS0}$, $\overline{RAS1}$ and $\overline{CAS}$ into high-impedance state.
CLK	Input	System clock.
$\overline{REFREQ}$	Input/Output	External refresh request as input. As output it indicates an internal refresh request.
$\overline{RAS0}$	Output	Row address strobe for memory bank 0.
$\overline{RAS1}$	Output	Row address strobe for memory bank 1.
$\overline{CAS}$	Output	Column address strobe.
RDY	Output	Ready signal for synchronizing slow memories that cannot guarantee microprocessor access time requirements.
TWST,FS0,FS1	Input	Strap inputs that determine the mode of operation and the internal refresh rate.

the use of DRAMs that are slower than expected by the microprocessor. When inactive (low), this output indicates that the memory is not ready. A memory access takes three or four clock cycles, depending on the mode of operation of the controller chip. One additional "wait" state (lasting one clock cycle) can be inserted, during which the RDY signal is low. When the refresh requests are generated internally, it is possible that a refresh cycle is in progress when the microprocessor requests a read or a write cycle. In this case, the refresh cycle proceeds to completion, which is indicated to the microprocessor by a low RDY signal, before the read/write access is performed. The mode of operation of the TMS4500 is selected by the strap inputs TWST (Timing/Wait Strap), FS0 and FS1 (Frequency Select 0 and 1).

A timing diagram for the controller access cycle is given in Figure 8.17. The values of the parameters shown correspond to TMS4500A-15, which is one of several versions of this chip. All values are stated in nanoseconds.

In order to implement the desired 256K-byte memory, we will connect the 4464 DRAM chips into two banks of four chips each. Each bank, consisting of 64K 16-bit words (128K bytes), will be accessed by a separate $\overline{RAS}$ signal.

There are a number of ways of designing the access circuitry to meet the timing requirements of the 68000, TMS4500, and 4464 chips. A possible arrangement is given in Figure 8.18. It shows the connection of all pins relevant to the operation of the memory, except for power supply and the strap inputs of TMS4500. The latter are assumed to be connected to select the internal refresh requests at the rate that meets the maximum refresh period of the 4464 DRAM.

Addressing information is latched into the controller chip by a negative transition of the 68000 address strobe signal, $\overline{AS}$. The same signal is also used to activate

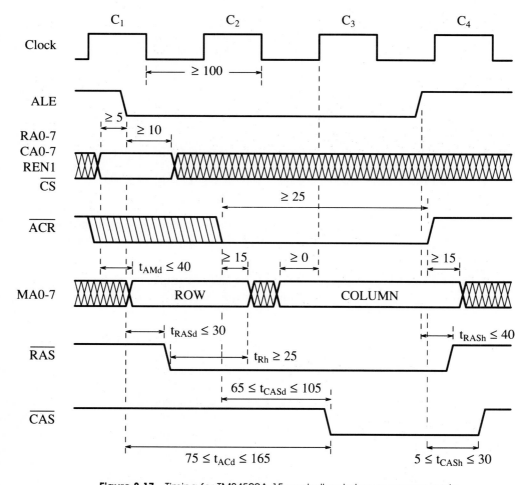

Figure 8.17 Timing for TMS4500A-15 controller during an access cycle.

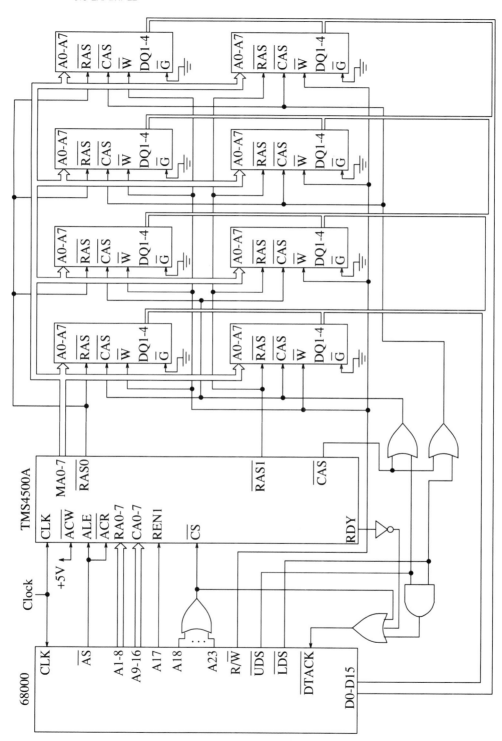

Figure 8.18 A dynamic memory connected to a 68000 microprocessor.

the read access control input, $\overline{ACR}$. The other access control input, $\overline{ACW}$, is not used because only one of these inputs is needed to enable $\overline{CAS}$ and in the case of the 68000 microprocessor both read and write cycles are started by an active $\overline{AS}$ signal. Address line A17 is connected to REN1 and, thus, is used to select a memory bank. When low, it selects the upper bank by means of $\overline{RAS0}$, and when high, $\overline{RAS1}$ enables the lower bank. Note that the memory is selected when address lines A18 to A23 are all zero.

As explained in Section 6.10, the 68000 microprocessor uses separate data strobe timing signals for the upper and lower bytes of a 16-bit word. Our memory is organized such that an active $\overline{UDS}$ (Upper Data Strobe) signal selects the four DRAM chips on the left, which correspond to the high-order byte. Similarly, the low-order byte is selected by $\overline{LDS}$ (Lower Data Strobe). This is accomplished by gating the $\overline{UDS}$ and $\overline{LDS}$ signals with $\overline{CAS}$ which is the signal that enables data transfers to or from a DRAM chip.

The remaining part of the access circuitry is devoted to the generation of the data acknowledge signal, $\overline{DTACK}$. From Figure 6.56, it is seen that during a read operation the 68000 expects valid data to be present on the data lines no later than t_{dd} after $\overline{DTACK}$ is activated. However, if $\overline{DTACK}$ is activated early in the read cycle, the microprocessor will not accept the data from the data lines of the bus until the end of state S6, which occurs during the fourth clock pulse. Our example is based on the assumption that the DRAM is fast enough to produce the data within this period, thus $\overline{DTACK}$ is activated as soon as either $\overline{UDS}$ or $\overline{LDS}$ is observed to be low. There is one case where $\overline{DTACK}$ should not be generated early. This happens when a read request occurs during a refresh cycle. While the refresh cycle is in progress, the controller places the RDY output into a low state, which in turn keeps $\overline{DTACK}$ high.

We should now check that the proposed circuit meets the timing requirements of the chips used. We will consider only the read cycle parameters in this example. Let us assume a 6-MHz clock and a corresponding 68000 microprocessor, known as 68000G6. The read cycle timing for this chip is as presented in Figure 6.53, except that some of the delays are longer than those shown in the figure. The parameters for the TMS4500A-15 DRAM controller are shown in Figure 8.17, and the DRAM used is TMS4464-10 as specified in Table 8.2.

A number of timing constraints must be met. For example, the address setup and hold times of the memory chip must be satisfied by the controller circuit, and the memory must respond to read requests within the time expected by the microprocessor. These constraints must be met even when the microprocessor issues a read request while a memory refresh cycle is in progress.

First, let us consider the row address setup and hold times for the DRAM. The minimum row address setup time depends upon the length of time during which the address is valid before the address strobe, $\overline{AS}$, becomes active. The address strobe delay, t_{as} in Figure 6.53, is guaranteed not to be less than 35 ns for the 68000G6 chip. The controller delays of interest are found in Figure 8.17. The time needed by the controller to generate a valid row address and present it to the memory, t_{AMd}, may be as long as 40 ns. This reduces the available row address setup time.

However, the desired setup time is increased by t_{RASd}, the time by which $\overline{RAS}$ is delayed with respect to the arrival of the address strobe (recall from Figure 8.18 that the $\overline{AS}$ bus line is connected to the ALE input of the controller, which determines the timing of the $\overline{RAS}$ signal). The t_{RASd} delay in Figure 8.17 has only its maximum value specified. Yet, when considering the address setup time, we are interested in its minimum value. When this value is not stated explicitly in manufacturers' data sheets, it is reasonable to assume that the minimum value is one third of the maximum given, as already discussed in Subsection 6.2.2. Hence, we will use $t_{RASd(min)} = 10$ ns. Therefore, the minimum row setup time is

$$
\begin{aligned}
t_{ASR(min)} &= t_{as(min)} - t_{AMd(max)} + t_{RASd(min)} \\
&= 35 - 40 + 10 \\
&= 5 \text{ ns}
\end{aligned}
$$

This satisfies the requirement $t_{ASR} \geq 0$, given in Table 8.2.

The row address hold time guaranteed by the controller is

$$t_{Rh} \geq 25 \text{ ns,}$$

which exceeds the minimum value of $t_{RAH} = 15$ ns. Note that the setup and hold times for the input address to the TMS4500A-15, which require minimum values of 5 and 10 ns, respectively, are easily met by the corresponding delays in the microprocessor.

Next, let us check if the data read from the DRAM will be placed on the data bus early enough. The 68000G6 expects data to be ready not later than $t_{dd} = 120$ ns after $\overline{DTACK}$ is asserted. However, as already mentioned above, during normal read cycles it will not accept the data from the bus until the end of the fourth clock pulse in the read cycle, even if $\overline{DTACK}$ is asserted early as we have done in this example. The DRAM responds with valid data within $t_{RAC} = 100$ ns of $\overline{RAS}$ or $t_{CAC} = 60$ ns of $\overline{CAS}$ going low, meeting the more stringent of these two requirements. From Figure 8.17, it is apparent that when ALE and $\overline{ACR}$ go low at the same time, the activation of $\overline{CAS}$ may be delayed by as much as

$$t_{CASd} - t_{RASd} = 105 - 30 = 75 \text{ ns}$$

from the time $\overline{RAS}$ becomes active. Since the activation of $\overline{CAS}$ governs the response of the DRAM, valid data will be placed on the bus no later than

$$t_{CASd} + t_{gate} + t_{CAC} = 105 + 15 + 60 = 180 \text{ ns}$$

after the ALE and $\overline{ACR}$ signals go low. We have included 15 ns for the delay of the OR gate through which $\overline{CAS}$ must propagate. Since the microprocessor does not read the data lines until more than two clock cycles after asserting the address strobe, $\overline{AS}$, which drives the ALE and $\overline{ACR}$ inputs, it is obvious that with 166 ns per clock cycle there is no difficulty in producing the data from the DRAM in plenty of time. Similar arguments apply to the evaluation of the write cycle.

Another timing requirement arises when a read (or write) request arrives while a refresh cycle is in progress. In this case, an active RDY signal will hold $\overline{DTACK}$ high until the refresh task is completed. A timing diagram illustrating this situation is given in Figure 8.19. Note that in addition to the four clock cycles needed for normal access, three "wait" clock cycles are inserted to account for the refresh. This means that data will not be valid when the microprocessor reaches its fourth clock cycle. Now, the timing for the data bus should be considered carefully. Figure 8.20 shows the requirements for the data bus, $\overline{\text{DTACK}}$, and clock signals, with delay values corresponding to the 68000G6 microprocessor. When $\overline{\text{DTACK}}$ goes low at least $t_q = 25$ ns before a negative edge of the clock, the microprocessor will accept the data from the bus at the following negative edge, and expects a data setup time, t_{dsr}, of at least 25 ns. An alternative constraint is that the data delay time from an active $\overline{\text{DTACK}}$ must be less than $t_{dd} = 120$ ns. Since the DRAM timing is known relative to the clock, it is more convenient to work with the timing requirements given in Figure 8.20.

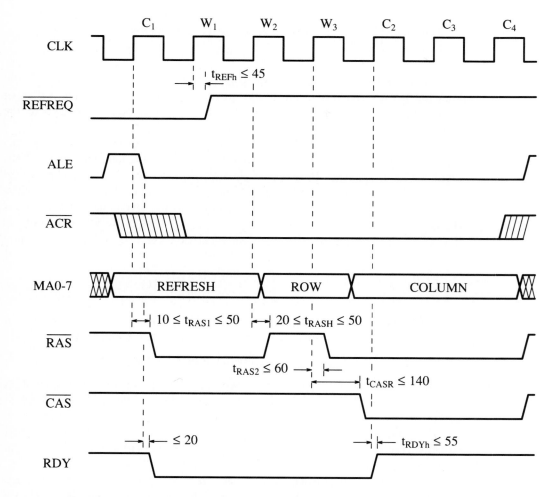

Figure 8.19 Timing for TMS4500A-15 controller during a refresh followed by a read cycle.

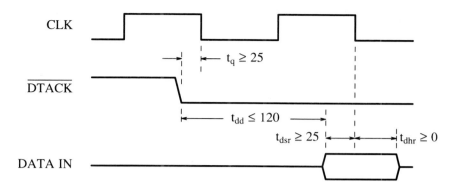

Figure 8.20 Data input timing for the MC68000G6 microprocessor.

From Figure 8.19 it is seen that RDY is asserted within t_{RDYh} = 55 ns of the positive edge of the C_2 clock pulse. This signal passes through an inverter and an OR gate, which add a maximum delay of 30 ns. Thus, $\overline{DTACK}$ is activated within 85 ns of the positive edge of C_2. Since this is just the maximum value for this delay, $\overline{DTACK}$ may be activated sooner. The negative edge of C_2 occurs 166/2 = 83 ns after its positive edge. Thus, $\overline{DTACK}$ may become active either before or after the required t_q = 25 ns of the negative edge of C_2. If it occurs before, the microprocessor will read the data from the bus at the negative edge of C_3. Otherwise, it will wait until C_4. Therefore, we should make sure that the data are valid, with sufficient setup time, at the negative edge of C_3. The controller activates $\overline{CAS}$ within t_{CASR} = 140 ns of the rising edge of the W_3 clock pulse. Thus, the DRAM will put valid data on the bus within

$$t_{CASR} + t_{gate} + t_{CAC} = 140 + 15 + 60 = 215 \text{ ns}$$

of the rising edge of W_3. This corresponds to 215 − 166 = 49 ns past the rising edge of C_2, meeting easily the requirements in Figure 8.20 for data to be read by the microprocessor at the end of the C_3 pulse.

Finally, it is important to verify that the minimum pulse width of the $\overline{RAS}$ signal in the high state is met. This parameter has an important function, known as "precharging," in preparing the DRAM chips for an access or refresh cycle. From Table 8.2, we see that this time, t_{RP}, must be at least 90 ns. It can be determined as

$$t_{RP(min)} = t_{SH(min)} + t_{RASd(min)} - t_{RASh(max)}$$

where t_{SH} is the time that the $\overline{AS}$ signal is positive, which is 180 ns for 68000G6. Taking the other values from Figure 8.17, we have

$$t_{RAS} = 180 + 10 - 40 = 150 \text{ ns}$$

which exceeds the required minimum.

All of our calculations have shown that the main timing requirements for a read operation are met easily. Obviously, the same consideration must be given to write and refresh cycles.

In this example we used a fast memory that allowed us to comfortably meet timing requirements of the chosen microprocessor. For cost reasons one may use slower chips. In other situations one may want to operate a microprocessor system at the highest possible speed. In all cases it is essential to ensure that all timing aspects are evaluated properly.

8.9 CACHE MEMORY

Performance of a microprocessor system is heavily dependent on the time needed to access information stored in its memory units. An electronic main memory is used, because when random locations are accessed it is much faster than secondary storage devices such as magnetic disks. Yet, there is a need for magnetic disks and tapes, because their cost per bit is much lower. In a typical application, programs and data are kept in the secondary storage and brought into the main memory at execution time. The microprocessor spends most of its time operating on the information available in the main memory, and refers to the slow secondary storage only infrequently.

While the main memory is very fast compared to secondary storage, it can still be slow with respect to the speed at which a microprocessor can operate. A substantial improvement can be achieved if an even faster memory unit, called a *cache,* is inserted between the main memory and the microprocessor, as shown in Figure 8.21. The faster the memory the more expensive it tends to be. Hence, the cache is likely to be much smaller than the main memory. The cache is often implemented as a physically distinct unit. But, the VLSI technology has advanced to the point where it is possible to incorporate a cache within the microprocessor chip itself, which reduces the time needed to access the information in the cache.

8.9.1 Locality of Reference

The microprocessor uses the cache as if it were the main memory, expecting to find the desired instructions and data there. If this information is not present in the cache, it must be brought there from the main memory. This suggests that the entire scheme will be effective only if most of the time the required information is found in the cache. Indeed, this is the case, for the following reason. Computer programs are written in a way that results in object code where instructions are most often executed in consecutive order. Moreover, loops consisting of several instructions are encountered frequently. A loop is usually executed several times, hence the microprocessor is likely to spend considerable time executing instructions in the loop, before it proceeds to other instructions. Now, if the instructions that constitute a loop are available in the cache, then their repeated execution can be achieved at a faster rate in view of the short access time of the cache. This phenomenon, that the

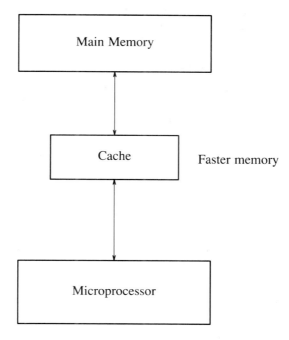

Figure 8.21 Use of a cache memory.

execution of a typical program involves frequent accessing of memory locations whose addresses are close together, is known as *locality of reference*.

There is a time overhead associated with transferring instructions from the main memory to the cache. The first time a given instruction is needed by the microprocessor, it may not be found in the cache. It has to be fetched from the main memory and written into the cache, as well as being forwarded to the microprocessor. Thus, the first time an instruction loop is executed, the time required may be the same as when the cache is not there. But, during the second and subsequent times through the loop, the execution will be much faster as the instructions can be read directly from the cache.

Another useful possibility arises from the observation that instructions are most often executed in the order in which they are stored in the memory. When a particular instruction is fetched from the main memory, it is beneficial to copy a block of words that contain the word being addressed from the memory into the cache. This action increases the likelihood that subsequent requests for instructions generated by the microprocessor will refer to information already in the cache.

The property of locality of reference is applicable mainly to memory references involving instructions. References to data are far less likely to involve contiguous memory locations or the same address repeatedly. For this reason, a cache used in a microprocessor system is often limited to storing instructions. Access requests for instruction operands are sent directly to the main memory in the usual manner.

Many modern microprocessors, such as the 68020 and the 68030, incorporate an on-chip instruction cache. The 68030 has an on-chip data cache as well.

8.9.2 Cache Structure and Accessing

Suppose that the main memory is partitioned into a number of blocks of equal size. A particular program that is being executed may occupy many blocks, exceeding the capacity of the cache. Blocks are transferred into the cache as required. If the cache is full, one of the old blocks must be overwritten. This implies that the cache should have labeling information associated with its entries, indicating which main memory blocks are currently available. The location of a new block in the cache and the choice of a block to be removed can be decided in different ways. Let us consider the scheme used with the 68020 microprocessor. The 68020 chip includes an instruction cache of 64 long words (256 bytes). Its organization is shown in Figure 8.22.

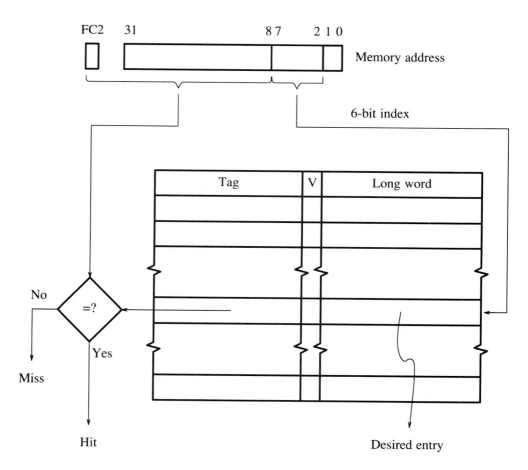

Figure 8.22   Instruction cache organization in the 68020 microprocessor.

A 6-bit index, consisting of address bits b_{7-2}, is used to select one of the 64 long words. (Note that bits b_{1-0} identify a byte within a long word.) The high-order 24 bits of the address and the function-code bit FC_2 form a *tag* field, which is used to identify an entry in the cache. One other bit, called the *valid* bit, V, is used to indicate whether the cache entry is valid. Since the low-order 8 bits of an address are used to identify a byte within a 256-byte block, a *direct mapping* is established between a 32-bit memory address, A, and the location A mod 256 in the cache. Thus, long words at memory addresses 0, 256, 512,...can be brought only into cache location 0, while long words at addresses 4, 260, 516,...can be only in cache location 4, and so on.

When a new long word is brought into the cache, the valid bit is set and the tag portion of the memory address is stored in the tag field of the corresponding entry. Later, when the microprocessor fetches an instruction by issuing its memory address, the tag portions of the presented address and the selected entry in the cache are compared. If they are the same and the valid bit is set, a *hit* is said to have occurred and the desired instruction is immediately available from the cache. Otherwise, in the case of a *miss*, the required instruction is fetched from the main memory and copied into the cache. One of the bits used in the tag comparison is the FC_2 bit. It is a part of a 3-bit function code, presented in Table 6.5, providing a distinction between the User and Supervisor modes in the operation of the 68020 microprocessor. Hence, an instruction loaded into the cache in the Supervisor mode cannot be used in the User mode, and vice versa. This is necessary because the FC_2 bit may be used in a 68020 system to select two physically distinct memories, one containing the user data and programs and the other containing the supervisor data and programs. Each of these memories may be addressed using all of the 32 address lines. Therefore, a single 32-bit address may actually refer to different physical memory locations, depending on whether the system design utilizes the function code signal in the memory selection hardware.

The main purpose of the valid bit, V, is to indicate whether the contents of a particular word in the cache match the contents of the corresponding location in the main memory, as determined by the tag bits. Clearly, this is not the case when power is turned on or when a new program is brought into the main memory from the secondary storage. Hence, the valid bits of the cache are cleared when the microprocessor is reset. It is also possible to clear the valid bits under software control.

A small variation in the cache structure is seen in the 68030 microprocessor [6]. The 68030 contains an instruction cache and a data cache, each capable of holding 256 bytes of information organized in a long-word format. The structure of both caches is similar to that used in the 68020 microprocessor. However, instead of viewing the cache as 64 independent blocks of 4 bytes (1 long word per block), the 68030 cache is viewed as 16 blocks of 16 bytes (four long words per block). This has the advantage of requiring fewer tags to be held in the cache, because a single tag is associated with all four words of a block.

Figure 8.23 shows the organization of the 68030 data cache. Note that only 16 tag locations are needed. A particular block location is selected by memory address

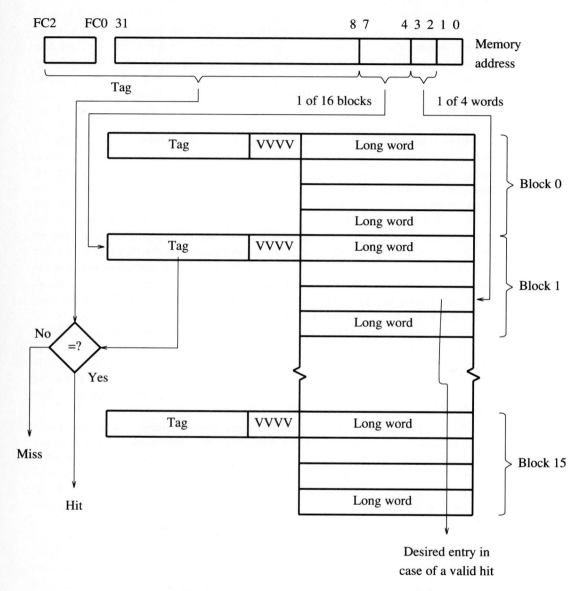

Figure 8.23 Data cache organization in 68030 microprocessor.

bits 4 to 7. One of the four long words in the chosen block is specified by address bits 2 and 3, while bits 0 and 1 identify the desired byte in a long word. The tag bits for this cache consist of the 24 high-order address bits and the function code bits FC_0 to FC_2. The inclusion of all three function code bits in a tag is the only structural difference between the data and instruction caches of the 68030. The instruction cache uses only the FC_2 bit as a part of the tag field, as was the case in the 68020 example in Figure 8.22.

The tag field in Figure 8.23 is followed by four valid bits. Each of these bits corresponds to one of the four long words of the associated block.

There exists one difference in the way the instruction and data caches are handled. The procedure for reading information from either cache is the same, but the writing process is not. It is never necessary to change the contents of the instruction cache, except by bringing new instructions from the main memory when a miss occurs. In contrast, the data cache entries may be altered during the execution of a program, where the result of an operation has to be written into a memory location that has its image resident in the cache at that time. It is possible to write the new information in the cache only, leaving the main memory unaffected. Later in the program, the block containing the altered data may have to be overwritten by new information from the main memory. At that time the altered block must first be stored in the main memory before it is overwritten by new data. An alternative approach is to arrange for the new data to be written both into the cache and in the main memory at the same time. Thus, the cache and the main memory are guaranteed to have identical information at all times. When it becomes necessary to load a new block into the cache, the data already in the cache may be safely discarded. This is called a *write through* approach. It is the approach used with the 68030 data cache.

Partitioning the cache into blocks of four long words has the advantage of reducing the number of tags, as mentioned above. But, it has another potentially more significant advantage. When a miss occurs, it is necessary to bring into the cache four new long words, rather than one. These long words are found in consecutive locations in the main memory. Hence, they can be read from the main memory using the burst mode described in Section 6.10.3. This mode enables data to be transferred at the highest possible rate. If the main memory is sufficiently fast, the four long words can be transferred in four successive clock cycles.

8.9.3 Alternative Mapping Schemes

The approach illustrated in Figure 8.23 uses one tag per block containing four long words. It still implies a direct mapping between the main memory and the cache, as was the case in the example of Figure 8.22. The direct mapping scheme is restrictive. For example, two blocks in the main memory whose addresses differ by 256 map onto the same cache block. A program that requires instructions or data from these two blocks will result in each block being continuously overwritten, even when the rest of the cache is empty. The difficulty can be avoided if the mapping scheme is altered so that both memory blocks are not necessarily mapped onto the same cache location. One way of achieving this is to provide space for two blocks in each block location in the cache. For example, suppose that the cache in Figure 8.23 is arranged so that it consists of eight *sets* of two blocks each, rather than 16 sets of one block. Let a given memory address, A, map into the cache address A modulo 128. Since there is space for two blocks in each set, the desired block location may be in either of the two halves of the set. A separate tag field, containing memory address bits 7 to 31, is needed in conjunction with each block location. Thus, when a reference to a particular set in the cache is made, the cache access circuitry checks both tags to determine which of the two blocks contains the desired information. When a miss occurs, the new block is brought in from the main memory to replace one of the two blocks in the set. Ideally, the block to be overwritten should be the one least likely to be needed again. The cache control circuitry should use a replacement strategy that attempts to achieve this goal. For example, a strategy that has been shown to be very effective is to replace the block that has been least recently used. To do so, the cache control circuitry needs to keep track of cache access history. A simpler, and quite effective strategy is to replace one of the blocks at random.

The type of cache organization where several blocks are associated with each set is said to use the *block-set-associative mapping* scheme. It is possible to carry the idea of using more than one block per set to an extreme where all blocks in the cache are considered to be a part of one set. This is known as the *associative mapping* scheme, where the location of a desired block can be determined by examining all tags in the cache. In this case, the tag for each block comprises all the address bits needed to specify the location of the block in the main memory. For example, if each block consists of 4 bytes, as in the cache of Figure 8.22, the tags must include the most-significant 30 bits of the memory address. However, if each block has 16 bytes, as in Figure 8.23, the tags must include only the most-significant 28 bits of the memory address. It is apparent that the associative mapping scheme is not very useful if the tags have to be examined one at a time. Its practical value stems from the fact that it is possible to design circuits that allow comparison of the tag field of a given memory address with all tags in the cache at the same time. An interesting use of such an "associative" cache will be discussed in Section 8.10.

It is apparent that the direct mapping scheme, while being restrictive, makes the cache easy to implement. The associative mapping scheme is very flexible, but costly to realize. A compromise between these two extremes is the block-set-associative scheme, which can be found in many commercial computers.

8.9.4 Performance

Using a cache that is faster than the main memory is the key to increasing the performance of a microprocessor system. It is particularly attractive to place a cache on a microprocessor chip, because accesses to memory locations within such a chip take less time than those that involve the external memory with the associated bus transfers. For example, the 68020 microprocessor uses three clock cycles for an external read bus cycle, but it needs only two clock cycles to access an instruction from the cache.

The effectiveness of a cache depends upon the likelihood that the desired information will be found in the cache. The percentage of memory accesses where a hit occurs is often referred to as the *hit ratio* or *hit rate*. It depends on the size of the cache and on the type of the program being executed. A quantitative assessment of the improvement achievable with a cache can be obtained by measuring the execution time of typical programs. One study of the 68020 microprocessor has shown that the performance gain resulting from the instruction cache is about 30% [7].

We have considered the concept of a cache memory and its implementation in Motorola microprocessors. Cache memories have become a standard way of improving the performance of computers. We have seen that they can be organized in a number of ways. Detailed discussion of the possible schemes can be found in several books on computer organization [8–11].

While a cache improves the performance of a microprocessor system by making the main memory appear faster than it actually is, it has little impact on the writing of software for such systems. In fact, the cache is essentially transparent to the programmer. In the next section we will discuss the virtual memory concept, which exploits some ideas presented in conjunction with the cache memory to make the main memory appear to the programmer to be larger than its actual physical size. The virtual memory provides considerable flexibility in the development of software, particularly in multiuser environments.

8.10 VIRTUAL MEMORY

Many microprocessor applications require execution of large programs. Ideally, a complete program is loaded in the main memory of a microcomputer at the time of execution. The size of such a program is constrained by two factors: the size of the addressable space and the size of the main memory. The addressable space can be very large. We have seen that the 68000 and 68020 microprocessors can address 16M ($=2^{24}$) and 4G ($=2^{32}$) locations, respectively. A main memory capable of holding this many bytes is expensive to implement. For economical reasons, typical microcomputers are equipped with a smaller main memory. However, it is possible to make the memory appear to the programmer larger than its physical size. This "virtual memory" concept is discussed in this section.

From a programmer's point of view it is convenient to have the main memory appear equal to the addressable space of the microprocessor. This can be accomplished even if the physical memory provided in a microcomputer is much smaller. The basic idea is quite simple. A program resides in the secondary storage, typically on a magnetic disk. If it exceeds the size of the main memory, it can be executed by bringing smaller parts of it into the memory at execution time. The transfer between the disk and the main memory is done automatically, requiring no intervention by the programmer. The addressable space may be regarded as a *virtual memory* that is divided into sections, called the *pages,* of fixed size. The physical main memory can hold a certain number of such pages, depending on the sizes of both the memory and the pages. A particular program may occupy many pages, which are brought into the main memory as required during the process of execution. Because a page is loaded into the main memory only when needed, the term *demand paging* is often used to describe such a scheme.

When the contents of a virtual memory page are read from the disk, they may be loaded into any page location in the physical memory. Figure 8.24 illustrates a possible situation. Suppose that the virtual memory consists of 2^{32} bytes, as is the case

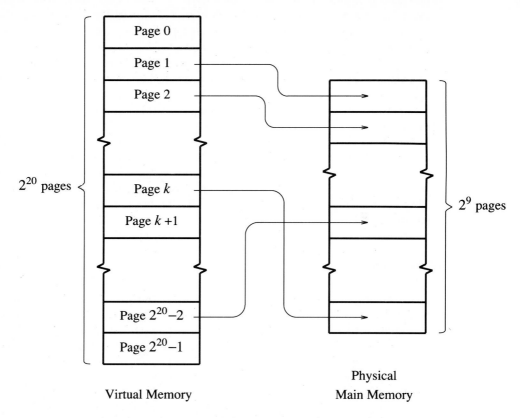

Figure 8.24 Page arrangement in a virtual memory system.

for the 68020 microprocessor. Assume that the size of each page is 4K (=2^{12}) bytes. Thus, there are more than a million pages, 2^{20} to be exact, in the virtual memory. Let the physical main memory be 2M bytes (=2^{21}) in size. It can hold 512 (=2^9) pages. The possible locations in the main memory that can hold a page are often called *page frames*. Since a particular page can be brought into any page frame, it is necessary to devise an address translation mechanism, which can be used to translate a *virtual address* specified by the program into a *physical address* that indicates where the requested information is to be found in the main memory.

When the microprocessor wishes to access a certain location in its addressable space, there arise two possible cases. If the page that contains this location happens to reside in the main memory, then the access is made in the usual way, after address translation. However, if the desired page is not in the main memory, it must first be brought there from the disk.

Before considering the details of the virtual memory scheme, we should question its conceptual workability. It is apparent that the amount of time spent on transferring pages between the disk and the physical memory is of great significance. The frequency of page transfers should be low. This is indeed the case, because the locality of reference phenomenon, explained in Section 8.9, suggests that much of the execution is likely to involve instructions that are already in the physical memory. We should also note that transferring an entire page can be done efficiently using the DMA techniques described in Chapter 7.

The operation of a virtual memory may be described as follows. The virtual addresses issued by the microprocessor are interpreted by a special unit, usually called the *Memory Management Unit* (MMU), as shown in Figure 8.25. The MMU incorporates an *address translation table.* For each page that is currently resident in the main memory, the table gives the corresponding physical address. Thus, if a virtual address received from the microprocessor involves a page that is already in the physical memory, the MMU generates a corresponding physical address and sends it to the memory. The data are transferred directly between the microprocessor and the physical memory. If the virtual address involves a page that is not in the physical memory, then a *page fault* is said to have occurred. The page that contains the desired location must be transferred from the disk to the main memory. This is done by having the MMU send a high priority interrupt to the microprocessor. The microprocessor suspends the execution of the current program and saves its state to allow later continuation of the program from the point of interruption. Then, it uses the operating system routines to determine the disk location of the desired page and brings the page into the main memory using the DMA technique. When this transfer is completed, the entries of the address translation table are updated, and execution of the interrupted program is resumed at the point where the page fault occurred. The required location is then accessed in the main memory.

Let us consider the process of address translation, from virtual to physical addresses, in some detail. We should note that the term "logical address" is sometimes found in the literature to denote a virtual address. This is the address generated by a program instruction and sent by the microprocessor to the MMU. Figure 8.26 depicts a simple translation scheme. We have assumed that 32-bit virtual

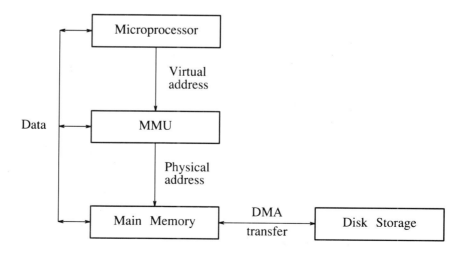

Figure 8.25 Virtual memory organization.

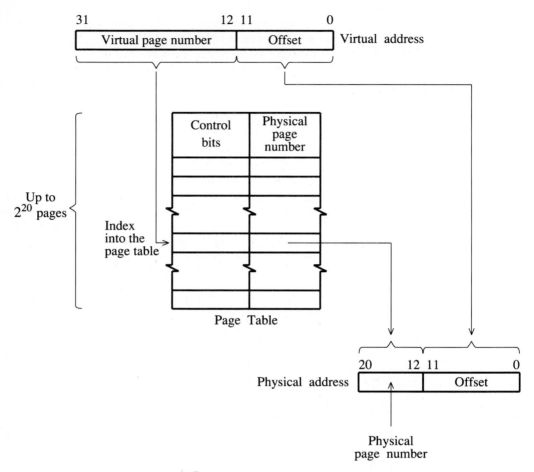

Figure 8.26 Virtual address translation.

addresses are generated by the microprocessor, the main memory has 2M bytes, and the pages are 4K bytes in size. Hence, the least-significant 12 bits of an address are an offset that specifies the location of a particular byte within a page. The offset bits are the same in the virtual and physical addresses. The most-significant 20 bits of a virtual address denote one of 2^{20} pages, and are referred to as a *virtual page number*. The main memory can hold up to 2^9 pages, identified by a 9-bit page frame number. These 9 bits and the 12-bit offset constitute the physical address.

Each entry in the address translation table, also called the *page table,* corresponds to a virtual page and contains a page frame number that indicates the location of the page when that page is present in the main memory. The virtual page number is used as an index into the page table, that points at the desired entry. In addition to the page frame number, each entry contains some control information, needed for controlling access to that page and for managing transfers between the main memory and the disk. One important component of the control information associated with a page is whether that page is resident in the main memory. It is also important to have an indication of whether the contents of the page have been changed since the page was read from the disk. A modified page must not be removed until its contents are first copied back on the disk. When the main memory is full and a new page is to be brought in, one of the old pages must be removed. The page to be removed may be chosen at random. This strategy works reasonably well in practice. But, an attractive alternative is to remove the page that was least recently used, on the assumption that it is least likely to be needed again soon. The latter scheme can be implemented by recording in the page table when a given page was used last.

Associating some protection information with each page allows the implementation of a security mechanism. In a simple case, some pages may be designated as "read only," while others may also be written into. More complex protection may also be provided. If some bits are included in the virtual address to denote a protection level of the request for access to a given page, then access will be permitted only if the protection level of the request is greater than or equal to that of the page.

Efficient operation of the virtual memory scheme is contingent upon the ability to ascertain quickly whether the desired page is in the main memory. This requires fast translation of a virtual address into a physical address, which suggests that the page table should be resident within the MMU. But, the table is likely to be too large for this approach to be feasible, because the MMU is normally implemented as a single IC chip. An alternative is to keep the page table in the main memory and have the MMU access it during every memory reference requested by the microprocessor. This is not attractive from the performance point of view, because the main memory is now accessed twice for each processor request. A more practical alternative is to use the concept of a cache, introduced in Section 8.9. We can keep the entire page table in the main memory and keep a copy of the active entries in the MMU.

Let us call the table in the MMU the "address translation cache," and the complete table in the main memory the "main page table." The address translation cache needs to hold only the information about some of the pages that are present in the main memory. When the MMU receives a virtual address from the microprocessor, it checks whether or not this address matches one of the entries in its translation cache. The simplest way of doing this would be to compare the virtual address with the addresses in the cache one by one, until either a matching address is found or the end of the cache is reached. However, the processing speed requirements demand that the entries in the address translation cache be examined quickly. Even though this table is not very large, it is impractical to have to search through it in sequential order. A good solution is to use an associative cache, which was introduced in Section 8.9. Figure 8.27 illustrates the use of an associative address translation cache. Note that it is the virtual page numbers stored in the cache that are compared with the equivalent portion of the virtual address provided by the microprocessor.

A disadvantage of an associative memory is its high cost. For this reason, the address translation cache is likely to contain the information about only a fraction of the pages present in the physical memory. For example, it may have 32 or 64

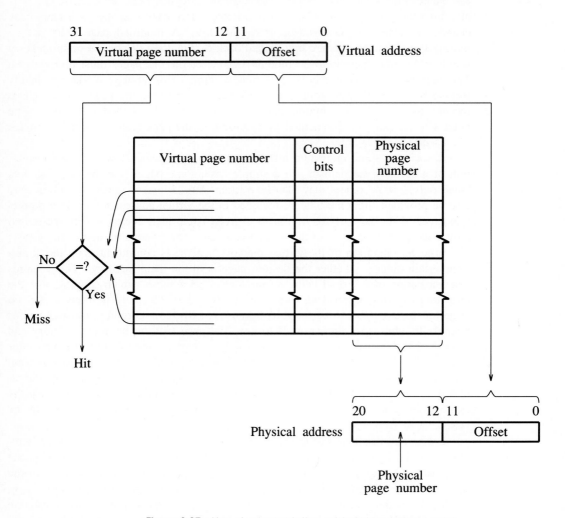

Figure 8.27 Use of an associative address translation cache.

entries, while there may be several hundred pages in the physical memory. If a reference to a page is not found in the address translation cache, this does not necessarily mean that the page is not present in the physical memory. In this case, the main page table must be consulted. If it is found that the desired page is indeed in the physical memory, a corresponding entry is made in the address translation cache, because the page is likely to be referenced again in the near future. The new entry is made in the address translation cache by overwriting an earlier entry, preferably one that corresponds to the page least recently used. In order to implement such a scheme, an indication of the time of the last usage should be recorded in the control field of each entry.

We have focused our discussion on the basic aspects of the virtual memory concept in general and address translation in particular. More complex schemes are found in commercial products. They usually support "multitasking" environments, where more than one user may be active at a time and the microprocessor is performing several tasks concurrently. This may involve using more than one page table and having to change from one table to another as a switch is made from one task to another. The page tables may be large, so that some of them may have to be kept on the disk and brought into the main memory only when needed. A more comprehensive discussion of virtual memory techniques can be found in references [8–12].

A good example of a powerful MMU chip is Motorola's MC68851 [13,14], which is intended for use with the 68020 microprocessor. It includes a 64-entry associative address translation cache. The entries in the cache are replaced using a version of the least recently used replacement strategy. In addition to providing the address translation mechanism, the cache contains control bits that are used to define a privilege level for access control to the main memory. These bits can be used to implement a protection mechanism. Beside the main distinction between the Supervisor and User modes, there can be up to eight levels of privilege defined for individual users.

The MC68851 allows several separate address translation tables to be used for different protection levels. It interacts with the main memory extensively, to access these tables. It can execute specialized instructions in a manner similar to the way the 68020 microprocessor executes its instructions. Hence, it is considered to be a coprocessor for the 68020 microprocessor. The concept of coprocessors will be discussed in Chapter 10.

The development of the VLSI technology has reached the point where it is feasible to include the MMU circuitry within the microprocessor itself. This is the case with the 68030 microprocessor [6, 15]. As we saw in Section 8.9, the 68030 also includes on-chip instruction and data caches. The 68030 operates in a manner where cache accesses and MMU translation are done in parallel. This is possible because the virtual addresses are used as tags in the cache. When the execution unit generates the virtual address of an operand to be read, the access circuitry in the data cache determines whether or not this operand is in the cache. Simultaneously, the MMU circuits translate the virtual address into a physical address. In the case of a hit, the operand is read from the data cache. But, when the cache access results in a miss, the physical address needed to access the main memory is immediately available to be placed on the address pins (assuming that a page fault has not occurred).

Including the MMU circuitry on the microprocessor chip may seem to suggest that users of systems based on such chips have to use the virtual memory capability whether they wish to do so or not. This is not necessarily the case. The MMU function in the 68030 can be disabled either by software or external hardware. Thus, it is possible to use a 68030 based microcomputer with or without a virtual memory scheme, or even to employ a different external MMU.

Before concluding this section, we should point out one important problem that arises in a virtual memory system when a page fault occurs. A microprocessor accepts an external interrupt after the instruction being executed at the time of interruption is completely finished. When a page fault occurs in a virtual memory system, a situation analogous to an urgent interrupt arises, except that it must be responded to before completing the instruction at hand, so that a new page may be loaded from the disk. The difficulty lies in the fact that the microprocessor cannot conclude the execution of the current instruction, because it requires the information just requested from the memory (e.g., the next word of a multiword instruction). There are two possible solutions. The first is to abandon the execution of the instruction completely and reexecute it correctly when the new page becomes available in the main memory. This approach requires that any registers or memory locations that may have been changed by the partially completed instruction must be restored to their state at the time the execution of that instruction began. The second solution is to save the state of the microprocessor at the point the instruction was interrupted, and to resume execution at this point later. The latter scheme is used in the 68020 and 68030 microprocessors. In either case, sufficient information about the state of the microprocessor must be saved to enable proper resumption of the suspended program. An interesting discussion of the demands that a virtual memory imposes on a microprocessor and possible ways for providing the required support in the architecture of the microprocessor can be found in reference [15].

Virtual memory support has become a common feature of 32-bit microprocessors. For smaller microprocessors, the addressable space may not be large enough, in comparison with the main memory provided, to warrant implementation of such schemes.

8.11 CONCLUDING REMARKS

This chapter has covered the essential aspects of main memory organization. It should enable the reader to understand the operation of commonly used memory circuits and the way in which they can be used to provide the main memory of a microcomputer.

We have seen that large memory units can be realized using a variety of memory IC chips, organized in array-like structures. The main memory can be of the RAM or ROM type, or a combination of both types.

RAM units can be constructed with static or dynamic circuits. Static memories are simpler to implement, but their cost is high. Dynamic memories lose their contents after a few milliseconds, hence they must be refreshed at regular time intervals. Their attraction lies in their lower cost, which is due to the simpler structure of the basic memory cells. Since fewer transistors are needed to implement each cell, dynamic memory circuits can be packed more densely on a chip, thus leading to ICs that have greater storage capacity than equivalently-sized static chips. They also consume less power. The cost of the refreshing circuitry is relatively small.

Both static and dynamic memory units lose their contents if the power supply is turned off. Such memories are referred to as being volatile. If it is necessary to retain some information permanently, then a portion of the main memory can be implemented with ROM chips. This may be the case with the initialization software, needed to make a microcomputer operational after the power is turned on or in the event of unexpected failures. A microprocessor system dedicated to a particular task may have most of its software resident in ROM storage. Typical examples include telephone line switching equipment and electronic game machines.

In place of ROM chips it is possible to use functionally equivalent PROM, EPROM, and EEPROM chips, which can be programmed easily by the user. They are attractive during engineering development and in applications where occasional changes in the memory contents need to be made. PROMs cannot be reprogrammed, but EPROMs and EEPROMS can be erased and then reprogrammed.

The speed of operation of a typical main memory is lower than the speed of a microprocessor. The performance of a microcomputer can be improved if a cache memory is interposed between the main memory and the microprocessor. The most common form of cache memory found in microprocessor systems is a small cache housed within the microprocessor chip itself.

The cost of the main memory limits its physical size to a fraction of the addressable space, especially in the case of 32-bit microprocessors. A virtual memory scheme may be used to make the physical memory appear to the programmer as being equal in size to the addressable space. Until recently such schemes have been in the domain of large computers, but are now being incorporated into microprocessor systems.

8.12 REVIEW QUESTIONS AND PROBLEMS

REVIEW QUESTIONS

1. What activates the word-select line in a memory chip?

2. Why is it useful to provide an "enable" input on a memory IC?

3. What are the advantages and disadvantages of multiplexing address pins on memory ICs?

4. What is the difference between static and dynamic RAMs?

5. In what applications should one use static rather than dynamic RAMs?

6. Describe the operation of a typical dynamic RAM controller circuit.

7. What happens if a read/write access is requested by a microprocessor while a dynamic RAM is in the midst of a refresh cycle?

8. Why is it useful to have different types of "read only" memory chips, i.e., ROM, PROM, EPROM, and EEPROM?

9. How can the "locality of reference" phenomenon be exploited in a microprocessor system?

10. What are the advantages and disadvantages of cache memories?

11. Could the cache concept be usefully exploited if the same technology is used to implement both the cache and the main memory, e.g., using CMOS circuits of the same type?

12. Is it useful to differentiate between an instruction and a data cache? Would it be better to use a common cache for both the instructions and the data?

13. Describe the possible address mapping schemes for use with cache memories. Why does a cache memory improve the performance of the 68030 microprocessor more than the performance of the 68020 microprocessor?

14. What are the advantages and disadvantages of the virtual memory scheme?

15. What information must be recorded in each entry of the page table in a virtual memory system?

16. How can the virtual address translation mechanism be implemented in order to avoid making a typical memory access much slower than an access in a microprocessor system that does not use virtual memory?

PRIMARY PROBLEMS

17. Give a design, similar to that in Figure 8.5, for a 64K-byte memory using 16K × 1 RAM chips.

18. Give a design for a 128K-byte memory unit, constructed with 64K × 1 ROM chips, which could be used with a 68000 microprocessor.

19. What is the minimum number of pins that a 256K × 1 ROM chip must have? Address multiplexing is not used with this chip.

20. Estimate the time overhead caused by the necessity to refresh a 4M-byte dynamic memory that may be used with a 68000 microprocessor that generates memory access requests every 400 ns. The memory is implemented using 256K × 1 chips, having access cycle time of 250 ns. Assume that refresh cycle time is also 250 ns.

21. The setup times t_{rs} and t_{cs} in Figure 8.11 must be provided for by the circuit of Figure 8.12. What delays in this circuit determine the desired setup times? Give an expression for each of t_{rs} and t_{cs} in terms of these delays.

22. Could a dynamic memory be used without refresh circuitry? Suggest an application where this might be feasible.

23. What address range is assigned to each of the two memory banks in Figure 8.18, where a bank is selected by either $\overline{\text{RAS0}}$ or $\overline{\text{RAS1}}$?

24. The circuit in Figure 8.18 is driven by a 6-MHz clock. Let the strap inputs of the TMS4500 controller be connected so that the clock signal is divided by 61 to generate the refresh request signal. During the refresh cycle, both $\overline{\text{RAS0}}$ and $\overline{\text{RAS1}}$ outputs are activated simultaneously. How frequently is each memory location refreshed? Is the minimum refresh requirement of the DRAM met?

25. Determine if the circuit of Figure 8.18 can be used with the TMS4464-20 DRAM. See Table 8.2 for the timing requirements of this chip.

26. Suppose that the DRAM chips used in the circuit of Figure 8.18 are too slow to produce valid data on the data bus to provide the setup time required by the

microprocessor. Assuming that proper operation could be achieved if the DTACK signal is activated one clock cycle after RDY goes high, modify the circuit to obtain this behavior. Note that the controller will insert one wait state, during which RDY will be low.

27. A "6809-like" microcomputer (which uses the 6809 instruction set, but does not have the same synchronous bus) has a small cache memory consisting of eight locations. Each location in the cache consists of a 13-bit tag field and an 8-bit data field. Main memory locations are mapped into the cache locations using the direct mapping scheme. When a miss occurs on a read operation, the requested word is read from the main memory. It is loaded into the cache and sent to the CPU at the same time. (This is known as the *load through* scheme.) Consider the following loop for adding numbers in a list:

```
LOOP      ADDA      ,X+
          DECB
          BHI       LOOP
```

Assume that before this loop is entered, registers A, B, and X contain 0, 5 and 2506_{16}, respectively, and that five 8-bit numbers are stored in the main memory starting at location 2506_{16} as follows: 23, 47, 3, 21, and 14. The loop starts at memory address LOOP = 1000_{16}.

(a) Show the contents of the cache at the end of each pass through the loop.

(b) Assume that the access time of the main memory is $2t$ and that of the cache is t. Ignoring the time taken by the microprocessor between memory cycles, calculate the execution time for each pass through the loop.

(c) Repeat part b for the case where only instruction words are stored in the cache. Data operands are fetched directly from the main memory and not copied into the cache.

(d) In what situations is it likely to be useful to have a cache for instructions only, rather than a cache that stores both instructions and data?

28. Consider the virtual memory scheme depicted in Figure 8.24, to be used in a 68020 microcomputer. Assume that at a given time the virtual pages 5, 6, 7, 304A, 304B, F0020, and F002C are resident in the main memory, in page frames 2A, 2B, 3, 1AC, 1F2, F4, and F5, respectively. What virtual addresses, used to fetch a 32-bit operand, associated with these pages will cause page faults? What main memory addresses hold the information associated with virtual addresses 5000, 304B2AA, and F002C005?

29. Give some arguments for determining the page size in a virtual memory system. Should the pages be small or large? What is likely to happen if the page size is too small? What if it is too large?

ADVANCED PROBLEMS

30. The main memory of a 6809 microcomputer consists of 48K bytes of RAM, occupying addresses in the range 0 to BFFF, and 12K bytes of EPROM in the range D000 to FFFF. The RAM is implemented with 8K × 8 static chips, and the EPROM with 4K × 8 chips. Show the circuitry needed to construct this main memory and connect it to the 6809 microprocessor bus. Assume that the memory chips are fast enough to meet the timing requirements of the 6809 bus.

31. Repeat Problem 30, using a single 16K × 8 EPROM chip to implement the read only memory. Your circuit must ensure that only the 12K locations corresponding to the address range D000 to FFFF respond to read requests.

32. Determine if the circuit of Figure 8.18 can be used with an 8-MHz clock. See Table 6.10 for the timing characteristics of the 8-MHz version of the 68000 microprocessor.

33. A 6809 microprocessor system requires 48K bytes of random access memory using DRAM chips. Assume that the chip used has characteristics similar to the 4464 chip discussed in Section 8.8, except that its size is 16K × 8 bits. Assume also that the TMS4500A-15 controller of Figure 8.18 is used to provide the access circuitry for the DRAM chips. Design a circuit that can be used to implement the desired memory.

34. Consider a program segment represented by Figure 8.28. It contains a loop consisting of instructions occupying memory addresses 1100 to 1180. Suppose that the program is executed on a 32-bit microprocessor that includes a cache such as the one depicted in Figure 8.22. Assume that the time needed to fetch 32 bits of

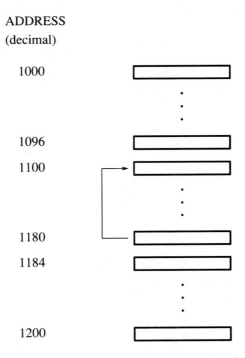

Figure 8.28 Structure of a program segment for Problem 34.

an instruction from the cache is t, while the time needed to fetch the same from the main memory is $4t$. Ignore the time needed to access operands and manipulate data.

(a) What is the total time needed to fetch the instructions of this segment if the loop is executed 30 times?

(b) How much time would be needed without the cache?

(c) Show the contents of the tag-bit field of the cache at the end of the first pass through the loop.

(d) What is the minimum number of times the loop should be executed to make the total time in a cache system shorter than the time in a system without a cache?

(e) How much improvement would be gained with the cache if the entire segment (from address 1000 to 1200) constituted an "outer" loop that is to be executed 10 times?

35. Repeat Problem 34, assuming that the program segment in Figure 8.28 occupies memory locations 1000 to 1500, and the loop within it occupies locations 1100 to 1376.

36. In a virtual memory system, the execution of an instruction may be interrupted by a page fault. What "state information" should be saved to enable later resumption of this instruction, after the new page has been brought into the main memory (which requires execution of other instructions to set up the DMA transfer)? Would it be simpler if the interrupted instruction was abandoned and reexecuted in its entirety? How could this be done?

37. Consider the following task. A 512×512 array of 32-bit integers is to be manipulated using a 68020 microcomputer that has virtual memory capability. Each column of the array is to be "normalized" by finding the largest element in the column and then dividing all elements in the column by this maximum value. Assume that each page contains 2K bytes, and that 0.5M bytes of the main memory are allocated for storing data during this computation. Assume also that it takes 50 ms to bring a page from the disk into the main memory when a page fault occurs.

(a) How many page faults would occur if the array data is stored in the virtual memory in column order?

(b) How many page faults would occur if the data is stored in row order?

(c) Estimate the total time needed to execute the task, for parts a and b.

8.13 REFERENCES

1. Glasser, L.A. and D.W. Dobberpuhl, *The Design and Analysis of VLSI Circuits*, Reading, MA: Addison-Wesley, 1985.

2. Weste, N., and K. Eshraghian, *Principles of CMOS VLSI Design — A Systems Perspective*, Reading, MA: Addison-Wesley, 1985.

3. Millman, J., and A. Grabel, *Microelectronics*, 2nd ed., New York: McGraw-Hill, 1987.

4. Burns, S.G., and P.R. Bond, *Principles of Electronic Circuits*, St. Paul, MN: West, 1987.

5. *MOS Memory Data Book*, Dallas: Texas Instruments, 1984.

6. *MC68030 Microprocessor Data Sheets*, Motorola, Inc. 1987.

7. MacGregor, D. and J. Rubinstein, "Performance analysis of MC68020-based systems," IEEE Micro, Dec.1985, pp. 50-70.

8. Hamacher, V.C., Z.G. Vranesic, and S.G. Zaky, *Computer Organization*, 2nd ed., New York: McGraw-Hill, 1984.

9. Baer, J.L., *Computer Systems Architecture*. Computer Science Press, 1980.

10. Mano, M.M., *Computer System Architecture*, 2nd ed., Englewood Cliffs, NJ: Prentice-Hall, 1982.

11. Hwang, K. and F.A. Briggs, *Computer Architecture and Parallel Processing*. New York: McGraw-Hill, 1984.

12. *MC68851 Paged Memory Management Unit User's Manual*. Motorola, Inc., 1986.

13. Cohen, B., and R. McGarity, "The design and implementation of the MC68851 paged memory management unit," *IEEE Micro*, Apr. 1986, pp. 13–28.

14. Holden, K., D. Mothersole, and R. Vegesna, "Memory management in the 68030 microprocessor," *VLSI Systems Design*, Feb. 1987, pp. 88–94.

15. Furht, B., and V. Milutinovic, "A survey of microprocessor architectures for memory management," *IEEE Computer*, Mar. 1987, pp. 48–67.

Chapter
9

Microcomputer Peripherals

Peripheral devices serve two primary purposes in a microprocessor system. They carry out the input and output operations needed for communication between the system and its environment, and they provide mass storage facilities for storing programs and data. Input and output devices include video terminals, printers, plotters, and modems. These devices are needed to enter programs and data into a microcomputer and to receive the results. The most commonly used mass-storage devices are magnetic disks and tapes. They provide nonvolatile media for storing large amounts of information. This chapter discusses the essential aspects of such peripheral devices.

9.1 VIDEO TERMINALS

Perhaps the most common computer peripheral is the video terminal. It consists of an input device, which is a keyboard for entering textual information, and an output device in the form of a display screen. A typical keyboard has 68 keys for numbers, upper- and lower-case letters, punctuation marks, and special symbols such as "$"

and "%." Video terminals are connected to a microprocessor system via a serial link. They may use synchronous or asynchronous transmission. Data is encoded using a standard character code, with ASCII and EBCDIC being the most commonly used (see Appendix A). The display screen is usually a cathode ray tube (CRT). However, some terminals use flat screens, such as liquid-crystal displays or plasma panels.

A simple video terminal transmits and receives characters one at a time, without performing any other function, and is sometimes called a "dumb" terminal. Other terminals provide a variety of functions that simplify the way in which the user interacts with the computer. When a character is received from the computer it is displayed following the text that is already on the screen. The place on the screen where a new character is entered is called the cursor position. Usually it is marked with a flashing underline sign or displayed in reverse video, that is, with bright background when the rest of the screen is dark, or vice versa. Most modern terminals allow cursor addressing. Upon receiving appropriate commands from the computer, the cursor may be moved to any point on the screen. The text that is already displayed is moved up, down, or to the side as appropriate. Such features are useful in conjunction with full-screen editors, which will be discussed in Chapter 12. Some terminals may also contain sufficient memory to store a few pages of text, and allow the user to view and possibly edit the contents of these pages.

Many microprocessor applications can benefit substantially if the terminal used can display graphical information. It is always possible to display simple graphs on any terminal by repeatedly printing some appropriate character, such as "*," at different places on the screen. This leads to low-resolution pictures that may be useful in some applications. In other cases, as in computer-aided design applications, the ability to produce high-resolution graphics is very important. Many terminals that offer such capabilities are available. The picture produced on the screen consists of an array of dots, called *pixels*, whose brightness and color can be changed under software control. Low-cost graphics terminals offer a resolution on the order of 700 $\times$ 700 pixels, with a single color and one level of brightness for each pixel. More sophisticated terminals have twice this resolution and allow several colors and many levels of brightness for each pixel.

The keyboard is a convenient tool for entering textual material. However, more flexible input devices are needed when dealing with graphical information. Devices such as a *joystick*, a *track ball*, or a *mouse* have been developed for this purpose. *Joysticks* are commonly encountered in video games. By tilting a stick in some direction, the computer is instructed to move the cursor on the screen in the same direction. A *track ball* is a small ball installed beside the keyboard, which can be rolled in any direction to indicate to the computer in which direction and by how much the cursor should be moved. The mouse is most commonly encountered in use with personal computers and in computer-aided design systems. It consists of a conveniently shaped—mouse-like—object that can be held by hand and moved over a flat surface. A set of wheels and gears inside the mouse are used to send messages to the computer to indicate the direction of motion and the distance traveled. Alternatively, a mouse containing optical sensors is placed over a checkerboard pad. As the mouse is moved, electrical pulses are sent to the terminal to enable it to count the number of squares traveled in the x and y directions. Most graphics terminals have provisions for attaching one of these devices, as a convenient means for moving the cursor on the screen.

9.2 MAGNETIC DISKS

Magnetic disks are by far the most commonly used form of on-line mass storage. They vary widely in size, speed of operation, and cost. The storage capacity of a single disk ranges from a few hundred kilobytes to several gigabytes. The speed, as represented by the data transfer rate, also spans a wide range. A small floppy disk drive transfers data at a rate of 40 Kbyte/s, while a large hard-disk drive may be capable of data rates as high as 10 Mbyte/s.

The storage medium of a disk is a thin circular platter covered with a magnetic material on one or both surfaces. Data are recorded serially on concentric tracks, which are usually divided into sectors, as shown in Figure 9.1. An index mark, which can be sensed either mechanically or optically, provides a reference point to identify the beginning of each track. The disk is mounted on an appropriate support and drive mechanism, which causes it to rotate past a read/write head, as shown in Figure 9.2. The read/write head can move radially to access any of the storage tracks on the disk surface. When both sides of the disk are used, a second head is introduced underneath the disk, requiring a more elaborate mounting and drive mechanism.

The disk may use a flexible plastic substrate on which a thin layer of magnetic material is deposited. Such disks are called "floppy disks." A floppy disk is a low-cost removable medium, capable of storing up to several megabytes of data. It is very popular in personal computers and small computer systems for distribution of software and storage of users' programs.

Because of mechanical difficulties, floppy disk drives are limited in both their capacity and data transfer rates. Higher performance drives use a disk made of a hard substrate. A single disk unit may use several platters organized as shown in Figure 9.3, with a dedicated read/write head for each surface. In this case, corresponding tracks on all surfaces are said to form a cylinder. Because all tracks in a cylinder are accessed in parallel, data transfer rates are much higher than in disks with a single platter.

In a widely used type known as a Winchester disk, the storage medium and the entire drive mechanism are sealed in an air-tight enclosure. This construction enables very close mechanical tolerances to be maintained and creates a dust-free environment. As a result, Winchester disks provide not only large storage capacities and high data rates, but also reliable operation and long life.

9.2.1 Disk Characteristics

The nature of magnetic recording and the organization of data on a disk lead to a number of attributes that influence the way in which disks are used in computer systems. Data are recorded serially on each of the tracks. Only one cylinder is

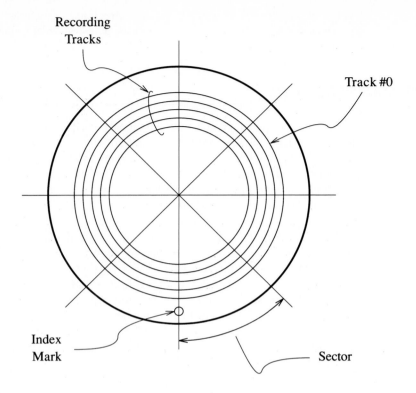

Recording Tracks

Track #0

Index Mark

Sector

Figure 9.1 A disk platter.

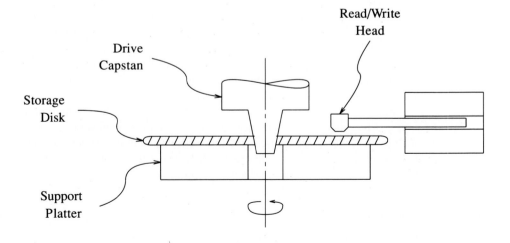

Read/Write Head

Drive Capstan

Storage Disk

Support Platter

Figure 9.2 An example of the drive mechanism for a floppy disk.

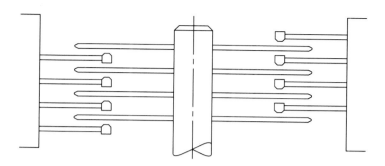

Figure 9.3 A multiple-platter disk unit.

accessed at any given time, depending on the position of the read/write heads. In order to access data on another cylinder, the heads must be repositioned. The operation of moving the heads to reach a particular cylinder is called a *seek*. No data can be read or written during a seek operation.

Serial transmission of data was discussed in Subsection 6.7.1. It was pointed out that certain encoding schemes are needed to incorporate the clock and data into a serial stream characterized by frequent state transitions. Magnetic recording uses similar encoding schemes. These schemes require that long strings of bits, usually on the order of several hundred, be transferred without interruption. By dividing tracks into sectors, as shown in Figure 9.1, the organization of data on a disk is compatible with this requirement, where a single sector is used to store a block of data that is either read or written in a single continuous operation.

Disk sectors can be accessed at random. In order to access a given sector, the read/write head is moved to the appropriate track. Then, the disk control circuitry waits for the required sector to come underneath the read/write head, at which point it starts transferring data. The delay involved in reaching a given sector consists of the time needed to move the read/write head, called the *seek time*, and the time spent waiting for the required sector to reach the read/write head. The latter is called the *rotational delay*, or *disk latency*. When sectors are accessed at random, the average rotational delay is the time required for the disk to rotate through 180°.

Controlling a disk drive requires attention to a large number of details. Most of the tasks are carried out by a special circuit called a *disk controller*, which receives commands from the microprocessor and issues the appropriate signals to the disk drive. The degree of sophistication of the disk controller varies substantially from one disk system to another. Typically, a single controller can control several disk drives.

The operation of dividing a disk into sectors is called *disk formatting*. It involves recording special patterns to mark the beginning of each sector. Due to difficulties in the disk manufacturing process, it is not possible to guarantee that the storage medium is completely free from defects. As a result, a few sectors may not be suitable for recording data. These sectors are identified by the manufacturer during testing, and their locations are recorded in a table, which is stored on the disk.

At the time of installation, this table is read by the computer and used in the process of disk formatting. As a part of the formatting procedure, individual sectors are assigned contiguous logical addresses, which may be different from their physical addresses. At all times, either the microprocessor or the disk controller maintains a mapping between the logical and physical sector addresses.

The information recorded in one sector is organized as shown in Figure 9.4. The preamble at the beginning of the sector consists of a few bits whose sole purpose is to enable the clock generator in the read circuitry to become synchronized so as to read the remainder of the recorded data correctly. At the end of the sector, a number of bits are used for error-checking purposes. They consist of a code that is computed based on the bit pattern stored in the data field. The error-check code represents an extension of the idea of parity presented in Subsection 6.7.1. It enables most errors that may occur during read or write operations to be detected. If one or more bits are recorded in error and the sector containing the error is read, the error-check code computed by the read circuitry will be different from that found at the end of the sector. Some error-check codes are sufficiently powerful that they enable the correct information to be computed from the erroneous data read from the disk. Others only provide an indication that an error has occurred, thus alerting the disk controller to take appropriate action.

Some of the errors introduced during a disk read or write operation are transient in nature. They may be caused by electrical noise or mechanical vibrations, and will not necessarily be repeated in subsequent attempts. These are called *soft errors*. Whenever new information is recorded on a sector, the controller reads it back to verify that it has been recorded correctly. If an error is detected the controller repeats the operation until the error disappears. However, if the error persists through several attempts, it is declared to be a *hard error*. Such errors may be due to a manufacturing defect in the disk surface, or they may be a result of damage caused by foreign objects or a malfunction of the head mechanism. When a hard error is detected, the sector involved is added to the damaged sector list. The operating system software must be informed of this occurrence to take appropriate action.

The preamble and error-check code occupy a certain amount of the available disk space. As such, they represent overhead. There is also overhead due to the sector markers recorded during the disk formatting process. The percentage of the

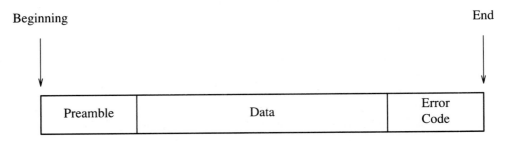

Figure 9.4 Organization of data in a disk sector.

recording surface lost to overhead increases as the sector size decreases. From this point of view, sectors should be made as large as possible. However, large sector sizes require large storage buffers in the control circuitry that handles data as the data are being transferred between the disk and the main memory. Also, since a sector is the smallest block of data that can be accessed by the microprocessor, a large sector would be wasteful when a short data file is stored. When program and data pages are transferred frequently between the disk and the main memory, the page size and the performance of the system are influenced by the sector size. Sector size should be chosen taking into account all these factors.

9.2.2 Magnetic Disk Replacement Technologies

There are several recording technologies that compete with magnetic disks for mass storage applications. They include magnetic bubbles and optical disks. *Magnetic bubble* devices use a thin wafer of a magnetic material in which information is deposited in the form of tiny regions called bubbles. The bubbles are moved about during read and write operations by current-carrying conductors deposited on the surface of the wafer and by a magnetic field created by surrounding coils. Magnetic bubble devices have the advantage of no moving parts, but are limited both in their storage capacity and speed. They are suitable for some specialized applications, such as in portable equipment.

Optical disks are very similar to the familiar compact disk (CD) audio players. The storage medium consists of a disk covered with a thin layer of aluminum. Information is recorded by burning tiny holes into the aluminum film, using a powerful laser beam. A laser beam of much lower intensity is used to read the recorded information. The beam is reflected off the surface of the disk to an optical detector, which can sense the presence or absence of the holes. Optical disks offer both high capacity and high speed. However, they are restricted to being a read-only storage medium, because the recorded information cannot be erased to store new information. Optical disks that use an erasable magnetic storage medium are available in research laboratories, but have not yet become a commercial reality.

Over the years, magnetic disks have been challenged by many technologies. However, through developments such as floppy disks and Winchester disks, as well as continuous improvements in the mechanical and electrical design of disk drives, magnetic disks continue to be by far the most widely used devices for random-access mass storage applications.

9.3 MAGNETIC TAPES

A *magnetic tape* is a compact medium that can hold large amounts of information. It is widely used for off-line storage and for transferring data from one computer to another. A tape is characterized by being a serial access device. In order to read a particular item on the tape, the tape drive must start at the beginning of the tape and skip past all recorded information up to the desired item.

Magnetic tapes are either 12.7 mm (1/2 in.), 6.35 mm (1/4 in.) or 3.175 mm (0.15 in.) wide, and they come in a variety of lengths. As in the case of audio recorders, digital magnetic tape recorders use either a reel-to-reel or a cassette configuration. Cassette tapes are more convenient to use, but have limited storage capacity.

Magnetic recording techniques for tapes are similar to those used for disks. Data and clock information is encoded into a serial stream, using one of the schemes mentioned in the discussion of serial transmission in Chapter 6. Data are organized in *records*, where a single record may contain several hundred bytes. While reading or writing information, a record must be accessed in a continuous, uninterrupted operation, to enable the clock recovery circuitry to function properly. Records are placed on tape one after the other, separated by blank gaps to enable the read circuitry to identify the beginning and end of each record. Since the information stored on a tape usually consists of files (see Chapter 12), the records that constitute one file are enclosed between two special markings called *file marks*. A file mark is a short record that contains a unique bit pattern. The first record following a file mark may be used as a file header that contains some identifying information about the contents of that file and the number of records it contains. Also, the first file on a tape may be a directory of various files stored on that tape. This organization is helpful when searching for a particular file stored on the tape.

The importance of error checking in magnetic recording was pointed out in the previous section. Error-check code fields similar to those used on disks are incorporated in each tape record. The operation of a tape drive is controlled by a tape controller, which is responsible for generating and checking the error codes. The controller accepts commands from the microprocessor to perform such functions as read one record, write one record, or verify that a particular record has been recorded correctly. In order to control tape motion, commands such as skip a given number of records forward or backward, move to the next file mark, or rewind tape are used. Some tape drives are capable of reading recorded data while moving forward or backward.

During normal operation, a tape drive reads one record of information, then stops at the following record gap. When it receives a command to read the next record, it must first set the tape in motion, and accelerate it to full speed before it can begin reading the next record. Hence, the record gap must be sufficiently long to accommodate both the braking and the acceleration periods. In order to minimize the size of the record gap, elaborate mechanisms are used to enable the tape drive to reach full speed quickly. Many tape drives incorporate a vacuum system that helps thread the tape through the read and write heads and reduces the mechanical problems associated with controlling the tape's motion.

STREAMING TAPE

Many applications in which tape drives are used do not require the ability to stop at the end of a record and start before the beginning of the next record. For example, tapes are often used for backing up the data stored on a disk. The procedure used involves copying the entire contents of a disk, or major portions of it, onto magnetic tape. Data are transferred at such rates that the tape need not be stopped at the end of each record. A type of tape drive known as a *streaming tape drive* uses a simple drive mechanism, which requires considerably longer time for braking and acceleration than conventional tape drives. It is suitable for applications in which large amounts of information are transferred without interruption, as in the case of backing up a disk. The recording format and organization of data on the tape is still the same as on conventional tapes. The main difference is that when a streaming tape drive receives a command to stop after reading a given record, it cannot stop within the gap following that record. Hence, to resume operation, the tape should be rewound, and operation started from the beginning of the tape. This mode of operation is obviously not suitable for applications that involve frequent starting and stopping.

9.4 MODEMS

Video terminals provide the simplest means of interaction between a user and a microcomputer. They are usually connected to the microcomputer via a serial link, as discussed in Section 6.7. When a terminal is situated near the computer, the connection is made by a simple cable joining the serial ports in the two devices. The signals transmitted along the link consist of voltage levels representing logic values 0 and 1. If the cable is not too long, typically less than 30 meters, it is possible to use ordinary driver and receiver circuits at both ends.

However, for long serial links, signal transmission becomes more complicated. A very long cable attenuates and distorts the transmitted signals and is likely to result in unreliable transmission. The characteristics of long cables are such that better results can be achieved if the DC voltage levels representing logic signals are translated into sinusoidal waveforms (AC signals). A device that performs the desired translation is called a *modulator*. At the receiving end of the transmission link, the reverse function is performed by a *demodulator*, which restores the signal to its original form. A device that can perform both of these functions is called a *modem*. It is used as an interface beween the digital serial port of a microprocessor or a video terminal and an analog transmission link. In general, the term "modem" refers to any device that encodes the transmitted information to suite the characteristics of the transmission link. For example, in the case of a link consisting of a fiber optic cable, binary data may be encoded by turning a laser beam on and off. The device that performs this function is called an *optical modem*.

Modems are needed when connections are made through telephone links. The telephone network is extensive and convenient to use. Telephone cables are available in most locations where computer equipment is likely to be installed. However, most telephone links are intended to carry voice signals. Hence they are designed to transmit efficiently sinusoidal signals with frequencies in the range from 300 to 3600 Hz. These analog lines are not suitable for transmission of DC signals. Thus, they can be used to transfer digital data only if the data are encoded as AC signals. We should note that higher-grade links, capable of transmitting signals with frequencies much above the audio range, are also readily available from the telephone companies. Of course, their cost is higher.

9.4.1 Modulation Techniques

The modulation process involves the transmission of a carrier signal that is modified in some way to represent binary data. The carrier is a uniform sinusoidal waveform, whose frequency is chosen according to the characteristics of the link. It may be modified, or *modulated*, using one of three basic techniques: amplitude, frequency, or phase modulation.

Modulation techniques are illustrated in Figure 9.5. Part (a) of the figure shows a binary signal. Parts b, c, and d show a carrier that has been modulated by this binary signal using the modulation schemes indicated. In each case, there are several cycles of the carrier signal for each transmitted bit. In amplitude modulation, two different signal amplitudes are used to represent 0 and 1. In frequency modulation, two different frequencies are used instead. The third possibility, phase modulation, is to represent the binary signal values by changes in the phase of the carrier signal, as shown.

9.4.2 Speed of Transmission

One of the parameters of interest to the user of a communication link is its bit rate, that is the number of bits transmitted per second. Another parameter that is often quoted is the *baud* rate. This is the rate at which the transmitted carrier changes its amplitude, frequency, or phase. In each of the schemes shown in Figure 9.5, one change occurs for each transmitted bit. Hence, the bit rate is equal to the baud rate. But, it is possible to use higher-level encoding, where several carrier amplitudes, for example, may be used. In this case, each change in the amplitude of the carrier represents more than one bit of information. Thus, the bit rate will be higher than the baud rate.

Typical bit rates used in peripheral equipment connected to microprocessor systems are 300, 600, 1200, 2400, 4800, 9600, and 19200 bits/s. We should caution the reader that the term "baud rate" is often mistakenly used to refer to the bit rate. The baud rate is of concern to the modem designer. From the user's point of view, only the bit rate is of interest. When a video terminal is connected to a computer via a 300 b/s link, the transmission of a screen full of text may take several seconds. At 1200 b/s, the operation becomes much more acceptable, while at 9600 b/s changes to the displayed text appear to take place quickly and smoothly.

9.4.3 Half- and Full-Duplex Transmission

Communication between a video terminal and a computer involves transferring information in both directions. If transmission can take place in only one direction at a time, the scheme is referred to as being *half duplex*. However, if data can be sent in both directions at the same time, the transmission link is said to be *full-duplex*.

It is obvious that half-duplex transmission can be provided on a single link. It is also possible to have a single link operate in a full-duplex mode. Consider, for example, the frequency modulation technique of Figure 9.5(c). Let the binary signal

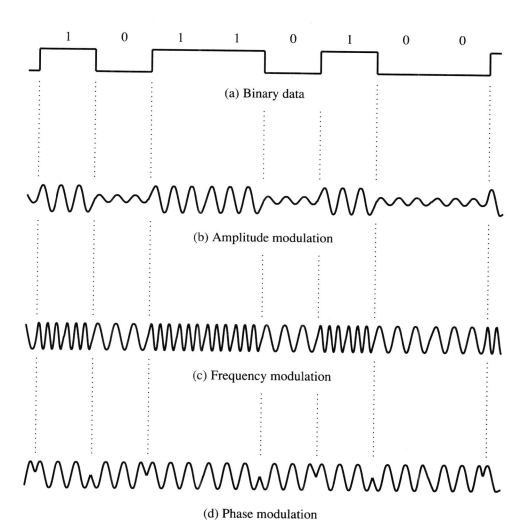

Figure 9.5 Modulation techniques.

be transmitted in one direction using a frequency of 1075 Hz to represent a 0 and 1275 Hz to represent a 1. We can make use of two other frequencies, such as 2025 and 2225 Hz, to represent 0s and 1s transmitted in the opposite direction. Thus, we have created two "channels" on a single physical link, one for each direction of transmission, where both channels can be active simultaneously.

Most commercially available modems can support either full-duplex or half-duplex transmission. Operation in the half-duplex mode is recommended when the transmission link is noisy.

9.4.4 Synchronous and Asynchronous Transmission

Transmission of data over a serial link can be synchronous or asynchronous, as discussed in Subsection 6.7.1. Asynchronous transmission is used predominantly at low speeds, while synchronous transmission is favored at higher speeds. Modems can be designed to function with either of these schemes. In the case of synchronous transmission, the clock signal may be generated by the modem and used to control transmission. Also, the receiving modem may perform transmission clock recovery as a part of the demodulation process.

9.4.5 Interface Requirements

A modem is connected to a digital I/O port on one side and to a communications link on the other. Both connections are subject to certain interface requirements. On the digital side, the most commonly used interface is called RS-232-C, which will be described in Chapter 11. On the side of the communication link, the modem may be connected directly to a dedicated transmission cable. It may also be connected to the switched telephone network, either directly or via an acoustic coupler. When an acoustic coupler is used, a telephone connection is esablished by the user by dialing the desired number. More sophisticated modems include autodialing capability and can be connected directly to the telephone line.

9.5 PRINTERS

Printers are used in microcomputer systems to produce hard copies of output data or text. A large variety of printers are available commercially. They vary in size, speed, type of printing mechanism, character shapes, and so on. Following is a short introduction to the most important characteristics of printers. Since we are primarily interested in printers from a user's point of view, we will consider only those aspects that are observable by the user.

9.5.1 Impact Versus Nonimpact

There exist two basic types of printing mechanisms. In the first, called *impact* printer, the image is formed by striking heads that press a printing ribbon against the paper. The speed of such printers is limited by the motion of mechanical parts. Despite this limitation, printing speeds exceeding 1000 lines per minute can be achieved.

The second type is called *nonimpact* printers. Typical examples of this type are electrostatic, thermal, and optical printers. The latter, also called laser printers, produce very high quality images. The speed of nonimpact printers is dependent on the technology used. They require fewer mechanical parts, which often leads to faster operation. However, the complexity of the process is a significant factor. For example, the process used in laser printers tends to limit the speed to 10 to 20 pages per minute. The speed and quality of a printer are directly related to its price.

considerations. Careful attention to these issues is needed throughout the design process, not as an afterthought. Noise problems are often time consuming to identify and costly to fix after a circuit board has been manufactured.

10.1 A SINGLE-BOARD COMPUTER

An example of a small microcomputer is shown in Figure 10.1. It consists of the following components:

1. 6809 microprocessor
2. 48K RAM
3. 12K EPROM
4. Two serial interfaces (ACIAs)
5. Two general-purpose parallel interface adapter chips (PIAs)
6. A bidirectional buffer

The entire system is housed on a single printed-circuit board, as shown in Figure 10.2. The bidirectional buffer is intended to provide the drive capability needed to extend the bus outside the board. This makes it possible to add peripherals, such as

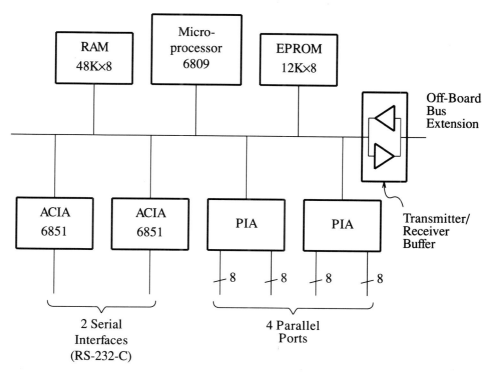

Figure 10.1 A single-board microcomputer based on the Motorola 6809 microprocessor.

Figure 10.2 A photograph of a printed-circuit board that houses the microprocessor system of Figure 10.1.

a disk drive, or to connect the board to other devices in various test and control applications. Most of the components in this system have been introduced in earlier chapters. In what follows, we will concentrate on issues relating to their interconnection and packaging in order to obtain a complete operational unit.

10.1.1 Microprocessor

The 6809 microprocessor chip was discussed in Section 6.9. It incorporates an on-chip oscillator, which generates the two clock signals, Enable (E) and Quadrature (Q), needed on the 6809 bus. The frequency of the oscillator is determined by a crystal, which should be connected to the Crystal (XTAL) and External Crystal (EXTAL) pins in Figure 6.50. The crystal used for this purpose is a quartz crystal with metallic electrodes deposited on opposite sides. The dimensions of the crystal are accurately controlled during manufacturing to provide a particular frequency of oscillation. Crystal-controlled oscillators produce a clock signal that has an accurate and stable frequency over a wide range of operating conditions.

For a microprocessor that does not have an on-chip oscillator, an external circuit is needed to generate the clock signal. An example of a crystal-controlled oscillator circuit is shown in Figure 10.3. A flip-flop is connected to the output of the oscillator, to divide the frequency of the oscillator's output signal by two. This arrangement guarantees that the high and low phases of the waveform at CLOCK have equal duration, because each is equal to one complete period of the oscillator's output signal.

10.1.2 Bus

Data transfer among various components in Figure 10.1 takes place over a common bus, whose signal lines are defined by the microprocessor chip. The 6809 bus signals were presented in Table 6.1. In addition, the 6809 and most chips on the board require a ground connection and a power supply voltage of +5 volts. Other power supply lines, such as +12 V and –5 V, should be added to the bus, if needed. As explained in Subsection 7.6.2, a Direct Memory Access Valid Memory Address (DMAVMA) signal should also be added if DMA transfers are to be supported.

Of course, not all bus signals are needed by every component in the system. Bus signals are routed only to the chips that use them. As a result, bus signals do not appear on the board as a single set of parallel lines that span the entire length of the bus. Some of the issues involved in board layout and wire routing will be discussed in Section 10.2.

One of the most important considerations in bus design is the question of electrical loading. Each chip output connected to a bus line must be capable of handling

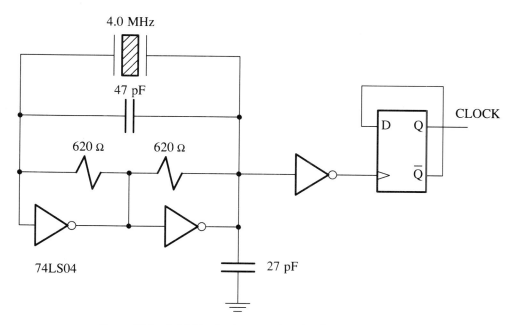

Figure 10.3 A 2-MHz clock generator circuit using a 4-MHz crystal.

sufficient current to satisfy the static, or steady state, requirements of all components connected to that line. Also, chip pins, whether input or output, and their interconnecting wiring represent a certain capacitive load. The entire capacitance associated with a bus line must be either charged or discharged every time the signal on the line changes state. Each chip output must be able to supply sufficient current to handle this dynamic load, that is, to charge and discharge the bus capacitance within the acceptable limits of signal rise and fall times. As the number of chips connected to a bus line increases, both the static and the dynamic loads increase. If the output drive capability of a chip is exceeded, the chip must be connected to the bus via a special driver that has the required drive capability.

MOS and CMOS devices have very high input impedance. Their input current seldom exceeds 2 microamperes (μA). An output that can supply 100 μA is capable of supporting 50 such devices, when only the static load is considered. Hence, if all the chips used are of the MOS or CMOS type, static current requirements on a single board are easily met. Input current requirements for TTL chips are considerably higher. An MOS output can seldom drive more than two low-power Schottky TTL loads. If several such loads are connected to a bus, high-current drivers must be used.

In an all-MOS system, capacitive loading is likely to be the factor that limits the number of chips that can be connected to a common bus. The number of pins that can be driven by a given output depends upon the drive capability of the chip and the acceptable rise and fall times of the output signal. The data sheet for a given chip usually specifies the timing delays for each output assuming that the load capacitance will not exceed a certain value, such as 50 or 100 picofarads (pF). Since the input capacitance per pin is usually in the range of 5 to 15 pF, an output specified for a maximum load capacitance of 100 pF can only drive about 10 other pins without exceeding its delay specifications.

In making loading computations, the designer should refer to the data sheets for the chips used. Exact values vary considerably from one chip to another, and even from one pin to another on the same chip. As an example, specifications for a few signal lines of the MC68B09 microprocessor chip are given in Table 10.1. In the low-voltage state, the microprocessor maintains a voltage of 0.5 V or less while sinking up to 2 milliamperes (mA) of load current on each address line. It can source a maximum of 145 μA at an output voltage of 2.4 V or more, when in the high-voltage state. (A gate output is said to act as a current sink when the direction of current flow is into the gate. When current flows from the gate to the bus, the gate output is said to act as a source.) For the microprocessor to meet its delay specifications, which were given in Table 6.2, the total capacitive load on each address line should not exceed 90 pF.

A DMA interface connected to the microprocessor bus drives the address lines when it becomes the bus master. At that time the microprocessor address outputs will be placed in the high-impedance state. Ideally, they are disconnected from the bus, and should present no load. However, they will in fact draw up to 100 μA in input current, which should be added to the static load seen by other drivers connected to the address lines. When calculating the capacitive load, 12 pF should be allowed for the microprocessor chip.

Table 10.1 ELECTRICAL SPECIFICATIONS FOR SOME SIGNALS OF THE MC68B09 MICROPROCESSOR CHIP

Signal	Parameter	Min.	Max.	Unit
$A_0 - A_{15}$	Output low voltage (for output current $\leq$ 2 mA)	—	0.5	V
	Output high voltage (for output current $\leq$ 145 μA)	2.4	—	V
	Input current when in high-impedance state	—	100	μA
	Capacitance	—	12	pF
	Capacitive drive capability	—	90	pF
$\overline{\text{IRQ}}$	Input low voltage	– 0.3	0.8	V
	Input high voltage	2.0	V_{cc}	V
	Input current	—	2.5	μA
	Capacitance	—	10	pF
$\overline{\text{RESET}}$	Input low voltage	– 0.3	0.8	V
	Input high voltage	4.0	V_{cc}	V

The parameters associated with the address pins are typical of the characteristics of the outputs of this chip, where other outputs may differ primarily in their current-handling capability. The table gives the electrical parameters of the interrupt-request input, $\overline{\text{IRQ}}$, as an example of input pins. The microprocessor considers $\overline{\text{IRQ}}$ to be in the low-voltage state whenever the voltage is less than 0.8 V. Since chip outputs generate a voltage of 0.5 V or less in the low-voltage state, there is a safety margin of 0.3 V. That is, the system can tolerate electrical noise, provided it does not cause the voltage on any line to change by more than 0.3 V. This is called the noise margin of the system. The minimum input voltage at $\overline{\text{IRQ}}$ is specified as –0.3 V. An input voltage lower than this value may cause electrical damage to the chip.

Most inputs and outputs of TTL and TTL-compatible ICs have low and high voltage thresholds that are very close to those quoted in Table 10.1 for the address and $\overline{\text{IRQ}}$ lines. The $\overline{\text{RESET}}$ input of the MC68B09 microprocessor chip differs in one respect. Its minimum high-state voltage is 4.0 V, instead of 2.0 V. The reason for this difference will become apparent from the discussion in Subsection 10.1.5.

10.1.3 System Organization and Address Assignment

Main memory and I/O interface chips are connected to the microprocessor bus in a straightforward manner. The data, R/$\overline{\text{W}}$ and clock (E) lines are connected directly to the corresponding pins of various ICs. In order to allow interrupts to be used in I/O operations, the interrupt-request outputs of the ACIA and PIA chips should be connected to the interrupt-request lines. We may choose to use $\overline{\text{IRQ}}$ for the ACIA and one of the PIAs, and $\overline{\text{FIRQ}}$ for the second PIA. This arrangement would provide two parallel ports, ports A and B of the second PIA, that are served at higher priority via the fast interrupt-request input of the microprocessor.

The chip-select inputs of various components on the board should be connected to the address lines via an appropriate address decoder. The available address space may be allocated to individual devices in a variety of ways. The only constraint in making address assignments is that the location of the interrupt-vector table, which occupies the address range FFF2 to FFFF, is fixed. If these addresses are chosen to be a part of the RAM address space, the interrupt vectors would reside in the RAM. Hence they may be changed by program instructions at any time, thus providing flexibility in the preparation of interrupt-service routines. On the other hand, being in RAM, the interrupt vectors may be corrupted as a result of errors. Also, since the contents of the RAM are destroyed when power is turned off, the vector table must be initialized by the software before interrupts can be used. Placing the entire interrupt-vector table in RAM is inconvenient, because the contents of at least one vector must be valid at all times. This is the Reset vector at address FFFE, which determines the address of the first instruction to be executed whenever the microprocessor is reset, including at power-up time. The Reset vector should be placed in a ROM or an EPROM. Alternatively, it may be obtained from mechanical switches that enable the user to enter a starting address manually.

Some form of a ROM is used in a microprocessor system to store programs that are to remain permanently in the main memory. In a general-purpose system, the ROM is likely to contain a program responsible for initialization of the system and for communicating with the user, as will be discussed in Chapter 12. In our example system, we will use an EPROM to hold the permanent programs. We will place it at the top of the addressable space, so that it will include the interrupt-vector table. A suitable arrangement for the addressable space based on this choice is shown in Table 10.2. The 48K-byte RAM is allocated addresses 0 to BFFF. The EPROM (12K bytes) occupies the range D000 to FFFF, leaving 4K bytes, at addresses C000 to CFFF, for other devices. On-board and off-board devices have been allocated 1K bytes each, in the range C800 to CFFF, and 2K bytes, from C000 to C7FF, are left unassigned.

When assigning addresses to devices, it is prudent to choose values that simplify the design of the address decoder. For example, the 1K bytes assigned to off-board devices have been chosen as C800 to CBFF. This range is uniquely defined by the six most significant address bits, A_{15-10}, having the pattern 110010. The remaining address bits, A_{9-0}, which can have any value in the range 000 to 3FF, will be used outside the board to select a particular address within the assigned range. The last column in the table gives the binary address patterns that can be used to identify each range of addresses. An entry marked with an "x" means that the corresponding line can be either 0 or 1, and may be treated as a "don't care" when designing the address decoder. For example, a signal that selects the RAM may be derived as follows:

$$\text{Select-RAM} \quad = \overline{A}_{15} + A_{15}\overline{A}_{14}$$

$$= \overline{A}_{15} + \overline{A}_{14}$$

Each ACIA requires only four locations of the address space (see Table 5.1). Since ACIA#1 has been assigned addresses starting at location CC00, the address decoder needs to recognize the binary pattern

$$1100\ 1100\ 0000\ 00xx$$

Recognition of this pattern requires 13 address bits, A_{15-2}, to be examined. When the address space available is only sparsely occupied by actual devices, as is the case with the 1K bytes allocated to on-board devices, we may simplify the address decoder by ignoring some address lines. For example, if A_2 is not examined by the address decoder, that is, if we treat this address bit as a "don't care," ACIA#1 would be selected for addresses in the range CC04 to CC07, in addition to its assigned range, CC00 to CC03. Its status register, for example, may be examined by reading location CC01 or CC05. If we ignore line A_5 instead of A_2, the status register would appear at locations CC01 and CC21. In the design of our example board, we will choose to ignore three address bits, A_2, A_3, and A_4, thus causing ACIA#1 to appear in the address range CC00 to CC1F. We will also ignore these bits when decoding the addresses of ACIA#2 and the two PIAs, which require four locations each (see Table 5.2).

Table 10.2   ADDRESS ASSIGNMENT IN THE 6809 BOARD EXAMPLE

Device	Address Range (Hex)	A_{15-12}	A_{11-8}	A_{7-4}	A_{3-0}
RAM	0000–BFFF	0 x x x	x x x x	x x x x	x x x x
		+ 1 0 x x	x x x x	x x x x	x x x x
Unassigned	C000–C7FF	1 1 0 0	0 x x x	x x x x	x x x x
Off-board	C800–CBFF	1 1 0 0	1 0 x x	x x x x	x x x x
On-board	CC00–CFFF	1 1 0 0	1 1 x x	x x x x	x x x x
ACIA#1	CC00–CC1F	1 1 0 0	1 1 0 0	0 0 0 x	x x x x
ACIA#2	CC20–CC3F	1 1 0 0	1 1 0 0	0 0 1 x	x x x x
PIA#1	CC40–CC5F	1 1 0 0	1 1 0 0	0 1 0 x	x x x x
PIA#2	CC60–CC7F	1 1 0 0	1 1 0 0	0 1 1 x	x x x x
EPROM	D000–FFFF	1 1 0 1	x x x x	x x x x	x x x x
		+ 1 1 1 x	x x x x	x x x x	x x x x

Based on the address assignments in Table 10.2, Figure 10.4 gives an appropriate address decoder, whose outputs realize the logic expressions given in Table 10.3. The signal called Other is equal to 1 when neither the RAM nor the EPROM is selected. Off-Board becomes active if the address is in the range assigned to off-board devices, and Input/Output Enable (I/OEN) becomes active if the address corresponds to one of the serial or parallel I/O chips on the board. One of these chips is selected by the 2-to-4 decoder, depending on address bits A_5 and A_6. Figure 10.5 shows how ACIA#1 may be connected to the bus, with one of its

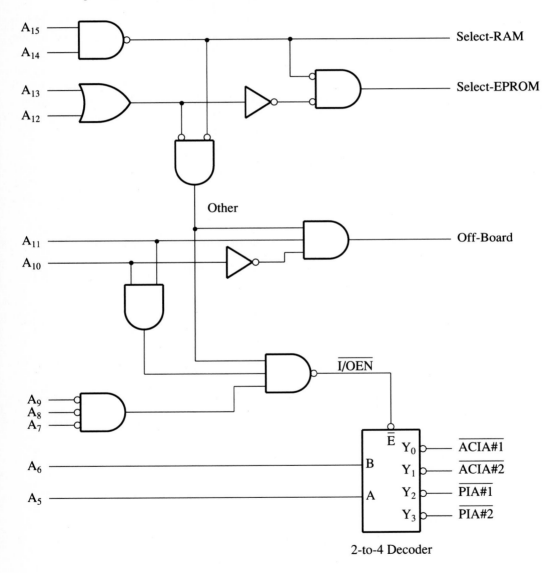

Figure 10.4   Address decoder for the 6809 board example.

Table 10.3 LOGIC EXPRESSIONS FOR THE ADDRESS DECODER OF FIGURE 10.4

Signal	Logic Expression
Select-RAM	$\overline{A}_{15} + \overline{A}_{14} = \overline{A_{15}A_{14}}$
Select-EPROM	$A_{15}A_{14}(A_{13} + A_{12})$
Other	$\overline{Select\text{–}RAM} \cdot \overline{Select\text{–}EPROM}$
	$= A_{15}A_{14}\overline{(A_{13} + A_{12})}$
Off-Board	$A_{15}A_{14}\,\overline{A}_{13}\overline{A}_{12}\,A_{11}\overline{A}_{10}$
$\overline{I/OEN}$	$Other \cdot (A_{11}A_{10})(\overline{A}_9\overline{A}_8\overline{A}_7)$

chip-select inputs connected to the corresponding output of the decoder. Its second chip-select input is connected to $\overline{DMAVMA}$ (Direct Memory Access Valid Memory

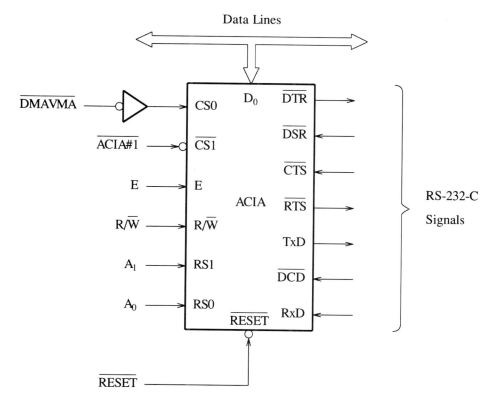

Figure 10.5 Connection of ACIA #1 to the microprocessor bus.

Address) line to prevent erroneous selection during the transfer of bus mastership. The $\overline{\text{DMAVMA}}$ signal performs the function of the Master-Active signal described in Subsection 7.6.1, and may be derived from the Bus Available (BA) and Bus Status (BS) outputs of the 6809, as was explained in Subsection 7.6.2. The serial data and RS-232-C control signals of the ACIA should be connected to appropriate RS-232-C drivers and receivers, as will be explained in Subsection 11.4.1.

10.1.4 Bus Buffer

The bidirectional buffers in Figure 10.1 transfer data between a device on the board and another outside the board. A control circuit is needed to enable the appropriate direction of transmission through each buffer, depending on the requirements of a given bus transfer. The direction of transfer is not necessarily the same for all bus lines. For example, during a read operation involving an off-board device, the buffers on the address lines should point outward, to transmit the address to the external devices. Meanwhile, the buffers on the data lines should point inward, so that the data transmitted by the addressed device can be read by the microprocessor.

When designing the control circuit for the buffers, all possible bus transfers should be considered, including read and write operations and DMA transfers between the main memory, which is on-board, and an off-board device. In a system in which expansion of the main memory by means of off-board memory chips is possible, transfers involving the external memory should also be taken into account. Table 10.4 gives the required buffer direction for the address, data, and R/$\overline{\text{W}}$ lines for all the cases relevant to the board of Figure 10.1. The table does not include the case of a DMA device on board, because none exists. All DMA transfers are assumed to be between an off-board device and the on-board main memory.

In order to generate the signals that control the bidirectional buffers, we need to determine whether the addressed device is on-board or off-board. This information is available from the address decoder of Figure 10.4. We also need to identify DMA

Table 10.4 CONTROL OF BUS BUFFERS FOR VARIOUS TYPES OF BUS TRANSFERS

Operation	Addressed device	Buffer Direction		
		Address	Data	R/$\overline{\text{W}}$
Microprocessor Read	On-board	O	O	O
Microprocessor Read	Off-board	O	I	O
Microprocessor Write	On- or off-board	O	O	O
DMA Read	On-board	I	O	I
DMA Write	On-board	I	I	I

Note: I, inward; O, outward.

operations, because according to Table 10.4, DMA transfers should be treated differently from read and write operations initiated by the microprocessor. It was pointed out in Subsection 7.6.2 that the 6809 activates BS and BA, to indicate to a device that it may start using the bus. The BS and BA signals are maintained in the active state until the device completes its data transfer and drops the bus request. Thus, any data transfer that takes place while BS = BA = 1 is a DMA transfer.

We now have all the information needed to control the bidirectional buffers. Figure 10.6 shows how the control logic may be implemented. Each buffer consists of two tri-state drivers pointing in opposite directions, as shown. A direction control input enables one driver or the other, so that the buffers point outward when this input is set to 1, and inward when it is set to 0. The reader should verify that this circuit causes the bidirectional buffers to behave as specified in Table 10.4.

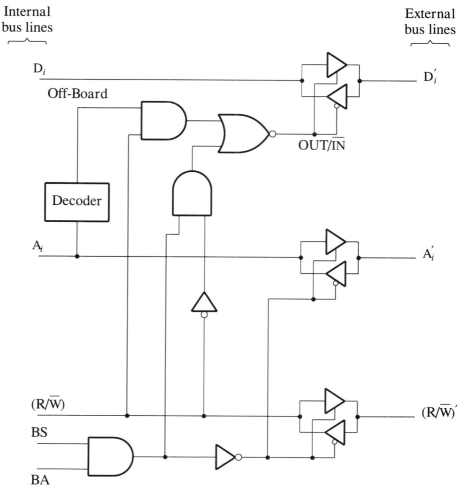

Figure 10.6 A circuit for controlling the bidirectional buffers of Figure 10.1.

10.1.5 Initialization

When power is applied to the board, the microprocessor must start its operation in an organized manner. The internal circuitry of the microprocessor must be placed in a well-defined initial state, so that it can start to fetch and execute instructions. A similar requirement applies to most programmable interfaces, such as a general-purpose interface or a DMA interface. All peripheral devices should assume an inactive state in which they do not transmit any data or interrupt the microprocessor. For example, the data direction register of a PIA chip should be initialized to set all port lines as inputs. This would prevent two devices whose outputs are connected together from transmitting on the same wire at the same time. (See Problem 30 in Chapter 5 for an example of two PIAs whose outputs are connected together.)

The initialization requirements are met by activating the $\overline{\text{RESET}}$ line on the microprocessor bus. Each chip that requires initialization at power-up time has a reset input that, when activated, performs the internal operations needed. All such inputs should be connected to the $\overline{\text{RESET}}$ line.

The $\overline{\text{RESET}}$ line may be activated by the circuit shown in Figure 10.7. The voltage on the $\overline{\text{RESET}}$ line is equal to the voltage across the capacitor. Before power is applied, the capacitor is in a discharged state, and point A is at 0 V. The capacitor will be charged through resistor R when power is turned on, causing the voltage at point A to start rising slowly from 0 toward +5 V. For a short time, the voltage at point A will be below the switching threshold of all the gates connected to the $\overline{\text{RESET}}$ line. Hence, the $\overline{\text{RESET}}$ line will be in the active state, and will cause all chips to be reset. When the capacitor voltage reaches the minimum voltage for the high state, the $\overline{\text{RESET}}$ line becomes inactive, and normal operation begins. The duration of the reset period is determined by the *time constant* of the circuit, which is the product RC. In addition to being activated at power-up time, the $\overline{\text{RESET}}$ line

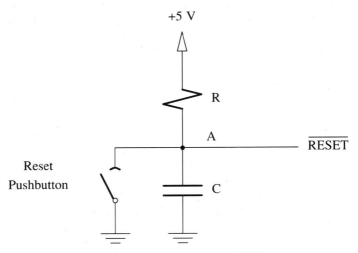

Figure 10.7 Generation of the $\overline{\text{RESET}}$ signal.

may be set to the low-voltage state by activating the reset pushbutton, which will discharge the capacitor. After releasing the pushbutton, the capacitor starts to charge slowly, as before.

The electrical characteristics of the $\overline{\text{RESET}}$ input of the microprocessor were given in Table 10.1. Its minimum high-state voltage is 4.0 V, instead of the usual 2.0 V for other inputs. As the voltage across the capacitor in Figure 10.7 increases, the microprocessor will be the last chip to exit the reset state and begin normal operation. Thus, when it begins executing instructions and sending commands to other devices on the bus, those devices will be ready to respond. If this order of ending the reset period is reversed, errors may result. For example, assume that a PIA chip detects an active state at its $\overline{\text{RESET}}$ input after the microprocessor begins executing instructions. The microprocessor may initiate a write operation to the data direction register to change one of the ports to an output port. However, because its $\overline{\text{RESET}}$ is still active, the PIA will ignore the write operation and maintain that port as an input port. This situation cannot arise when the microprocessor is the last to exit the reset state.

10.2 ELECTRICAL AND PACKAGING CONSIDERATIONS

Having completed the design of the logic circuitry needed for our single-board computer, it is time to start construction of the actual board. A few considerations that should be kept in mind during this phase will be discussed in this section.

10.2.1 Circuit Boards

Two methods of construction are commonly used, wire-wrapping and printed circuits. On a wire-wrapped board, connections are made by means of thin insulated wires (size AWG 30), where the stripped ends of the wires are wrapped around 1-inch long pins on the underside of the board, using a special tool. The pins have a square cross-section to ensure good electrical contact and reduce the probability of the wires unwinding. Wire-wrap pins are attached directly to the board or to sockets in which the chips are mounted. They may be gold plated to improve the durability of electrical contact with the wires.

The wire-wrapping technique is well suited to the construction of prototype boards. A skilled technician can complete the construction of a board in a relatively short time, working directly from circuit diagrams. Design changes are easily implemented by unwrapping the undesired connections and replacing them with new ones.

For volume production, printed-circuit boards are by far the most widely used technology for construction of computer boards. Chips may be mounted directly on the board or plugged into sockets soldered to conductive pads on the board. Connections are in the form of thin copper traces on both sides of the board. The manufacturing process begins with a board consisting of an insulating material sandwiched between two thin sheets of copper. A material known as "photoresist" is spread over the surface of the board. Then, the interconnection pattern, which is drawn on a transparent sheet, is transferred photographically to the photoresist, and the board is chemically etched so that only the desired copper traces remain. Holes are drilled at pin locations, and the IC chips or sockets are inserted through them. The pins may be soldered to the traces on the board manually, but in a mass production environment an automated process known as "wave soldering" is used.

External connections to a printed-circuit board are often made via an *edge-connector,* which consists of an array of traces in the form of narrow fingers at the edge of the board. The traces act as electrical contacts when the edge of the board is plugged into an appropriate female connector. Edge connectors provide a simple, low-cost scheme for making external connections and for mechanically supporting the board in a card cage. However, they may be a source of trouble in the long term, particularly when boards are inserted and removed often during normal operation and service. A variety of pin and socket connectors are available for mounting on printed-circuit boards. They provide a more reliable, but more expensive, means for making external connections.

We now turn our attention to the question of how wires and circuit traces should be laid out on a board. Ideally, the chips should be organized such that all connections are short, with as few crossovers as possible. This is a challenging task, which is very tedious when done manually. Several software packages are currently available for this purpose as part of computer-aided design facilities.

In the case of wire-wrapped boards, there is no particular problem with crossovers, because insulated wires are used. However, it is still advisable to organize the chips in a way that minimizes the length of interconnections. Each wire should run along the straight line between the two points it interconnects. It is particularly *not* advisable to run wires in neat bundles parallel to the sides of the board, because this arrangement maximizes the electrical coupling between the wires. Also, all wires should be kept as close to the surface of the board as possible. The reasons for these precautions will be explained shortly.

In addition to keeping wires short, several other factors should be kept in mind during the layout and construction of circuit boards. We will discuss some of these below.

10.2.2 Power Supply

The power supply connection for TTL and MOS circuits consists of +5 V and ground leads. Some may require additional supply voltages, such as +12 V and −5 V. The power supply voltage must be within a small tolerance of its nominal value. Typically, the allowable tolerance is ±0.5 V, less in some cases, depending upon the types of chips used.

Close examination of the operation of a TTL or an MOS logic gate reveals that there are sharp current pulses on the power supply lines coincident with state transitions at the gate output. The inductance of the power supply lines combined with these current pulses results in voltage pulses appearing on the +5 V and ground wires. Hence, the power supply voltage may momentarily reach values outside the allowable range. Also, the power supply lines become a medium through which these voltage pulses are distributed as undesirable electrical noise to other components on the board. The cumulative effect of noise from various sources can lead to a malfunction. For example, it may cause a flip-flop to change its state erroneously.

Design of the power distribution network on the board should ensure that the effect of the current pulses mentioned above is minimized. The first step in this direction is to make the resistance and, more importantly, the inductance of the power supply wires as low as possible. Both the resistance and the inductance of these wires are reduced by using thicker and wider traces. The inductance is also influenced by the area of the loop encompassed by the power supply current as it flows from the power source to a given chip and back on the ground trace. The area of this loop should be minimized.

Another important technique for reducing power supply noise is the use of *bypass capacitors*. These are small capacitors connected between the +5 V and ground traces on the board, very close to the IC chips. The current pulses resulting from state transitions will be mostly absorbed by the capacitors, causing very little change in the supply voltage. In other words, the capacitors act as filters for the noise induced on the power lines as a result of logic switching. The function of a bypass capacitor is illustrated in Figure 10.8, where the inductance shown is that created by the long power-supply lines. A current pulse emanating from the chip finds a "shortcut" through the capacitor, and is not forced to flow through the line inductance. A larger capacitor diverts a larger portion of the current pulse away from the line inductance, and thus reduces the amplitude of the resulting noise voltage.

If a bypass capacitor is to perform its function properly, it must have very low inductance. This means that the capacitor must have short leads, must be mounted physically close to the chip, and must have an internal construction that produces low internal inductance. Many capacitor types, such as ceramic disk capacitors, are

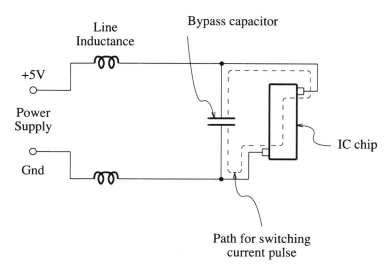

Figure 10.8 Use of a bypass capacitor.

available for this purpose. It is a common practice to mount a capacitor in the range 0.01 to 0.1 µF near every chip on a microprocessor board. A few larger capacitors may also be strategically located on the board, such as near external power-supply connections, to further reduce noise in the power distribution network.

10.2.3 Ground Plane

We mentioned above that power supply traces, +5 V and ground, should be made wide in order to reduce power line inductance. There is another equally important reason why these lines should cover as wide an area as possible. This is related to the reduction of crosstalk between wires and noise coupled into the circuit from external sources.

The term *crosstalk* refers to the fact that electrical signals on a wire induce similar signals, usually of lower amplitude, on nearby wires. These effects can generally be reduced by avoiding situations in which a given pair of wires run in close physical proximity for a long distance. Wherever possible, wires should cross each other at right angles. However, because of the limited space on a circuit board and the large number of interconnections, it is inevitable that parallel wires will be present.

A significant amount of crosstalk is caused by capacitive coupling, which is illustrated in Figure 10.9. Three traces are in close physical proximity, two signal wires, A and B, and a ground trace. Because of their proximity, the two signal wires have a mutual capacitance C_{AB}. The capacitance between wire B and ground is C_{BG}. A signal on wire A, such as a high-to-low transition, can be regarded as emanating from a signal source between wire A and ground. The two capacitors act as a voltage divider, causing an induced voltage on wire B given by

$$V_B = V_A \times \frac{C_{AB}}{C_{BG} + C_{AB}}$$

Hence, the induced voltage, V_B, can be reduced by keeping the two wires far apart, to reduce C_{AB}, and by increasing the capacitance between the signal wires and

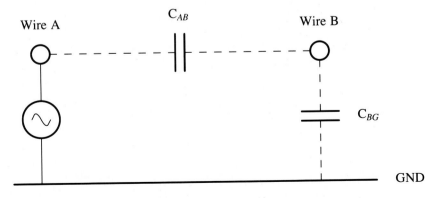

Figure 10.9   *Capacitively coupled crosstalk.*

ground, C_{BG}. Capacitance to ground can be increased by enlarging ground traces on the board and by keeping signal wires or traces as close to the ground traces as possible. A similar argument applies to capacitively coupled noise from sources external to the board.

It was mentioned earlier that, for high-frequency signals, bypass capacitors provide a low-impedance path between the +5 V and ground return traces. Hence, all power supply traces serve as ground for the purpose of reduction of crosstalk. Together they form the *ground plane*. One of the objectives in the design of circuit boards is to maximize the area of the ground plane and to place all signals as close to it as possible.

Because of its importance in reducing various types of noise on a circuit board, many construction techniques provide a ground plane in the form of a solid copper sheet covering the entire area of the board. An excellent example of this approach is the four-layer printed-circuit board. Instead of the usual two sheets of copper, four sheets are used separated by insulating layers. The middle two serve as the power supply lines, +5 V and ground, while the outer two are used for circuit interconnections. The two parallel sheets in the middle act as power supply leads that have very low inductance. At the same time, they form an excellent ground plane. All circuit traces on the outside are very close to the ground plane, while being easily accessible for testing and servicing. Interconnection of signals on the two sides of the board is still possible, via holes drilled through the intermediate layers.

10.2.4 Timing Signals

Signals that control the timing of data transfers, such as the clock signal on a synchronous bus or the handshake signals of an asynchronous connection, require special attention. Any unwanted pulses, spikes, or improper transitions are almost certain to lead to unreliable operation, if not outright failure. Timing lines are particularly vulnerable to various causes of signal degradation, because they are usually long lines connected to a large number of chips. For example, the clock signal on a synchronous bus is needed by most chips that participate in data transfers over the bus. Since the clock line serves more chips, it carries higher currents than other signal lines. This means that it is not only vulnerable to interference, but it is also a significant source of crosstalk for neighboring lines.

The recommendations given earlier with regard to current handling capability of line drivers, proximity to the ground plane, and routing of signals should be strictly adhered to when dealing with timing lines. This will usually lead to satisfactory results, particularly when relatively low-speed TTL and CMOS components are used. In some cases, additional precautions are recommended. For example, in the absence of a well-defined ground plane, a twisted pair of wires may be used for the clock signal, where one wire serves as the hot wire and the other as a ground return. In this case, the ground-return wire becomes a part of the ground plane on the board. In addition to guaranteeing that the hot wire is physically close to the ground plane, twisting provides a form of a shield that further reduces crosstalk and susceptibility to interference. It is particularly effective when long wires are involved. An even more robust arrangement can be achieved by using a thin coaxial cable. The inner conductor carries the signal, and the outer sheath provides the ground return.

More elaborate precautions may be needed when dealing with high-speed circuitry, harsh environments, or applications requiring a high degree of reliability. Such precautions may involve a detailed analysis of current flow in the circuit, particularly on the ground plane, in order to improve the board layout. They may also involve the use of shielding and power line filtering, both to protect the microcomputer board from external influences, and to reduce the effect of the board on other system components.

There is another issue that must be taken into consideration when designing the interconnections on or off a circuit board. This is the question of signal propagation over the transmission lines formed by the interconnecting wires. Because of its importance, we will discuss this issue in some detail in the following section.

10.3 TRANSMISSION LINE EFFECTS

Two wires that form the path of an electrical signal constitute a *transmission line*. Consider a logic gate driving a load, R_L, through a long pair of wires, as shown in Figure 10.10. (The point at which the ground-return wire connects to the driver gate should not be confused with the tri-state control input as it appears, for instance, in Fig. 10.6.) We normally think of a signal transition, such as a low-to-high transition at the output of the gate, as an event that appears simultaneously at all points of a wire. In fact, such signals travel on the transmission line at speeds close to the speed of light. The signal can be regarded as a wave that propagates on the line in much the same way as a disturbance propagates on the surface of water. When the signal reaches the end of the line, the load acts as a boundary and reflects the signal back on the line. The reflected wave travels toward the beginning of the line, where it may again be reflected. Reflections continue until the signal levels at both ends of the line reach their steady-state values.

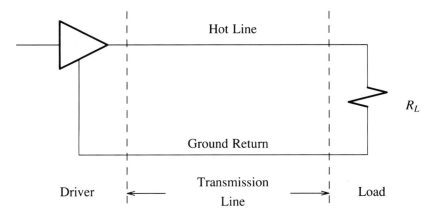

Figure 10.10 Transmission line connecting a driver gate to a load.

To understand transmission line phenomena we will examine the voltage and current flow on the line when a voltage step is applied. For the purpose of this discussion, let us replace the driver gate in Figure 10.10 by a DC voltage source, V_S, in series with an impedance R_S and a switch, as shown in Figure 10.11. The transmission line that connects the load, R_L, to the source can be thought of as consisting of a large number of very short sections. The two wires that form a section Δx can be represented by an inductor ΔL and a capacitor ΔC, as shown in Figure 10.12. In this simplified model, we have ignored the effect of the resistance of the wires and the losses associated with the insulating dielectric. When the switch is closed, a voltage step of magnitude V_1 is applied to the line. As a result, a current i begins to flow in the inductor of the first section of the line. As the current charges the capacitance of that section, the rise in the capacitor voltage, v, causes current to flow in the next section, and so on. Thus, the voltage step propagates from one section to the next.

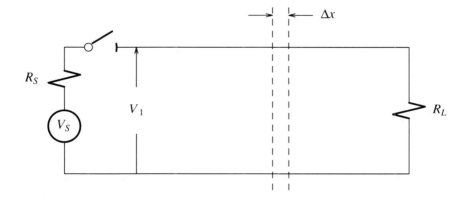

Figure 10.11 A load connected to a DC source via a transmission line.

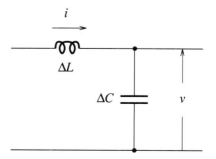

Figure 10.12 Equivalent circuit for the section Δx in Figure 10.11.

Exact expressions for the voltage and current on the line as the wave propagates can be obtained by examining the limiting case when the length of the sections approaches 0. The details of the analysis are outside the scope of this book. Interested readers are referred to books on electromagnetic field theory, such as [1,2]. For our purposes, it suffices to give the main results. The voltage step propagates down the line without degradation. That is, its amplitude and rise time remain unchanged. A current step, I_1, is associated with the voltage step, and the relationship between the two is given by

$$\frac{V_1}{I_1} = \sqrt{\frac{L}{C}}$$

where L and C are, respectively, the inductance and capacitance of the line per unit length. Thus, the transmission line appears to the propagating voltage and current wave as a resistive load having a resistance R_0 given by

$$R_0 = \sqrt{L/C} \tag{1}$$

which is known as the *characteristic impedance* of the line. The speed, s, with which the wave propagates on the line is given by

$$s = \frac{1}{\sqrt{LC}} \tag{2}$$

A snapshot of the voltage and current distribution on the line shortly after the voltage step is applied is shown in Figure 10.13. The figure shows the instant when the wavefront has reached a distance d from the beginning of the line.

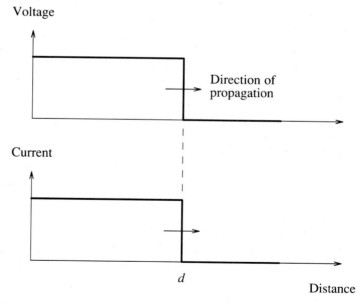

Figure 10.13 Voltage and current distribution over the line in Figure 10.11 shortly after the switch is closed.

The values of L and C are determined by the physical characteristics of the transmission line wires (shape, size, and spacing) and by the dielectric constant, ε_r, of the insulating medium. (ε_r is the permittivity of the medium relative to that of free space.) For example, consider a circuit board with signal traces having the dimensions shown in Figure 10.14. Assume the board to be made of a material known commercially as G-10, which has a dielectric constant, ε_r, of 4.7. The capacitance and inductance per unit length of the signal traces on this board are $C = 70$ picofarads per meter and $L = 0.4$ microhenry per meter, respectively. Substituting in the equation for R_0, we obtain

$$R_0 = \sqrt{\frac{0.4 \times 10^{-6}}{70 \times 10^{-12}}} = 76 \text{ ohms}$$

The characteristic impedance increases for thinner conductors, lower permitivity, and larger spacing between the signal traces and the ground plane. For typical board arrangements, it is in the range of 35 to 150 ohms. The speed of propagation over the line in Figure 10.14 is given by

$$s = \frac{1}{\sqrt{0.4 \times 10^{-6} \times 70 \times 10^{-12}}} = 1.89 \times 10^8 \text{ meters per second}$$

The value of s is determined primarily by the dielectric constant of the surrounding insulating medium. In fact, if the space around the conductors is totally filled with an insulator having uniform characteristics, the speed of propagation becomes independent of the conductor geometry, and is given by

$$s = s_0 / \sqrt{\varepsilon_r}$$

where s_0 is the speed of light in free space, 3×10^8 m/s (30 cm/ns or about 1 ft/ns). For the line mentioned above this equation yields $s = .46 \, s_0$, or 1.38×10^8 m/s.

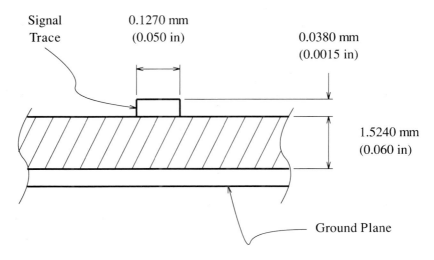

Figure 10.14 Example of a signal trace on a printed circuit board.

This value is somewhat lower than the actual speed of propagation, because the dielectric in Figure 10.14 occupies only part of the space around the conductors.

10.3.1 Reflections

Consider now what happens when the wavefront in Figure 10.13 reaches the end of the line. As the capacitor of the last section of the line is charged, current begins to flow in the load, R_L. In the special case where $R_L = R_0$, all the current I_1 eventually flows through R_L, creating a voltage V_L equal to V_1, the voltage to which the capacitor has been charged. Hence, no further changes in voltage or current occur. The system has reached a steady state. In this case, that is when the load impedance is equal to the characteristic impedance of the line, the load is said to *match* the line.

If the load impedance does not match the line, a steady state cannot be reached at the moment the wave reaches the load, because the voltage to current ratio associated with the incident wave cannot be sustained by the load. Instead, the voltage step, V_L, observed across the load will be different from V_1. The difference, $V_L - V_1$, is a voltage step that must now propagate toward the source, to charge the line capacitance to the new value, V_L. The original incident wave can be thought of as having been reflected off the mismatched load. The situation is illustrated in Figure 10.15. A portion I_L of the incident current I_1 flows through the load, such that $V_L = I_L R_L$. Current continuity requires that the current component, I_2, of the reflected wave be equal to $I_1 - I_L$, as shown. Hence, the reflected wave consists of a voltage step $V_L - V_1$ and a current step $I_1 - I_L$, which must satisfy the constraint imposed by the transmission line impedance; that is,

$$V_L - V_1 = R_0(I_1 - I_L)$$

Since $V_1 = I_1 R_0$ and $V_L = I_L R_L$, we obtain

$$V_L = \frac{2R_L}{R_L + R_0} V_1$$

Hence, the voltage step of the reflected wave is given by

$$V_2 = V_L - V_1 = \frac{R_L - R_0}{R_L + R_0} V_1$$

The ratio of the amplitude of the reflected to the incident wave is known as the *reflection coefficient*. We will denote it by ρ_L. Therefore,

$$\rho_L = \frac{R_L - R_0}{R_L + R_0} \tag{3}$$

Clearly, when $R_L = R_0$ the reflection coefficient is equal to 0, which is the case of a matched load where no reflection occurs. When the reflected wave reaches the source, it may be reflected again. The reflection coefficient, ρ_S, at the source is given by

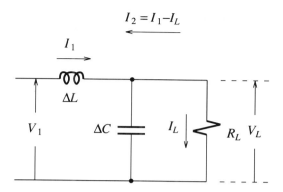

Figure 10.15 Voltage and current distribution at the load.

$$\rho_S \;=\; \frac{R_S - R_0}{R_S + R_0} \tag{4}$$

The above discussion indicates that the initial voltage step, V_1, causes a wave that travels along the line and, as a result of reflections, bounces back and forth between the two ends. Since the reflection coefficient at each end is less than one, the wave eventually dies down, and the voltage at any point on the line reaches the steady-state value. Since we have assumed a DC source, the inductance and capacitance of the line have no effect on the steady-state conditions. Therefore, the steady-state voltage is given by

$$V_{steady-state} \;=\; \frac{R_L}{R_S + R_L}\,V_S \tag{5}$$

We have yet to determine the value of the initial voltage step, V_1. This value is easily obtained from the observation that the voltage and current associated with any wave traveling on the transmission line are related by the characteristic impedance, R_0, of the line. At the moment the switch in Figure 10.11 is closed, the transmission line appears to the source as a load having that impedance. Therefore,

$$V_1 \;=\; \frac{R_0}{R_S + R_0}\,V_S \tag{6}$$

As an example, assume that the printed-circuit trace in Figure 10.14 connects a source having $V_S = 5$ V and $R_S = 200$ ohms to a load $R_L = 1000$ ohms. The initial voltage step, V_1, is given by

$$V_1 \;=\; \frac{76}{200 + 76} \times 5 \;=\; 1.38 \text{ V}$$

The reflection coefficients at the two ends of the line are given by Equations 3 and 4, which yield

$$\rho_L = \frac{1000 - 76}{1000 + 76} = 0.86$$

$$\rho_S = \frac{200 - 76}{200 + 76} = 0.45$$

Hence, the voltage steps associated with successive reflections will have the amplitudes

$$
\begin{array}{lclcl}
V_2 & = & 0.86 \times 1.38 & = & 1.19 \text{ V} \\
V_3 & = & 0.45 \times 1.19 & = & 0.54 \text{ V} \\
V_4 & = & 0.86 \times 0.54 & = & 0.46 \text{ V}
\end{array}
$$

and so on. Let P be the propagation delay from one end of the line to the other. The waveform of the voltage across the load will be as shown in Figure 10.16, where $t = 0$ corresponds to the instant the initial voltage step is applied. At $t = P$, the voltage step, V_1, arrives at the load and is reflected, causing the load voltage to rise to 2.57 V ($= V_1 + V_2$). A round trip later, at $t = 3P$, the wave that has been reflected off the source V_3, arrives and is again reflected. As a result, the load voltage increases by an amount $V_3 + V_4$ to 3.57 V. As reflections continue, the load voltage approaches its steady-state value of

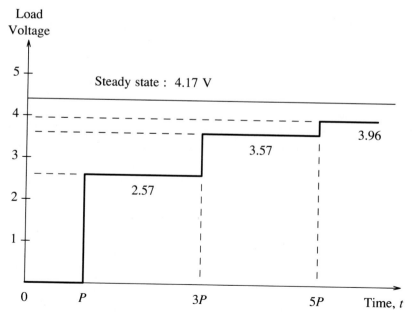

Figure 10.16 Voltage waveform at the load for $V_S = 5$ V, $\rho_S = 0.45$, and $\rho_L = 0.86$.

$$5 \times \frac{1000}{200 + 1000} = 4.17 \text{ V}$$

In the example above, both R_L and R_S are larger than the characteristic impedance of the line. If the termination resistance is smaller than R_0, Equation 3 indicates that the reflection coefficient becomes negative. That is, the voltage step associated with the reflected wave will be opposite in sign to that of the incident wave. Consider the same load and transmission line, this time driven by a source having $R_S = 40$ ohm. In this case, $V_1 = 3.28$ V and $\rho_L = 0.86$ as before, but $\rho_S = -0.31$. Hence, $V_2 = 2.82$ V, $V_3 = -0.87$ V and $V_4 = -0.75$ V. Because of the alternating signs of the voltage steps associated with successive reflections, the load voltage oscillates as shown in Figure 10.17, until it settles to its final value of 4.81 V.

In general, an oscillatory waveform results if ρ_S and ρ_L have opposite signs, that is, if $R_S < R_0 < R_L$ or $R_L < R_0 < R_S$. The load waveform has the staircase appearance of Figure 10.16 if $R_0 < R_S, R_L$ or $R_0 > R_S, R_L$. When either the source impedance or the load impedance matches that of the transmission line, a single voltage step appears at the load. These results are summarized in Figure 10.18. We should add that when $R_L = R_0$, no reflections occur at either end of the line. When $R_S = R_0$, the steady state will be reached after only one reflection at the load. Thus,

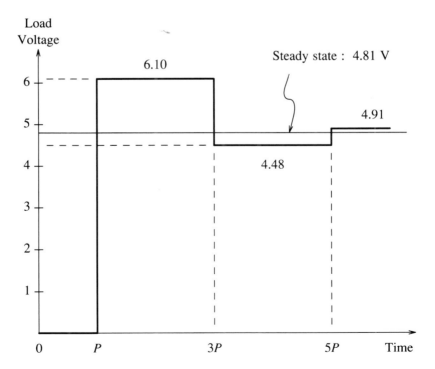

Figure 10.17 Voltage waveform at the load for $V_S = 5$ V, $\rho_S = -0.31$, and $\rho_L = 0.86$.

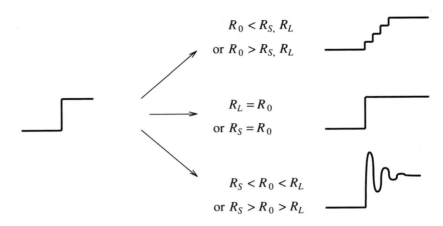

Figure 10.18 Load voltage waveforms for different values of termination impedance.

the voltage at all points of the line other than at the load end will reach its final value in two steps.

10.3.2 Graphical Representation

The amplitudes of the reflected waves can be obtained by a simple graphical procedure. Graphical analysis is particularly useful when dealing with logic circuits, which are highly nonlinear devices that cannot be replaced by a single constant resistance as in the previous discussion. We will describe the graphical procedure by applying it first to the example given above.

The relationships between voltage and current at the source and load terminals are represented by the dark lines in Figure 10.19. We have assumed the positive direction of current flow to be from the source toward the load. The transmission line imposes the condition $\Delta v = R_0 \Delta i$ for any wave traveling in the forward direction, where Δv and Δi are the voltage and current steps, respectively, associated with the wave. For a wave traveling toward the source, a positive voltage step will cause current to flow in the negative direction. Hence, for that wave, $\Delta v = -R_0 \Delta i$.

Initially, the conditions on the transmission line are represented by point A in the figure. When the switch in Figure 10.11 is closed, the operating point at the source end of the transmission line jumps to point B, which satisfies the conditions imposed by both the source and the forward wave. When the wave represented by point B arrives at the load, it is reflected. The new operating point must satisfy the characteristics of the load and a backward traveling wave. Recall that the load voltage and current are equal to the algebraic sum of the voltage and current components, respectively, of the incident and reflected waves. Thus, by drawing a line with slope $-R_0$ from point B we obtain point C, which represents the conditions after the first reflection. We continue the process to obtain points D, E, and so on.

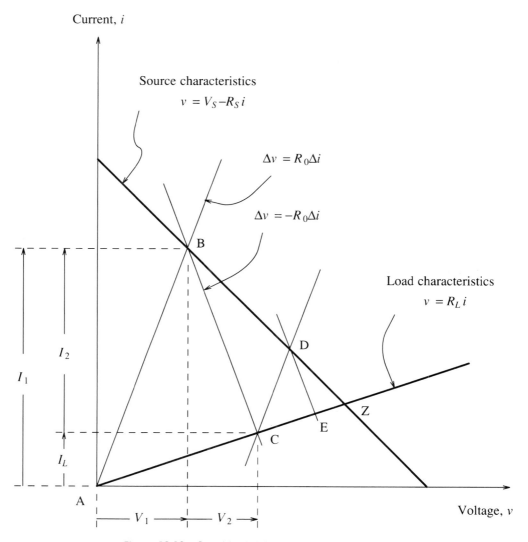

Figure 10.19 Graphical determination of wave amplitudes.

Clearly the operating point at either end of the line eventually converges to point Z, which represents the steady-state conditions.

The graphical procedure enables us to deal with nonlinear drivers or loads. For example, consider two low-power Schottky (LS) TTL logic gates connected via a twisted pair of wires, as shown at the top of Figure 10.20. The voltage-current characteristics at the output of the driver in the high- and low-voltage states and at the input of the receiver are given in the figure [3, 4]. When the driver output is in the low-voltage state, the operating point will be point A. This means that the voltage on the line will be about 0.2 V, and a small amount of current will flow from

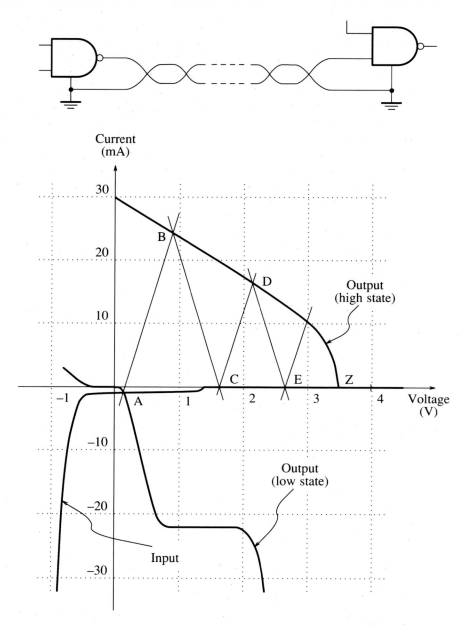

Figure 10.20 Reflections on a transmission line connecting two TTL-LS gates.

the receiver to the driver (a TTL input acts as a current source and a driver acts as a current sink when in the low-voltage state). As the gate output changes from low to high, the input voltage at the receiver follows the path A, C, E, etc., until it reaches the steady-state point, Z. We have assumed that the transmission line has a characteristic impedance of 30 ohms. The resulting waveform at the input of the receiver is shown in Figure 10.21. It is left as an exercise for the reader to draw the waveform for a high-to-low transition.

It is instructive to note the effect of the current sourcing capability of the driver on signal delays. The high-state threshold at the input of the receiver in Figure 10.20 is 2 V. The input voltage exceeds this value only when point E is reached, that is, after three end-to-end propagation delays. For a higher-current driver, point B moves up and to the right, and point C moves to the right. When the current-handling capability of the driver is sufficiently high, the 2-V threshold would be exceeded after only one propagation delay. A good indicator of the ability of a given driver to provide the required current is the value of its short-circuit current, which is the output current when the driver's output is connected directly to ground. The output curve in Figure 10.20 has a short-circuit current of 30 mA. The typical value for TTL-LS components is about 60 mA. However, the guaranteed minimum value under worst-case operating conditions of temperature and supply voltage is only 20 mA.

10.3.3 Design Considerations

Let us now consider the effect of transmission line phenomena on the design and operation of microcomputer boards. The duration of individual steps in Figures 10.16 and 10.17 is equal to twice the end-to-end propagation delay. The effect of these steps is significant when this delay is comparable to or larger than the rise or

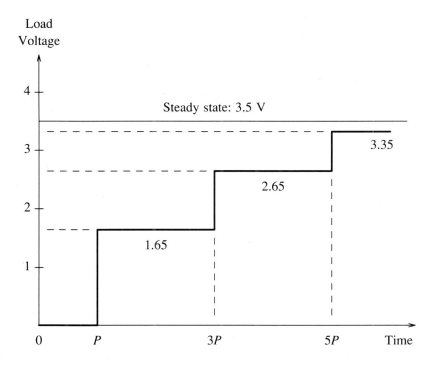

Figure 10.21 Voltage waveform at the input of the receiver in Figure 10.20.

fall time of the signal. For example, a high-capacitance line may have a propagation speed of 0.1 s_0, or 3 cm/ns. The round-trip delay for a 30-cm length is 20 ns. If the rise and fall times are in the range 10 to 20 ns, the waveforms of Figures 10.16 and 10.17 would be clearly visible when observed on the screen of an oscilloscope. For faster devices, transmission line effects should be taken into account for much shorter lengths.

Figure 10.21 indicates that one result of transmission line reflections is to increase the effective propagation delay over a bus. Also, since the drivers and loads on different lines of a bus are not necessarily identical, different numbers of reflections will be needed to reach the switching threshold. Hence, transmission line effects also increase the bus skew. Let us allow two round trips for the signal to reach a stable state, which amounts to 40 ns for the bus mentioned in the preceding paragraph. This value should be added to any circuit delays at the sending and receiving ends when making the propagation delay calculations discussed in Subsection 6.2.2.

There is an important factor that should be kept in mind when estimating the speed of propagation on a bus line. A typical bus line has many loads connected to it, not just one at the end. Detailed analysis of this case is quite complicated. It suffices to say that the distributed capacitive load results in a significant reduction in the characteristic impedance and the speed of propagation on the bus. Since a certain amount of capacitance is associated with each device input or output connected to the bus, the distributed load can be regarded as causing an increase in the line's capacitance per unit length. The associated reduction in the characteristic impedance and speed of propagation can be estimated from Equations 1 and 2. For heavily loaded bus lines, such as for the backplane buses to be discussed in Chapter 11, the reduction factor may be as high as 5 or more.

When dealing with the data and address lines on a bus, correct operation can be achieved provided that sufficient delay is allowed for transmission line reflections to die down. Timing lines are a different matter. On a synchronous bus, for example, various devices use the clock line to determine when to take specific actions, such as placing data on the bus or loading data into an internal buffer. Thus, the state of the signal on the clock line must be well defined at all times. Figure 10.22 illustrates a possible situation that might arise as a result of transmission line reflections. A driver and a receiver gate are connected as shown. The figure gives the waveforms at points B and C resulting from a low-to-high transition at point A. The dashed line represents the position of the switching threshold of the receiver. Voltages above that threshold are interpreted as 1, and produce a high output. Voltages below the threshold produce a low output. As a result of reflections, three transitions appear at the output of the receiver, point C. The unwanted transitions may cause false data to be received, a counter to be advanced, or a flag to be cleared prematurely, thus leading to a system failure.

It is often necessary to reduce transmission line reflections, either to increase the speed of operation of the system or to eliminate the problem depicted in Figure 10.22. From Figure 10.18, it is clear that this can be achieved by arranging for the load impedance, R_L, to be equal to the characteristic impedance of the transmission line, R_0.

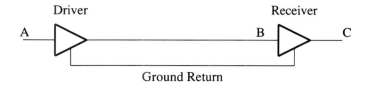

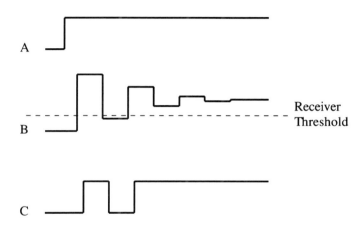

Figure 10.22 Effect of transmission line reflections on the received waveform.

The input impedance for most logic gates is very high compared to the characteristic impedance of transmission lines normally found on microprocessor circuit boards. The latter is usually on the order of 100 ohms. Hence, separate resistors, called *termination resistors*, should be attached at the end of a transmission line to provide the desired matching. Since the current that flows through this added resistive load must be supplied by the line driver, proper termination of a transmission line comes at the expense of increased power dissipation and the need for line drivers that have higher current handling capacity.

Often, it is not desirable to simply connect a single resistor from a signal line to ground, because too much current would be required to pull the line up to the high-voltage state. Instead, the arrangement shown in Figure 10.23(a) may be used. The termination impedance seen by the transmission line is the parallel combination of the two resistors R_1 and R_2, whose values should be chosen to provide the desired matching load. At the same time, the ratio of the two resistors determines the voltage at that point when all drivers on the bus are turned off. For an open-collector line, this voltage must correspond to the high-voltage state on the line, which means it must be above 2.4 V for TTL-compatible buses. As an example, a matching

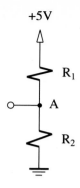

(a) Passive

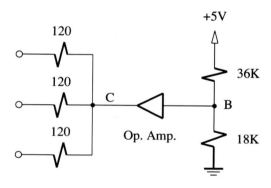

(b) Active

Figure 10.23   Line termination.

termination that maintains a voltage of about 3 V on a 120-ohm open-collector line is obtained by choosing R_1 and R_2 such that

$$\frac{R_2}{R_1 + R_2} \times 5 = 3$$

and

$$\frac{R_1 R_2}{R_1 + R_2} = 120$$

These two equations yield $R_1 = 200$ ohms and $R_2 = 300$ ohms. In practice, it is customary to use higher resistor values, to reduce the current handling requirements for bus drivers. For example, the VMEbus, which will be presented in Section 11.2, uses $R_1 = 330$ ohms and $R_2 = 470$ ohms. The same values are recommended for terminating all bus lines, including those that are not of the open-collector type, because TTL drivers can sink much more current than they can source. That is, it is much easier for a TTL driver to pull the line down than up. For the Multibus, described in Section 11.3, the values $R_1 = 220$ ohms and $R_2 = 330$ ohms are used on clock lines. Most other lines are terminated in a single pull-up resistor of either 1K or 2.2K ohms. Each of these choices represents a compromise between the need to provide perfect impedance matching and the desire to keep current levels and power dissipation low.

The termination arrangement of Figure 10.23(a) draws a significant amount of current at all times. When designing for very low power dissipation, the active termination circuit shown in Figure 10.23(b) is preferable. A voltage divider produces the desired bias voltage at point B. This divider draws only a small amount of current from the supply because the resistors used are large. The voltage at point B is transferred to point C by an operational amplifier, which will supply whatever current is needed to maintain the voltage at point C at the same value as at point B. Thus, the output of the operational amplifier constitutes a voltage source having the desired bias voltage. A bus line may be terminated by connecting it to point C via an appropriate resistor. Several lines can be terminated using only one bias circuit, as shown.

In concluding the discussion of transmission lines, we should note that it is not practically possible to attain perfect matching of lines in the environment of a microprocessor system. Circuit traces on a printed-circuit board are likely to vary in size and spacing from one point to another. Hence, the line impedance is not uniform. Also, with a bus line, which normally feeds many chips, it is very difficult to provide proper termination at all load points. A long bus line is usually terminated only at the two ends.

10.4 MULTIPROCESSOR SYSTEMS

In this and previous chapters, we discussed microprocessor systems that comprise a microprocessor connected via a common bus to the main memory and several I/O peripherals. Such systems are referred to as *uniprocessor systems*, because they contain a single processor that controls the operation of all other devices. The main memory and the peripheral devices make no decisions on their own. They simply carry out the commands they receive from the microprocessor.

Some of the peripherals in a uniprocessor system may be quite sophisticated, perhaps even containing their own microprocessors. For example, a disk drive may be controlled internally by a microprocessor whose job is to attend to the details of head movement and the recording and retrieval of data. Peripherals that use a microprocessor in this way may provide many functions and a high level of performance. Such a microprocessor constitutes an integral part of the peripheral device. Its existence is totally transparent to the rest of the system in which the I/O device is used. A system that has one microprocessor functioning as a central processing unit may continue to be called a uniprocessor system, even if some of its devices contain internal microprocessors that control their operation.

In order to increase the computing capability of a uniprocessor system, more powerful microprocessors and faster buses and memory devices are needed. This approach has lead to many of today's large computers. An alternative approach is to use several microprocessors in the same system, and divide the computational tasks among them. The main difficulty in designing such systems is finding efficient ways for a large number of microprocessors to share the workload. Currently available commercial systems make effective use of a relatively small number of microprocessors working in parallel. However, systems comprising much larger numbers of microprocessors will soon be available, and will likely constitute the supercomputers of the future.

In what follows, we will discuss briefly some of the ways in which more than one microprocessor can be incorporated in a single system.

10.4.1 Coprocessors

The first step in attempting to expand a uniprocessor system is to consider attaching a second microprocessor to the same bus, as shown in Figure 10.24. Both processors obtain program instructions and data from the main memory, via the bus. Hence, access to the main memory is likely to become the bottleneck in this arrangement. Significant improvement in performance can only be achieved if the tasks assigned to the second processor require relatively infrequent access to the main memory. For example, the second microprocessor can be dedicated to computationally intensive tasks, such as performing floating-point arithmetic. As can easily be appreciated from the example in Section 4.9, a single floating-point operation involves a large number of steps. A dedicated special-purpose processor can be designed to perform floating-point operations using its internal registers for temporary storage and requiring access to the main-memory only to obtain the initial operands, and later to deposit the result.

A special-purpose processor used as described above is often called a *coprocessor*. It can be regarded as expanding the capabilities of a general-purpose microprocessor by providing certain functional capabilities that would otherwise require considerable programming effort and execution time. Many modern microprocessors have associated coprocessor chips, with floating-point and I/O coprocessors being the most commonly encountered examples.

10.4.2 General-Purpose Multiprocessor Systems

Coprocessors can increase the capability of a microprocessor system significantly. However, they are only suitable in situations in which specific, computationally intensive tasks can be identified and assigned to a processor specifically designed for that task. In a true multiprocessor system, we wish to use several general-purpose microprocessors that can share various computing tasks without interfering with each other. The main drawback of the organization of Figure 10.24 is that the system's

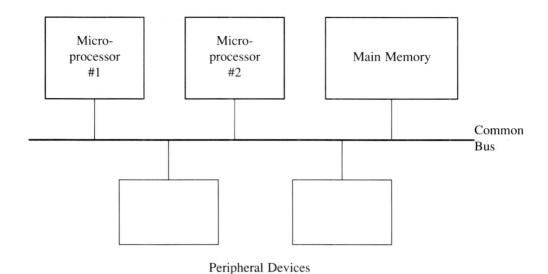

Figure 10.24 A simple two-processor system.

performance is limited by the bandwidth, or data transfer capability, of the common bus and the main memory. In an attempt to increase the memory bandwidth, let us consider the more general organization shown in Figure 10.25. The main memory is

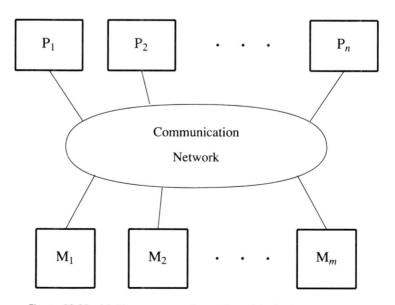

Figure 10.25 Multiprocessor system using global memory modules.

divided into m modules, M_1 to M_m, which can be accessed independently by n microprocessors, P_1 to P_n. Also, instead of the single bus of Figure 10.24 a communication network is used. Because any memory module can be accessed by any processor, the modules are said to constitute a *global memory*. For the system of Figure 10.25 to provide higher performance than that of Figure 10.24 it must be possible for several processor-memory transfers to proceed in parallel. At the same time, the delays through the communication network must be very small. This is difficult to achieve.

A more flexible organization is shown in Figure 10.26. Each processor is provided with a *local memory*, LM_i, attached to the processor's bus in the usual manner. The bus is attached via an interface circuit to the communication network that provides access to the global memory modules, GM_j. In order to make effective use of this system, programs and data should be organized such that, most of the

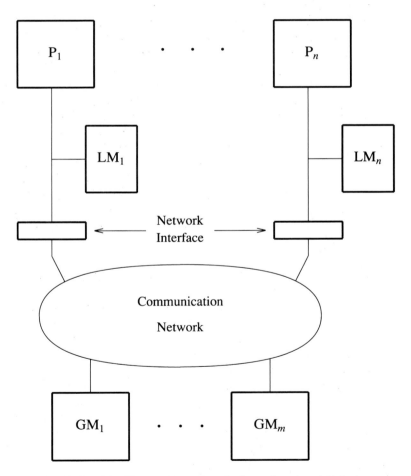

Figure 10.26 A multiprocessor system that uses local and global memory.

time, each processor uses data that are available in its local memory. Much less frequently a processor may access one or more of the global memory modules, which contain data shared by all processors.

Many practical systems are based on the organization of Figure 10.26, where the communication network is implemented in the form of a bus. In the next chapter, we will describe two standard buses, VMEbus and Multibus, that incorporate special features to make them suitable for the multiprocessor environment. Other configurations are also being developed for the communication network to enable many processors to access global memory modules simultaneously, thus increasing the total processing capability of the system.

A detailed study of multiprocessor systems is beyond the scope of this book. However, before concluding this discussion we will comment on two important aspects. The first relates to the general question of allocation of work to individual processors. The second is a specific problem related to the need for synchronization of tasks that run in different processors and require access to common data.

LOAD SHARING

Consider a multiprocessor system used in a large library. The system keeps track of the library inventory as new books are added and deleted from the collection, of book loans to determine which books are on the shelves, of overdue books, and so forth. Many terminals are connected to the computer system to enter enquiries and updates, record book loans and returns, and carry out various administrative functions related to the running of the library.

How should this enormous workload be divided among different microprocessors in a multiprocessor system? There are many possibilities. We can assign a processor to look after all the work related to one section of the library, such as science, history, fiction, and so forth. Alternatively, one processor may be used to keep track of inventory, another to record loans, and another for members' records and the issuing of library cards. In yet another approach, the processors may be regarded as a large pool of computing resources that are assigned to tasks upon demand. In this case, some mechanism must be implemented to select free processors and direct them to handle new tasks, irrespective of the nature of those tasks.

No matter which of the above approaches is chosen, it is clear that certain data, such as records of the permanent collection of the library, must be accessible to many, perhaps all, processors. Other data are more local in nature. For example, the details of a request for a book search need only be known to the processors involved in responding to that particular request. Also, processors must be able to communicate with one another. When a loan is made or when a book is added or lost, this information must be made available to all processors concerned. The hardware organization of Figure 10.26 offers the main features needed. The global and local memories enable the processors to function independently, while sharing a common data base. The communication network provides a path for exchanging messages and accessing the global memory.

Perhaps the most difficult aspect of the design of a multiprocessor system is the design of the algorithms and software needed to share the workload among the processors and to guarantee correctness of the entire operation. In our library example, it is important to ensure that no request is forgotten, or serviced twice. Also, we must prevent two processors working independently from loaning the same book to two different people.

The discussion above illustrates some of the difficulties involved in the design of multiprocessor systems. There are several other important issues that should be addressed. For example, it is desirable to achieve a balanced sharing of the workload among different processors. Bottlenecks can easily develop when a particular resource, such as the communication network or some global memory module, is heavily used. Also, situations in which processors spend long idle periods waiting for other processors to complete certain tasks should be avoided. Some applications, such as the library example appear well suited to implementation on a multiprocessor system, because of the multiplicity of tasks that need to be carried out in parallel. There are many other applications where the division of tasks among processors is less clear. Dealing with such issues is, and will continue to be, an important topic of research.

In the remainder of this section, we will discuss one problem that every multiprocessing system must be able to handle correctly. This is the issue of task synchronization.

TASK SYNCHRONIZATION

A fundamental problem in the design of a multiprocessor system is that of task synchronization. When two or more tasks require access to a common data item, it is often necessary to guarantee that only one task is allowed access at a given time. For example, consider two requests to reserve a given book in the library system mentioned above. Obviously, only one person should be allowed to reserve the book. Once the book is assigned to a person, other requests should either be rejected or should be placed in a queue of requests for that book. To guarantee correct operation when the requests are handled by different processors, access to the part of the database that contains the pertinent records must be granted to only one processor at a time. This restriction on access is not as easy to implement as it may seem. To illustrate the difficulty, let us consider the problem in more detail.

A suitable mechanism for guaranteeing exclusive access to common data is based on the notion of semaphores, as used to control railway lines. A train can enter a particular segment of a line only when the associated semaphore indicates that there are no other trains on that segment. As soon as the train enters the segment, the semaphore is changed to the "busy" position. It returns to the "free" position when the train exits the segment. The same concept can be implemented in software to control access to a particular data structure residing in the global memory. The sequence of instructions in Figure 10.27 tests a binary variable called SEMAPHORE. If SEMAPHORE = 1, the data structure is not available, and the processor must try again. Otherwise, if SEMAPHORE = 0, it is set to 1 and the processor proceeds with the task at hand, which may involve many data transfers between the processor and the area in the global memory that the semaphore protects.

It appears that if every processor wishing to access the same data structure checks the semaphore in the manner described above, no two processors can have the

```
WAIT       Test          SEMAPHORE
           Branch_if_≠0   WAIT
           Move           #1,SEMAPHORE
           Begin protected task
           . . .
```

Figure 10.27 Use of a semaphore.

right of access at the same time. Unfortunately, closer inspection reveals that this is not the case. Consider two processors A and B, each requiring access to the same data structure, and each attempting to test the associated semaphore. Let processor A complete execution of the first instruction in Figure 10.27, and obtain the result SEMAPHORE = 0 . But, before executing the instruction that sets the semaphore to 1, processor B begins to access the same memory module. It tests the state of the semaphore and it too obtains SEMAPHORE = 0. Not knowing what happened, each processor proceeds to set the semaphore to 1 and access the data structure.

In order to alleviate this problem, it must be possible for each processor to test and set the semaphore without any chance of other processors gaining access to the same global memory module between the two operations. For example, if the communication network consists of a single bus, processor A can be guaranteed exclusive access to the global memory during that period by preventing processor B from becoming the bus master. Such a restriction requires cooperation between the software and the bus arbitration hardware. The Test-and-Set (TAS) instruction of the 68000 microprocessor is intended for use in such situations. It reads the contents of a memory location, sets the condition codes accordingly, so that they may be tested later by a conditional branch instruction, then writes a 1 in that location. The TAS instruction may be used as shown in Figure 10.28. In order to ensure that no other device can gain access to the main memory between the test and set parts of the instruction, the 68000 carries out the two transfers using the indivisible read-modify-write cycle described in Subsection 6.10.1. For the general case of Figure 10.26, the control hardware of the communication network must be able to implement a read-modify-write cycle as an indivisible transaction. The two standard buses described in Chapter 11 have special features to support such operations.

In the absence of a read-modify-write capability, either in the microprocessor or in the communication network, some hardware feature must be incorporated in the system to offer an alternative facility. For example, the instructions in Figure 10.27 may be preceded by a command to the control hardware to acquire and retain access

```
WAIT       TAS    SEMAPHORE
           BNE    WAIT
           Begin protected task
           . . .
```

Figure 10.28 A semaphore mechanism using the Test-and-Set instruction of the 68000 microprocessor.

to the required memory module. Then, another command to release the module should be issued after the test and set operations have been completed.

We have considered just a few issues pertinent to multiprocessor systems. There are many other issues involved. The complexity of multiprocessor systems is much greater than that of uniprocessor systems. Interested readers should consult one of the specialized books in this area [5–8].

10.5 CONCLUDING REMARKS

In the previous chapters we have discussed many facets of microprocessor system hardware. This chapter addressed the issue of "putting it all together." An example microprocessor board was presented, to illustrate such aspects as address decoding, generation of clock signals, and system initialization at power-up time.

The chapter also discussed several electrical circuit considerations that need to be taken into account in the design of a microcomputer. Particular attention was paid to the sources of noise encountered in logic circuits, including crosstalk, transmission line reflections, and noise on the power supply lines resulting from gate switching. The need for and importance of a ground plane was discussed in some detail.

Microprocessor systems in which several microprocessors share the processing load were considered. One way of sharing the load is through the use of coprocessors to perform specialized tasks such as floating-point computations. An alternative approach is to use a number of identical processors and connect them to the main memory via a high-speed communication network, which may take the form of a backplane bus. Some of the issues that need to be addressed in the design of such systems, particularly the question of task synchronization, were introduced.

In the next chapter, we will examine two standard backplane buses, VMEbus and Multibus. We will also discuss other interconnection standards for connecting peripheral devices to a microcomputer. In particular, the RS-232-C and RS-449 serial interconnection standards will be presented, as well as the Small Computer System Interface (SCSI) bus used for connecting disks, tapes, and other high-speed peripherals.

10.6 REVIEW QUESTIONS AND EXERCISES

REVIEW QUESTIONS

1. What causes dynamic loading to be higher than static loading of a logic gate? What happens as the dynamic load increases?

2. Can a high dynamic load lead to errors? Explain.

3. Inspect the data sheet of any digital IC. What parameters indicate the capability of this chip to handle dynamic loads?

4. If the capacitive load driven by a gate exceeds the value quoted in the data sheet, does this necessarily lead to failure? Why?

5. What is the noise margin in a digital circuit?

6. What happens to the noise margin if the loading limits of a logic gate are exceeded? How does this problem manifest itself?

7. Why is the high-voltage state threshold on the $\overline{\text{RESET}}$ input of the 6809 microprocessor higher than the standard TTL threshold of 2 V?

8. For a microprocessor that does not have the feature mentioned in Problem 7, what would you do to avoid the danger mentioned in Subsection 10.1.5?

9. What are the advantages and disadvantages of putting the interrupt vectors of a microprocessor in read only memory?

10. What is the purpose of the bypass capacitors? Instead of installing a large number of small capacitors, is it equally effective to use a single capacitor equal in value to the sum of the small ones? Why?

11. What is the ground plane and what functions does it serve?

PRIMARY PROBLEMS

12. Some microprocessors require two clock signals that are 90° out of phase. Give a logic circuit that can be used to generate a second clock signal, CLOCK2, that is 90° out of phase with CLOCK in Figure 10.3.

13. Design a RAM circuit for the 6809 microcomputer of Section 10.1, using 8K × 8 static RAM chips. Modify or augment the address decoder of Figure 10.4 as you see fit.

14. Repeat Problem 13 for the EPROM section of the microcomputer. The EPROM chips available to you have an organization of 4K × 8.

15. Repeat Problem 13 using 16K × 1 static RAM chips.

16. Assume that the inputs of the memory chips present the same electrical load as the $\overline{\text{IRQ}}$ line in Table 10.1. Memory outputs represent a capacitive load of 10 pF each, and sink 10 μA to ground when in the high-impedance state. Calculate the total static and dynamic load on the address and data lines in the two circuits of Problems 13 and 15. Compare these loads to the drive capability of the microprocessor and comment on the results.

17. Calculate the characteristic impedance of a transmission line that has an inductance and capacitance per unit length of 0.6 μH/m and 60 pF/m, respectively.

18. Consider a 500-mm bus line on the backplane of a card cage. The line has a capacitance to ground of 120 pF/m and a speed of propagation of 180 mm/ns. Board connectors are spaced 20.32 mm apart, and introduce a capacitance to ground of 8 pF each. The transceiver gate on each board adds another 15 pF per pin, on average. For a fully loaded bus, estimate the effective speed of

propagation and the period of the oscillatory noise signal resulting from transmission line reflections.

19. Consider a bus line terminated at both ends by the termination network of Figure 10.23(a). What is the current sinking capability of a gate output needed to pull the line down to 0.4 V?

20. Consider the circuit shown in Figure 10.29, and assume the D input of the flip-flop to have the same characteristics as $\overline{\text{IRQ}}$ in Table 10.1. The impedance of the ground return path between the flip-flop and the driver gate is Z. Show the waveform of the effective clock signal seen by the flip-flop resulting from the pulse at point A, assuming that the signal at the clock input of the flip-flop is in the high-voltage state. (The effective clock signal is the voltage between points C and G'.)

21. An I/O device has eight internal registers and is assigned the address range FC20 to FC27. Where else will it appear in the address space if A_6 and A_5 are ignored by the address decoder?

22. Why is a $\overline{\text{DMAVMA}}$ signal needed on the 6809 bus but not on the 68000 bus?

23. Complete Table 10.4 to show the case of a DMA device on the board that may access the main memory either on or off the board.

24. In order to extend the 6809 bus off board, what type of buffers would you suggest for each of the following lines: DMA/Bus Request (DMA/BREQ), BS, and BA? Why?

25. Prepare a table similar to Table 10.4 for the 68000 bus lines $\overline{\text{AS}}$, $\overline{\text{DS0}}$, and XACK.

26. A high-speed, 8-bit wide communication link is established between two microprocessor boards, using a PIA chip on each board. Both PIA ports are used, one for transferring data in each direction. The same software runs on both processors.

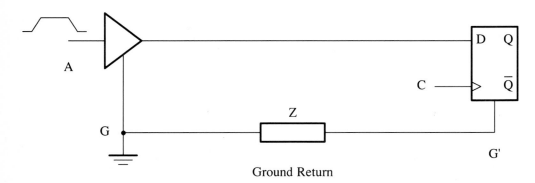

Figure 10.29   Problem 20.

How would you connect the PIA ports and control lines, and which of the PIA modes would you use on each port? Can you guarantee that the two ports connected to each other can never be programmed as outputs at the same time, including at power-up time? Can this guarantee be extended to cases where programming errors occur?

27. A state transition at the output of a logic gate causes the current pulse shown in Figure 10.30 to appear on the power-supply lines. What size bypass capacitor is needed to limit the resulting voltage pulse to 250 mV?

28. Reduction of crosstalk requires signal wires and printed-circuit board traces to be placed close to a ground plane. What effect does this have on the transmission line impedance and the speed of propagation?

29. The traces on a circuit board have inductance L and capacitance C per unit length given by: $L = 0.3\ \mu\text{H/m}$ and $C = 75\ \text{pF/m}$. Calculate the characteristic impedance and speed of propagation on this line.

ADVANCED PROBLEMS

30. The interrupt vectors in the microcomputer described in Section 10.1 reside in an EPROM. Suggest a software mechanism that makes it possible to move all vectors except Reset and NMI to locations in RAM, starting at address 0.

31. The circuit in Figure 10.6 uses the output of the address decoder to determine whether the memory being addressed is on board or off board. In the case described in Problem 23, how would you determine whether a particular DMA transfer has been initiated by a device on the board or off the board? Modify the control circuits in Figure 10.6 to accommodate the case of on-board DMA devices.

32. Discuss the differences between the effect of electrical noise on an analog circuit, such as an audio amplifier, and on a digital circuit.

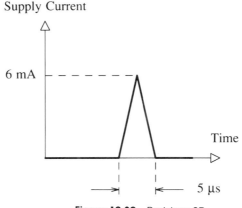

Figure 10.30 Problem 27.

33. When an analog circuit is built in a sloppy manner, the effect is usually immediately apparent during testing (oscillations, audible noise, and so on.) This is not always the case with digital circuits. Why? Does this mean we do not need to be as careful when designing digital circuits?

34. "We need to worry about transmission line effects in digital circuits only when designing systems that use a high clock rate." Is this statement true? Explain by giving some examples.

35. Give an example to show that a poor ground return can lead to errors, even in a circuit operating at a low clock rate.

10.7 REFERENCES

1. Kraus, J.D., *Electromagnetics.* New York: McGraw-Hill, 1953.
2. Hayt, W.H., *Engineering Electromagnetics.* New York: McGraw-Hill, 1981.
3. *The TTL Data Book,* Vol. 2. Dallas: Texas Instruments, 1985.
4. *Supplement to TTL Data Book,* Vol. 3, Dallas: Texas Instruments, 1984.
5. Hwang, K. and Briggs, F.A., *Computer Architecture and Parallel Processing.* New York: McGraw-Hill, 1984.
6. Satyanarayanan, M., *Multiprocessors: A Comparative Study.* Englewood Cliffs, NJ: Prentice-Hall, 1980.
7. Weitzman, C., *Distributed Micro/Minicomputer Systems.* Englewood Cliffs, NJ: Prentice-Hall, 1980.
8. *Multiprocessors and Parallel Processing,* P.H. Enslow, Jr., Ed., Wiley-Interscience, 1974.

Chapter
11

Interconnection Standards

In this chapter we present some of the commercial interconnection standards that have evolved in the computer field. These standards are intended to facilitate the interconnection of equipment manufactured by a wide variety of companies. Without such standards, computer users would be left with the unpleasant and expensive task of having to design a specialized interface for each new device added to a system.

Many commercial standards are simply "de facto standards"; that is, they are the result of a widespread popularity of a particular bus or interconnection scheme. Others have acquired a more official status, after being sanctioned by various government bodies and industrywide associations. In the area of interconnections, four institutions play a particularly important role in the development of commercial standards; namely,

- American National Standards Institute (ANSI)
- Electronics Industry Association (EIA)
- International Telegraph and Telephone Consultative Committee (Comité Consultatif Internationale Télégraphique et Téléphonique—CCITT)

• Institute of Electrical and Electronics Engineers (IEEE)

ANSI and EIA are concerned with standards within the United States. CCITT is an international organization formed by the telephone companies in various countries. IEEE is an international scientific society that has no direct link to any government or industry.

This chapter is not intended to provide a comprehensive coverage of standards, nor is it meant to study the details of those presented. Its main aim is to provide the reader with an appreciation for the variety and complexity of the interconnection schemes used in microprocessor systems, through the study of a few examples. The reader is already familiar with the concept of a bus as a medium for interconnecting microprocessor system components. The data transfer, arbitration, and interrupt facilities needed were discussed at length in Chapters 6 and 7, together with examples of the 6809 and 68000 buses. In this chapter we introduce the concept of a backplane bus, which forms the backbone of a microprocessor system, and describe two relevant standards, the VMEbus and the Multibus. Two commonly used standards for serial communication, called RS-232-C and RS-449, are also discussed, followed by a discussion of some peripheral buses, particularly the Small Computer System Interconnection (SCSI) bus.

11.1 BACKPLANE BUS

Single-board microcomputers are compact and inexpensive, but limited in their functional capability. They have a main memory of a certain size and a few serial and parallel interfaces. A single-board microcomputer may also incorporate specialized circuitry, such as analog-to-digital and digital-to-analog converters to handle analog inputs and outputs. But, the types and number of such devices are determined at the time the board is designed.

A general-purpose microcomputer—a personal computer, for example—requires more flexibility than can be afforded by a single-board arrangement. Mass storage must be provided, usually in the form of a disk drive. During the lifetime of a general-purpose system, many additional components are likely to be needed—an expanded main memory, I/O interfaces to printers, plotters or modems, larger or additional disk drives, etc. All such devices should be connected to the microprocessor bus. An arrangement that provides the required flexibility is to extend the microprocessor bus lines outside the microprocessor board and use them to attach other boards that have the required interface circuitry. Typically, a bus runs along the back of a cage, with appropriate connectors into which circuit boards may be plugged, as illustrated in Figure 11.1. A bus that interconnects circuit boards in this manner is usually called a *backplane* or a *system bus*. The backplane wires may be either physically separate wires or they may themselves be traces on a printed circuit board that constitutes the back of the cage.

In the design of the microprocessor board presented in Chapter 10, a bidirectional buffer was used to extend the microprocessor bus outside the board (see Fig. 10.1). The outer side of this buffer may be connected directly to the backplane bus,

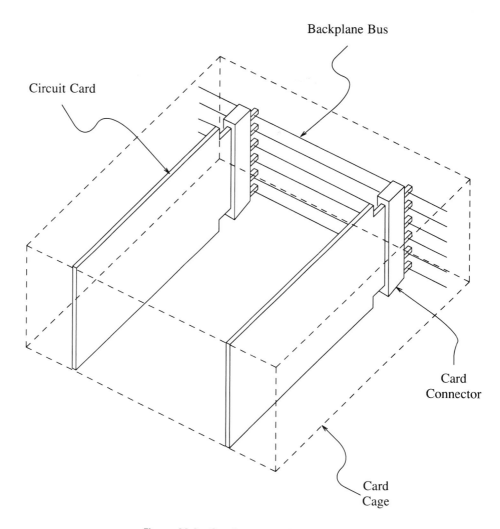

Backplane Bus

Circuit Card

Card
Connector

Card
Cage

Figure 11.1 Card cage and backplane.

in which case all lines of the backplane bus would be functionally identical to their counterparts on the microprocessor board. This arrangement has one major drawback. There are many microprocessors on the market. Even those available from the same manufacturer may have different buses. When the backplane bus is chosen to match the bus signals of a particular microprocessor, then other circuit boards can be used only if their bus interfaces are designed for that microprocessor.

Ideally, the structure of the backplane bus should be agreed upon by all manufacturers, thus forming an industrywide standard. Because of the wide range of functional capabilities that the backplane bus is expected to support in different microprocessor systems, and because of the highly competitive nature of the marketplace, such an agreement has not taken place; nor is it forthcoming. Instead, several standards have emerged, and have met with varying degrees of support among manufacturers of microcomputers and peripherals. The definitions of some such buses have been studied, modified slightly and published by the IEEE as recommended standards [1].

The use of a backplane bus is illustrated in Figure 11.2. The microprocessor board contains a microprocessor, some local memory and peripheral device interfaces interconnected by a local bus, which is the microprocessor's own bus. The local bus is connected to the backplane bus via a protocol-translation circuit. Other circuit boards connected to the backplane bus may contain additional memory and I/O interfaces. When the microprocessor initiates a read or a write operation on its local bus, the protocol-translation circuit transfers the address, data, and control information between the two buses. While doing so, it changes the encoding and/or timing of various signals, according to the needs of each bus. The memory connected to the backplane bus is usually referred to as *global memory*, because it is accessible to all devices connected to that bus. Microprocessors may or may not be allowed to access the local memory on boards other than their own.

In the arrangement of Figure 11.2, the protocol-translation operation is completely transparent. Devices connected to the backplane bus are assigned addresses within the address space of the microprocessor. The microprocessor issues read and write requests in the normal manner over its local bus. Depending upon the address

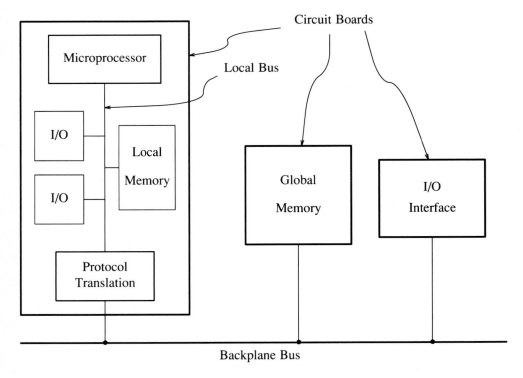

Figure 11.2 Use of a backplane (system) bus.

used, some of these requests are relayed over the backplane bus to another system component, such as the global memory module. When the reply is received, it is forwarded to the microprocessor. Thus, the only difference between local and global memory is that the latter appears slower, because of the delays introduced by the protocol-translation circuit and the backplane bus.

We will describe two examples of backplane bus standards that are in common use—the VMEbus and the Multibus—primarily to provide a complete example of the functional capabilities needed on a system bus. Some of the related issues regarding electrical drive requirements and mechanical considerations will also be discussed. Sufficient information is presented to enable simple interface circuits to be designed. However, the complete specification contains many details regarding timing, special cases, electrical circuit considerations, etc., which are not given here, and with which the hardware designer must be familiar. Interested readers should consult the relevant specification documents.

The two sections dealing with the VMEbus and the Multibus are independent. The reader may choose to study only one of the two buses, then proceed to study serial interconnection schemes without loss of continuity.

11.2 VMEbus—IEEE STANDARD 1014

The VMEbus is a sophisticated backplane bus intended for high-performance microprocessor systems. It has several advanced features that make it suitable for use in both single-processor and multiprocessor systems.

The VMEbus is a successor of the VERSAbus, which was designed by Motorola as a part of their EXORmac 68000 development system. The VERSAbus specifications were modified slightly to use a European circuit-board standard known as Eurocard. The modified version was later jointly adopted by Motorola, Mostek, and Philips/Signetics as a standard bus for some of their product lines, under the name VMEbus [2]. It has also been adopted as IEEE Standard 1014 [3].

Mechanically, the VMEbus specifications call for a card cage that can be mounted on a standard 482.6 mm (19 in.) rack. The cage is equipped with slots at 20.32 mm (0.8 in.) spacing, numbered A1, A2, etc. to hold the circuit boards. A full-width cage accommodates up to 20 circuit boards, and a half-width cage accommodates 9. A single row of 96-pin connectors is mounted on a circuit board that runs along the back of the cage to provide the backplane interconnections for Euro-card circuit boards, which are 100 mm high and 160 mm deep. A double-height cage may also be used. It provides two rows of connectors and accommodates double-height cards, which are 233 mm high and have two 96-pin connectors each. Standard-size cards may be mounted in a double-height cage, using only the upper-row connectors.

There are several options available for configuring a system based on the VMEbus. The bus may have 8, 16, or 32 data lines and may use 16, 24 or 32-bit addressing. The standard configuration, which uses one connector only, has 16 data lines and uses 24-bit addressing. An extended bus provides a 32-bit data path and 32-bit addressing, and requires both connectors of a double-height card. Sixteen-bit addresses are used only for I/O devices, in a system that has a separate I/O address space.

The signals on the VMEbus are divided into four groups, as follows:

1. Data transfer bus (DTB)
2. DTB Arbitration
3. Priority Interrupt
4. Utilities

The DTB comprises the data, address, and control lines used for transferring data. The devices that use the DTB fall into one of two categories, depending upon their role during data transfer operations. Those capable of initiating read and write requests on the bus are called *masters*. They include microprocessor boards and direct memory access (DMA) devices. Devices that are addressed by the masters during data transfer operations are called *slaves*. At a given time, there can only be one active master on the bus. The DTB arbitration lines are used by different masters to send requests for the use of the bus to an arbiter, which is responsible for the transfer of the bus mastership from one master to another.

The priority interrupt lines enable prioritized interrupt requests and acknowledgments to be exchanged between devices. The utilities lines provide support functions, such as power-on reset and detection of failures of the AC power supply. Each of the bus-line groups will be examined in the subsections below. We will first give a brief description of the electrical specifications of the bus and the types of drivers needed.

The VMEbus uses TTL-compatible signals. The bus lines are terminated at each end of the backplane circuit board, as shown in Figure 11.3. This termination is used on all lines, except those connected between devices to form a daisy chain. The 330- and 470-ohm resistors provide a termination impedance of 194 ohms. They also form a voltage divider that produces a voltage of 2.94 V when no other device is driving the bus. Since the high-state threshold for TTL inputs is 2.0 V, all lines appear in the high-voltage state, unless driven to the low-voltage state by some device. To pull a bus line to 0.5 V, a device must be able to sink a current of

$$2 \times \frac{2.94 - 0.5}{0.194} = 25.2 \text{ mA}$$

The factor of 2 in this equation is due to the fact that the line is terminated at both ends. This current is in addition to the current needed by all the logic gates connected to the line.

Interface circuits connected to the bus use different types of drivers on different lines, depending on the nature of the signal and the number of inputs and outputs connected to the line. A standard TTL totem-pole driver (TP) is used on a line that has only one sender and one receiver attached to it. A high-current version (HC) is needed when there are multiple receivers. For example, the clock line on the bus requires a TP-HC driver. Lines that have multiple drivers require either tri-state (3ST) or open-collector (OC) drivers. Because of the line terminations, these lines

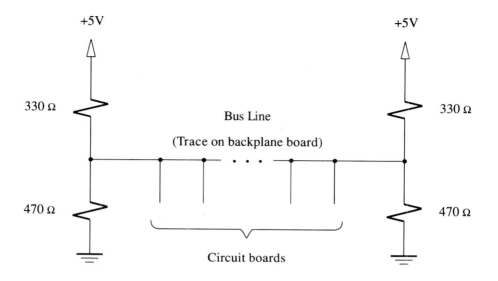

Figure 11.3 Termination of VMEbus lines.

will always be in the high-voltage state when none of the drivers connected to them is enabled. However, the state of a 3ST line is unpredictable for the short period during which a particular driver is being enabled or disabled. For this reason, OC drivers are used on critical timing lines, such as those that carry the data-strobe signals, where false transitions may lead to errors. Tri-state drivers are used on lines where no particular significance is attached to signal edges, e.g., on address and data lines. Open-collector drivers must also be used on lines that may be activated (driven low) by more than one driver simultaneously, as in the case of an interrupt-request line where the desired state is the logical OR of the states of all the drivers connected to the line.

11.2.1 Data Transfer Bus

The data transfer bus comprises the lines used during read and write operations, which are summarized in Table 11.1. They include the address and data lines and the control lines needed to carry timing signals and information such as the operand size and the direction of the transfer. For each line, the table gives the mnemonic designation, a brief functional description, the type of driver needed, and the number of bus connector pins involved. A line whose name ends with an "*" is active when in the low-voltage state.

Data transfer over the VMEbus can take place in 8-bit (byte), 16-bit (word), or 32-bit (long word) quantities. The number of address bits used may be 16 (short), 24 (standard), or 32 (extended), where short addresses are intended only for addressing I/O devices. Six additional lines may be used to carry an "address-modifier" code, whose role will be described later. In the remainder of this section we will refer to

Table 11.1   VMEbus DATA TRANSFER LINES

Designation	Function	Driver	Number
A01-A31	Address	3ST	31
AM0-AM5	Address modifier	3ST	6
D00-D31	Data	3ST	32
WRITE*	Indicates write operation	3ST	1
AS*	Address strobe	3ST-HC	1
DS0*, DS1*	Data strobe for low (odd) and high (even) byte	3ST-HC	2
LWORD*	Indicates a long word	3ST	1
DTACK*	Data acknowledge	OC	1
BERR*	Bus error—indicates that a requested transfer cannot take place	OC	1
		Total	76

32-bit quantities, but the discussion applies equally to addresses and data of any length.

Read and write requests on the VMEbus are differentiated by means of the WRITE* line, which is activated for write operations only. The timing of a data transfer is controlled by asynchronous handshake signals similar to those used on the 68000 bus (see Subsection 6.10.1). The master activates the Address and Data Strobe signals, AS*, DS0*, and DS1*, and the slave responds by activating the Data-Acknowledge signal, DTACK*. In the case of an error, the bus error line, BERR*, is activated instead of DTACK*. The bus address space is organized as illustrated in Figure 11.4. Bytes can have any 32-bit address, A_{31-0}. Word addresses must always have $A_0 = 0$. When $A_1 = 0$ they are said to be even word addresses, and when $A_1 = 1$ they are odd word addresses. The high-order byte of a word has the same address as that of the word, and the low-order byte has the next-higher address. Long word addresses must have $A_1 = A_0 = 0$. This organization is essentially the same as that of the 68000 microprocessor, with the exception that the 68000 allows long words to have any even address. Only the high-order 31 bits of an address are transmitted on the address lines of the VMEbus, A31 to A01. As in the case of the 68000 bus, separate data strobe signals, DS1* and DS0*, are provided for even and odd bytes, respectively. Both data strobe lines are activated for word transfers. A third line, LWORD*, is activated by the bus master for long word transfers.

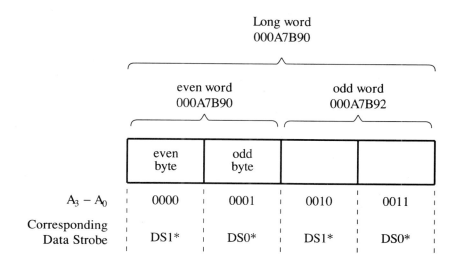

Figure 11.4 Example of addressing and available data lengths on the VMEbus.

Irrespective of the address, data are always placed on the data lines such that the least-significant data bit is on line D00. A byte transfer always uses lines D07 to D00 and a word transfer uses lines D15 to D00. Hence, a 32-bit device, such as a memory board, must incorporate appropriate multiplexers to shift byte or word data to the low-order data lines, to enable byte or word transfers to take place. Figure 11.5 shows how a 32-bit memory may be connected to the bus to enable byte, word, and long-word read operations. A multiplexer, MPX, selects either the high-order or low-order word of data, depending on the state of address bit A_1. Then, either the even or the odd byte of that word is selected for transmission on data lines D07–D00, as determined by the data strobe lines. $\overline{EB}$ is activated for an even byte transfer (only DS1* is active) and $\overline{OB}$ for an odd byte (DS0* is active). For word operations (both DS0* and DS1* are active), $\overline{W}$ and $\overline{OB}$ are activated, and for long-word operations (LWORD* is active) $\overline{LW}$ is also activated. Table 11.2 gives the desired behavior for the four signals $\overline{OB}$, $\overline{EB}$, $\overline{W}$, and $\overline{LW}$, and Figure 11.6 shows how they may be generated. A multiplexing circuit similar to that in Figure 11.5 is needed for write operations.

Shifting data to the low-order section of the bus may seem to be an unnecessary complication, but it is needed to simplify the connection of 8 and 16-bit devices. For example, a disk drive that transfers data 16 bits at a time should be connected only to lines D15 to D00. With the multiplexing circuitry of Figure 11.5, DMA transfers from a 32-bit memory board to the disk can take place in a straightforward manner. Similarly, an 8-bit device should be connected to lines D07 to D00.

Recall that the approach taken in the design of the 68000 bus is slightly different, where byte transfers are allowed on either half of the 16-bit data bus. An odd

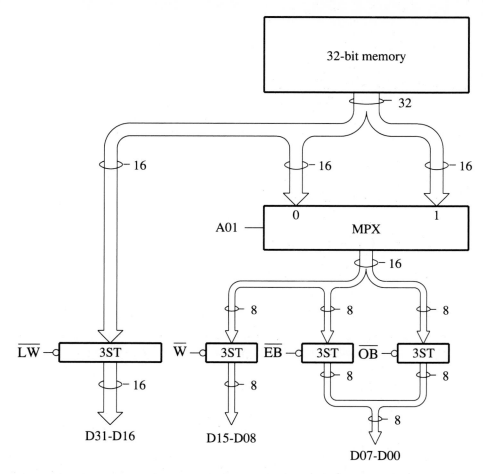

Figure 11.5 An example of the multiplexing needed when connecting a 32-bit memory to the VMEbus.

Table 11.2 DESIRED BEHAVIOR OF THE CONTROL SIGNALS OF THE DRIVERS IN FIGURE 11.5

Transfer	$\overline{\text{LW}}$	$\overline{\text{W}}$	$\overline{\text{EB}}$	$\overline{\text{OB}}$
Long word	0	0	1	0
Word	1	0	1	0
Even byte	1	1	0	1
Odd byte	1	1	1	0

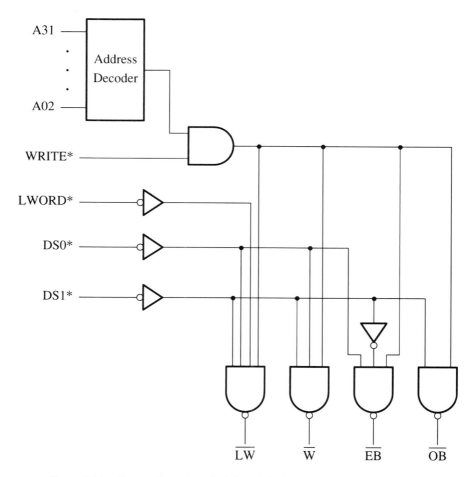

Figure 11.6 Generation of control signals for the tri-state drivers in Figure 11.5.

data byte is transferred on D_{7-0} and an even byte on D_{15-8}. To simplify the connection of 8-bit devices, the 68000 provides a special instruction, MOVP, which transfers successive data bytes using only one half of the data bus. When a 68000 microprocessor communicates with an 8-bit device via the VMEbus, the MOVP instruction is not needed, because a multiplexing arrangement similar to that of Figure 11.5 will be incorporated in the bus-protocol translation circuit of Figure 11.2.

DATA TRANSFER TIMING

The signal exchange between a master and a slave during a read operation over the VMEbus is illustrated in Figure 11.7. The master places new information on the address and address-modifier lines and, in the case of a long-word transfer, activates LWORD*. After a short period (minimum 35 ns) to allow for bus skew and address decoder delays at the slave, the master activates the address strobe line, AS*. Then, it activates either DS0* or DS1*, or both, as appropriate. The master cannot activate the data strobe signals until DTACK*, which may have been active in a previous operation, has returned to the inactive state. It must also ensure that the WRITE* signal has been in the inactive (high) state for at least 35 ns.

Upon receiving the data strobe, the selected slave places the requested data on the data lines, then activates DTACK*. In response, the master loads the data into an internal buffer, removes the address information, and deactivates its strobe signals. Bus specifications require the slave to guarantee that the information on the data lines is valid before it activates DTACK*. It must also guarantee that the delay, t_{ack}, between the time it receives the data strobe and the time it activates DTACK* is not less than 30 ns. This delay forces the master to keep a valid address on the

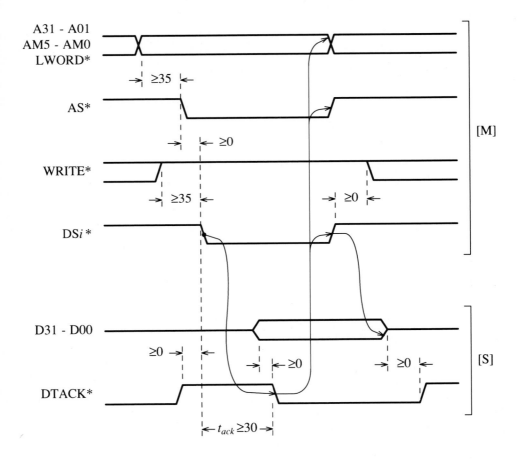

[M] Lines driven by master

[S] Lines driven by slave

Figure 11.7 A read operation on the VMEbus. Delays are given in nanoseconds.

bus for at least that long, to ensure that other devices have a chance to interpret the address properly. The slave maintains valid data on the bus until the data strobe returns to its inactive state, at which point the slave places its data drivers in the high-impedance state, then deactivates DTACK*.

A write operation proceeds in the same manner, except that the data are transmitted by the master. Its timing details are shown in Figure 11.8. The master activates WRITE* and places the data on the data lines. Then, after at least 35 ns, it activates the data-strobe signals. This delay allows for bus skew, and provides some setup time for the input buffers at the slave. Note that the master must ensure that DTACK* has reached its high state following the previous operation before turning on the tri-state drivers on the data bus.

The master may combine a read transfer and a write transfer into a single indivisible read-modify-write cycle. This cycle is needed to facilitate task synchronization in multiprocessor systems, as discussed in Subsection 10.3.2. Its implementation on the VMEbus is exactly the same as on the 68000 microprocessor bus (see Subsection 6.10.1). After completing the read operation, the master maintains AS*

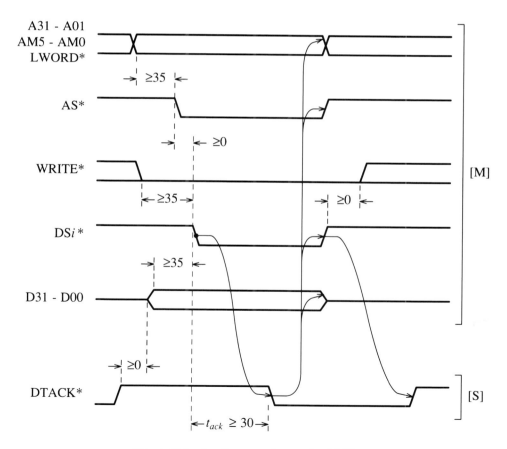

Figure 11.8 A write operation on the VMEbus.

in the active state, activates WRITE*, puts new data on the data lines and issues a second data strobe. The cycle terminates when AS* returns to the inactive state at the end of the write operation. No other device can use the bus between the read and write operations, because the bus mastership can be transferred to a new device only while AS* is inactive, as will be explained in Subsection 11.2.2.

A few comments are in order about the timing of data transfers over the VMEbus. The bus designers have chosen to use the DTACK* signal to guarantee that no two devices have their data bus drivers enabled at the same time. In a read operation, Figure 11.7, the slave enables its drivers only after it receives a data strobe, which is not issued by the master unless DTACK* is inactive. At the end of a transfer, the slave ensures that its data drivers have been turned off before returning DTACK* to the inactive state. These two conditions guarantee that a slave will not enable its drivers until the device transmitting data in the immediately preceding transfer has turned off its drivers. Similar conditions apply in a write operation.

Another useful feature in the timing diagrams of Figures 11.7 and 11.8 is that the master has the responsibility for inserting most of the delays needed for correct operation. In particular, the address and data strobe signals are delayed slightly to allow for bus skew and for address decoding and setup requirements in the slave. In a read operation, the master must also compensate for the fact that, due to bus skew, it may receive an active state on DTACK* slightly before the data is valid. As a result of these constraints on the master, it is possible to use simple circuitry in most slave interfaces. If the delays needed in a particular slave exceed those allowed for by the master, the slave must introduce additional delays internally.

Finally, let us examine the way in which a read or a write cycle ends. When the master receives an active state on DTACK*, it removes all its signals from the bus. It is important to consider when these signals change state as seen by various devices on the bus. Because of bus skew, the address lines may change their state before the address strobe. Also, in a write operation, the data may become invalid before the data strobe returns to the inactive state. Slave interfaces must ensure that such conditions will not lead to false selection or to incorrect data being stored. The device selection circuitry should examine the address lines only at the falling edge of AS*, because once DTACK* is activated the information on the address lines is unreliable. Also, in the case of a write operation, the data buffer in the interface should be isolated from the data lines before activating DTACK*, by using edge-triggered flip-flops, for example.

ADDRESS MODIFICATION

The VMEbus is intended for use in systems in which there may be several microprocessors acting as masters and several memory and I/O devices, which normally act as slaves. Some I/O devices may become bus masters during DMA transfers. The address-modifier lines provide flexibility in managing the operation of such a system. For example, it may be desirable to partition the system such that some devices respond only to certain processors. The desired partitions may be created by arranging for each processor's interface to place an identifying code on the address-modifier lines. All slaves examine the address-modifier lines, and respond only to the codes assigned to them. A given device may respond at different addresses for different address-modifier codes. These codes may also be used to implement privileged access. A memory or I/O device will respond only to access requests that are accompanied by an address-modifier code indicating that the request has a sufficiently high level of privilege.

Some address-modifier codes have been assigned certain interpretations as a part of the specification of the VMEbus. Others are left to be defined by system designers and users. Among those that form a part of the bus specification are codes that distinguish between Supervisor (privileged) and User (nonprivileged) access, and between program and data access. The address-modifier codes are also used to indicate whether the address lines carry an extended (32-bit), a standard (24-bit), or a short (16-bit) address. A short address may be used in a system that has an I/O address space separate from that of the main memory. This is the case when the microprocessors used have special instructions for I/O operations, as is the case with microprocessors in the Intel family. A microprocessor that uses memory-mapped I/O, such as 6809 and 68000, has a single address space for all devices on its bus.

Another feature that can be selected by the address-modifier code is a sequential access cycle. This is a bus cycle in which a contiguous block of data is transferred to or from the main memory. The entire transfer appears as a single bus cycle. It is initiated by the master by issuing a normal read or write request, and placing the sequential-access code on the address-modifier lines. All devices that have sequential access capability record the address transmitted by the master in an internal address counter. Then, the memory board selected by that address proceeds with the requested read or write operation in the normal manner. At the rising edge of DSi *, which marks the end of the transfer operation, all participating slaves increment their internal address counters by 1, 2, or 4, for byte, word, and long-word transfers, respectively. Meanwhile, instead of terminating the transfer by deactivating AS*, the bus master maintains AS* in the active state, and requests another transfer by issuing a new data strobe. This time, no new address is placed on the address lines; the participating slaves use their internal address counters instead. At the end of the transfer, the address counters are once again incremented. Data transfers at sequentially increasing addresses continue in this manner until the master returns the address strobe line to the inactive state.

Note that although a single slave—a memory board in this case—is involved in a given transfer operation, all slaves that have sequential access capability record the starting address and increment it at the end of each transfer. This action is necessary because as the address is incremented it may cross the boundary from one memory board to another.

A sequential-access cycle differs from a sequence of separate read or write cycles at sequential addresses in one respect. When a sequential-access cycle is used, availability of the bus is guaranteed for the entire block to be transferred without interruption. According to the bus arbitration protocol, a new master can acquire the bus mastership only after the current bus master has released the address strobe line, AS*. If separate cycles are used, a higher-priority master may acquire the bus mastership halfway through the block transfer, thus making the total time needed to complete the transfer unpredictable.

11.2.2 DTB Arbitration

A backplane bus is often shared among several masters. Some arbitration schemes that can be used to schedule the use of the bus in such circumstances were introduced in Chapter 7. We will now briefly describe the arbitration process on the VMEbus.

The signals for bus arbitration are summarized in Table 11.3. Four bus request signals, BR0*-BR3*, are available for masters to request the bus mastership in four priority groups. The bus arbiter, which must reside on the circuit board plugged into slot A1, responds by activating one of four bus-grant signals. The grant lines are connected in a daisy chain fashion to all devices in each group. At each device a grant signal appears on two pins, one for input and one for output, having the names BGiIN* and BGiOUT*, respectively. The chain is formed by connecting the output signal at slot Aj to the input signal at slot A(j+1). The bus-grant outputs of the arbiter are connected to the BGiIN* pins at slot A1. At any board location where a given grant signal is not used, the BGiIN* pins must be jumpered to the corresponding BGiOUT* pins, to enable the grant signal to propagate to other masters downstream. Note that the arbiter produces the grant signal on the input side of its board to ensure that this signal will be seen by all bus interfaces on the chain, including those that may be on the same board as the arbiter. Because of the daisy chain organization, the driver connected to BGiOUT* on one board drives only a single receiver connected to BGiIN* on an adjacent board. Hence, these lines require standard totem-pole drivers.

The VMEbus specifications define three types of bus arbiters. The simplest accepts bus requests on BR3* only, and generates a grant on BG3IN*. In this case all masters on the bus are connected into the BG3IN/BG3OUT daisy chain.

The second type uses all four request lines and generates grant signals on the corresponding grant lines. It treats simultaneous requests on a fixed priority basis, where BR3* has the highest priority and BR0* the lowest priority. If the arbiter receives a request whose priority is higher than that of the current bus master, the

Table 11.3   ARBITRATION SIGNALS ON VMEbus

Designation	Function	Driver	Number
BR0*-BR3*	Bus Request	OC	4
BG0IN*-BG3IN*	Bus grant daisy chain inputs	TP	4
BG0OUT*-BG3OUT*	Bus grant daisy chain outputs	TP	4
BCLR*	Bus Clear—a request to the current master to release the bus.	TP-HC	1
BBSY*	Bus Busy	OC	1
		Total	14

arbiter activates BCLR*, to request that the current bus master stop using the bus. While the current bus master is not obliged to respond within any particular period, it should release the bus as soon as it reaches a point in its operation where it is safe to do so. As soon as the bus becomes free, the arbiter signals the higher-priority master to begin using the bus.

The third type of bus arbiter uses the round-robin or rotating-priority scheme described in Subsection 7.7.1. After granting a request at level n, level $n-1$ (mod 4) is given the highest priority. Of course, devices connected to the same bus-request line continue to have a fixed-priority ordering based on their electrical distance from the arbiter.

Let us now examine the way in which the bus mastership is passed on to a new master. The bus busy, BBSY*, and address strobe, AS*, signals play important roles in this operation. Bus busy is an open-collector line, which is inactive unless it is pulled to the low-voltage state by some device. Its main function is to indicate to the bus arbiter and to other masters that some device is currently using the bus. The current bus master maintains BBSY* in the active state as long as it wishes to transfer data, releasing it only when it is ready to relinquish the bus. The arbiter cannot generate a bus grant unless BBSY* is in the inactive state.

The sequence of events for transferring the bus mastership from one master, A, to another, B, is illustrated in Figure 11.9. Before t_0, master A is using the bus. The low periods on line AS* indicate data transfer cycles generated by device A.

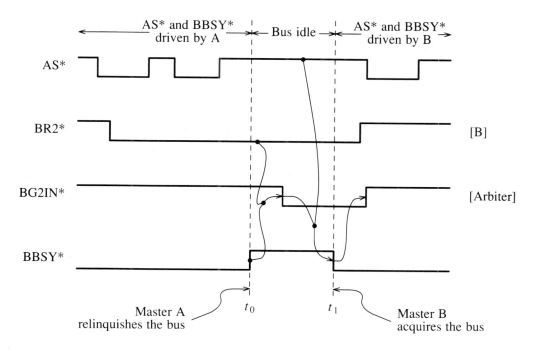

Figure 11.9 Transfer of bus mastership from master A to master B on the VMEbus.

After completing its last data transfer cycle, device A releases BBSY*, thus indicating to the arbiter that it may grant the bus to another master. At this time there is an outstanding request from master B on bus request line #2. Hence, the arbiter issues a bus grant at level 2, which propagates along the BG2IN*/BG2OUT* chain until it reaches device B. We have used two arrows emanating from the BBSY* and BR2* signals to point to the activation of BG2IN*. This is intended to show that the arbiter issues the grant in response to the request on BR2*, but only after BBSY* becomes inactive. Similarly, Master B activates BBSY* and assumes the bus mastership when its BG2IN* input becomes active; however, it must first check that the AS* line is in the inactive state, to ensure that the previous master has turned off all its output drivers. After becoming the bus master, device B drops its bus request and may begin transferring data over the bus. The arbiter deactivates the bus-grant signal in response to BBSY* becoming active.

The idle period between t_0 and t_1 in Figure 11.9 is the time taken by the arbiter to select the highest-priority request and by the bus-grant signal to propagate to the first device requesting the bus on the daisy chain. According to the specifications of the VMEbus, this period may overlap the last data-transfer cycle of master A. Instead of waiting until that transfer cycle is completed, device A may release BBSY* after it has activated AS* for the last time, as shown in Figure 11.10. When device B receives the bus grant, AS* may still be in the active state. In this case device B must wait until AS* becomes inactive, before activating BBSY* and beginning to transfer data.

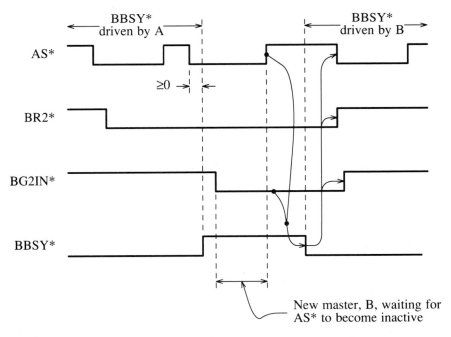

Figure 11.10 Transfer of bus mastership with arbitration occurring during the last data transfer cycle.

11.2.3 Interrupts

Before introducing the interrupt scheme used, we should reiterate that as a standard for a backplane bus, the VMEbus is not intended to support any particular microprocessor, even though it has many similarities to the 68000 bus. In fact, a multiprocessor configuration may accommodate several different types of microprocessors within a single system, each with its own way of handling interrupts. The backplane bus provides a communications channel through which processors and their peripherals exchange interrupt-request and acknowledgment signals. We will refer to any processor that is capable of receiving and processing interrupt requests as an *interrupt handler*, and to devices that issue interrupt requests as *interrupters*.

The VMEbus lines used for interrupts are given in Table 11.4. There are seven interrupt-request lines, IRQ1* to IRQ7*. All seven lines may be connected to a single interrupt handler, in which case IRQ7* is treated as the highest-priority and IRQ1* the lowest-priority line. Alternatively, the responsibility for servicing interrupts may be distributed among several interrupt handlers. For example, one microprocessor may service interrupt requests on lines IRQ1* to IRQ3* while another microprocessor services lines IRQ4* to IRQ7*. In this case, IRQ3* will have the highest priority in the first group and IRQ7* the highest priority in the second group. No explicit priority hierarchy is defined for lines serviced by one interrupt handler relative to those serviced by another. However, as will be explained below, an interrupt handler needs to acquire the bus mastership before it can accept an interrupt request. Hence, if simultaneous requests arrive on lines IRQ3* and IRQ4* in our example, the request that will be serviced first is the one whose handler is connected to a higher-priority bus-request line.

A single interrupt-acknowledge line, IACK*, is used for acknowledging the interrupt requests received on any of the seven interrupt-request lines. The particular line being acknowledged is identified by transmitting a 3-bit code on address lines A03 to A01. In order to resolve multiple requests on the same request line, the acknowledgment signal is connected to all interrupters in a daisy chain fashion, similar to that used for the bus-grant signal. However, one important difference is that while the bus grant originates at the arbiter, which always resides at slot A1, the interrupt-

Table 11.4   INTERRUPT-HANDLING SIGNALS ON VMEbus

Designation	Function	Driver	Number
IRQ1*–IRQ7*	Interrupt-request	OC	7
IACK*	Interrupt-acknowledge	3ST or OC	1
IACKIN*	Interrupt-acknowledge daisy chain input	—	1
IACKOUT*	Interrupt-acknowledge daisy chain output	TP	1
		Total	10

acknowledge signal may be generated by any interrupt handler. Hence, the interrupt-acknowledge signal must be sent to the beginning of the daisy chain, from where it can propagate sequentially through all device interfaces. The arrangement used is illustrated in Figure 11.11, which shows the first six slots of the bus. Interrupter devices are assumed to be plugged into slots A1, A4, and A5, and an interrupt handler is plugged in slot A2. The device in slot A6 is assumed to be a microprocessor that is capable of generating interrupt requests (to other microprocessors) and of servicing interrupts. That is, it can function either as an interrupter or as an interrupt handler. Slot A3 is either empty or has a device that does not use interrupts. The IACK* line is connected to all slots to form a common bus line. It may be activated by any interrupt handler, using an open-collector driver.

In addition to IACK*, each bus device has an interrupt-acknowledge input, IACKIN*, and output, IACKOUT*, which are connected to form a daisy chain, as shown. At slot #1, IACK* is connected to the beginning of this chain. At any slot where the interrupt-acknowledge signal is not used, the IACKIN* and IACKOUT* pins are connected together to maintain continuity of the chain. The mechanical specification of the VMEbus provides jumper plugs for this purpose. At other slots, interrupters must propagate the interrupt-acknowledge signal downstream, unless

1. The device has an interrupt request pending on one of the interrupt-request lines, IRQ1* to IRQ7*.
2. The code on address lines A03–A01 matches the number of that interrupt-request line.

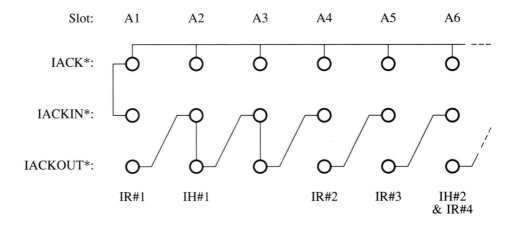

Figure 11.11 Pin interconnections for the interrupt-acknowledge signal.

In this case, the interrupter blocks the propagation of the interrupt-acknowledge signal and identifies itself to the interrupt handler microprocessor by transmitting an ID code on the data lines. The ID code may be a byte, a word, or a long word. This information will normally be used by the interrupt handler to locate the corresponding interrupt vector in its main memory, as in the case of the 68000 microprocessor. Since the nature and use of the ID information differs from one microprocessor to another, it is not given any particular interpretation by the VMEbus specifications.

The process of transferring the interrupter's ID is initiated and controlled by the interrupt handler in essentially the same manner as a normal read operation. Before issuing an interrupt acknowledgment, the interrupt handler must first acquire the bus mastership, if it is not already the bus master at the time it decides to acknowledge an interrupt request. It then starts an interrupt-acknowledge cycle, which is illustrated in Figure 11.12. The figure shows the case where an interrupt request is raised on line IRQ4* by interrupter IR#2 of Figure 11.11. The interrupt handler responsible for line IRQ4* acknowledges this request by transmitting the interrupt-level code, 4 in this case, on the address lines and activating IACK*. Then, it activates the address and data strobes, as in a read operation. All interrupters that are requesting service examine the state of the IACK* line at the time AS* becomes active. An active state on IACK* indicates that this is an interrupt-acknowledge cycle instead of a data transfer cycle. Hence, the interrupters decode address lines A03 to A01 to determine the level of the interrupt being acknowledged. Those not requesting an interrupt at level 4 simply pass the acknowledgment downstream, but they must wait at least 40 ns after DSi* has become active before activating their IACKOUT* outputs.

The active state on IACK* propagates to slot A1 and appears at the IACKIN* input of IR#1. Assuming that IR#1 is not requesting an interrupt, it waits for 40 ns after DSi* becomes active, then it activates its IACKOUT* output. The interrupt acknowledgment propagates down the daisy chain until it reaches IR#2, which responds by transmitting its identifying code on lines D07 to D00 and activating DTACK*. Note that the interrupter checks both IACKIN* and DSi* before transmitting its ID code. This may appear unnecessary, because IR#1 is not allowed to activate IACKOUT* until DSi* becomes active. However, an interrupter in slot A1 may receive an interrupt acknowledgment before DSi* becomes active. For the sake of generality and interchangeability, interrupters are required to check both signals. Interrupter #2 maintains its IACKOUT* output in the inactive state, thus terminating the propagation of the interrupt acknowledgment.

The bus cycle is completed in the same way as a normal read cycle. All devices on the daisy chain must deactivate their IACKOUT* outputs as soon as AS* becomes inactive. The interrupter that received service, IR#2 in the case of Figure 11.12, may remove its interrupt request either at the end of the acknowledgment cycle or later when one of its registers is accessed by the interrupt handler. This flexibility is intended to accommodate the wide variety of peripheral ICs on the market. For example, the ACIA turns off its $\overline{IRQ}$ output when its data register is accessed by the microprocessor, as was explained in Subsection 7.2.1.

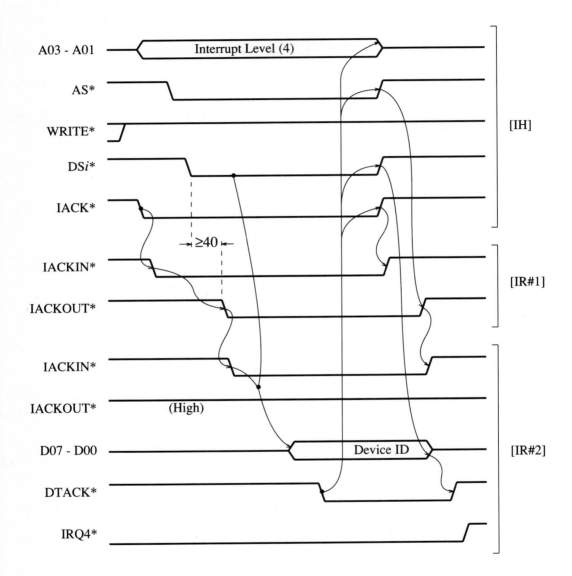

IH: Interrupt handler responsible for IRQ4*.
IR#1: Interrupter in slot A1 of Figure 11.11.
IR#2: Interrupter in slot A4 of Figure 11.11.

Figure 11.12 Timing of an interrupt-acknowledge cycle on the VMEbus.

11.2.4 Utilities and Power Supply

The bus lines described so far carry the signals needed to perform the main functions of the bus, which are to enable various devices to exchange data and interrupt signals. In a microprocessor system, a number of bus lines are needed to coordinate system operation at the time power is turned on, to detect failures, etc. Table 11.5 lists the utility and power supply lines on the VMEbus. A brief description of the utility lines is given below.

- SYSCLK: This line carries a 16 MHz clock, which is independent of all other bus signals. It is provided for use by interface circuits that require a clock signal for any purpose. An example of the use of a clock signal to introduce a delay was given in Subsection 6.10.1 and Figure 6.58. SYSCLK should not be confused with the clock signal used to control data transfers on a synchronous bus. As pointed out earlier, the VMEbus uses an asynchronous handshake between the data strobe and acknowledgment signals.

- ACFAIL*: The power supply circuit in a computer system generates the DC supply voltages needed from the AC power. This circuit incorporates a large storage capacitor capable of maintaining the DC voltages within their specified limits for several milliseconds after the AC power is interrupted. In order to guard against loss of data following a power failure, the system software should use this "grace period" to save any data stored in volatile RAM devices

Table 11.5 UTILITY LINES ON THE VMEbus

Designation	Function	Driver	Number
SYSCLK	System clock	TP-HC	1
ACFAIL*	AC power failure	OC	1
SYSRESET*	System Reset	OC	1
SYSFAIL*	System failed self test	OC	1
+5V	Power supply lines		6
+12V	Power supply line		1
−12V	Power supply line		1
+5VSTDBY	+5V supply for devices requiring battery backup		1
GND	Ground		12
Reserved	Reserved for future extensions		1
		Total	26

on non-volatile devices, such as disks. The ACFAIL* line is used to alert various devices on the bus when the AC power is lost and at least 4 ms before the DC supply voltages fall below their specified values. The way in which the information on the ACFAIL* line should be used is not a part of the VMEbus specifications. It is the responsibility of the system designer to use this information to start the task of saving volatile data and shutting down various operations gracefully. Perhaps an active state on ACFAIL* could cause an interrupt in the microprocessor responsible for maintaining the system's integrity, to inform it of the imminent loss of power.

- SYSRESET*: In previous chapters, we encountered several examples of the need for reset signals in microprocessors and interface circuits. The SYS-RESET* line should be connected to the reset inputs of all devices on the VMEbus. It should be activated by the power supply circuit whenever power is turned on, to place all devices in their proper starting state. It may also be activated by an operator via a switch.

- SYSFAIL*: A reliable computer system is one that operates for a long time without failure. Another, and perhaps more important, aspect of system reliability is that the system should not appear to operate properly while in fact it is producing incorrect results because of some internal malfunction. Because of the complexity of modern microcomputer systems, elaborate diagnostic tests are carried out often on individual components to ensure system integrity. These tests are usually run at the time power is turned on, before normal system operation begins. They may also be carried out at regular intervals while the system is running, under control of the system software. The SYS-FAIL* line provides a common communications medium for all devices on the VMEbus to report the result of their diagnostic tests. Following a system reset, any device that has a self-testing capability activates the SYSFAIL* line and begins running its self-diagnosis test. If it passes the test, it releases SYS-FAIL*. Hence, SYSFAIL* will remain active if any device detects a failure. The system software should check the state of this line before beginning, or resuming, normal operation.

11.2.5 Interface Example

The VMEbus introduces more constraints than those encountered in the simple interface circuits discussed in Chapter 6. Consider an input interface. According to the discussion in Subsection 11.2.1, the circuit that implements this interface is required to meet the two conditions:

1. It must determine whether it is being addressed at the falling edge of AS*.
2. It must guarantee a minimum delay of 30 ns after receiving DSi* before it activates DTACK*.

Because of these two requirements, the interface cannot be implemented as a combinational circuit, as was the case in the example of Figure 6.6. Memory elements are needed to control various stages of the data transfer operation. In other words, the interface must be implemented as a *sequential machine*. The timing diagram in Figure 11.7 indicates that we may represent the operation of the interface circuit by the state diagram in Figure 11.13. Initially, the interface is in the Idle state. When it receives an address strobe while its address is on the address lines it moves to the Selected state, where it waits for a data strobe. After receiving the data strobe, it waits for a short period, represented by the Delay state. Then, it enters the Response state and sends an acknowledgment. When the data strobe becomes inactive the interface returns to the Idle state. During a read operation (WRITE* is inactive), the interface should transmit the requested data as soon as it can after it receives the data strobe. If it can do so in less than 30 ns, the delay period may be set to its minimum value of 30 ns. Otherwise, the delay period should be extended as needed, to ensure that valid data will appear on the data lines before DTACK* becomes active.

A logic circuit that implements the slave interface described above is shown in Figure 11.14. The address strobe sets the Selected flip-flop if the device is being addressed. The address transmitted by the master is valid 35 ns before AS* is activated. However, because of bus skew, the slave is guaranteed an address setup time of not more than 10 ns. If more time is needed, a short delay, D1, may be inserted as shown. In choosing the value of D1, it must be remembered that the address cannot be assumed valid later than 30 ns after AS* becomes active. When the data strobe is activated and the requested transfer is a read operation, Enable-D* is activated (set low), and after a delay D2, the Response latch is set to 1 to activate DTACK*. The value of D2 should be chosen to control the delay between the data strobe and acknowledgment as described above. The acknowledgment latch is reset as soon as one of the two data strobes returns to the inactive state.

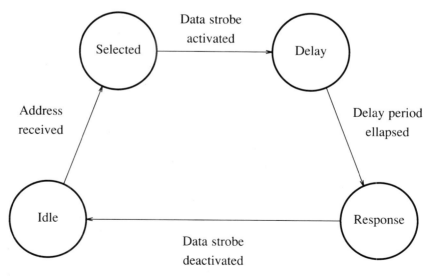

Figure 11.13 A state diagram representing the operation of a VMEbus interface circuit.

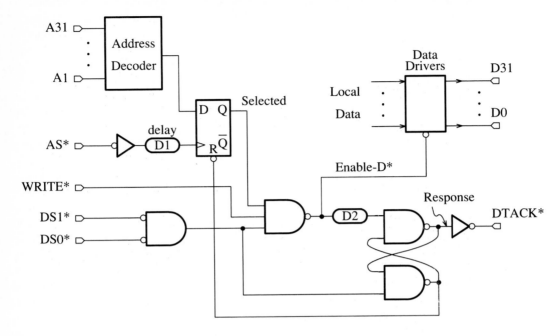

Figure 11.14   An input interface circuit for the VMEbus.

In this circuit, Enable-D* is activated only after the data strobe becomes active. The reader should consider the possibility of activating this signal as soon as the device is selected (Selected=1). Will that violate any of the rules of the bus protocol?

11.2.6 Bus Bandwidth

The term *bus bandwidth* is often used to refer to the maximum rate at which data can be transferred over a bus. For a synchronous bus on which one data transfer takes place in every clock period, the bandwidth is simply the clock frequency times the width of the bus. For example, a 16-bit synchronous bus that operates with a clock frequency of 4 MHz has a bandwidth of 8 Mbytes/s. The bandwidth is more difficult to assess for asynchronous buses, because some of the delays within a bus cycle are variable. These delays depend on the physical distance between the devices involved and on the internal delays within each device. Hence, the bus bandwidth may be computed based on typical or worst-case values, depending on the intended use.

In order to estimate the bandwidth of the VMEbus, we need to consider the propagation delays introduced by the bus lines during the handshake between the master and the slave. We also need to determine how soon after one transfer ends another can begin. To examine the propagation delays, it is convenient to draw a timing diagram in which each signal appears twice, once as seen by the master and once as seen by the slave. The timing diagram of Figure 11.7 redrawn in this manner is shown in Figure 11.15, where P is the propagation delay on the bus. Because of bus skew, the value of P may vary from one bus line to another. For example, although the minimum address setup time introduced by the master is 35 ns, the bus specifications guarantee a value of only 10 ns at the slave. Also, while the slave transmits its data before it activates DTACK*, the data may arrive at the master as much as 25 ns after the acknowledgment. The master must compensate for this delay by waiting a period t_d, equal to 25 ns plus whatever the master needs for setup time, before loading the data into its input buffer and deactivating the data strobe.

The master cannot issue a new data strobe until it sees DTACK* return to the inactive state. Another constraint that affects the minimum separation between cycles is that once the data strobe signals return to the inactive state, they must remain in that state for a minimum of 40 ns before they can be activated again. A similar constraint applies to the address strobe. These constraints are needed to ensure that the pulses on the strobe lines can be seen by all devices, including those not involved in the transfer. For example, all masters need to monitor AS* because of the role it plays in the transfer of bus mastership. Also, several devices may need to monitor the data strobe to increment their internal address counters during block transfers.

Let T be the time interval between two successive data strobes. To obtain the minimum value of T, let us assume that each device requires at least two gate delays to respond to any signal. Therefore

$$T_{min} = 2P + t_{ack} + t_d + Max\ (40, 2P + 4G)$$

where G stands for one gate delay. For two devices that are 25 cm apart, and if we assume a propagation speed of 3 cm/ns (see Subsection 10.3.3), we obtain $P = 8$ ns. Let $G = 10$ ns, $t_{ack} = 30$ ns, and $t_d = 30$ ns. In this case, $T_{min} = 130$ ns, which means that the two devices can exchange data at a rate of about 7.7 transfers every microsecond. Since the bus can carry 4 bytes in parallel, this represents a bandwidth of 31 Mbytes/s.

The value of 31 Mbytes/s should be regarded as an upper bound rather than a value that can be expected in practice. We have used minimum values for bus delays such as address setup time. Designers are likely to use somewhat larger values to account for variations in component characteristics with temperature, age, etc. Also, the delays discussed here are in addition to any delays introduced on the circuit boards connected to the VMEbus. For example, consider the cycle time seen by the microprocessor in Figure 11.2, when transferring data to or from a global memory module. The duration of the cycle will be determined by the sum of the delays on the microprocessor board, the system bus and the memory board. Clearly, the rate of such data transfers will be considerably lower than 31 Mbytes/s.

Another factor that affects the memory access time is that the microprocessor board may have to contend for bus mastership before transferring data. If another device happens to be using the bus, the microprocessor will be forced to wait. This waiting period increases the average access time of the global memory as seen by the microprocessor. We should also note that the time overhead introduced during the transfer of bus mastership (see Figs. 11.9 and 11.10) can significantly affect the

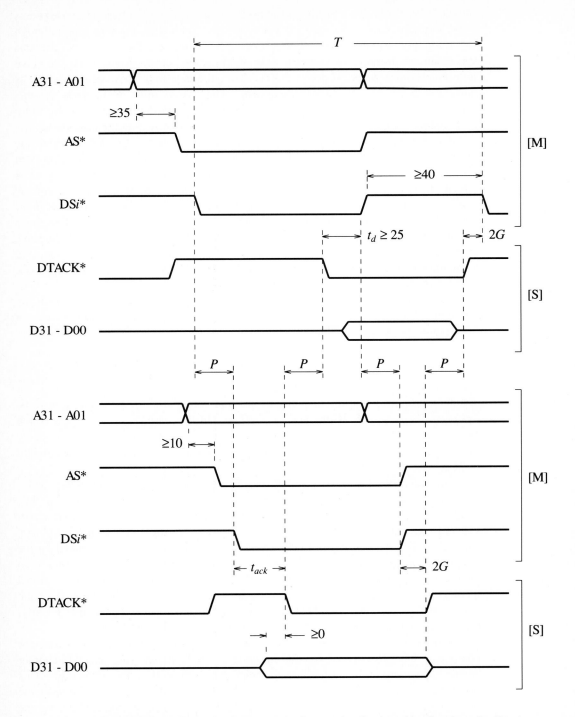

Figure 11.15 VMEbus signals as seen by the master (top) and by the slave (bottom).

aggregate rate of data transfer over the system bus, if the bus mastership is transferred frequently from one device to another.

Since the system bus is not involved in data transfers between the microprocessor and its local memory, these transfers can be significantly faster than those between the microprocessor and the global memory. Hence, it is advantageous to provide as much memory as possible on the circuit board that houses the microprocessor. Also, in organizing the software, the local memory should be used to store instructions and data that are accessed frequently. The allocation of memory to programs and data to achieve this objective is one of the functions of the operating system software.

11.2.7 Review

The VMEbus provides many features to support single- and multiple-microprocessor systems. It enables 8, 16, and 32-bit data transfers and uses up to 32 bits of address. Data transfers are controlled by an asynchronous handshake between data strobe and acknowledgment signals. The address-modifier lines enable certain organizational features to be easily implemented, such as partitioning the system among various masters, or restricting access to certain devices based on the master's level of privilege. An indivisible operation consisting of a read followed by a write transfer to the same address is defined to support the test-and-set operations needed in multiprocessor systems.

Bus arbitration on the VMEbus requires an arbiter that resides on the circuit card at slot #1. Four bus-request and four bus-grant lines are provided to implement either fixed or rotating priority. Priority arbitration may take place while the bus is idle, or during the last data transfer cycle of the current bus master. The arbiter can request the current bus master to release the bus if there is a pending request from a higher-priority device.

Seven interrupt-request lines are available, The responsibility of servicing requests on these lines may be shared among several interrupt handlers. When an interrupt request is accepted, the interrupt handler must first acquire the bus mastership. Then, it issues an interrupt-acknowledge signal to request the interrupter to send its ID information.

The VMEbus has several utility lines, including a clock signal, a reset line, and a line to indicate a failure in the AC power source. Another utility line is provided to facilitate the implementation of self-diagnostic tests on various devices connected to the bus.

11.3 MULTIBUS—IEEE STANDARD 796

The Multibus was introduced by Intel in the mid 1970s to serve as a backplane bus for various Intel products that use the 8085 and 8086 microprocessors [4]. It gained wide acceptance, and is currently being used in many systems including ones that do not use Intel microprocessors. There are two versions in existence, Multibus and Multibus II, where Multibus II is an expanded version of the basic Multibus, intended to support 32-bit microprocessors. The Multibus has also been issued as IEEE Standard 796 [5,6] and Multibus II as Standard 1296 [7].

The Multibus has 86 lines, which comprise 16 data lines, 20 address lines, 26 control lines, and 20 lines for power supply and ground. The remaining four lines are reserved for future use. They may be assigned functions in newer versions of the bus to be announced in the future. Multibus circuit cards are 304.8 mm (12.00 in.) long and 163.8 mm (6.45 in.) wide. Each board has two edge connectors that protrude a further 0.8 mm (0.30 in.) at the back. The first edge connector, called P1, consists of 86 fingers, 43 on each side of the board, and carries the Multibus signals. The second, P2, is an auxiliary connector that carries four additional address lines and several utility lines.

We will discuss the Multibus by examining its basic functional capabilities, starting with the way in which data transfers take place. Devices that are capable of initiating read and write requests are called *masters*, and those addressed by a master are called *slaves*. At a given time, only one master can be the current bus master and request data transfer operations. If another master wishes to transfer data, it must first request and acquire the bus mastership, using the arbitration scheme to be described in Subsection 11.3.3.

The Multibus uses TTL logic levels. The tables in the following sections give the type of drivers needed to generate each signal. Most lines use a three-state driver (3ST). Open-collector drivers (OC) are used on lines that may be driven by more than one device at the same time. Signal lines that are driven by one device only use standard totem-pole TTL drivers (TP) when there is only one receiver on the line, and a high-current version (TP-HC) on lines that drive multiple receivers.

11.3.1 Data Transfer

The lines used for transferring data are given in Table 11.6. There are 20 address lines, which define a 1-Mbyte addressable space, and 16 data lines. These are the available signals when only the main connector, P1, is used. The auxiliary connector, P2, provides four additional address lines, A20* to A23*, which expand the addressable space to 16 Mbytes. The Multibus uses the asynchronous handshake technique described in Subsection 6.8.2 to control data transfers. Separate handshake signals are provided for the main memory and for I/O devices. Also, instead of using a R/W line to indicate the direction of a transfer, different handshake signals are provided for read and write operations. The bus master initiates a data transfer operation by activating one of the command lines, either the memory read command, MRDC*, or the memory write command, MWTC*. Similar command lines are available for I/O transfers. In all cases, the addressed device responds using the transfer acknowledge line, XACK*. We will adopt the convention that signal names ending with an "*" are active when in the low-voltage state. This convention is consistent with the nomenclature of the IEEE 796 Standard. In Intel's publications, a "/" is used instead of an "*."

Addresses on the Multibus are organized as shown in Figure 11.16. The low-order byte of a word has an even address, which is the same as the address of the word. The high-order byte has the next higher address. A single bus cycle may be

Table 11.6 DATA TRANSFER SIGNALS ON THE MAIN CONNECTOR (P1) OF THE MULTIBUS

Designation	Function	Driver	Number
A0*-A19*	Address lines	3ST	20
D0*-D15*	Data lines	3ST	16
BHEN*	Byte high enable	3ST	1
MRDC*	Memory read command	3ST	1
MWTC*	Memory write command	3ST	1
IORC*	I/O read command	3ST	1
IOWC*	I/O write command	3ST	1
XACK*	Transfer acknowledge	3ST	1
INH1*-INH2*	Inhibit another slave at the same address	OC	2
LOCK*	Lock out access for other masters	3ST	1
		Total	45

used to transfer either one byte or one word of data. There are two ways of transferring single bytes over the bus. The first allows 16-bit and 8-bit devices to coexist on the same bus, while the second is intended for systems that have only 16-bit devices.

Eight-bit devices use data lines D7* to D0*. Hence, to enable a 16-bit device and an 8-bit device to exchange data, all byte transfers take place on these lines. The high-order data lines, D15* to D8*, are used only during word transfers. The least-significant address bit, A0*, and Byte High Enable, BHEN*, indicate whether a given transfer involves an even byte, an odd byte, or a word, where BHEN* is activated during word transfers only. Figure 11.17 illustrates how a 16-bit memory may be connected to the data lines. When the memory is selected, the multiplexer MPX selects either the low-order or the high-order byte for transmission over lines D7* to D0*, depending upon whether the requested location has an even or an odd

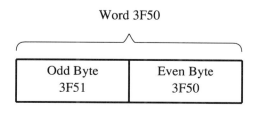

Word 3F50

Odd Byte 3F51	Even Byte 3F50

Figure 11.16 Byte and word addresses on the Multibus.

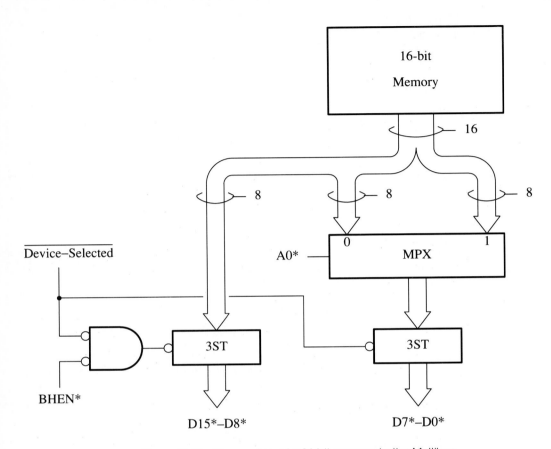

Figure 11.17 Connection of a 16-bit memory to the Multibus.

address, respectively. The high-order byte of the memory is also connected to lines D15* to D8* via an appropriate driver, which is enabled only during word transfers, i.e., when BHEN* is active. A similar multiplexing arrangement is needed for write operations.

The scheme described above enables 8-bit and 16-bit devices to communicate in a straightforward manner. The Multibus specifications define a second mode in which even bytes are transferred on lines D7* to D0* and odd bytes on lines D15* to D8*. This mode avoids the need for the multiplexer in Figure 11.17. However, it can only be used when all devices connected to the bus are 16 bits wide.

A read operation from the main memory is illustrated in Figure 11.18. The bus master places an address on the address lines and activates BHEN*, if appropriate. After at least a 50-ns delay, to account for bus skew and address decoder delay at the slave, it activates MRDC*. In response, the main memory places the requested data on the data lines and activates XACK*. The master loads the data into its input buffer and deactivates MRDC*, after waiting at least 20 ns to compensate for bus

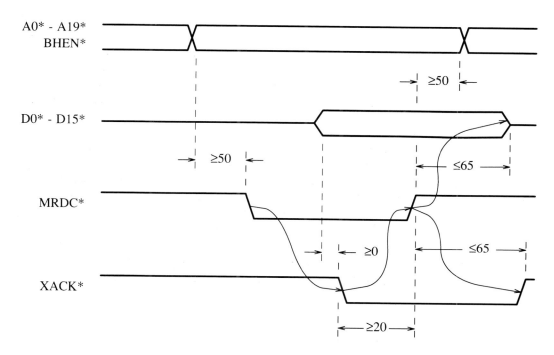

Figure 11.18 Timing of a read operation on the Multibus.

skew. It maintains a valid address on the bus for at least 50 ns after MRDC* has become inactive. Then, it may place a new address on the bus to start another access cycle. In this way, all devices on the bus are guaranteed to see a valid address as long as MRDC* is active, even after allowing for bus skew. The slave must maintain valid data until MRDC* becomes inactive, after which point it must turn off its data drivers and deactivate XACK* within 65 ns. The 65-ns limit guarantees that the slave will not interfere with the subsequent bus cycle.

In the case of an I/O operation, the same sequence of events takes place as in Figure 11.18, except that MRDC* is replaced by IORC*. It is possible for the same address to refer to a main memory location and to an I/O device register. One or the other will be selected depending upon which command is used. The separation of the I/O address space from the address space of the main memory can be utilized by microprocessors that have separate I/O instructions, such as those in the Intel family. For microprocessors that use memory-mapped I/O, which include both the 6809 and the 68000, only the memory command signals are used.

Write operations proceed as illustrated in Figure 11.19. The master places the data on the data lines at the same time as it transmits the address information. Then, it activates the appropriate write command. The addressed device accepts the data and responds by activating XACK*.

An interesting feature of the Multibus is the "inhibit" facility. It enables one device, B, to occupy an address assigned to another device, A. When the common

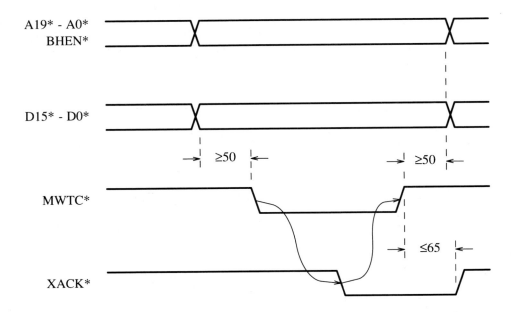

Figure 11.19 Timing of a write operation on the Multibus.

address appears on the bus, device B activates one of the inhibit lines, thus instructing device A not to participate in the requested operation. An example of the use of this facility is given in Figure 11.20. Device A is a 16K RAM, occupying the address space 0C000 to 0FFFF. Suppose that it is desired to assign a small portion of this space, say 0C020 to 0C027, to an I/O device—device B in the figure. In this case, the address decoder of the RAM should respond to all addresses in the range 0C000 to 0FFFF, except 0C020 to 0C027. A decoder that behaves exactly in this manner would be somewhat complicated. More importantly, the presence of device B in the system would have to be anticipated at the time the interface for the RAM is designed. A more flexible arrangement is shown in Figure 11.21. The RAM interface responds to all addresses in the range 0C000 to 0FFFF, except when the INH1* line is active. When an address corresponding to device B appears on the address lines, it activates the inhibit line, via an open-collector gate, thus preventing the RAM from responding. The open-collector arrangement allows several devices to use the INH1* line to inhibit device A. Thus, a third device, C, may be added at a later time and assigned a different portion of A's address space. The Multibus has two inhibit lines, making it possible to arrange devices in a two-level hierarchy. For instance, three devices, A, B, and C, may be arranged such that device B inhibits device A using INH1*, and device C inhibits both A and B, using INH2*.

The last signal in Table 11.6 is LOCK*. It is intended for use in indivisible data transfer operations, and will be described in Subsection 11.3.3.

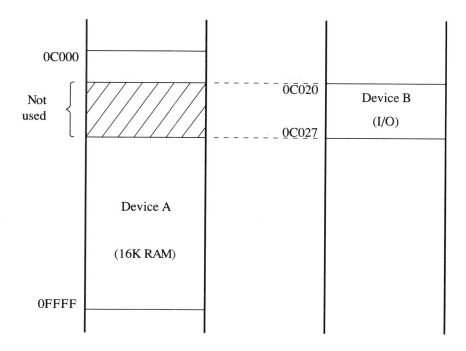

Figure 11.20 Address space assignment made possible by the inhibit feature of the Multibus.

11.3.2 Interrupts

The Multibus supports both vectored and nonvectored interrupts. It has eight interrupt-request lines, and one interrupt-acknowledge line, as shown in Table 11.7. The interrupt-request lines provide eight levels of priority, with INT0* having the highest priority. Recall that the Multibus is a backplane bus, which serves as a medium for connecting memory and I/O devices to the bus of a microprocessor. The extent to which the lines and functions defined in the standard are used in a given system depends on the needs of the particular microprocessor that controls the system. A 6809 microprocessor, for example, requires only two lines to transfer interrupt requests from I/O devices to its $\overline{IRQ}$ and $\overline{FIRQ}$ inputs. All interrupt lines of the Multibus are defined to be level-triggered. No edge-triggered line is provided. Hence, the $\overline{NMI}$ line of the 6809 microprocessor can only be used locally, that is on the circuit board housing the microprocessor. Since the 6809 does not normally read an interrupt-vector address from I/O devices, the interrupt-acknowledge facility of the Multibus is not needed. A 68000 system, on the other hand, can use six interrupt lines. In this case, INT0* to INT5* may be connected to $\overline{IRQ1}$ to $\overline{IRQ6}$ of Figure 7.30. The interrupt-acknowledge signal issued by the 68000 can be sent to the I/O devices over the INTA* line. However, the signaling scheme for transferring the interrupt-vector address on the Multibus is slightly different from that of the 68000

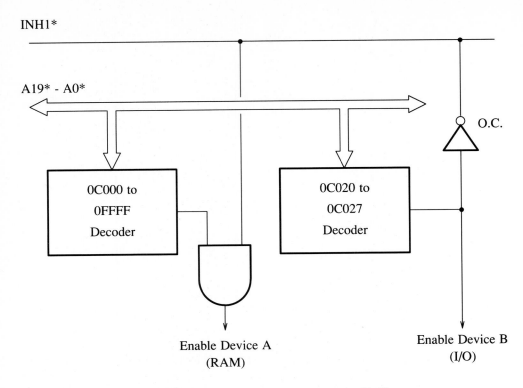

INH1*

A19* - A0*

O.C.

| 0C000 to |
| 0FFFF |
| Decoder |

| 0C020 to |
| 0C027 |
| Decoder |

Enable Device A
(RAM)

Enable Device B
(I/O)

Figure 11.21 Use of the inhibit facility of the Multibus.

bus. Some intervening circuitry on the processor card is needed to translate one bus
protocol into the other.

The transfer of an interrupt-vector address over the Multibus is illustrated in
Figure 11.22. When the microprocessor accepts an interrupt request, it issues a pulse
of fixed width on the INTA* line. This pulse alerts all devices on the bus that an
interrupt-acknowledge cycle is in progress, and that the microprocessor is about to
request the interrupt vector address. No device is allowed to change the state of the

Table 11.7   MULTIBUS INTERRUPT SIGNALS

Designation	Function	Driver	Number
INT0*-INT7*	Interrupt-request lines, INT0* has the highest priority.	OC	8
INTA*	Interrupt-acknowledge	3ST	1
		Total	9

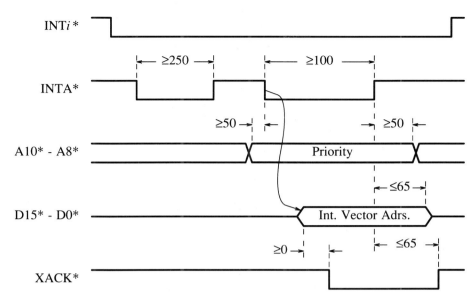

Figure 11.22 Transfer of interrupt vector address over the Multibus.

INT*i* * lines until the interrupt-acknowledge cycle is completed. The processor places the priority level of the highest-priority request on address lines A10* to A8*, then activates the INTA* line a second time. The device having the priority indicated on the address lines places an 8-bit interrupt-vector address on D7* to D0* and activates XACK*. The cycle is terminated in the same manner as a memory read cycle, with INTA* replacing MRDC*.

11.3.3 Bus Mastership

The Multibus provides two arbitration mechanisms for transferring the bus mastership from one master to another—a serial-priority scheme and a parallel-priority scheme. The same input and output pins on a given master board are used in both cases. The particular scheme used is determined by the way in which these pins are wired on the backplane. Table 11.8 summarizes the functions of the signals involved.

The serial priority scheme uses a variation of the daisy chain concept described in Subsection 7.3.3. No bus controller is needed to generate a bus-grant signal in response to bus requests. Instead, a signal called Bus Priority propagates through the devices, which are connected in a daisy chain fashion as shown in Figure 11.23. An active state at the Bus Priority input, BPRN*, of any device indicates that the device has permission to begin using the bus when the current user stops using it. When a device wishes to use the bus and its BPRN* input is active, it deactivates its Bus Priority output, BPRO*, indicating to the devices downstream that they cannot use the bus. Otherwise, it transfers the signal received at BPRN* to BPRO*. Since the

Table 11.8   MULTIBUS ARBITRATION SIGNALS

Designation	Function	Driver	Number
BPRN*	Bus priority input	—	1
BPRO*	Bus priority output—used only in serial arbitration	TP	1
BREQ*	Bus request—used only in parallel arbitration	TP	1
BUSY*	Bus busy—activated by current bus master	—	1
BCLK*	Bus clock—used for synchronization of arbitration logic	TP-HC	1
CBRQ*	Common bus request	OC	1
		Total	6

first device in the chain has its BPRN* input permanently connected to ground, this device will be able to acquire the bus mastership whenever it wishes.

The daisy chain connection determines the relative priorities of the devices that wish to acquire the bus mastership. However, it does not indicate when the highest priority device may actually begin to transfer data over the bus. The BUSY* line provides this information. It is an open-collector line that is pulled down to the active state by the current bus master, and released only after the current bus master has turned off all its bus drivers. Hence, a device that has received permission to use the bus through the daisy chain must wait until BUSY* becomes inactive. Then, it should reactivate BUSY* and begin using the bus.

The Bus Request output, BREQ*, is not used in the serial arbitration scheme. An internal bus-request signal in each device initiates the procedure for acquiring the bus mastership, but this signal need not be broadcast to other devices. Instead, the

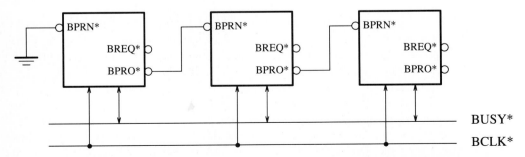

Figure 11.23   Daisy-chain connection for serial priority arbitration on the Multibus.

BPRO* output of the device that wishes to use the bus is placed in the inactive state, to withdraw the permission for bus mastership from all devices downstream. Should the current bus master be one of these devices, it should promptly release the bus after completing the transfer cycle, or indivisible sequence of bus cycles, that it is currently involved in.

The serial arbitration scheme described above has the advantage of not requiring a bus allocation controller for generating a bus-grant signal, as is the case for the conventional daisy chain scheme described in Subsection 7.6.2. The latter represents *centralized control*, because it uses a single device to control bus allocation. The Multibus scheme uses *decentralized*, or *distributed*, *control*. Unfortunately, distributed control gives rise to a serious timing problem, which we will describe briefly together with the way it is resolved on the Multibus.

Consider a period during which the bus is not being used by any device. The BUSY* signal is inactive and all BPRN* inputs are active. Any device wishing to acquire the bus mastership can activate BUSY* and begin transmitting data immediately. Once BUSY* becomes active, no other device can use the bus. Clearly, a conflict arises if, while the bus is idle, two devices decide at the same time to acquire the bus mastership. Seeing BUSY* inactive and their BPRN* inputs active, both devices begin to use the bus. In order to avoid this conflict, the Multibus specifications require all events related to bus acquisition to be synchronized with the common clock signal, BCLK*, shown in Figure 11.23. The transfer of the bus mastership from one device, A, to another device, B, of higher priority is illustrated in Figure 11.24. Initially, device A is using the bus. Its priority input is active, and it keeps BUSY* active and its priority output inactive. When device B needs to use the bus it deactivates its priority output, synchronizing its action with the falling edge of BCLK*, as shown at time t_0. This change in state propagates down the priority daisy chain until it reaches the BPRN* input of device A. Realizing that it has lost the permission to use the bus, device A promptly turns off its bus drivers and releases BUSY*, also synchronizing its actions with the falling edge of BCLK*. During the clock period between t_1 and t_2, device B detects the inactive state on BUSY*. Hence, at the next falling edge of the clock, it activates BUSY* and assumes the bus mastership.

We should now reexamine the case of two devices, A and B, simultaneously deciding to acquire the bus mastership. Both devices deactivate their BPRO* outputs at the beginning of the same clock period, and begin to monitor the BUSY* signal. Assuming that BUSY* is inactive, each device concludes that it can use the bus; however, it must wait for the next falling edge of the clock before doing so. During that period, the inactive state at the BPRO* output of device B reaches the BPRN* input of device A. As a result, device A is forced to abort its attempt to acquire the bus mastership. Obviously, for this scheme to work reliably the clock period must be longer than the longest propagation delay on the priority daisy chain.

The number of bus masters that can be accommodated on the serial priority arbitration chain is limited because of the delay constraint mentioned above. For a large number of masters, parallel arbitration may be used instead. A system that uses this scheme has a parallel priority arbitration circuit, housed on the circuit board that constitutes the backplane of the card cage. The Bus Request outputs, BREQ*, of all

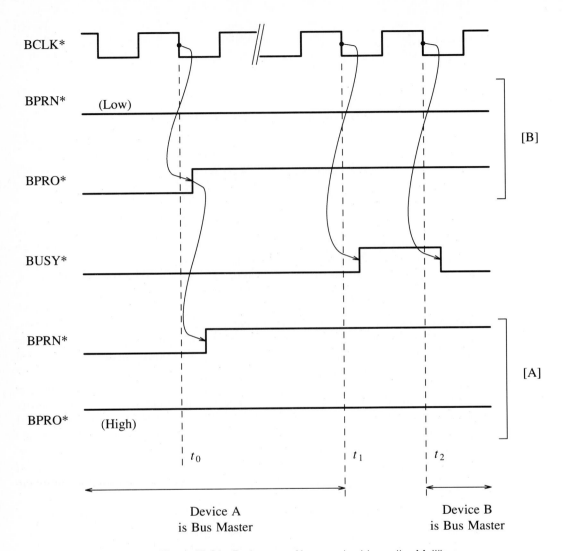

Figure 11.24 Exchange of bus mastership on the Multibus.

masters are connected to that circuit. Also, the Bus Priority inputs, BPRN*, are connected to the outputs of the priority circuit, instead of the daisy chain arrangement. The Bus Priority outputs, BPRO*, are not used. A device wishing to acquire the bus mastership activates its BREQ* output. During each clock period, the arbitration circuit selects the highest priority request and activates the corresponding BPRN* signal, synchronized with the falling edge of BCLK*. In turn, the selected device proceeds to acquire the bus mastership in exactly the same manner as before.

We should once again note that from the point of view of the devices on the bus, the serial and parallel arbitration schemes appear identical. They differ only in the way the request and priority signals are used on the backplane. In the serial scheme, the BPRO* output of one device is connected to the BPRN* input of another, and BREQ* outputs are ignored. In the parallel scheme, the BPRO* outputs are ignored and BREQ* and BPRN* are connected to a parallel arbitration circuit.

Irrespective of the method used for priority arbitration, a low-priority device can acquire the bus mastership only if higher-priority devices voluntarily relinquish the bus mastership. A high-priority device has no way of knowing whether a lower-priority device is waiting to use the bus. Hence, it must give up the bus mastership periodically, usually after each data-transfer cycle, then contend for getting it back whenever it is ready for another transfer. In this way, all devices get a chance to transfer data. However, releasing the bus and getting it back introduces unnecessary delays when no other device is waiting to use the bus. The Multibus has one signal, called Common Bus Request, CBRQ*, which is used to inform the current bus master whether other devices are waiting to use the bus. Any device wishing to acquire the bus mastership should activate CBRQ*. When this signal is inactive, the current bus master may retain the bus mastership even when it has no data to transfer. Thus, the delay associated with regaining control of the bus is avoided when no other devices need to transfer data.

We should point out one other Multibus feature intended to support indivisible transfers, such as a read-modify-write operation (see Subsection 6.10.1). The signals mentioned above are sufficient to support such an operation, because the bus master is not obliged to release the bus between the read and the write parts of the operation. After completing the read cycle, it simply continues to activate BUSY*, and starts the write operation. However, consider the system shown in Figure 11.25. Two processors, A and B, access a common memory, each via its own Multibus. When processor A initiates a read-modify-write cycle, it is guaranteed exclusive access to its Multibus until both the read and the write transfers are completed. However, the two transfers appear as independent requests to the common memory. The memory control circuitry may allow processor B to access the memory between these two transfers, thus defeating the purpose of the read-modify-write operation. In order to guarantee that such an operation remains indivisible, the memory control circuit must be informed of this necessity. None of the bus signals discussed so far provides the required information. The BUSY* signal is not suitable, because it is activated at other times, when processor B need not be locked out. The Multibus has a special signal called LOCK*, which is intended for use during indivisible operations. While it is not needed on the Multibus itself, it can be activated by a master wishing to perform an indivisible operation to communicate this fact to other devices. The memory control circuit should monitor the LOCK* signal, and provide exclusive access to one processor whenever this signal is active.

11.3.4 Utilities and Power Supply

The utility and power supply lines available on the main Multibus connector are given in Table 11.9. The INIT* line provides a common reset signal, which should be connected to the reset inputs of all devices on the bus. This line should be

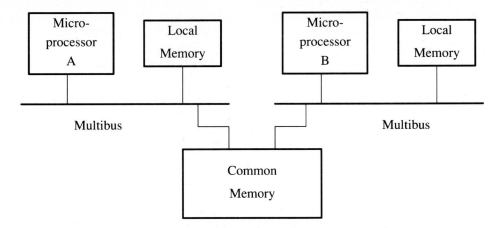

Figure 11.25   Two Multibus systems that communicate via a common memory.

activated at power-up time to reset all devices and cause the controlling microprocessor to start the system's operation.

The Constant Clock line, CCLK*, is a continuously running 10 MHz clock, which is independent of all bus signals. It is provided for use in interface circuits whenever a clock signal is needed. An example of its use is given in the next section.

The auxiliary Multibus connector, P2, carries several signals to coordinate the actions of various devices on the bus under abnormal conditions. For example, an

Table 11.9   MULTIBUS UTILITY AND POWER SUPPLY LINES

Designation	Function	Driver	Number
INIT*	Initialization—resets the system to its initial state	OC	1
CCLK*	Constant clock	TP-HC	1
+5V	Power supply lines		8
+12V	Power supply lines		2
−12V	Power supply lines		2
GND	Ground return lines		8
RFU	Reserved for future use		4
		Total	26

open-collector line called ACLO (for AC power low) is normally kept inactive (low) by the power supply circuit. It has a pull-up resistor connected to a standby power supply, such as a battery. If the AC power is interrupted, this line becomes active. The power supply circuit continues to maintain the DC power supply lines within their specified values for at least 3 ms after ACLO becomes active. This period should be used to save volatile data on nonvolatile storage devices, such as disks, and to shut down the system's operation gracefully. For details of this and other auxiliary signals, interested readers should consult the Multibus specifications.

11.3.5 Example

We will now present a simple example of the circuits needed to control data transfer over the Multibus. The main purpose of this discussion is to examine the way in which the timing specifications in Figures 11.18 and 11.19 influence circuit design. Consider a Multibus slave board that houses several 8-bit I/O devices. The control signals involved in data transfers are the read and write commands, MRDC* and MWTC* (assuming memory-mapped I/O), and the transfer acknowledgment, XACK*. The way in which these signals are used on the board is illustrated in Figure 11.26. A bidirectional data buffer connects the Multibus data lines to the data bus on the board. The state of this buffer is controlled by the two signals $\overline{OE}$ (Output Enable) and T (Transmit). The T input controls the direction of the buffer. When T=1, data are transferred from side A to side B, and vice versa when T=0. All buffer outputs are in the high-impedance state when $\overline{OE}$ is inactive. The address decoder on the board (not shown) activates $\overline{\text{Device-Selected}}$ when one of the devices on the board is addressed. After one of the two command signals becomes active, either Read or Write will become active. During a read operation T is set to 1, to transfer the data on the local bus to D7* to D0*. Otherwise, T is equal to 0, causing the data received on D7* to D0* to be transferred to the local bus.

The tri-state driver for XACK* is enabled as soon as either a read or a write command is received. However, XACK* remains inactive, because the shift register is normally held in the cleared state, with all its Q_i outputs equal to 0. The $\overline{\text{Clear}}$ input of the shift register is deactivated when either a read or a write command is received, allowing the Constant Clock, CCLK*, to shift a 1 into the shift register. After three clock pulses, Q_2 becomes equal to 1, thus activating XACK*. As soon as the master removes the read or write command, the data buffer and XACK* driver outputs are disabled and the shift register is cleared.

The amount of delay introduced by the shift register during a read operation should be chosen such that XACK* becomes active after valid data have been placed on the Multibus data lines, taking into account both the access time of the I/O device register and the propagation delay through the bidirectional buffer. For a write operation, the shift register delay determines the duration of the active state on the signal Write-Pulse. By terminating Write-Pulse at the time XACK* is activated, rather than when MWTC* becomes inactive, we provide hold time at the data inputs of the I/O device, in addition to the 50 ns introduced by the master (see Fig. 11.19).

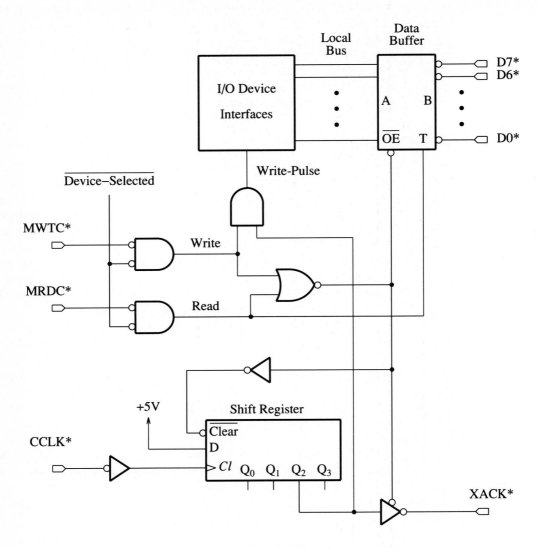

Figure 11.26 A Multibus interface example.

The handshake lines on many computer buses use open-collector drivers. The Multibus uses three-state drivers on these lines, to obtain slightly faster rise and fall times. However, significantly longer delays are introduced when enabling or disabling a three-state driver. For this reason, the XACK* driver in Figure 11.26 is enabled as soon as a command is received, long before the time at which XACK* is to be activated. It is important that the driver not introduce any glitches on the line while it is being enabled. Some commercially-available tri-state drivers, such as the 8098 manufactured by Intel, have been designed with this requirement in mind.

11.3.6 Bus Bandwidth

The bus bandwidth was defined in Subsection 11.2.6. It was shown that an upper bound for the rate at which two devices that are 25 cm apart on the VMEbus can transfer data is 31 Mbytes/s. For purposes of comparison, let us estimate the bandwidth of the Multibus.

To determine the period, T, between two successive data transfers, a complete read cycle is shown in Figure 11.27. This diagram is the same as the timing diagram in Figure 11.18 with the signals shown twice, once at the master and once at the slave. It should be compared to Figure 11.15 for the VMEbus. The main difference between the two is that the Multibus guarantees a valid address on the bus as long as one of the command lines is active. As a result, a larger separation is needed between two successive cycles. Also, the Multibus uses a slightly larger allowance for bus skew and address decoding.

According to Figure 11.27, The minimum value for T may be obtained from the equation

$$T_{min} = 2P + t_{ack} + t_d + Max\ (100, 2P + 2G).$$

For the Multibus, $t_{ack} \leq 2G$ and $t_d \geq 20$ ns. Hence, for $P = 8$ ns, the minimum value for T is 156 ns, compared to 130 ns for the VMEbus. Because the Multibus is only 2 bytes wide, the maximum data transfer rate between the devices is about 13 Mbytes/s.

To reduce power dissipation and the need for high-current drivers, the Multibus specifications do not call for line terminations in the same way as the VMEbus. Instead, most lines have a single pull-up resistor placed at some point on the backplane. The value of this resistor is 2.2 kohm for the address and data lines, 1 kohm for the command lines, and 500 ohm for the XACK* line. The amplitude of the ringing caused by transmission line reflections may be sufficient to delay the settling of the signals on the address and data lines by as much as two roundtrip delays (see Subsection 10.3.3). To allow for these delays, the minimum address and data setup times shown in the figure should be increased by about 40 ns. In this case, the minimum value for T becomes 234 ns, which corresponds to a bandwidth of 8.5 Mbytes/s.

11.3.7 Review

The Multibus, or IEEE 796 bus Standard, is intended for use as a backplane bus in microprocessor systems. It supports 8- and 16-bit devices. By restricting byte transfers to the eight low-order data lines, a 16-bit device may exchange data with an 8-bit device.

The Multibus can support several masters, which may be microprocessor boards or direct memory access (DMA) devices. Either a serial or a parallel arbitration scheme may be used to enable the masters to contend for the bus mastership at any given time. The serial scheme does not require a central controller. It is a distributed arbitration scheme that uses a daisy chain to define a fixed priority ordering for all bus masters. In the parallel scheme, an arbitration circuit that resides on the

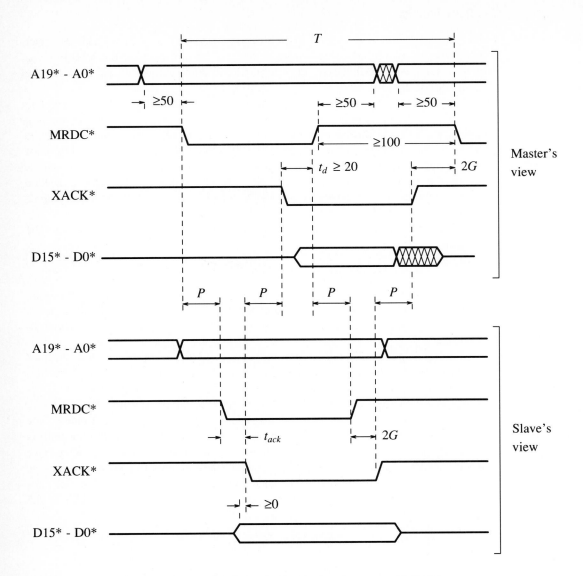

Figure 11.27 Data transfer cycle on the Multibus.

backplane circuit board receives requests for bus mastership and grants the bus to the highest-priority device. All bus arbitration signals are synchronized with a common clock signal, to avoid race conditions that could arise as a result of the distributed nature of the arbitration scheme.

Eight interrupt lines, which are assigned a fixed priority ordering, enable peripheral devices to send interrupt requests to the microprocessor that controls the operation of the system. If there are several microprocessors on the bus, one of them must be selected to handle all interrupts. It is responsible for generating an interrupt-acknowledge signal and receiving the interrupt vector from the interrupting device.

The main Multibus signals are available on a single edge connector on each circuit board. A second connector provides four additional address bits and several other signals for auxiliary functions, such as informing various devices about an imminent loss of power.

11.4 DEVICE INTERFACE STANDARDS

The VMEbus and the Multibus standards describe backplane buses that interconnect various microprocessor system components, such as microprocessor boards, main memory boards, and peripheral interface circuits. The place of a backplane bus within a system was illustrated in Figure 11.2. Another point in a system where it is useful to define a standard for interconnection is shown in Figure 11.28. This is the point at which an I/O device, such as a video terminal, a modem, or a printer is connected to its interface circuit. Several commercial standards exist that define the signals exchanged across this interface. They use either serial or parallel transmission of data.

Serial transmission is used for connecting many peripherals to microcomputers, because it offers flexibility and low cost. A serial transmission cable, whether it is a

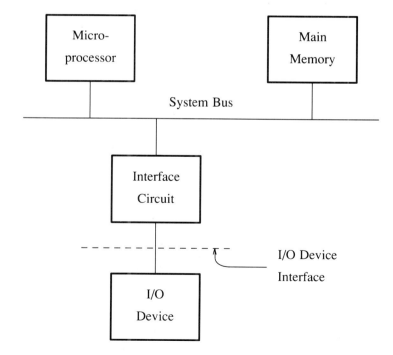

Figure 11.28 Connection point specified by an I/O interface standard.

coaxial cable, a fiber optic cable or a simple telephone cable, is easily routed through the wiring ducts in a building. Also, serial links enable data to be transmitted to remote locations via the telephone network. Whenever long links are involved, or when the telephone network is used, a modem is used at each end (see Chapter 9). The main schemes used for encoding the transmitted data and timing information on serial lines were discussed in Subsection 6.7.1.

In this section, we will discuss the most commonly used standard for interconnecting serial transmission links. It is known in North America as RS-232-C [8], and internationally as CCITT Recommendation V.24. We will then introduce a modified version called RS-449 [9], which can be used for transmission over longer distances and at higher speeds. Each of these standards defines a connector for interconnecting computers and their peripherals via serial transmission links. The connector comprises a set of signals used for transmitting and receiving data and for controlling the transmission link. Each standard defines the electrical characteristics of the signals, the functions they perform, and the mechanical specifications of the connector. The RS-232-C interface uses a 25-pin connector, and the RS-449 a 37-pin connector. In many applications, only a few of these pins are needed. However, the standard connector is usually used to maintain industrywide compatibility. In the case of RS-449, an additional nine-pin connector is also defined for use in certain situations, as will be explained later.

We will also present an example of a parallel interconnection scheme commonly known as a Centronix interface. This is not a standard that is sanctioned by any official body in the same way as the RS-232-C. Nevertheless, it has been adopted by many manufacturers. The connector it describes is intended for connecting printers and printing terminals to computers.

11.4.1 RS-232-C Standard

The RS-232-C Standard divides equipment into two categories: data terminal equipment (DTE) and data circuit-terminating equipment (DCE). Data terminal equipment comprises all data processing equipment, including computers, video terminals, printers, etc. The second category includes any device, such as a modem, whose primary function is to provide a communication path between two devices in the DTE category. The RS-232-C connector is intended to connect a DTE to its associated DCE, as illustrated in Figure 11.29. Modems are needed when the terminal is far away from the computer, in a different building, or even in a different city. The cable connecting the two modems is typically leased from the telephone company. It is also possible to connect the modems over the switched (dialed) network.

Modems transmit data by modulating a carrier signal, as was explained in Section 9.4. Before data transmission can begin, each modem must turn on its carrier. When both carriers are turned on, a *full-duplex* communication channel exists between the two modems, allowing data transmission to take place in both directions at the same time. The role of the RS-232-C signals is to enable the DTE at each end to control the operation of its associated DCE, to establish a communication channel with the remote DTE. After establishing a connection, the two DTEs can exchange

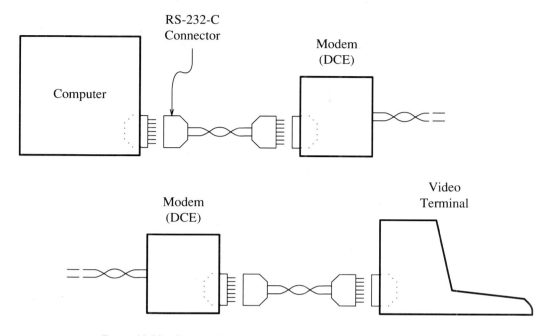

Figure 11.29 Connection of a video terminal to a computer via a modem.

data, while continuing to monitor the communication channel to detect abnormal conditions, such as loss of the carrier.

The signals defined by the RS-232-C Standard are given in Table 11.10, grouped according to their functions. The first two columns in the table give the designation used by the two Standards organizations, EIA and CCITT. The third column gives the number of the connector pin used for each signal, where pin numbering on a male connector when viewed from the pin side is as shown in Figure 11.30.

Table 11.11 provides further clarification for the functions of some of the RS-232-C signals commonly used to control modems. The RS-232-C interface also includes three timing signals. In the case of synchronous data links, a common clock signal may be needed by the DTE and its associated modem. The clock for transmitted data may be generated by the DTE and sent to the modem using signal DA, or generated by the modem and sent to the DTE using signal DB. The clock for the received data is recovered by the modem and sent to the DTE using signal DD.

On some communication links, two channels are established for each direction of transmission, using two different carriers. The first is the primary channel used for data transmission. The second, which is usually of a lower speed, is called an auxiliary or a secondary channel, and is used for transferring control information between the two ends. The last five signals in Table 11.10 are used to carry the data and control information associated with the secondary channel.

Table 11.10 FUNCTIONAL DESCRIPTION OF RS-232-C SIGNALS

Designation		Pin No.*	Source	Functional Name	Abbreviation
EIA	CCITT				
AA	101	1		Protection ground	PG
AB	102	7		Signal ground	SG
BA	103	2	DTE	Transmitted data	TD
BB	104	3	DCE	Received data	RD
CA	105	4	DTE	Request to send	RTS
CB	106	5	DCE	Clear to send	CTS
CC	107	6	DCE	Data set ready	DSR
CD	108.2	20	DTE	Data terminal ready	DTR
CE	125	22	DCE	Ring indicator	RI
CF	109	8	DCE	Data carrier detected	DCD
CG	110	21	DCE	Signal quality detector	
CH or CI†	111 or 112	23	DTE or DCE	Data signal rate selector	
DA	113	24	DTE	Transmitter clock	
DB	114	15	DCE	Transmitter clock	
DD	115	17	DCE	Receiver clock	
SBA	118	14	DTE	Secondary transmitted data	
SBB	119	16	DCE	Secondary received data	
SCA	120	19	DTE	Secondary request to send	
SCB	121	13	DCE	Secondary clear to send	
SCF	122	12	DCE	Second data carrier detected	

*   Pins 9, 10, 11, 18, and 25 are not used.

†   This signal is generated by the DTE in some installations, and called CH, and by the DCE in others, and called CI.

Table 11.11   COMMONLY USED MODEM CONTROL SIGNALS AND THEIR FUNCTIONS

Signal	Functional description
DSR	Activated by a modem when it is powered and connected to the line, and when it has gone off hook in the case of a switched connection.
RI	Used by a modem having a telephone answering capability and connected to a telephone network. It is activated to inform the DTE that a ringing signal has been received.
DTR	Activated by DTE when ready, e.g., power is turned on and communication software is running. In the case of a switched connection, it indicates that the DTE is willing to accept incoming calls and that the modem should go off hook when the telephone rings.
RTS	Activated by DTE when it wants to start transmission. DCE should establish a transmission channel by transmitting the carrier.
CTS	Activated by DCE to indicate that it has a channel ready for transmission (carrier is on). DTE may start sending data.
DCD	Activated by DCE when it has detected an incoming carrier.

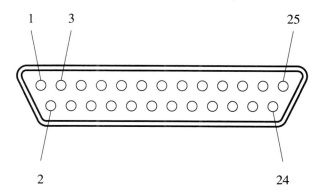

Figure 11.30   Pin numbering on an RS-232-C male connector viewed from the pin side.

ELECTRICAL SPECIFICATIONS

An RS-232-C connection uses *single-ended* transmission. In this arrangement, the signal ground terminals (pin 7) of the DTE and the DCE are connected together, and one wire—the hot wire—is used to carry the signal. Single-ended transmission is illustrated in Figure 11.31. The ground connection provides a path for the return current for all signals exchanged by the two ends. This simple arrangement is suitable for low bit rates and short distances. The RS-232-C interface is not recommended for cables longer than 15 m (50 ft) or for transmission speeds higher than 20 kb/s. However, cables of up to 100 m in length and more have been used successfully, when high reliability is not essential.

The voltage levels used in RS-232-C connections are summarized in Figure 11.32. On the data lines, (BA, BB, SBA and SBB) the sender generates a voltage between −5 and −15 V for a 1 and between +5 and +15 V for a 0. It must have sufficient current handling capacity to produce these voltages across a load impedance of 3000 ohms. The receiver interprets any voltage less than −3 V as a 1 and higher than +3 V as a 0, thus providing a noise margin of 2 V. It should be capable of withstanding any voltage between −25 V and +25 V without electrical damage. The same voltage levels are used for the control signals, with a positive voltage indicating an ON, or active, state and a negative voltage for an OFF, or inactive, state. We should note that the terms *mark* and *space* are sometimes used to refer to the 1 and 0 states, respectively, on the data lines. These terms are a carry-over from older transmission technologies used in teletypewriters and telegraph equipment, where mechanical on/off switches were used. An open loop (no current flow) represented a 0, or a space condition, and a closed loop (current flow) represented a 1, or a mark condition.

An important cautionary note is in order regarding the way in which ground connections are handled when long transmission cables are involved. Of course, the signal grounds at the two ends of a transmission link must be connected to provide a

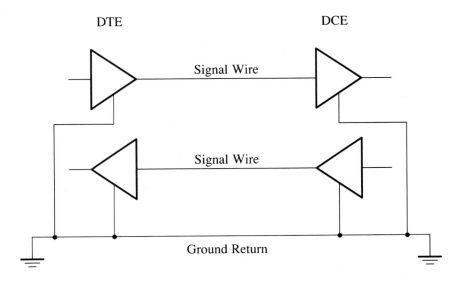

Figure 11.31 Single-ended transmission.

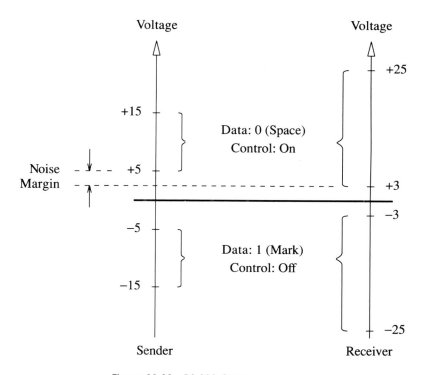

Figure 11.32 RS-232-C voltage levels.

path for the return current. However, the signal ground should be connected to the protective ground at one point only in any system, as shown in Figure 11.33. The protective ground is usually the ground pin on the AC mains connector. The reason for this precaution is illustrated in Figure 11.34, which shows the signal ground connected to protective ground at both ends of a transmission link. Ideally, the protective ground wires (green or bare wires) of the power distribution network should not carry any current. If that were the case, the two grounds in Figure 11.34 would be at the same voltage. However, in reality, protective ground wires carry both DC and AC current components, due to capacitive and inductive coupling to other wires, differences in earth potential at different locations, etc. As a result, the connections to the two grounds in Figure 11.34 result in the introduction of a noise source, as shown. Since the ground return wire has a nonzero impedance, Z, the noise voltage is added to the signal voltage seen by the receiver. The noise voltage introduced in this way can be sufficiently large to cause errors, or even to cause damage to the equipment, particularly during weather storms. (Audiophiles should be familiar with this problem, where a characteristic hum sound is heard if the same mistake is made when wiring a signal source, such as a tuner, to the input of an amplifier.)

For these reasons the signal ground should be connected to the protective ground at one point only in any system, in order to avoid the creation of ground loops. We

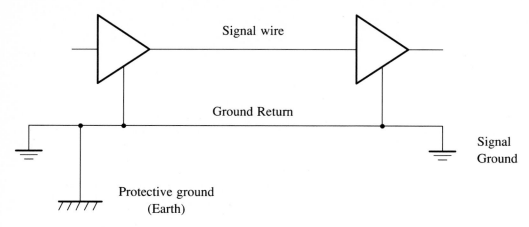

Figure 11.33   Signal and protective ground.

should add that this precaution eliminates any problem arising from DC voltage differences between the two ends, but it does not completely eliminate AC noise components. There will always be a certain amount of capacitive coupling between the signal ground and the protective ground at each end, because metallic enclosures and any other exposed conductive objects are usually connected to the protective ground for safety reasons. Hence, high-frequency ground noise components will continue to affect the receiver even when signal and protective grounds are not physically connected. This is one of the factors that limits the performance of single-ended transmission schemes.

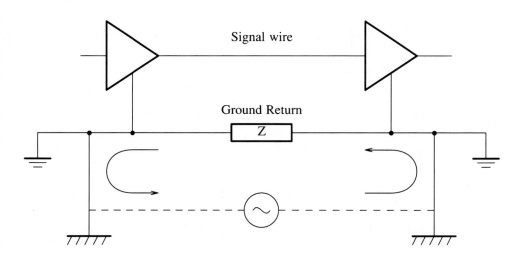

Figure 11.34   Noise introduced by a ground loop.

CONNECTION EXAMPLES

An example of the way in which an RS-232-C interface may be designed is shown in Figure 11.35. An ACIA chip is used for the transmission and reception of serial data. The TTL voltage levels at each ACIA output are converted to the RS-232-C levels shown in Figure 11.32 by an appropriate driver, such as the MC1488 chip. Similarly, an RS-232-C receiver, MC1489, is used for the input signals. By connecting the modem-control signals of the RS-232-C interface to the corresponding inputs and outputs of the ACIA, the state of these signals can be placed under software control via the Command and Status registers of the ACIA, as was explained in Subsection 5.3.1. The receiver section of the ACIA is disabled when Data Carrier Detect ($\overline{\text{DCD}}$) is inactive. In a system in which it is not anticipated that this signal will be used, $\overline{\text{DCD}}$ should be set permanently in the active state.

Details of the connection of the driver and receiver chips are illustrated in parts (b) and (c) of Figure 11.35. The driver requires a positive and a negative power supply to generate the voltage levels of Figure 11.32. A small capacitor is used to limit the rate at which the output voltage changes during high-to-low and low-to-high transitions. The RS-232-C specifications require this rate, which is often called the *slew rate* of the driver, not to exceed 30 V/μs, to reduce the amount of high-frequency noise components that the connecting cable introduces in the environment. A typical RS-232-C receiver requires only a single +5 V power supply, to produce TTL levels at its output, and has an input threshold of about 1 V for high-to-low transitions and 1.25 V for low-to-high transitions. The difference between these two thresholds is called *hysteresis*, which provides some immunity to noise. An additional feature found in some receivers is the availability of a "response control" input, marked R in Figure 11.35(c). By connecting a small capacitor between this input and ground, the input bandwidth of the receiver is reduced so that it can reject high-frequency noise components. The R input may also be used to adjust the receiver's threshold voltages in some applications.

When the distance between the terminal and the computer is small, the modems in Figure 11.29 are not needed. The terminal may be connected directly to the serial port of the computer, provided attention is paid to the way in which the RS-232-C signals are used by each device. Both the computer and the terminal are DTEs, according to the standard's terminology. Hence, they should be connected such that each appears as a DCE to the other. The main difference between the DTE and the DCE, as far as the RS-232-C signals are concerned, is the pin on which data are transmitted. The DTE sends data to the DCE on pin 2 and receives data on pin 3. This means that pin 2 at the computer end of the interconnection cable should be connected to pin 3 at the terminal end, and vice versa, as shown in Figure 11.36. The two signals CTS (Clear To Send, pin 5) and DCD (Data Carrier Detected, pin 8) have no significance for the video terminal. At the computer end, they should be set permanently in the active state, to make the terminal appear to the computer to be ready for transferring data. They are shown connected to DTR (Data Terminal Ready, pin 20). In this way, they will become active as soon as the computer activates DTR to indicate that it is ready to transmit and receive data. Alternatively, a fourth wire in the interconnecting cable may be used to connect CTS and DCD on

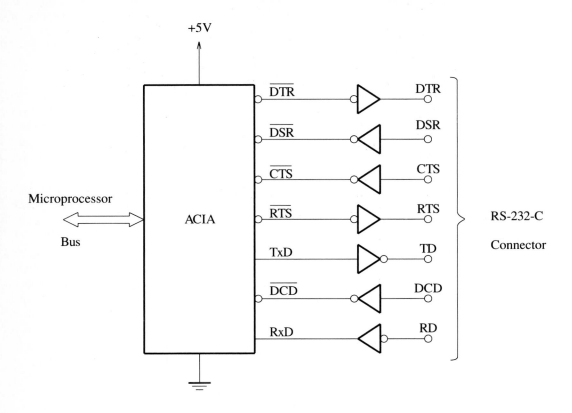

(a) Connection to ACIA

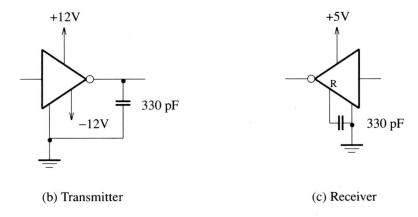

(b) Transmitter

(c) Receiver

Figure 11.35   Handling of RS-232-C signals.

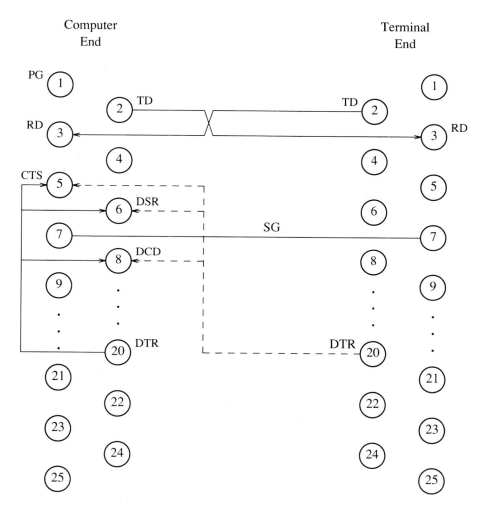

Figure 11.36 Direct connection of RS-232-C signals of a video terminal and a computer.

the computer side to DTR on the terminal side, as indicated by the dashed connection in the figure. This arrangement enables the computer to determine whether the terminal is ready, because the terminal keeps DTR in the active state whenever it is turned on and ready for operation.

Another example of the use of the RS-232-C signals is shown in Figure 11.37. In this case, a printer is connected to a computer via a serial port. Printers often have an internal buffer capable of storing a number of characters, typically in the range 256 bytes to 2K bytes. The computer may send characters at a higher speed than they can be printed, in which case the characters are stored in the buffer. The main feature that the connection in Figure 11.37 provides is that it enables the printer

Computer
End

Printer
End

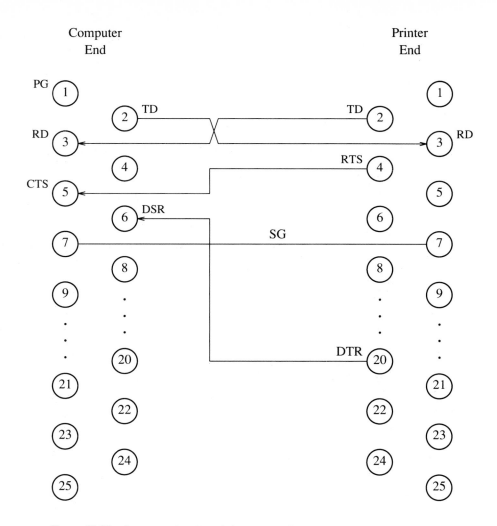

Figure 11.37 An example of a printer connection using the RS-232-C signals.

to stop the flow of data when its buffer is full. It may do this in one of two ways, using a so-called soft handshake or hard handshake. In the soft handshake, when the printer buffer is nearly full, perhaps at about the 75% point, the printer sends a control character to the microprocessor over the input data line (pin 2 of the printer interface). The character most frequently used to stop the flow of data is called X-OFF, which is the ASCII character DC3 (13_{16}). As printing progresses and the buffer approaches the empty point, e.g., when the buffer is less than 25 percent full, the printer sends another control character informing the microprocessor that it may send more data. The character sent in this case is called X-ON (DC1 or 11_{16}). Note that the printer sends X-OFF before its buffer is completely full, because in an interrupt driven system there is likely to be some delay before the microprocessor recognizes the character and stops sending data. Similarly, it sends X-ON before its buffer is completely empty.

The two characters X-OFF and X-ON may be used to control data flow to devices other than printers. For example, they may be sent from the keyboard of a video terminal while perusing a long file. They enable the user to stop and start the printing of data on the screen at any time. They may be sent from the keyboard as CTRL S and CTRL Q, respectively. On some terminals, a special key called Hold Screen is provided which sends X-OFF when activated. It sends X-ON when activated a second time.

The soft handshake uses only the data lines of the RS-232-C interface, and requires the printer to be able to send data to the microprocessor. In the case of a hard handshake, the DTR line of the printer is connected to the DSR input on the microprocessor side. The connection between TD of the printer and RD of the microprocessor is not needed. Instead of sending X-OFF, the printer deactivates DTR, as shown in Figure 11.38, to stop the flow of data.

Whether a soft or a hard handshake is used, the printer may use its RTS output to indicate that it is ready to receive data. For example, the printer may not be ready to receive data immediately after power is turned on. In this case, it delays the activation of RTS until it has completed its internal initialization functions, or until it has been placed *on-line* by the operator. This line is shown connected to the CTS input on the computer side in Figure 11.37, so that it may be tested by the software.

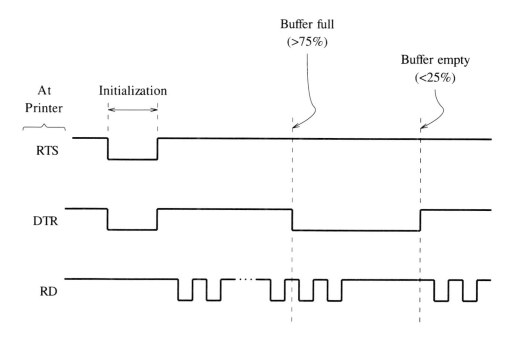

Figure 11.38 Printer buffer control using the RS-232-C signals.

11.4.2 RS-449 Standard

The main limitation to the performance of RS-232-C connections results from the use of the single-ended transmission scheme depicted in Figure 11.31. One source of noise in this arrangement has already been discussed in conjunction with Figure 11.34. There are other noise sources resulting from various forms of coupling between a transmission link and its environment. Crosstalk due to capacitive coupling was discussed in Subsection 10.2.3. Also, the loop created by the signal wire and ground return in Figure 11.31 may pick up interfering signals as a result of inductive coupling with nearby current-carrying conductors. The effect of all these factors is much more pronounced at higher frequencies. Since high-speed transmission requires a receiver with a wide bandwidth, a significant noise component passes through the receiver, eventually swamping the desired signal.

A powerful technique that leads to a significant reduction in all the noise components mentioned above is known as *differential* transmission, illustrated in Figure 11.39. Two signal wires are used to carry the signal current, with the ground connection only providing a common voltage reference for the transmitter and the receiver circuitry. The signal is transmitted in the form of a voltage difference between the two signal wires. The average voltage between the signal wires and ground, called the *common mode* voltage, has very little effect on the receiver's output, ideally none. Every attempt is made to *balance* the transmission link, by making the two signal wires identical, both electrically and mechanically, so that the noise voltages induced on them are the same. Since the receiver is sensitive only to the voltage difference between the two wires, interfering noise will be canceled out.

There remains one important source of noise pick-up in Figure 11.39, which is the loop created by the two signal wires. Any voltage induced in this loop by other

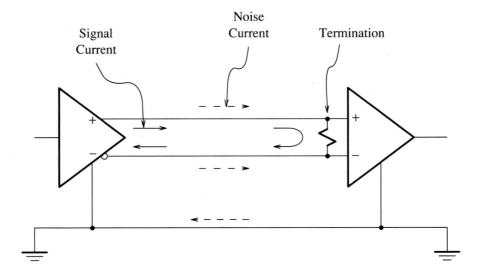

Figure 11.39 Differential transmission.

current-carrying conductors in close proximity will appear as a difference voltage at the input of the receiver. Hence, the area of this loop should be minimized. An effective way to do this, which also helps in balancing the line, is to twist the two wires together tightly. The two wires are sometimes enclosed in a conductive shield to further isolate them from their surroundings.

As in the case of RS-232-C, the RS-449 Standard provides a primary channel and a secondary channel. The primary channel uses a 37-pin connector. Ten of the signals exchanged through the connector are called Category I signals, and are transmitted using the differential scheme of Figure 11.39. Hence, they use two pins each. The remaining signals are called Category II signals. They use a scheme called *unbalanced differential transmission*, in which the sender is single-ended and the receiver is differential, as shown in Figure 11.40. Since the driver is single-ended, only one hot wire is needed for each signal. The signal ground at the DTE is connected through the Send Common line to the negative inputs of all Category II receivers in the DCE, and vice versa. This arrangement provides better performance than the single-ended arrangement of Figure 11.33, because the ground-loop noise source shown in Figure 11.34 will not contribute to the received voltage. However, unbalanced transmission is considerably less effective in rejecting other sources of noise than the balanced differential scheme of Figure 11.39. As a result, the signals included in Category II are those that change state only infrequently and do not require fast receivers.

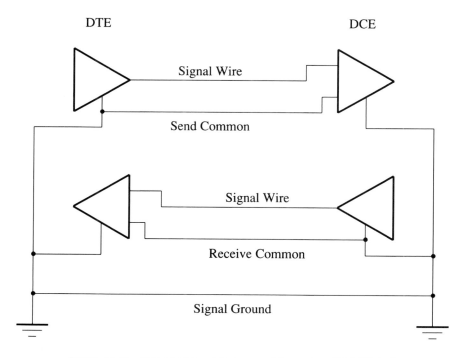

Figure 11.40 Connection of Category II lines of an RS-449 interface.

Table 11.12 gives the RS-449 signals and their RS-232-C equivalents, whenever they exist. Different names are used in the two standards to avoid confusion. The functional differences between the signals in the two standards are minor, and will not be discussed further. The signals given in the table are those used in the primary channel. They are available on the 37-pin connector of an RS-449 interface. When a secondary channel is needed, an additional nine-pin connector is used. It accommodates the secondary channel control signals equivalent to those in Table 11.10. All secondary channel signals are in Category II.

As a result of using differential transmission for Category I signals, an RS-449 interface enables data transfer rates of up to 2 Mb/s, at a nominal maximum distance of 10 m. At 100 kb/s it may be used for distances up to 1 km. Longer distances or faster speeds are also possible, depending on the environment and the type of cable used. The termination resistor in Figure 11.39, which has a value of 100 ohms, is needed only for transmission speeds higher than 20 kb/s. The electrical characteristics of the drivers and receivers are specified in a separate EIA Standard known as RS-422 [10] for Category I signals and RS-423 [11] for Category II signals.

Before concluding this discussion we should point out that the definition of the RS-449 signals has been carefully planned to enable interconnection between RS-449 and RS-232-C equipment. For example, a computer that has an RS-449 interface may be connected to an RS-232-C modem via a simple box that translates one set of signals into the other. Only wire interchanges and a few resistors are required for this task [12].

11.4.3 Centronix Parallel Interface

The interface described in this section is commonly found on printing devices. It was first used on printers manufactured by Centronix Corporation, and later adopted by many other manufacturers. However, there is no official standard by this name.

As in the case of the RS-232-C, the Centronix interface defines a connector and the signals it carries. The signals exchanged between the interface circuit and the printer comprise eight data signals and seven control signals, as shown in Figure 11.41. We have used an "*" to identify signals that are active when in the low-voltage state. The connector has 36 pins, which include a separate ground pin for each signal. The interconnection cable consists of a twisted pair for each signal—a hot wire and a ground wire.

A typical exchange over the Centronix interface is illustrated in Figure 11.42. When the microprocessor sends a byte of data to the printer interface circuit, the interface circuit sends the data to the printer together with a pulse on the Data-Strobe* line. In response, the printer activates the Busy line. When it is ready to accept another byte of data, it sends a pulse on the Acknowledge* line and deactivates Busy. This exchange is essentially the same as the pulse-mode handshake described in Subsection 5.4.3 and Figure 5.23. Figure 11.42 also shows the Printer-Ready status signal inside the interface circuit. Note that, in this case, Printer-Ready can be derived simply as the complement of the Busy signal.

As mentioned in Subsection 11.4.1, many printers contain a buffer capable of storing a number of bytes of data. Instead of asserting Busy during each transfer, the

Table 11.12   PRIMARY CHANNEL SIGNALS IN AN RS-449 INTERFACE

Signal Mnemonic	Pin(s)*	Category	Source	Functional Name	RS-232-C Equivalent
SG	19	—	—	Signal Ground	AB
SC	37	—	DTE	Send Common	—
RC	20	—	DCE	Receive Common	—
SD	4,22	I	DTE	Send Data	BA
RD	6,24	I	DCE	Receive Data	BB
IS	28	II	DTE	In service	—
IC	15	II	DCE	Incoming Call	CE
TR	12,30	I	DTE	Terminal Ready	CD
DM	11,29	I	DCE	Data Mode	CC
RS	7,25	I	DTE	Request to send	CA
CS	9,27	I	DCE	Clear to send	CB
RR	13,31	I	DCE	Receiver ready	CF
SQ	33	II	DCE	Signal quality	CG
NS	34	II	DTE	New signal	—
SF†	16	II	DTE	Select frequency	—
SR†	16	II	DTE	Signaling rate selector	CH
SI	2	II	DCE	Signaling rate indicator	CI
TT	17,35	I	DTE	Terminal timing	DA
ST	5,23	I	DCE	Send timing	DB
RT	8,26	I	DCE	Receive timing	DD
LL	10	I	DTE	Local loopback	—
RL	14	II	DTE	Remote loopback	—
TM	18	II	DCE	Test mode	—
SS	32	II	DTE	Select standby	—
SB	36	II	DCE	Standby indicator	—

*   Pin 1 is connected to the connector and cable shield, and should be connected to the protective ground at one end only. Pins 3 and 21 are spare.

†   Signals SF and SR share the same pin. Only one or the other is used in a given installation.

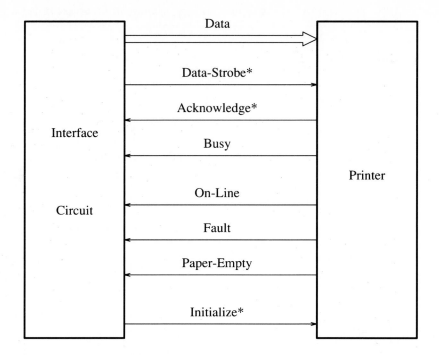

Figure 11.41 Centronix interface signals.

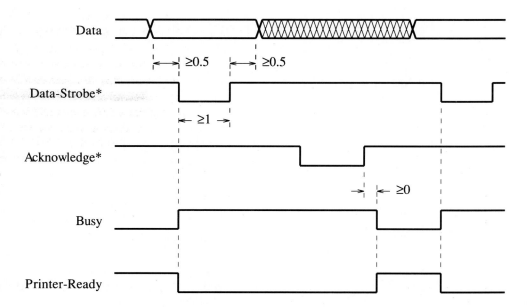

Figure 11.42 Signal exchange on the Centronix interface (delays are given in microseconds).

printer may accept data until its internal buffer is full. However, the Printer-Ready bit in the interface must still go to 0 during each byte transfer, as in Figure 11.42.

The printer sends three status signals to the interface circuit: On-Line (also called Select), Fault, and Paper-Empty. As implied by their names, they indicate that the printer is on line, that a fault condition exists (e.g., paper is jammed), and that the printer has run out of paper, respectively. Most printers have a control switch to allow the user to put the printer either in the on-line or in the local mode. When in the local mode, the printer cannot accept commands from the computer. This mode enables the user to attend to the printer, perhaps to load new paper or to advance the paper manually.

The last control signal in Figure 11.41 is called Initialize*. When activated by the interface circuit, it causes the printer to discard any data that may be in its storage buffer, and to perform any internal initialization procedures that may be needed to get ready for printing. The printer will activate Busy as soon as it receives the Initialize* signal, and will return it to the inactive state when it is ready to accept data.

The Centronix interface is an example of a parallel interface that is intended for a specific application area. Its popularity is a result of its simplicity and the fact that it can support a range of performance and functional capabilities.

11.5 PERIPHERAL BUSES

The previous section introduced some device-interface standards. In each case, a device connected to a microprocessor system requires its own connecting cable and interface circuit. It is often convenient to connect several devices to a common cable that is connected to the microprocessor bus through a single interface circuit. This approach gives rise to the concept of a *peripheral bus*, which is illustrated in Figure 11.43. The microprocessor communicates with the peripheral bus controller, which, in turn, communicates with the individual I/O devices.

The peripheral bus offers many benefits, particularly in a high-performance system. Its design may be tailored to meet the needs of a particular type of peripherals or application environment. The bus controller may itself be an intelligent device, perhaps containing a microprocessor dedicated to the job of controlling the peripheral bus and the devices connected to it. Such a controller would reduce the processing load of the main microprocessor. The devices connected to the peripheral bus may share the use of a DMA controller, as shown in the figure. Also, by using a peripheral bus, fewer devices need to be connected to the system bus. Since the system bus is often a bottleneck, the use of a peripheral bus can lead to a significant improvement in performance.

There exists several standards for peripheral buses. We will describe first the SCSI bus standard, which is used for connecting medium to high-speed peripherals such as disks, tapes, and printers. It is used extensively in microprocessor systems, including popular workstations such as SUN[1] and personal computers such as the

---

1. Product of Sun Microsystems, Inc., Mountain View, CA.

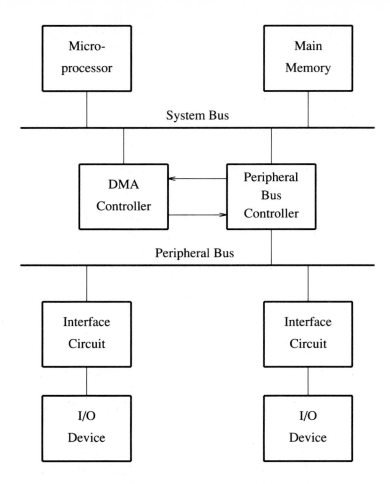

Figure 11.43 The use of a peripheral bus in a microprocessor system.

Macintosh.[2] Later, we will introduce briefly the IEEE-488 and the Computer Automated Measurement and Control (CAMAC) bus standards, which are used primarily in laboratories for connecting measuring instruments.

11.5.1 Small Computer System Interface (SCSI)

This standard is defined by the American National Standards Institute, under the designation X3.131 [13], for connecting microcomputer peripherals to a common peripheral bus. It is intended to enable data transfer rates of up to 4 Mbytes/s. The total length of a SCSI bus is limited to 6 m with single-ended transmission, and

---

2. Product of Apple Computer Inc., Cupertino, CA.

25 m with differential transmission. For short buses, data transfer rates of up to 20 Mbytes/s have been realized. Connection to the bus is via a 50-pin connector.

The speed of transmission and the functional features available on a SCSI bus make it suitable for connecting devices such as disks, tapes, and printers to a micro-computer. Also, several microcomputers may be connected to one SCSI bus. They can use it to communicate with each other as well as with the peripheral devices. In this case, a single disk drive may be accessed by two different microcomputers. Figure 11.44 shows a SCSI bus being used to connect two microcomputers, two disk drives, a magnetic tape drive and a high-speed printer. Each device is connected to the bus via an interface circuit, or controller, that implements the SCSI bus protocol. In total, up to eight controllers may be connected to a single bus. The bus protocol allows several devices to be connected to a single controller, as in the case of the disk controller in the figure.

Devices on the SCSI bus fall into one of two categories: *initiators* and *targets*. An initiator is a device that issues commands to other devices to perform certain operations. A target is a device that receives these commands and provides the

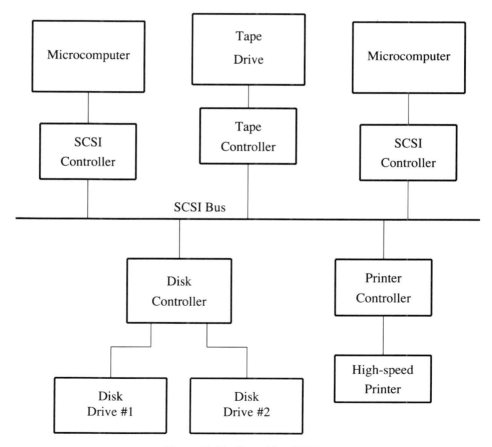

Figure 11.44 Use of the SCSI bus.

service specified. For example, each of the SCSI controllers connected to the micro-computers in Figure 11.44 must be capable of operating as an initiator, to enable the microcomputers to request data transfer operations to or from any of the peripheral devices. The disk, tape, and printer controllers are target devices that implement the commands they receive from an initiator. If the two microcomputers are expected to talk to each other directly, then their SCSI controllers must also be capable of operating as targets.

An initiator and a target communicate by first establishing a logical connection between them over the bus. Then, they use this connection to exchange information, which may be control information or blocks of input or output data. The control information is used by the initiator and target to exchange commands and status information and to coordinate their operation. For example, after establishing a con-nection with the disk controller in Figure 11.44, the initiator needs to specify one of the disk drives, whether it needs to read or write data, the disk address of the data, etc. The target performs the requested operation and transfers the data as specified, or it may indicate that an error has been detected.

Because of the nature of the operation of devices such as disks and tapes, several messages may be exchanged between the target and the initiator to carry out a partic-ular command. Furthermore, successive messages may be separated by long waiting periods, while a disk performs a seek operation or reads a particular sector. Connec-tions on the SCSI bus are designed to accommodate such behavior. An initiator begins by selecting a target. In order to do this, it must first request and obtain con-trol of the bus, and establish a logical connection with the target. Then, it exchanges as many messages as needed with the selected target to specify the desired service. The connection may now be suspended while the target performs the requested operation. During this time the bus is released for use by other devices. When the target is ready to respond to the initiator, it contends for control of the bus. When successful, it reactivates the suspended connection and forwards its response to the initiator. The process of suspending and later reactivating a connection may be repeated several times before a particular task is completed.

In what follows, we will describe the SCSI bus signals and the way they are used to exchange control information and input/output data between the initiator and target.

BUS SIGNALS

The SCSI bus comprises 18 signals, whose functions are summarized in Table 11.13. Recall that either single-ended or differential transmission may be used. In the single-ended case, all signals are asserted, or equal to 1, when in the low-voltage state. When referring to connector pins or bus lines, rather than logic signals, the signal names are preceded with a "–" sign to indicate this fact. For example, the line that carries the BSY signal is called –BSY. In the case of differential transmission, two lines are used, and are called –BSY and +BSY.

For ease of reference, the bus signals listed in the table are divided into several categories. There are nine data signals and nine control signals. The data signals are used for transferring up to 8 bits of information in parallel, plus a parity bit for error detection. The information being transferred may consist of I/O data. It may also

Table 11.13 THE SCSI BUS LINES

Category	Name	Function
Data	−DB(0) to −DB(7)	Data lines: used to carry one byte of information during an information transfer phase and device identification (address) during arbitration, selection, and reselection phases
	−DB(P)	Parity bit for the data bus
Phase	−BSY	Busy
	−SEL	Asserted during selection and reselection
Information type	−C/D	Control/Data: asserted during transfer of control information (command, status or message)
	−MSG	Message: indicates that the information being transferred is a message
Handshake	−REQ	Request: asserted by a target to request a data transfer cycle
	−ACK	Acknowledge: asserted by the initiator when it has completed a data transfer operation
Direction of transfer	−I/O	Asserted to indicate an input operation (relative to the initiator)
Other	−ATN	Attention: asserted by an initiator when it wishes to send a message to a target
	−RST	Reset

consist of one of three types of control information, called command, status, and messages. The type of information is specified by the two control signals C/D and MSG. A third control signal, I/O, indicates the direction of the transfer. One of the two devices involved in any data transfer must be an initiator and the other a target. A transfer is said to be an input or an output operation based on its direction relative to the initiator. Thus, when I/O = 1, indicating an input operation, data are transferred from the target to the initiator, and vice versa.

Individual byte transfers on the lines are always controlled by the target. The target drives the C/D, MSG, and I/O signals according to the type and direction of each byte transfer. The timing of the transfer is controlled by a handshake between the request and acknowledge signals, REQ and ACK. The target asserts REQ and the initiator responds by asserting ACK.

Operation of the SCSI bus proceeds in stages called *phases*. Starting from an idle state, called the *bus-free* phase, any device wishing to transfer data may request the use of the bus. The action of requesting the bus causes all device interfaces to enter the *arbitration* phase. The device that wins the arbitration starts a *selection* phase, to select another device with which it wishes to exchange data. The selection phase results in a logical connection being established between the two devices, and the bus enters the *information transfer* phase. At this point, the two devices may transfer as many bytes of information as they wish, where the information transferred may consist of an arbitrary mix of control information and data. Then, the bus returns to the bus-free phase. Throughout this process, the Busy (BSY) and Select (SEL) signals indicate which phase is currently in progress. We will describe each of these phases briefly, limiting our discussion to features that have not been encountered before, and which have been introduced because of the special needs of the SCSI bus. First, we will describe the addressing scheme used.

DEVICE IDENTIFICATION

The SCSI bus supports a maximum of eight devices, numbered 0 to 7, that can be addressed directly. Each device is associated with one of the eight data lines, DB(0) to DB(7), which are used as address, or ID, lines during the arbitration and selection phases. When a device wishes to identify itself, it asserts its associated data line. Similarly, to select another device, it asserts the ID line of that device.

Because only one line is needed to identify a device, it is possible to identify several devices at the same time. For example, when an initiator selects a target, it places its own ID on the line together with that of the target, thus enabling the target to know whom it is talking to. Also, during the arbitration phase, several devices may place their IDs on the bus simultaneously, as will be explained below. Of course, in this case each participating device must drive only its own ID line.

ARBITRATION

There is no single device that performs the function of the bus arbiter. Instead, a distributed arbitration scheme is used, in which all device interfaces participate. A device may request the bus only while the bus is free. This state is identified by the BSY and SEL lines being simultaneously in the inactive state for some minimum period. To request the bus, a device asserts BSY and at the same time asserts its ID on the data lines. No other device is permitted to transmit its ID to request the bus after it observes that BSY has been asserted. However, it is possible that while the bus is being used by some device, two or more other devices decide they need to transfer data. As soon as the bus becomes free, they all assert BSY and their ID lines, hence the need for arbitration.

Each device on the bus is assigned a fixed priority. A request on line DB(7) is regarded as having the highest priority and DB(0) the lowest priority. Each device independently examines the data lines to determine whether it has the highest priority. If so, it proceeds to the selection phase. Otherwise, it turns off its drivers and waits for the bus to become free again.

The arbitration mechanism described above is very simple, but it is fraught with danger. It is based on the assumption that all devices see the same information, and hence will reach the same decision as to which the highest-priority request is. To guarantee reliable operation, once a device detects an active state on the BSY line, it is not allowed to change the state of its ID line until the arbitration phase ends. Also, after each step in the arbitration process, sufficient delay is introduced to allow for propagation delays and transmission line effects. The sequence of events is illustrated in Figure 11.45, which also gives some representative values for the required waiting periods. The figure shows the case where devices 2 and 6 are waiting to use the bus. They must wait for BSY and SEL to become inactive and remain in that state for at least 800 ns. Then, they assert BSY and their respective ID lines. After a minimum of 2200 ns, each device examines the data lines to determine whether it has won the arbitration. This rather long period is needed to ensure that BSY has reached a stable state, that all devices have observed the asserted state on BSY, and that they have stopped changing the state of the ID lines.

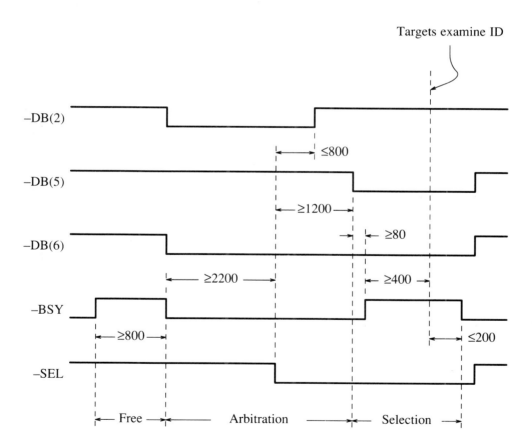

Figure 11.45 Sequence of events during arbitration and selection on the SCSI bus. (Delays in nanoseconds.)

Device 6 wins the arbitration cycle in the figure, and it asserts SEL. Device 2 is required to release all bus lines within 800 ns of asserting SEL. After asserting the SEL signal, the winning device waits for at least 1200 ns to allow other devices to clear the bus and the bus signals to settle. Then it enters the selection phase.

SELECTION

The devices contending for the bus during the arbitration phase may be either initiators or targets. Initiators request the bus to start a new I/O operation, while targets request the bus to resume a suspended operation. The selection phase is entered when the device that won the arbitration is an initiator. In the case of a target winning the arbitration, the bus enters the reselection phase, which will be described later.

Having gained control of the bus, the initiator selects one of the target devices. To do so, it asserts the SEL line and places the ID of the desired target, together with its own ID, on the data lines. In the case of Figure 11.45, we have assumed that device 6 is an initiator that wishes to establish a connection with device 5, which must be a target. After asserting DB(5), it waits for a short period to allow for bus skew, then releases the BSY line. The combination of an active state on SEL and an inactive state on BSY indicates that the bus is now in the selection phase. All targets wait at least 400 ns before examining the data line, to allow these lines to settle. Then, the selected target responds by asserting BSY. The initiator releases SEL and the ID lines, thus terminating the selection phase. At this point, the target is in control of the bus, and it may begin transferring information at any time.

INFORMATION TRANSFER

The selected target acts as the bus master during the information transfer phase. The initiator, which started this process by selecting the target, performs the role of a slave that responds to data transfer requests. Usually, when a target is selected, it begins by requesting a command from the initiator. It does so by starting a data transfer operation in which it sets I/O = 0, C/D = 1 and MSG = 0. The command the target receives from the initiator may specify that a block of data is to be read from one of the disk drives controlled by the target, for example. The target begins reading the requested data from the disk. As each byte becomes available, the target sends it to the initiator using an input data transfer (i.e., with I/O = 1 and C/D = MSG = 0). After transferring all bytes of the block, the target sends a command complete message (I/O = 1, C/D = 1 and MSG = 1). Throughout this process, the target maintains control of the bus by keeping the BSY line asserted. At the end, it releases BSY, thus disconnecting itself from the initiator and returning the bus to the free phase.

It was mentioned above that the target need not remain connected to the initiator until the requested operation is completed. A disk seek may be needed before the data can be read, or the target may read the entire block into an internal buffer before it begins sending it to the initiator. To allow other devices to use the bus during this period, the target may suspend its connection to the initiator after receiving the read command. First, it sends a single-byte message to the initiator, indicating that it is about to suspend the connection and that the initiator should save any information related to the command. Then, the target releases the BSY line. When it is ready to transfer the data it contends for the bus, and after gaining control, it reactivates the suspended connection using the reselection procedure described below.

RESELECTION

The reselection phase is essentially the same as the selection phase, with the roles of the target and the initiator interchanged. The signal exchange on the bus is illustrated in Figure 11.46, where the letters I and T identify whether it is the initiator or the target that is driving a given line. The target requests the bus, and when it wins an arbitration cycle, it asserts the SEL line, its ID line and that of the initiator, then releases BSY. The initiator responds by asserting the BSY line.

At this point, we encounter the main difference between the selection and reselection procedures. Recall that it is always the target that controls the data transfer phase. After reselecting the initiator, it needs to get back control of the BSY line. When it receives the initiator's response in the form of an asserted state on the BSY line, it too asserts BSY before releasing SEL. Then, it begins transferring data. The initiator should release BSY soon after SEL becomes inactive. But, this time BSY will remain held in the asserted state by the target.

The process or handing over control of the BSY line from the initiator to the target gives rise to a subtle, yet potentially dangerous problem. Just before the initiator releases BSY, this line is being held in the asserted state by both the target and the initiator. When the initiator turns off its driver, no change in state should be observed on the BSY line. However, careful analysis of current flow reveals the

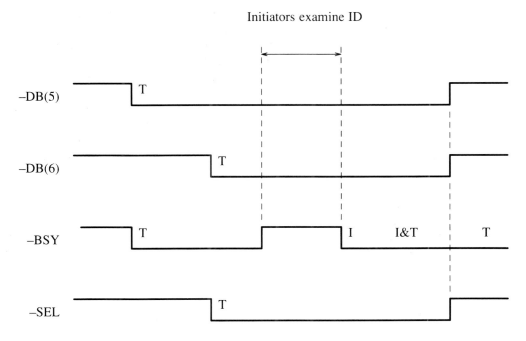

Figure 11.46 Target (5) reselects the initiator (6).

possibility of this line becoming inactive for a short time [14]. The length of this period is a function of the distance between the two devices and the total length of the bus. The pulse that may appear on the bus under these circumstances is commonly known as the "OR glitch," because the state of the BSY line is the logical OR of the states of the drivers connected to it. It is important to ensure that no device will interpret this narrow pulse as an indication of the bus being free. Hence, no device should assume that the bus is free until the BSY line has remained in the inactive state longer than the maximum possible width of the OR glitch. This phenomenon is the main reason for the 800-ns minimum duration of the bus-free phase shown in Figure 11.45.

COMMENTS

In concluding the discussion of the SCSI bus, it is instructive to reflect on some of its salient features and how they have been influenced by the intended application environment. The bus can be up to 25 m long and can be operated at data rates up to 4 Mbytes/s. It serves a wide variety of high-speed peripherals, such as tapes, disks, and printers. At the hardware level, the needs of this environment have affected the design of the bus in many ways. For example, an important characteristic of disks and tapes is that the timing of their operation is controlled by their mechanical motion. In the absence of large buffers in the device controller, data transfers over the bus must also be coordinated with the operation of the device. Hence, the function of controlling the information transfer phase on the SCSI bus is assigned to the target, rather than the initiator as might otherwise be expected. To start an operation, the initiator requests the bus. Then, when it gains control it selects a target and hands the bus over to that target.

Another characteristic of disks and tapes is the long delay between the time a request for access to a particular data block is received and the time data transfer can begin. The ability to suspend and later reactivate a connection over the bus makes it possible to use the bus for other purposes while waiting for the disk or tape to reach the desired block.

The ability to suspend a connection has farther reaching implications than the question of bus utilization. This feature implies that the initiator may send a command to read from or write into a particular disk block, without worrying about first sending a command to perform the seek operation. The target, which acts as the device controller, automatically generates a seek command, if needed, and when that is completed begins transferring data. This is but a simple example of many functions that the device controller may perform. By increasing the functional capabilities of the device controller, the initiator, and ultimately the microprocessor that generates the request, can be isolated from the details of the operation of the peripheral device. The microprocessor may issue a request to read a "logical" block of data. The requested block may reside on several, possibly noncontiguous physical disk sectors. When this request is relayed by the initiator to the target, the target translates it into a sequence of seek and read operations to access the data. The connection between the initiator and target may be suspended several times while these operations are being performed.

Many commercial device controllers designed for use with the SCSI bus offer functional capabilities of the type mentioned above. They are also capable of detecting disk errors and keeping track of defects on the storage medium. The elaborate structure of the control information that can be exchanged on the SCSI bus, consisting of commands, messages, and status information, provides the communication channel between the initiator and target needed to use these functional capabilities.

In addition to specifying the data transfer and arbitration schemes used on the bus, the SCSI Standard defines a number of messages that can be exchanged by the initiators and targets for different types of devices. The messages contain addressing information, status information, descriptions of the required operations, such as read, write, verify, seek, etc. Messages are provided for disk formatting, exchanging lists of defects on disk tracks, and other device specific functions. Many powerful commands, such as commands to search a certain area of a disk or tape for a particular data pattern, are also included in the Standard.

The SCSI bus has gained considerable popularity. Many microcomputers, disks, tapes and other peripherals are commercially available with a built-in SCSI interface. Such devices can be connected together in a straightforward manner by means of a SCSI cable.

11.5.2 Instrumentation Bus—IEEE 488

The IEEE 488 Standard defines an interconnection scheme that uses a parallel bus organization to connect several peripheral devices to a computer [15]. It is mainly intended for connecting laboratory instruments, such as oscilloscopes, network analyzers, voltmeters, etc. For this reason, it is called a General-Purpose Instrumentation Bus (GPIB). Since it was first defined and used by the Hewlett-Packard Company, it is also known as HPIB.

Figure 11.47 shows a GPIB used to connect a voltmeter, an oscilloscope, and a plotter to a microcomputer. While a microprocessor bus is usually limited to a few tens of centimeters in length, the GPIB may be up to 20 m long. Hence, the instruments can be conveniently located on a laboratory bench. The bus consists of 16 signal wires that comprise 8 data lines and 8 control lines, where the data lines may carry a device address or a byte of data.

The way in which devices are selected and data is transferred over the GPIB differs somewhat from the familiar read and write operations on a microprocessor bus. All data transfers are controlled by a single device called a controller, which is the microcomputer in the case of Figure 11.47. Devices that are capable of transmitting data over the GPIB are called *talkers*, and those that receive data are called *listeners*. A device that has both talk and listen capabilities is assigned two different addresses, one for each function.

Consider an experiment in which it is desired to obtain some readings from the voltmeter, use these readings to display a waveform on the oscilloscope, and produce a hardcopy record on the plotter. The required transfer of data takes place as follows. The microcomputer informs the oscilloscope and the plotter that they are to

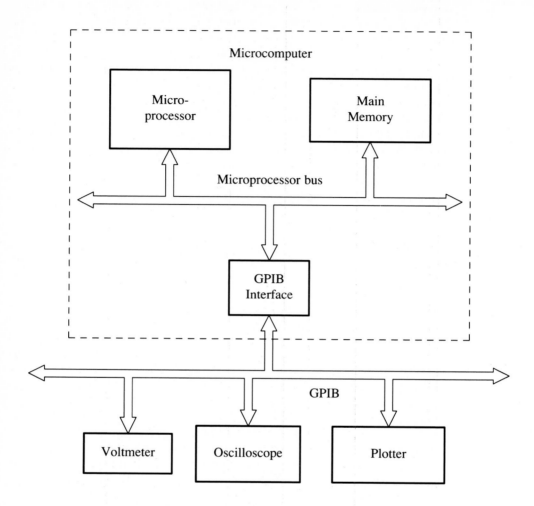

Figure 11.47 An example of a system that uses the GPIB.

receive data. To do this, it sends two successive control messages on the GPIB, the first containing the listen address of the oscilloscope and the second the listen address of the plotter. As a result, these two devices begin to listen to any data transmitted on the bus. Then, the microcomputer instructs the voltmeter to start transmitting data, by sending a third control message containing the voltmeter's talk address. The voltmeter may now begin to transmit any amount of data, one byte at a time, which both the oscilloscope and the plotter will receive. No addresses need be specified during data transmission, because the intended destinations were selected before the transfers began. Data transfer continues until the voltmeter sends a control message indicating "end of data." Alternatively, the microcomputer may terminate data transfer by issuing an "attention" message, which causes the talker to stop transmitting data. Throughout this exchange, the control lines of the bus are used to indicate the nature of the information on the data lines, and to distinguish between control messages and data messages.

In addition to data transfer, the GPIB controller can send control messages that act as a trigger, which would cause some device to begin performing a predefined operation. For example, a trigger message may be sent to a testing device to initiate a particular test sequence. There are also several ways for the controller to collect status information from all the devices on the bus, either sequentially or in parallel.

The GPIB has met with considerable success in the marketplace. Many laboratory instruments are available with a built-in GPIB interface, which enables direct connection to a GPIB cable. Special GPIB interface cards are available for many microcomputer systems and personal computers.

11.5.3 CAMAC Digital Interface System

This standard describes the interconnection system for Computer Automated Measurement and Control (CAMAC). It was originally developed for the measurement and control equipment used in the nuclear field, but it found applications in many other environments. The CAMAC scheme was adopted as an IEEE standard in 1975 [16].

In the CAMAC scheme, circuit boards are housed in card cages called *crates.* The standard defines a bus called a *Dataway,* which runs at the back of a crate to interconnect the circuit boards. It also defines a scheme for interconnecting several crates over either a *parallel highway* or a *serial highway.* Each crate has up to 25 mounting *stations,* numbered 1 to 25, where each station is provided with an 86-pin connector to access the Dataway. The interface and control circuitry for a given instrument is housed in a *module* or *plug-in unit,* which may occupy one or more stations.

Station 25 in each crate is called a *control station,* and the remaining 24 are *normal stations.* One of the modules mounted in a crate functions as a *crate controller.* It occupies the controller station plus at least one normal station, usually number 24.

The crate controller is responsible for controlling the operation of all other modules in the crate. It can read from or write into the registers of any module. It may also send a command to a module to enable it or disable it, or to perform a special function. In the remainder of this section, we will describe briefly some of the distinguishing features of the Dataway.

The Dataway carries 24 bits of data in parallel. Unlike the buses we have encountered so far, the data lines are not bidirectional. There are 24 lines for transferring data from any station to the Crate Controller (for read operations) and another set of 24 lines for carrying data from the Controller to other stations (for write operations). While transferring data or sending commands, the Crate Controller selects one or more normal stations to participate in the operation. There are 24 select lines provided for this purpose, one for each station. To select a station, the controller activates the corresponding line. Each station may have up to 16 internal registers. One of these is selected using a 4-bit subaddress, which is transmitted by the controller at the same time it activates the select line for the station.

During each Dataway transaction, the Controller transmits a command that specifies the type of action to be taken, such as write data, disable device, etc. The selected modules may respond by sending status information on separate lines provided for this purpose. In addition, each station has one dedicated line called "Look-At-Me," which it activates when it requires service from the Crate Controller.

The timing of data transfers over the Dataway is controlled by two *strobe* lines. After placing a command on the bus, the Controller sends a pulse on the first strobe line. A while later, it sends a pulse on the second line. In this way, a Dataway transaction is divided into two stages, during which different actions can take place. The two stages are somewhat similar to the two phases of a clock period on a synchronous bus.

The availability of several command and status lines, dedicated select and Look-At-Me lines, and separate data lines for read and write operations makes CAMAC modules very simple to design. Only a limited amount of decoding is needed, and correct timing is easily implemented by using the pulses on the two strobe lines. This approach is well suited to the instrumentation environment. It also reflects the constraints of the earlier stages of digital circuit technology, where gate count dominated the cost of a circuit module.

11.6 CONCLUDING REMARKS

This chapter introduced the concept of a system bus, or backplane bus, and how it may be used in microprocessor systems. Two examples of industry standards, the VMEbus and the Multibus, were discussed in detail to provide the reader with a complete picture of the hardware involved.

The backplane bus provides the backbone of a microprocessor system. However, it is limited in length and in the number of separate devices that can be connected to it. Several serial and parallel schemes are available for connecting various types of peripherals to a microcomputer. The RS-232-C and its successor, RS-449, standards were described. They are widely used with data terminals, modems, and other low to medium-speed devices. Three peripheral buses were also described, which provide special features to support certain types of devices. The GPIB and CAMAC buses are suitable for laboratory instrumentation and control applications, while the SCSI bus is used for connecting disks, tapes, and other high-speed devices.

This chapter concludes the presentation of the hardware features of microprocessor systems. Machine and assembly-level programming was presented in earlier chapters. In the following chapter, an overview of the software that controls the system's operation will be presented.

11.7 REVIEW QUESTIONS AND EXERCISES

REVIEW QUESTIONS

1. What is the difference between the system bus and the bus of a microprocessor? When can the same bus serve both functions?

2. Why is the protocol translation circuit in Figure 11.2 needed? Give a simple example to illustrate your answer.

3. The protocol translation circuit in Figure 11.2 is said to be transparent. What does this mean? Give an example of a situation in which this circuit is not transparent to the programmer. (*Hint*: consider buses having different address lengths.)

4. Indicate which of the following transfers is allowed on the VMEbus:

   (a) Read 4 bytes at A32C14

   (b) Write 2 bytes at 5DE641

   (c) Write 1 byte at 5DE641

   (d) Read 1 byte at 4F62A3C

   In each case, indicate which of the three lines DS0*, DS1*, and LWORD* will be activated and which data lines will be used to carry the data.

5. Figure 11.7 indicates that the slave must not deactivate DTACK* before setting its data lines in the high-impedance state. Why is this condition necessary?

6. There are three types of arbiters that may be used on the VMEbus. Is the design of the interface circuit of a device connected to the bus affected by the type of arbiter used?

7. What is the function of the BCLR* signal on the VMEbus? How would the absence of this signal affect the strategy used by a given master as to when to acquire and release the bus?

8. The VMEbus uses a daisy chain arrangement to propagate the interrupt-acknowledge signal. Does this place any constraint on the location of an interrupt handler relative to the interrupters it serves?

9. Based on the delay constraints in Figure 11.7, what are the shortest and longest possible durations for a read cycle on the VMEbus?

10. After completing a data transfer, a bus master should release the bus, unless it is going to carry out another data transfer immediately. When is such a strategy necessary? Considering the signals on the VMEbus, is there an alternative?

11. What is the role of the BBSY* signal in the arbitration process on the VMEbus?

12. When a VMEbus has several interrupt handlers, how is interrupt priority determined?

13. Consider a microprocessor board connected to the VMEbus. It was suggested in Subsection 11.2.4 that the ACFAIL* signal may be connected to one of the interrupt-request lines of the microprocessor. How would you handle the SYS-FAIL line?

14. The VMEbus requires the slave to remove the data before it deactivates the DTACK* line. On the Multibus, the slave must deactivate XACK* first. What difference does this make to the way the master loads the data into its input buffer?

15. Give one or two examples from your own experience where the inhibit facility of the Multibus could be useful.

16. Can you connect a device that does not transmit an interrupt vector to the Multibus? Does the INTA* line get activated in this case?

17. Consider the serial arbitration scheme of the Multibus. If the bus is free, what is the maximum delay from the time a device decides it needs to become the bus master until it can begin transferring data?

18. The BCLR* signal of the VMEbus and the CBRQ* signal of the Multibus are both used to inform the current bus master that another device wishes to use the bus. What is the difference in the way these two signals are used?

19. In Subsection 11.4.2, it was mentioned that unbalanced differential transmission is used with signals that "do not change state frequently." Why is it possible to use a lower-grade link on such lines?

PRIMARY PROBLEMS

20. Consider a 68000 microprocessor board connected to a VMEbus. Design a logic circuit to generate the address and data strobe signals on the VMEbus from the signals on the 68000 bus.

21. The VMEbus does not guarantee address hold time after AS* becomes inactive. Hence, in the presence of bus skew, a simple address decoder whose output is gated with AS* may lead to false selection. One way to handle this problem is to use an edge-triggered flip-flop, as was done in the example in Figure 11.14. Examine the timing constraints in Figure 11.7 carefully and suggest an alternative circuit.

22. Give a timing diagram showing two successive read cycles on the VMEbus involving two different devices. Calculate the minimum delay from the time one device disables its data drivers until the other enables its drivers.

23. Repeat Problem 22 for a read cycle followed by a write cycle.

24. One of the data registers in an I/O interface consists of an 8-bit buffer that has one enable input, E. Input data is transferred to the register's output while E = 1 and is locked in when E changes to 0. Show how the E input may be controlled

to enable data to be reliably written into the register from the VMEbus. Assume that the register requires a setup time of 65 ns and a hold time of 20 ns (both relative to the $1 \rightarrow 0$ transition at the E input).

25. Design a serial interface for the VMEbus using an ACIA chip. Show only those parts of the interface needed for read and write operations.

26. The multiplexing arrangement in Figure 11.5 is needed for read operations. Design a separate circuit that can be used to write bytes, words, and long words into the memory.

27. Give an alternative design for the circuit in Figure 11.5 using the 8287 bidirectional buffer (see Fig. 11.26) to enable both read and write operations.

28. Design an interface circuit similar to that of Figure 11.26 to connect an ACIA to the Multibus.

29. The role of the common bus request signal, CBRQ*, was explained in Subsection 11.3.3. Consider a situation in which this signal is not used. The current bus master completes a data transfer operation and relinquishes the bus mastership. Soon after that, it decides it has another data item to transfer. It requests and acquires the bus mastership and transfers the data. Assuming that no other devices were using the bus during that period, how much time could have been saved by using the common bus request feature of the Multibus?

30. Consider a 6809 microprocessor board connected to a Multibus. It is required to make the contents of the low-order byte of the Reset vector programmable via eight switches, which are installed on a different board. Show how you may use the inhibit facility of the Multibus bus for this purpose.

31. Consider an RS-232-C connection between a microprocessor and a modem. Examine Tables 11.10 and 11.11 carefully and suggest a possible sequence of signal exchanges while establishing a connection with another modem. Assume that the two modems are permanently connected to each other.

32. Repeat Problem 31 for the case of a modem with a telephone answering capability. (*Hint*: the microprocessor must first instruct the modem to answer the telephone.)

33. Draw a diagram to illustrate how capacitive coupling alone can introduce ground loop noise into a single-ended system.

34. Consider the RS-232-C interface in Figure 11.35. A modem with a telephone answering capability is connected to a 6809 microprocessor via this interface. The ACIA occupies locations CB00 to CB03, and the RI signal of the RS-232-C interface is available as the least-significant bit at location CB04. Write a 6809 program that accepts incoming calls and establishes a communication channel with the calling device (see Tables 11.10 and 11.11). The last instruction in your program should be a branch to a routine called TALK.

35. Two microprocessor systems are to be connected together via a serial link. Each system has an RS-232-C interface similar to that shown in Figure 11.35. Show how you may interconnect the RS-232-C pins of the two systems to provide the required link. Use all the signals shown in such a way that each system sees the same signaling sequences as if it were connected to a modem (i.e., do not set any signals permanently in the active state).

36. Consider the printer connection shown in Figure 11.37. Write a 6809 or a 68000 program to send data to the printer using the RS-232-C signals to control the flow of data to the printer buffer. Assume that these signals are connected to the microprocessor via an ACIA, as shown in Figure 11.35.

ADVANCED PROBLEMS

37. The 68000 microprocessor allows byte transfers on either the high-order or low-order data lines. Show how you may generate the address and data strobe signals on the 68000 bus from the signals on the VMEbus to support any byte or word transfer that may be initiated by the microprocessor.

38. Consider a VMEbus I/O interface that contains one ACIA and supports interrupt-controlled I/O. The interface generates an interrupt request on INT3* whenever the interrupt-request output of the ACIA is active. When it receives an interrupt acknowledgment it transmits the ID code $5B_{16}$. Give an appropriate design for the part of the interface responsible for handling the interrupt signals on the bus and for transmitting the ID code. How and when is the interrupt request removed from the bus?

39. Repeat Problem 38 for an interface that has two ACIAs and one PIA. Interrupt requests from the PIA, either IRQA or IRQB, should be transmitted at priority level 4, while those from the ACIAs should be at level 3. The ID codes are $5A_{16}$ for the PIA and $5B_{16}$ for the ACIAs. (*Hint*: Requests for a given level can simply be ORed together.)

40. Consider a case in which there are two pending requests for interrupts at the outputs of the two ACIAs in Problem 39. Give a possible sequence of events from the time the interrupt handler recognizes the interrupt request at level 3 until both ACIAs turn off their requests.

41. Figure 11.48 shows the control circuit for the serial priority network in the interface of a Multibus master. The circuit begins the process of acquiring the bus mastership when REQ is activated. After it has acquired the bus and activated BUSY*, it activates its Master output to indicate to the remainder of the interface circuit that this device is now the bus master. The bus mastership is relinquished when REQ becomes inactive. Suggest an appropriate design for the serial priority controller.

42. What changes, if any, would you make to the circuit of Problem 41 for the case of parallel priority? Can you define a circuit that works in both cases?

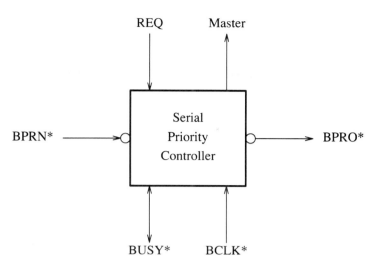

Figure 11.48 Problem 41.

11.8 REFERENCES

1. Dawson, W.K., and R.W. Dobison, "A framework for computer design." *IEEE Spectrum*, Oct. 1986, pp. 49-54.

2. *The VMEbus Specification.* Motorola, Oct. 1985.

3. *IEEE Standard for a Versatile Backplane Bus: VMEbus.* ANSI/IEEE Standard 1014-1987.

4. *Intel Multibus Specification.* Intel, 1981.

5. *IEEE Standard Microcomputer System Bus.* ANSI/IEEE Standard 796-1983.

6. Boberg, R.W., "Proposed microcomputer system 796 bus standard." *Computer*, vol. 13, Oct. 1980, pp. 89-105.

7. *A Full-Feature 32-Bit Backplane Bus.* IEEE Draft Standard 1296.

8. *Interface Between Data Terminal Equipment and Data Circuit-Terminating Equipment Employing Serial Binary Data Interchange.* Electronics Industries Assoc., Standard RS-232-C, Oct. 1969.

9. *EIA Standard RS-449: General Purpose 37-Position Interface for Data Terminal Equipment and Data Circuit-Terminating Equipment Employing Serial Binary Data Interchange.* Electronics Industries Assoc., Standard RS-449, Nov. 1977.

10. *Electrical Characteristics of Balanced Voltage Digital Interface Circuits.* EIA, Standard RS-422-A, Dec. 1978.

11. *Electrical Characteristics of Unbalanced Voltage Digital Interface Circuits.* EIA, Standard RS-423-A, Dec. 1978.

12. *Application Notes on Interconnection Between Interface Circuits Using RS-449 and RS-232-C.* EIA, Nov. 1977.

13. *Small Computer System Interface (SCSI).* ANSI Standard X3.131 (draft), 1985.

14. Gustavson, D.B., "Wire-OR logic on transmission lines." *IEEE Micro*, vol. 3, June 1983, pp. 51-55.

15. *IEEE Standard Digital Interface for Programmable Instrumentation.* ANSI/IEEE, Standard 488, 1978.

16. *IEEE Standard Modular Instrumentation and Digital Interface System (CAMAC).* IEEE Standard 583, 1975.

Chapter 12

Microprocessor System Software

In the preceding chapters we examined many aspects of microcomputer hardware and programming concepts at the assembly language level. We considered the structure of the hardware and the way in which it executes programs. An implicit assumption was that programs can be prepared and placed in the memory in some reasonably straightforward manner. The only software aids considered were the assembler programs, which assemble users' programs written in assembly language into machine level object code that can be executed directly.

A computer system cannot be used effectively without an appreciation of the available software resources. How can a desired program be prepared using the available text editing facilities? Having prepared the program, how can it be assembled, or compiled, then executed to produce the desired output? These tasks are performed with the help of software that resides in the computer. Some of this software is closely tied to the particular type of computer used. It enables the user to use the machine efficiently without being familiar with the details of the system. This software is referred to as the *system software*. In contrast to the system software, the term *application programs* is used to refer to programs that perform functions other than those needed for the operation of the computer system itself. Application programs may be written by an individual user to solve a problem or process data. They could also be programs that are more generally available to perform tasks such as bookkeeping, accounting, or playing games. System software facilitates the preparation and execution of application programs by hiding the individual characteristics of the system from the user.

An informal description of typical system software functions is confined to this single chapter, but this does not imply that software issues are less important than hardware ones. It is merely a consequence of the fact that this text is primarily a treatment of microcomputer organization, which emphasizes hardware aspects. Detailed discussion of software topics can be found in other textbooks, which the reader is encouraged to consult to obtain deeper understanding of system software.

12.1 SYSTEM SOFTWARE OVERVIEW

Microcomputer systems are found in many widely differing configurations, ranging from those that are dedicated to a single user or a single task to systems that accommodate many users or many tasks simultaneously. The supporting software differs according to the size and the nature of a system. In a single-task system it tends to be fairly small. In a multitask system, system software is often large as it must provide for efficient sharing of resources among several tasks.

12.1.1 A Simple System

Regardless of the size of the system, certain basic functions must be available. Let us begin by considering a simple microprocessor system, implemented on a single printed-circuit board similar to that discussed in Section 10.1. The system hardware consists of a microprocessor, the main memory, and a video terminal that serves as the input/output device. There is no secondary storage provided.

Assume that this system is used to prepare and run application programs written in assembly language. In order to do so conveniently, it is necessary to have a text editing program, an assembler program, and some means to facilitate the debugging and running of application programs. Since there is no secondary storage device provided, the desired software must be permanently resident in the main memory. Hence, a part of the main memory has to be of the read-only type.

Figure 12.1 illustrates the software elements that the user is aware of when dealing with an assembly language application program. The editor, assembler, and debugger are programs that can be used by the user as needed. Each of these programs performs a well-defined function, which is useful to a variety of different users interested in preparing and running various application programs. Such programs are referred to as *utility programs*. Since we have assumed that there is no secondary storage in this simple system, these programs are resident in the main memory at all times. In addition to these programs, the main memory must have space for the user's data entered during the editing process. Since editing is done on data organized in files, this space is usually called the *file buffer*. The assembler assembles a source program in the file buffer and leaves the resultant object code in the memory,

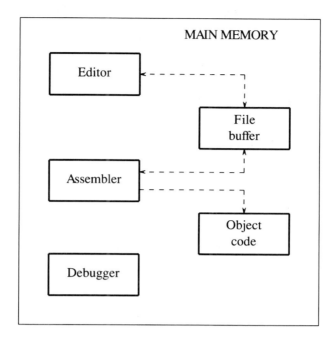

Figure 12.1 User's view of basic software.

ready for execution. The debugger is included to facilitate the development of new application programs, by providing a convenient means for finding errors in a program.

The user sees the blocks on the left side of Figure 12.1 as software modules associated with specific tasks. These tasks are well defined and essentially independent of each other. In order to perform a particular task, the user must be able to communicate with the corresponding module. Moreover, he must be able to invoke the use of the desired module and specify what is to be done. A common way of accomplishing this is to include another software module whose main function is to interpret the commands entered by the user. This module is called the *command interpreter*.

The user is primarily interested in interacting with the software modules in a computer. However, this interaction involves the use of hardware, as exemplified by the input/output transactions via a video terminal. Such I/O devices are controlled by special software modules, called *device drivers* or *I/O drivers*. All of this suggests that Figure 12.1 gives an oversimplified picture of the desired software. A more representative diagram is shown in Figure 12.2, which shows the software and hardware elements of our simple system. The figure indicates the relationships among the software modules, the hardware units, and the user. Let us first consider briefly each software module. Then, we will discuss how they interact.

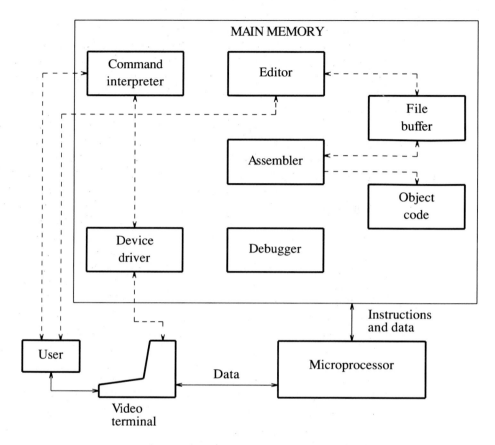

Figure 12.2 Software elements in a simple system.

- *Command Interpreter*: This is a program used to interpret the user's com- mands. The user interacts with the microprocessor system by typing com- mands on the terminal keyboard and observing the response displayed on the screen. The commands include requests for the use of other programs, such as the editor or the assembler. They allow the user to specify a desired action, such as to request that the object code be executed, or that a text file be displayed on the screen. The command interpreter program accepts user's commands and initiates the action specified. Should the user make an error in the command entered, the command interpreter recognizes this fact and displays an appropriate error message on the screen.
- *Editor*: The editor program is used to enter and edit textual information. This information may be the source code of an application program. It may also be the text of a document being entered into the computer as a part of a word pro- cessing application.

- *Assembler*: The assembler program is used to assemble the source code of a program written in assembly language into object code. We have discussed the characteristics of assemblers at length in Chapters 2, 3, and 4.

- *Debugger*: Few programs are ever written without some mistakes. When the user attempts to execute an erroneous program, the result is often unpredictable. Finding errors in such a program may not be easy. In order to simplify this task, a debugging facility is normally provided in microprocessor systems. We will discuss the debugging software in detail in Section 12.5.

- *Device Driver*: We have encountered the concept of *device drivers* in Chapters 3 to 7. These are the programs that transfer data to and from the registers in I/O device interfaces. In the system of Figure 12.2, a device driver is used to deal with the video terminal. The driver reads the characters entered on the keyboard and passes them on to other software modules for interpretation and processing. It also receives the data to be displayed from these modules and transmits it to the terminal.

Figure 12.2 shows that the terminal is connected to the system by means of a data link. The actual physical connection is likely to be done through an ACIA chip, as described in Section 5.3. The device driver must include all of the code needed to initialize the I/O interface registers and to handle read and write transfers as required. An example of this code was given in Figure 5.11.

One of the main reasons for including the device drivers in the system software is to reduce the need for the user to know all the details of the I/O interfaces in the system, and to simplify the problem of coping with the wide variety of I/O devices in existence. A separate device driver module is provided for each device connected to a microcomputer. It contains all the device-specific instructions needed to transfer data to or from that device. Hence, if a device is replaced or a new one is added, only the new device driver needs to be installed. The rest of the system software and any existing user programs remain unchanged. A user program, or a system software module such as the editor or the command interpreter, initiates an I/O operation by sending an appropriate request to the device driver. The driver performs the desired operation and informs the requesting program when it is finished. We will deal with the question of communication between two software modules in Section 12.6.

Physically, the user communicates with the microcomputer by means of the video terminal, which sends and receives data via a hardware link to the microprocessor. The read and write instructions that cause the data transfers are a part of the device driver. The driver also includes a mechanism for passing input/output data to and from other software modules.

The user is not aware of the existence of the device driver. Initially, the commands entered at the terminal are intended for the command interpreter, which means that there is a conceptual link between the user and the command interpreter. The data associated with this link passes through the hardware link between the terminal and the microprocessor, then by software via the device driver to the command interpreter. When the user requests an editing session, he communicates with the editor module, which also takes place using the device driver.

USE OF INTERRUPTS

It was pointed out in Chapter 7 that interrupts may be used to coordinate data transfers between the terminal and the microprocessor. The use of interrupts makes it possible to switch back and forth between two or more tasks. For example, a slightly more sophisticated system than the one shown in Figure 12.2 may allow the user to start editing a file while the assembler module is assembling a program or while some previous results are being printed by a printing module. In this system, the video terminal connection includes an interrupt request line, as shown in Figure 12.3. Interrupts are used to indicate the readiness of the terminal to participate in I/O transfers. Again, an ACIA chip is used to provide the I/O interface, as discussed in Section 7.4 or 7.5.

In a microprocessor system that supports a number of different interrupts, it is useful to provide one system software module that coordinates all interrupt-related activities. Such a module is called an *interrupt handler*. An interrupt request from the terminal is recognized by the interrupt handler, which in turn instructs the device

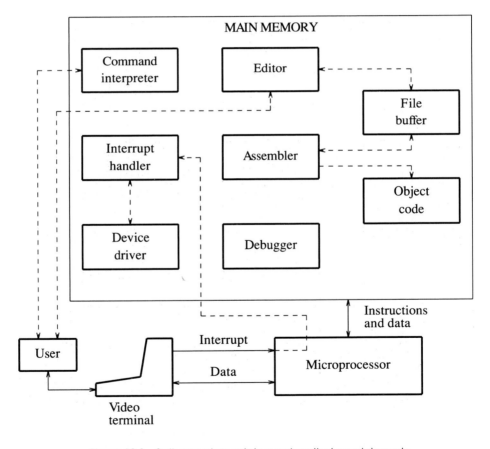

Figure 12.3 Software elements in a system that uses interrupts.

driver to take the appropriate action. In more complex systems there is a variety of I/O devices capable of raising interrupt requests. Moreover, the transfer of control between different software modules may be accomplished using software interrupts, as we will see in Section 12.6. All interrupt requests, regardless of their source, are received by the interrupt handler, which then initiates an appropriate response. In effect, the interrupt handler consists of a collection of interrupt service routines in the system. Each of these routines may make use of other modules to deal with specific requests.

The software elements described above are relatively simple and small in size; yet they are sufficiently useful to be used in many simple microprocessor systems. In particular, they may be found in equipment used for industrial control applications or in an educational laboratory. The editor and assembler programs are usually separate entities that the user calls when needed. The remaining modules, which provide the facilities needed for interacting with the user and for debugging and running of application programs, may constitute a single entity that is often called a *monitor*.

The monitor must include a routine that performs the initialization procedure at start-up time, when power is first turned on. Initialization involves actions such as setting various control registers in the I/O interfaces (see Chapters 5 and 6) and establishing a stack by loading a predetermined value into the stack pointer. The monitor normally resides in ROM and starts running automatically when power is turned on. At the end of the initilization process, the monitor is ready to accept commands from the user. It indicates its readiness by displaying a prompt symbol, such as $ or %, on the terminal screen.

The initialization procedure is often needed at times other than when the power is turned on. For example, it is needed when the user finds the machine in an unrecoverable state from which it is impossible to proceed. It is also needed if user's programs contain catastrophic errors that inadvertently destroy the contents of some important registers, perhaps the stack pointer or some control registers in the I/O interfaces. In such cases, having recognized that something has gone irrecoverably astray, the user can start over again by reinitializing the microcomputer system. Most systems have a switch or a key that can be activated to reset the machine.

12.1.2 Systems with Disk Storage

A great limitation of the system described in the previous subsection is that it does not have any secondary storage. This reduces the scope of the software that can be provided. Let us now consider a larger system, which has a magnetic disk storage and possibly a larger main memory. Such a system is representative of modern microcomputers that are used as workstations, personal computers, etc.

The system should make it easy to create, store and access a variety of files, organized in a suitably structured file system. The response of the computer to user's requests should be fast, even if the system is used by a number of users simultaneously via separate video terminals. These features can be provided only if powerful hardware is coupled with efficient system software.

Since both the system software and the user's files are likely to be extensive, most of this information has to be kept in the disk storage and brought into the main memory when needed. This can be done quickly, because the disk unit is almost always connected to the main memory using the DMA technique, as was explained in Chapter 7.

Figure 12.4 shows the software elements found in a typical system. The basic modules discussed in conjunction with Figure 12.3 are used as before. Other modules are included to enhance the capability of the system and to account for the existence of the disk storage. A disk driver is used to control transfers to and from the disk. The interrupt handler is more extensive, because of the increased number of interrupt sources. A software routine is needed to look after the management of the file system. An essential feature of the system in Figure 12.4 is its ability to move data back and forth between the main memory and the disk. A memory management module is included to control the organization of information in the main memory as it is brought from the disk. For example, if one of the virtual memory schemes described in Chapter 8 is implemented, this would be the responsibility of the memory management software.

The editor, assembler, and debugger modules of Figure 12.3 are not shown explicitly in Figure 12.4. They are included in the block labeled "utility programs." Other examples of utility programs are compilers and interpreters for high-level languages. Utility programs also include linkers and loaders, which will be discussed in Section 12.4.

System software is a term that refers to programs permanently "installed" in a computer system. Some of this software is essential for the operation of the system, and it in fact characterizes the system. It includes the monitor functions defined in the previous subsection, the file system management, the memory management, and the general control of the activities that take place in the computer system. A commonly used term to refer to such software is *operating system.* Utility programs are not considered to be a part of an operating system. They are independent programs that may be added at any time, even though they are often sold as part of an operating system package.

An operating system performs many functions, some pertinent to the internal operations in the system and others providing services to the user. The central part of the operating system consists of modules dealing with internal functions, such as the interrupt handler, the device drivers, the file system, and memory management routines. This part is often called the *kernel* of the operating system. Note that the command interpreter in Figure 12.4 is not included in the kernel, because its primary function is to interact with the user.

An operating system can be large, often too large to fit in the main memory. This is not a significant problem, because only relatively small portions of it are needed at any given time. The entire operating system is stored on the disk and the parts needed are brought into the main memory as required. However, some parts, which comprise the kernel, are needed very frequently and are normally kept in the main memory at all times. Ideally, the kernel should be small enough not to occupy a significant portion of the available main memory.

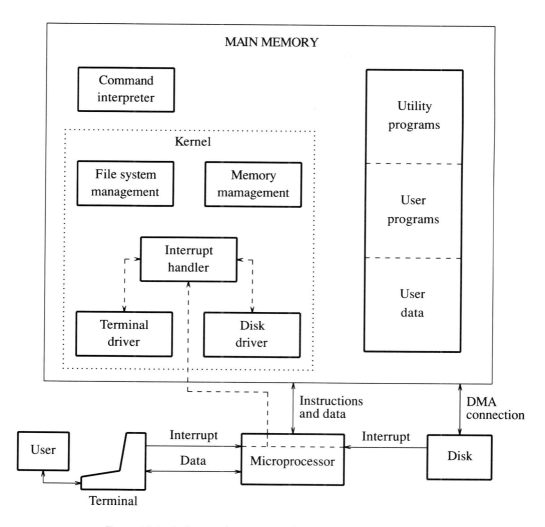

Figure 12.4 Software elements in a microcomputer with disk storage.

Numerous operating systems have been developed for microprocessor systems. Some have attained widespread use. We will pursue the topic of operating systems further in Section 12.6, and consider a specific example of the UNIX system in Section 12.7.

The subsequent sections of this chapter deal with the main features of several important utility programs. Our aim is to outline the functions of these programs, and to illustrate through them the interaction between the software and hardware in a computer system.

12.2 TEXT EDITOR

A basic task that a computer user should be able to perform is to write programs. This may be done by writing the statements that constitute a program on paper. Then, satisfied with the outcome, the statements may be typed on the terminal to enter the program into the computer. The text editor enables the user to enter the program text, to correct typing mistakes, and to make any changes in the text. Of course, not only programs but any text can be prepared with the aid of a text editor.

A *text editor,* or simply an *editor,* is a program that interacts with the user during the preparation of the text, typically via a video terminal. The user types the text and the commands to the editor on the keyboard. The editor responds by displaying on the terminal screen the input information, any other desired information that is already stored in the computer, as well as instructions to the user.

Modern editors display one screenful of text, and allow the user to augment or update any part of this text in a simple manner. They are called *full-screen editors.* Simpler editors, particularly those intended for operation with teletypewriter-like terminals, allow manipulation of only one line, the last line being displayed. Hence, they are called *line editors.* Line editors are simpler to implement, but are less convenient to the user. In contrast, full-screen editors provide simple means for working on any part of the displayed page.

When dealing with large amounts of information, it is necessary to handle it in an organized manner. A common way is to structure it in the form of files. A *file* is a collection of data treated by the system software as a single entity. It is assigned a unique name for reference. A file may be stored on a disk or a tape, identified by its name. The information stored in a file may be a program to be executed on a computer, data used by this program, English text, or anything else that is suitably encoded. Files may be long or short. A long file may comprise a chapter of a book, a whole book, payroll information, student records, etc. A short file may contain a frequently used mailing address, or perhaps even a single character.

A text editor program deals with text in the form of a file. To enter text into the computer, the user first opens a file and gives it a name that can be recalled later. This can usually be done with a simple command such as

Edit filename

where "Edit" is a command indicating to the operating system that the user wishes to edit a file, and "filename" is a suitable name chosen by the user. The operating system loads the editor program from the disk into the main memory and starts its execution. Now, any commands typed at the terminal will be received and acted upon by the editor. At this point, the desired text may be entered to be stored in the file. If the named file already exists, the new text is appended at the location specified by the user.

After entering the text, the user will want to make changes and corrections. The editor provides commands to allow insertion and deletion of symbols, words, and phrases, as well as moving portions of the text around. We will not pursue the details of the editing process. The reader has undoubtedly used an editor on some computer and is familiar with the nature of commands provided and the interactions involved.

Most editors operate on information held in the main memory of the computer. Often one finds the term "text buffer," "file buffer," or simply "buffer," associated with the part of the memory that holds the currently active text. When satisfied with the edited text, the user instructs the editor to copy the contents of the buffer on the disk, or possibly on some other secondary storage device such as magnetic tape. Using a buffer space in the main memory allows for much faster operation than would be possible if editing required frequent accesses to the file stored on the disk. The existence of a buffer also provides some measure of protection against unexpected results, or errors commited inadvertently. In such a case, it is only the information in the buffer that is affected. The original copy of the file on the disk is not changed until the user instructs the editor to update it.

Let us now consider the interaction between the software and the hardware during the process of editing a file, using the system of Figure 12.3. As mentioned earlier, the editor uses the device driver to send data to the terminal and to receive data from the terminal. In the case of a line editor, the editor sends or receives a complete line at a time. With full-screen editors, characters are passed from the driver to the editor one at a time, as they are entered by the user. The editor sends either a complete string of characters to be displayed, or a command to the terminal to insert a character at some point on the screen, assuming, of course, that the terminal is capable of performing such operations.

A line editor informs the operating system when it expects a new line. The operating system suspends the editor and waits for the input information. The user enters the characters by pressing keys on the terminal keyboard. When a key is pressed, the terminal sends an interrupt request to the microprocessor, which activates the interrupt handler. After determining the source of the interrupt, the interrupt handler calls the driver of the video terminal. The driver reads the character, places it in a buffer that will eventually contain the entire line. It may also echo the character back to the screen. Having done this, the driver is not needed until another interrupt request arrives, indicating the availability of the next character. While waiting for the next character, the operating system may activate another task (not related to this editing task) or it may simply do nothing. When the last character in the line has been received, which is the "carriage return" character, the operating system resumes the execution of the suspended editor and passes the input information to it.

In a full screen editor, each character is passed on to the editor as soon as it has been read. The editor then takes an appropriate action, depending upon its functional mode and the nature of the character received. Suppose that the received character is to be appended to a line of text the user is currently typing. Then, the editor places the character into the file buffer and issues a call to the operating system to display the character on the screen. Note that, in this case, it is the editor that instructs the driver to echo back the character. This is a very flexible mode of operation, where the editor takes some action as soon as any character is read from the keyboard. However, there is a price to pay for this flexibility, because control is passed from the driver to the editor and vice versa for each input character, in contrast to the line editor where control is passed only once for each line. The time overhead involved becomes a significant factor in heavily used computers. If a number of terminals are active simultaneously, the response time observed by the users may become unacceptably long.

From the discussion above, it is apparent that line editors place smaller demands on the run time of the machine. But, they are also less attractive from the user convenience point of view, compared to full-screen editors.

12.3 ASSEMBLERS, COMPILERS, AND INTERPRETERS

Computer programs are written either in a high-level language, such as PASCAL, FORTRAN, BASIC, or COBOL, or in the assembly language of the machine at hand. In order to execute such programs, they must be translated into machine code. We have already discussed assembly languages and the process of assembling programs into machine code in Chapters 2, 3, and 4. The desired translation is achieved by means of an assembler program.

A given assembly language applies only to the processor for which it is intended. The addressing modes and the instruction set available are those of the processor in question. This means, for example, that programs should be written in the 68000 assembly language only if they are to be executed on a 68000 microprocessor.

The programming examples presented in previous chapters are relatively small, and thus presented little difficulty. In more complex applications this may not be the case. Large assembly language programs can be tedious to write and difficult to decipher. It is much easier to write programs in a high-level language because the available instructions are powerful. A single high-level language instruction often requires several assembly language instructions to perform the same function. High-level languages are defined without reference to any particular machine. Hence, programs written in such languages can be run on a variety of computers. But, the source code written in a high-level language must first be translated into machine code, called the object code, by means of a *compiler* program, or simply a compiler. A different compiler is needed for each language and each type of computer. Thus, one speaks of a PASCAL compiler for the 6809 microprocessor, which translates PASCAL programs into 6809 instructions, or a PASCAL compiler for the 68000 microprocessor, which produces 68000 machine code. The two are, clearly, not the same.

The process of compilation, like the process of assembling programs, involves translating an entire source program into machine object code, which can then be executed. There exists another possibility for dealing with high-level language programs. Instead of a compiler, one may use a program called an *interpreter* to execute a program written in a high-level language. The interpreter examines each program statement and performs the required action immediately, before moving to the next statement. Interpretive processing does not yield object code that can be reexecuted. If a given source program is to be executed many times, it will have to be interpreted each time.

There are important tradeoffs when selecting between compilers and interpreters. The process of compiling a program can be a time consuming task. But, the result is a machine-coded version of the program that can be executed many times without having to be recompiled. Thus, when a given program must be run frequently, as is often the case with typical application programs, it is best to use it in the compiled form. On the other hand, if a source program is to be executed only once, the interpretive approach is attractive. Perhaps its key attraction lies in the fact that it facilitates debugging of programs. Since a program is interpreted one statement at a time, it is easy to tell the user where an error has occurred.

Assemblers, compilers, and interpreters are all utility programs that are included in the system software of a given computer.

12.4 LOADERS AND LINKERS

After the object code is generated by a compiler or an assembler, it is usually stored on a secondary storage device—in most cases a magnetic disk. In order to execute this code, it must be brought, or loaded, into the main memory of the computer. The desired transfer is accomplished by means of a *loader* program. In a simple case, the task of loading a program is straightforward. Suppose that a given program was written and its object code prepared knowing that it will be loaded into a particular part of the main memory, say starting at location 100. Then, the loader merely copies the object code from the disk into the specified part of the memory. The only information needed by the loader is the location of the object code on the disk, its size, and the destination location in the main memory. However, such is not always the case. Sometimes it is desirable to place the object code into a part of memory different from the one originally envisaged. Moreover, it may be convenient to piece together an object program from separately compiled segments. These possibilities have profound implications on the task of loading. We will discuss some commonly found facilities in this section.

The simplest loader is one that assumes that the object code was prepared in such a form that it can be copied directly into prespecified memory locations for execution. The loader performs the copying task using the information about the length of the object program and its location in the memory. This information is normally generated by the compiler and included in a header field that precedes the object code stored on the disk. A loader used to load object code of this type is referred to as an *absolute loader*.

There are other types of loaders, capable of handling more complex situations. Suppose that a given microcomputer can deal with two different programs at the same time, as will be discussed in Section 12.6. It executes a portion of one program, then switches to the second program. After executing a part of the second program, it reverts to the first one, and so on. Thus, the exact location of various segments in the main memory is determined only at the time of execution, and cannot be predicted when the segments are compiled. In order to deal with this situation, it must be possible to modify the object code to make it suitable for execution in the new location. For example, absolute addresses that appear in the object code must be changed. The process of modifying the object code to suit the new location is called *relocation*.

We have seen in Chapters 3 and 4 that it is possible to incorporate facilities for relocation in the design of a microprocessor. A simple mechanism, as illustrated by the 6809 and 68000 microprocessors, is to have an addressing mode where the addresses are generated relative to the current contents of the program counter. Then, programs that rely on this relative addressing mode can be located anywhere in the memory at execution time. However, while most of the code can be derived using relative addresses, there will inevitably be some addresses that will have to be specified as absolute addresses. The values of such addresses are referred to as *address constants,* which may have to be changed when a program segment is relocated. In order to enable the necessary changes to be made, a list of address constants and their places in the object code is usually included at the start of the file that contains the object code for a program.

All the modifications needed to relocate object code can be performed by the loader program. A program with this capability is called a *relocating loader.* It is more extensive, but also more useful than an absolute loader.

Our example microprocessors facilitate relocatability by providing the relative addressing modes. There are other ways of achieving this purpose. For instance, some computers have a special register called the *base register.* The contents of this register are incorporated in the computation of operand addresses such that a program can be relocated simply by changing the contents of the base register. In this case, the object code is generated assuming that it will be loaded in the main memory starting at address 0. However, it can be loaded and executed in a different place in memory, say starting at location 2000, by placing the value 2000 into the base register. The existence of base registers simplifies considerably the task of relocating programs and the complexity of the relocating loader. But, it also makes the processor structure more complex.

We should note that the loader itself must be resident in the main memory, if it is to be used to load other programs. This raises the question of how it got there in the first place. An obvious possibility is to implement a part of the memory as ROM, and place the loader into it, thus making it always available. Unfortunately, this approach requires a portion of the main memory to be dedicated to this function. An attractive compromise is to keep only a small loader in ROM—one that is just sufficient to bring the main loader from the disk into the main memory. In this type of arrangement, the loader in ROM is called a *bootstrap loader.*

Programs are often constructed from several separately produced pieces, possibly written by different people. Consider, for example, the case of a subroutine written by one person to be used by other people in their application programs. A user wishing to use this subroutine in a program may do so by appending the source code of the subroutine to that of the program. However, if the subroutine is rather extensive, it is not desirable to compile it each time it is to be incorporated into another program. Instead, it is possible to store the compiled object code of the subroutine on the disk and have it available for general use.

In order to link two object-code modules together coherently, it is necessary to adjust certain values in the code. For example, suppose that some names of variables or address constants in the source code are used in more than one module. Obviously, these values should be the same in all modules. Yet, since the modules were compiled separately, without specific knowledge of the details of other modules, it is likely that the values in question will be different. When the modules are linked into a single object-code program, all such values must be adjusted. This is accomplished by means of a *linker* program.

The linker can link object code modules if it knows exactly where the externally used names are found. A name is called *external* to a given module if it was defined in another module. A practical way of providing the linker with the required information is to include at the beginning of each module a list of external names and their locations in the module. In order to enable the assembler or compiler to produce this list, these names must be declared in the source program. For example, let the assembler directive EXT (for EXTERNAL) indicate that a given name is defined in another module. Also, let ENT (for ENTRY) be the assembler directive that indicates that a given name may be referenced in other modules. A possible use of these directives is illustrated in Figure 12.5. This is the program of Figure 4.15, which displays a one-line message on a video terminal. It is written in the form of a general purpose subroutine that may be called by other programs. The figure shows the declaration of externally defined names, used to refer to the locations of the message and the display subroutine. When this program is assembled, a list of the names specified in the EXT and ENT directives and their location in the program is included in the object code generated by the assembler, preceeding the program itself.

An interesting situation arises when an object program exceeds the size of the main memory available in a given microcomputer. Such programs can be executed by segmenting them into modules of suitable size, which can be brought into the memory as needed. They must be linked using a linker program. When a new segment is loaded into the memory, it displaces the code that belongs to one or more modules that have already been executed. It is customary to say that the new module *overlays* an old one. Of course, the reader will recall that another way of dealing with large programs is by means of a virtual memory, as discussed in Section 8.10, if the microcomputer includes this feature.

Loader and linker programs are important parts of system software. Sometimes, their functions are performed by a single program, which is then called a *linking loader*.

12.5 DEBUGGER

In Section 12.1 we saw that the monitor part of the system software provides an interactive link between the user and the computer system. It responds to user's commands by performing the tasks requested. It prompts the user to supply additional information when needed. It also sets up the hardware so that it can execute various programs.

When complex programs are written numerous errors may occur. Detection of these errors, which is referred to as program *debugging,* can be a difficult task, particularly for programs written in assembly language. Debugging of such programs can be facilitated with specialized utility programs called *debuggers*. A debugger

```
* Main program.
            EXT         DISPMESG        Name defined elsewhere.
            ENT         MESG            This name may be used in
*                                           other program modules.
MESG        DS.B        100             Define message space.
            .
            .                           Instructions that place a
            .                               message at MESG.
            JSR         DISPMESG        Call the display subroutine.
            .
            .
            .
```

(a) A part of the main program

```
* Subroutine to display a 1-line message.
            EXT         MESG            Name defined elsewhere.
            ENT         DISPMESG        This name may be used in
*                                           other program modules.
OUTBUF      EQU         $80A00          Address of output and
STATUS      EQU         $80A01              status registers.
DISPMESG    MOVEA.L     #MESG,A0        Initialize pointer.
WAIT        BTST.B      #1,STATUS       Wait for the display
            BEQ         WAIT                to become ready.
            MOVE.B      (A0)+,OUTBUF    Transfer a character to display.
            CMPI.B      #$0D,-1(A0)     Continue displaying if not CR.
            BNE         WAIT
            RTS
```

(b) Display subroutine

Figure 12.5 Specification of linking information.

allows the user to follow closely the execution of the object code. While it may have to change temporarily some of the object code to perform its function, such changes are transparent to the user. That is, from the user's point of view, the object code appears unaffected. In the discussion below, we will consider some of the most important features found in debuggers.

In order to find errors in a source program, it is useful to be able to

1. Examine the flow of execution of the object code
2. Stop the execution at specified points in a program and examine the values of various registers and memory locations

3. Modify the contents of registers and memory locations
4. Slow down the execution by single-stepping through the instructions, one at a time
5. Watch the value of a given variable or pointer as it changes during the execution

In general, the user should be able to observe, alter, and control various aspects of a running program.

A most direct way of examining the object code is to look at the contents of memory locations. This may be done with a command to the debugger in which the user specifies the address of the desired location and the debugger responds by displaying the contents of that location on the terminal screen. A new value may be deposited in that location via the terminal keyboard. Such commands give the user an editing capability that affects the object code directly. An extension of this type of examination facility is to have the debugger write out the contents of a larger part, or even all of the memory, on the screen or some other output device. The term *dump* is used to denote this function, which usually yields a printout of object code and associated data, which can then be laboriously studied.

To execute a program the address of the first instruction in the object code is loaded into the program counter. Then, the instructions are executed one after the other. As an aid in locating programming errors, it is convenient to have the debugger stop the execution of the program at some desired points. Such points are called *breakpoints,* and they are specified by giving their addresses to the debugger. The resultant effect is as follows. Execution of the program procedes normally until a breakpoint is reached. Then, control is transferred to the debugger program, which allows the user to examine the status reached. Execution is resumed when the user issues a continue command, and proceeds until the next breakpoint is encountered. Some debuggers allow the user to place a breakpoint within a program loop and specify how many times the loop should be executed before the breakpoint is activated. The convenience of breakpoints stems from the ease with which they can be set or removed, without having to modify the source code.

Sometimes it is convenient to observe status changes after executing each instruction. Single-stepping through a program in this manner is available as a feature in most debuggers. It can be used particularly well in conjunction with the capability to set breakpoints. The user can set a breakpoint in the vicinity of a troublesome part of a given program. Then, by single-stepping through the suspected segment, the problem may be identified.

A microprocessor may contain design features that facilitate the implementation of debugging schemes. For example, the 68000 microprocessor provides a trace facility using bit 15 of the status register, as mentioned in Section 4.1. When the trace flag, T, is set to 1, the processor is in the *trace* mode. In this mode, a software interrupt is raised at the end of execution of each instruction, and the debugging software is executed as a result. The trace feature may be used to realize single-stepping through a program. It may also be used to print out or display the contents of some processor registers or memory locations as the instructions of the program that is being debugged are executed. This information, which is often referred to as a *program trace,* can be very useful in locating errors that are otherwise difficult to discover.

A good debugger program is a valuable tool that can save the user many hours and much frustration in detecting errors in a program. Debuggers may be used with programs written in either a high-level or an assembly language.

12.6 OPERATING SYSTEMS

In Section 12.1, we introduced the concept of operating systems as being the means for providing various services needed to prepare and run application programs. Operating systems have evolved with the progress in computer technology. The increasing power of microprocessors has opened the door to sophisticated operating systems with characteristics that used to be the exclusive domain of large mainframes. Numerous operating systems have been developed. The popularity of some has reached the point where they have become de facto standards.

There are several types of operating systems for microcomputers. The simplest type is a *single-user* system. In its basic form, it permits one task to be active at a time. A more sophisticated version may permit more than one task to be active, in which case it is called a *multitasking* system. It enables the user to continue interactive work, for example, while previous results are being printed on a printer. The operating system coordinates the activities of two or more tasks, using interrupts and the control structures described in Chapter 7. A *multiuser* operating system is capable of supporting several users at a time. The number of users that can be accommodated depends upon the computing power of the microprocessor and the nature of user jobs. Such operating systems can be very complex. They provide each user with access to the system's resources, and they control the execution of users' programs to prevent them from interfering with one another.

We have already stated that the main purpose of an operating system is to make the computer easy to use in a variety of applications. It makes it possible to use the computer without knowing the intricacies of its hardware. It allows the user to use the computer resources as if they were simple logical entities. A multiuser system also allows each user to view the machine as if it belonged to him exclusively.

Operating systems differ widely in the functions they provide and the techniques used to implement them. In order to give the reader a rough indication of how an operating system functions, let us consider a commonly used way of handling various resources. Computers perform tasks by executing programs. Let us assume that a given task is achieved by a specific program, which can be scheduled to run whenever the task is needed. Such a task, as represented by a program being executed, is often referred to as a *process.* For example, when one of the utility programs in Figure 12.4 is being used it can be regarded as a process. Many systems allow one program, such as an editor, to be used by several users simultaneously. In this case, each use of the editor constitutes a different process. Thus, the concept of a process enables the operating system to identify each incident of use as a distinct entity that can be scheduled for execution and to which various system resources can be allocated.

The task associated with a process may require execution of several program modules, and vice versa; namely, a given program module may involve several processes. For instance, a device driver involves several tasks, such as initialization and read and write operations. A different process may be associated with each of these tasks. In fact, it is useful to think of an entire I/O task dealing with a specific I/O device as a process involving the relevant portions of the interrupt handler and the device driver. In this case, the process that reads characters from a terminal keyboard would include the "read" part of the device driver as well as the portion of the interrupt handler needed to decide what to do with interrupt requests from the terminal.

One should think of a process as a logical entity that can be created and later destroyed by the operating system. When the user logs into the system a new command interpreter process is created. If the user requests the use of the editor, a new editor process is created, and so on. Execution of the program that represents a given process may be started and stopped several times during the life of that process. When the user issues a command to the command interpreter to edit a file, the execution of the command interpreter is stopped, the editor process is created and the editor program is executed. We say that the command interpreter process is suspended and control is passed to the editor process. Later, when the editor requires some input/output transactions, it is suspended and control is passed to the desired I/O process. When the user leaves the editor, the editor process is destroyed, and the command interpreter process is resumed. A process is suspended when control is passed from it to another process, with the expectation that the suspended process will be resumed sometime in the future. When the task associated with the process is no longer needed, the process is destroyed.

The notion of processes is particularly useful when dealing with multitasking and multiuser operating systems. Microprocessors are fast enough and powerful enough to accommodate several users at a time. Two or more tasks can appear to be performed simultaneously if they are done fast enough so that the users do not experience an unusual time delay. A common way of implementing multitasking is to divide time into relatively short slots, typically on the order of several milliseconds, determined by a timer circuit that raises interrupts at regular intervals. The operating system kernel contains a scheduling routine which allocates a slot to each active process in turn. As a result, the processes appear to the users to be running simultaneously even though only a single instruction of one process is being executed at a given instant.

When several users are using the system simultaneously, there is at least one process active for each user. Moreover, a given process is likely to make use of other processes, such as the editor using the device driver to display a message on the terminal screen. All of this suggests that control is frequently passed from one process to another. There is an overhead associated with this transfer of control, because the status of an interrupted process must be saved, so that the process can be resumed later. The severity of this overhead depends on the particular implementation of the operating system and the hardware. The response of the machine as seen by the users depends mostly on the number of active users and the nature of their work, but is also influenced by the amount of switching between processes.

In order to get a better appreciation of how control may be passed from one process to another, let us consider the example of reading characters from a terminal keyboard and displaying text on the screen during the editing session, which has already been discussed in Section 12.2. A possible arrangement for the interaction between the editor and device driver programs may be as follows. The editor, like all utility programs, uses a software interrupt to indicate to the kernel that some action external to the editor is needed. The interrupt handler recognizes the occurrence of the interrupt and transfers control to the driver. The interrupt handler may be implemented as a general interrupt service routine and the driver as appropriate subroutines that are called when needed, as was suggested in Section 12.1. Thus, two types of interrupts occur in this example. Hardware interrupts from the terminal indicate its readiness to handle a character. These interrupts are serviced by the interrupt handler, which calls either the read or the write subroutine of the driver. Software interrupts come from the editor, indicating to the interrupt handler that another action involving the driver is needed, which may be either reading another character or displaying text on the screen.

Figure 12.6 illustrates a possible structure of the software for a 68000 microcomputer. We have assumed that the editor has prepared a string of characters to be displayed on the terminal screen. The string is stored in the main memory starting at a location called MESG, and it is terminated by a null character. The mechanism used to display the string is as follows. The editor passes the address of the string as a parameter in register A0, to be used by the device driver. Since the device driver provides several services, such as read and write operations, the editor must indicate which specific function is being requested. We have assumed that a simple code is used where 1 represents a write operation and 2 represents a read operation. This code is passed as a parameter in register D0.

In Chapter 7 it was explained that all interrupts in a 68000 system are implemented as "exceptions." One of the exceptions is caused by the TRAP instruction, which serves as a software interrupt. Since it is useful to have several different software interrupt instructions, the TRAP instruction actually comprises a set of 16 distinct instructions identified by a number in the range 0 to 15 in the operand field. Each of the 16 TRAP instructions is assigned a different exception vector location, as indicated in Table 7.4. These locations range from 80_{16} to BF_{16}, each location specifying a 32-bit address of the corresponding interrupt service routine. In Figure 12.6 we have selected the instruction

TRAP   #7

to serve as the software interrupt used by the editor when it requires an input/output task to be performed. This instruction is associated with the exception vector at location $9C_{16}$. Hence, during system initialization, which is performed when power is turned on, this location is loaded with the address of the relevant module of the interrupt handler, 5000_{16} in our example.

The interrupt handler includes the instructions needed to decide whether a read or a write operation is being requested, by examining the code in register D0. Then,

```
*  System initialization
                ORG          $9C              Load exception vector for
                DC.L         #CHK1               TRAP instruction.
                .
                .
                .


****************************************************************
*  Editor program
                .
                .
                .

                MOVEA.L      #MESG,A0         Parameters are string address
                MOVE.B       #1,D0              and type of operation.
                TRAP         #7               Software interrupt.
                .
                .
                .


****************************************************************
*  Interrupt handler
                .
                .
                .

                ORG          $5000
CHK1            CMPI.B       #1,D0            Check if write operation.
                BNE          CHK2
                JSR          INITW            Call the write routine.
                MOVE         #$2000,SR        Enable interrupts in 68000.
WAIT            TST.B        D0               Wait for the next interrupt
                BNE          WAIT               from the terminal.
                MOVE         #0,SR            Disable interrupts in 68000.
                RTE
CHK2            CMPI.B       #2,D0            Check if read operation.
                BNE          ERROR            Invalid request.
                JSR          INITR            Call the read routine.
                .
                .
                .

                ORG          $6000
DISPLAY         TST.B        (A0)             Check if the end of string
                BEQ          DONE               has been reached.
                JSR          OUTCHAR          Send a character to display.
                RTE
```

Figure 12.6 (continued next page.)

```
DONE        JSR       DISINT            Disable further interrupts.
            CLR       D0                Indicate the end of string.
            RTE
             .
             .
             .
```

****

```
* Device driver
             .
             .
             .

OUTACIA     EQU       $80A00            Addresses of output, command
CMNDR       EQU       $80A02               and control registers in
CNTLR       EQU       $80A03               the ACIA interface chip.
INTVEC      EQU       $64               Autovector for terminal interrupts.
INITW       MOVE.L    #DISPLAY,INTVEC   Use autovector interrupts.
            MOVE.B    #$1E,CNTLR        Initialize ACIA and enable
            MOVE.B    #7,CMNDR             transmitter interrupts.
            RTS
DISINT      CLR.B     CMNDR             Disable transmitter interrupts.
            RTS
OUTCHAR     MOVE.B    (A0)+,OUTACIA     Display one character.
            RTS
             .
             .
             .
```

Figure 12.6 An example of a system software task.

it calls the device driver to perform the required action. When it decides that a write operation is involved, it first requests the driver to initialize the interface of the terminal for an output operation. This is done using the scheme presented in Figure 7.34, assuming that the serial connection to the terminal is implemented using an ACIA chip, as discussed in Section 7.5. The driver initializes the ACIA to select the output mode and to enable transmitter interrupts using the autovector exception feature. The interrupt vector at the chosen autovector location, 64_{16}, is set to point at the appropriate entry in the interrupt handler. This entry is at location DISPLAY, which is the beginning of the section of the interrupt handler responsible for write operations.

After initializing the ACIA, the interrupt handler lowers the processor's priority to 0 to enable ACIA interrupts, assuming that this is the level at which ACIA interrupts are received. Once the interrupt vectors have been set and interrupts enabled, the output process proceeds independently. The interrupt handler waits for an interrupt request from the terminal indicating that it is free to accept a character for display. The arrival of an interrupt request from the ACIA causes the DISPLAY routine of the interrupt handler to be executed. This routine sends a character to the display by calling the output subroutine of the device driver. Before returning, it checks to see if the end of the string has been reached, in which case it disables the transmitter interrupts and indicates the end by clearing the contents of register D0. Clearing D0 causes the wait loop to be terminated, and the interrupt handler to return to the editor.

The reader should observe the separation of input/output functions between the interrupt handler and the device driver. The interrupt handler performs the general functions needed to deal with various interrupt requests. The device driver performs all device-specific functions. Thus, if a different device is to be used, it is only necessary to change the driver routine to meet the requirements of the device.

The example in Figure 12.6 is somewhat oversimplified, causing two of the subroutines in the device driver to be rather trivial. In practice, the interrupt handler and the device drivers may be considerably more complicated, dealing with a variety of other conditions. We have adhered to this simple structure to focus on the concept of passing control among the software modules in an operating system.

As a final observation with respect to Figure 12.6, it is apparent that our choice of waiting idly for successive interrupts from the terminal detracts from the processing efficiency of the system. Instead of waiting in the WAIT loop, the interrupt handler could pass control to another process that has been suspended. This must be done with care, to ensure that no computation is corrupted by unexpected transfers of control. For instance, the contents of registers used in a given interrupt service routine should be saved prior to entering the routine and restored when leaving it.

We have indicated briefly some of the most basic aspects of operating systems available for modern microcomputers. Extensive discussion of operating systems can be found in many books dedicated to this topic [1–7]. In the next section we will discuss briefly a few characteristics of the UNIX system, which has attained widespread use.

12.7 UNIX

UNIX is an operating system developed at Bell Laboratories that quickly became popular and widely used due to its many useful features, particularly its design which enables application programs to be completely machine independent and thus easily portable. That is, an application program developed to run under UNIX can be used without extensive modifications on computers of different manufacture, provided that these computers use the UNIX operating system. The program merely has to be recompiled for use on a different machine. Many application programs are available to run under UNIX.

UNIX has many features whose description is beyond the scope of this text, but is readily available in many books [7–9]. We will consider only a few of its features to give an idea of what to expect in an operating system of this type. UNIX is a multiuser system, which requires a user to log-in using a password, to provide protection for programs and data files. The file structure is flexible and easy to use. Files are edited using any of several text editing programs, and can be created, copied, removed, displayed, or printed, using simple commands.

Interaction between a user and UNIX normally takes place via a video terminal, using a command interpreter program known as the *shell*. The output produced in response to user commands is displayed on the terminal screen, unless specified otherwise. For example, if two files are concatenated to form a larger file by a command

$$\text{cat} \qquad \text{filename1} \qquad \text{filename2}$$

the resultant output will be displayed on the screen, but no new file will be created. A task specified in a command can use input data that comes from an input device, such as the terminal keyboard, or from a file. The output of a command may be redirected to a file or to a device. For example, in order to concatenate two files using the cat command and have the resultant file stored under the name filename3, the command

$$\text{cat} \qquad \text{filename1} \qquad \text{filename2} \qquad >\text{filename3}$$

may be used. The ">" symbol indicates that the output produced should be redirected to a file named filename3, instead of being displayed on the screen. If a file bearing the name filename3 is already in existence, it is deleted before a new file is created having the new contents.

It is also possible to use the output of one routine as the input for another routine. For instance, the command

$$\text{cat} \qquad \text{filename1} \qquad \text{filename2} \qquad | \text{ lpr}$$

will concatenate the two files and use the resultant file as input to the "lpr" command, which will print the file contents on the line printer. The bar symbol (|) denotes that the output of cat is to be connected to the input of the lpr routine. This feature is called a *pipe*.

The file system structure in UNIX is in the form of a tree. In order to provide a convenient organization the user can define *directories,* which may contain files or other subdirectories. Each directory is a node in a tree that may have any number of descendants.

Figure 12.7 gives an example of the file structure for a user called George. Rectangular nodes represent directories, while ellipsoidal nodes depict files. We have assumed that George's files pertain to three kinds of activities. First, he has application programs that are either being developed or used as functional programs that perform some specialized computation. Second, he is working on a manuscript for a book, making use of the UNIX editing and typesetting programs. Third, he uses electronic mail and retains some of the received messages for future reference. His root node is a directory called "george." This directory is referred to as the "home directory," which contains two other directories, "Progrms" and "Book," as well as a file for saving mail messages, labeled "mailbox." Directory Progrms contains two application programs, appl.1 and appl.2. The first consists of several modules stored as separate files in a directory called "appl.1." The second comprises a single file. Similarly, the directory Book contains subdirectories that correspond to chapters in the manuscript. Each chapter directory has files for various sections of that chapter.

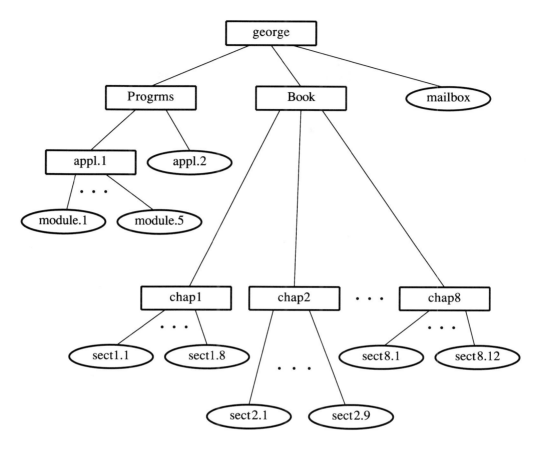

Figure 12.7 Tree data structure of the UNIX file system.

A given file is accessed by specifying the path that leads to it. For example, the file containing Section 2.1 of Chapter 2 of the manuscript is identified by

Book/chap2/sect2.1

The text of this file can be displayed on the screen using the command

cat   Book/chap2/sect2.1

Since it is likely that the user will make frequent references to files in a given directory, it is convenient to change the base reference point from the home directory to the directory desired. For instance, when working on various sections of Chapter 2 it is useful to use chap2 as the base reference point. This can be done by a "change directory" command

cd   Book/chap2

Having done this, the previous task of displaying Section 2.1 on the screen is accomplished simply by

$$\text{cat} \qquad \text{sect2.1}$$

A change to any directory can be made in a similar manner. A reference to the home directory is indicated by the symbol "~" (tilde). Thus, a change from directory chap2 to directory appl.1 is made by the command

$$\text{cd} \qquad \text{~/Progrms/appl.1}$$

New directories can be created with the "make directory" command. For example, suppose that George is currently working in directory Progrms and that he wants to create a new subdirectory called appl.3. He can do this with the command

$$\text{mkdir} \qquad \text{appl.3}$$

Directories can be removed in the same way with the "remove directory" command, rmdir.

A new file may be generated by copying an existing file, concatenating two or more files, or by using the editor to enter information into a new file. The latter case arises when one calls the editor to work on a nonexisting file. If in our example the current directory is chap1 and George wishes to add one more section to Chapter 1, he can do this by calling the editor with the command

$$\text{ed} \qquad \text{sect1.9}$$

Since no file exists in this directory with the name sect1.9, a new file is created. We should note that two different files can have the same name if they are in different directories.

We have considered just a few simple examples of UNIX features. There are many others. In a typical UNIX system there exists a wide variety of software tools, facilities for document preparation, as well as mail service through which the users of the system can communicate with each other.

12.8 CONCLUDING REMARKS

System software is a vital part of a microcomputer system. Without it the microcomputer would be difficult to use. The system software includes various utility programs and an operating system that provides a flexible and easy to use facility for handling files, interacting with I/O devices and invoking various systems functions. The operating system removes the need for the user to be thoroughly familiar with the implementation details of the computer hardware. It enables the user to view the hardware and software resources as logical entities that perform well-defined functions.

In this chapter we have attempted to give the reader a brief overview of system software, dealing only with the most general aspects. Our hope is that, having been told about the significance and usefulness of this software, the reader will consult a more specialized book that discusses these important topics in greater detail.

The choice of system software is largely dictated by the type of microprocessor used, the availability of the desired software and its cost. It is important that system software be properly supported. This means that someone, typically the originator or a distributor, is continuously involved in correcting errors that may be discovered from time to time and in producing enhancements. For this reason, it is not surprising that the most popular system software tends to be supported by large companies and institutions.

12.9 REVIEW QUESTIONS AND PROBLEMS

REVIEW QUESTIONS

1. What are the typical functions of system software?

2. What characterizes a file?

3. What tasks should a text editor be able to perform?

4. In what type of applications should one write a source program in the assembly language of a given machine?

5. What is the function of a compiler?

6. Explain the difference between a compiler and an interpreter.

7. What types of loader programs are used in microcomputers? What are their main features?

8. When is it necessary to use a linker program?

9. What must be done as a part of the hardware initialization process for a given microprocessor system?

10. What should a user expect from a debugger program?

11. In what situation would one use breakpoints? Indicate some examples where it would be useful to insert a breakpoint.

12. What is a program trace?

13. What are the functions of an operating system?

14. What is the function of a command interpreter?

15. What is an I/O device driver?

16. Explain the interaction between the device driver for a video terminal and a text editor program, during a typical editing session.

17. What is referred to as the "kernel" of an operating system?

18. Describe the notion of a "process," with respect to an operating system.

19. What is a "directory" in UNIX?

20. What can be accomplished with the "pipe" feature of UNIX?

21. What are the advantages of a tree file structure?

22. Why is it advantageous to have a widely used operating system?

23. How can a single microcomputer be used by several people simultaneously?

PROBLEMS

24. Is the program in Figure 3.21 relocatable? If not, how would you change it to make it relocatable?

25. Repeat Problem 24 for the program in Figure 3.33.

26. Is the program in Figure 4.20 relocatable? If not, what changes should be made?

27. A debugger program must be able to insert a breakpoint at a given address of the object code of the program that is being debugged. But, the object code must not appear to be changed by the debugger, as far as the user is concerned. A possible scheme for inserting the breakpoint is to replace temporarily the instruction at the specified address with a trap instruction (software interrupt), which will transfer control to the debugger when execution of the user's program reaches that point. To continue the execution of the program, the original instruction that was removed to make room for the trap must be restored.

    Write a 68000 routine capable of inserting a breakpoint, in response to a user's command that specifies the desired address. Execution of the user's program is to be resumed after the user issues a continue command by pressing the "c" key on the keyboard. Assume that execution stops at the breakpoint only once, even if the breakpoint happens to be in a program loop.

28. Repeat Problem 27 for a 6809 microcomputer.

29. Could one use a Call_subroutine instruction instead of a trap in the debugger of Problem 27? Explain.

30. Devise a scheme for inserting a breakpoint (see Problem 27) in a program loop, such that the breakpoint is activated each time through the loop. Write a 68000 program to implement your scheme. (*Hint*: It will be necessary to single-step through the instruction after the breakpoint.)

31. Can the task of Problem 30 be done on a 6809 microcomputer? What difficulties would be encountered? Suggest possible solutions using either software or hardware.

32. It is desired to insert a breakpoint after the CMPA.L instruction in the program of Figure 4.11. When the breakpoint is activated, the contents of all relevant registers are to be displayed on a video terminal. However, the breakpoint is not to be activated each time through the loop. Instead, the user may indicate how many passes through the loop should occur before the breakpoint is activated, using a suitable mechanism. Develop a scheme that meets this requirement and show the programming details involved.

33. It is desired to obtain a "trace" of the program in Figure 4.11, where the contents of all registers are to be printed at the end of execution of each instruction. Write a program capable of accomplishing this.

34. Propose a scheme that can be used for single-stepping through a 68000 program. The user causes the next instruction to be executed by pressing the "c" key on the keyboard. Show the details of a software routine that can be used to accomplish this feature.

35. Can single-stepping be implemented on a 6809 microcomputer in the same way as on a 68000 microcomputer? What problems would be encountered? Could these problems be overcome? Explain.

36. Write a routine that could be a part of an interrupt handler of an operating system that runs on a 6809 microcomputer. Show only what is needed for the interaction between a device driver for a video terminal and an editor program. Assume that this interaction takes place as suggested in Section 12.6, which was illustrated in Figure 12.6.

37. The example in Figure 12.6 includes a call to subroutine INITR, which is executed if the editor requests a read operation from the terminal. Write this subroutine and any other code necessary to perform the read operation, assuming that the editor expects only one input character.

38. Repeat Problem 37 assuming that a line editor is used, which expects one line of input characters.

12.10 REFERENCES

1. Tanenbaum, A.S., *Operating Systems: Design and Implementation.* Englewood Cliffs, NJ: Prentice Hall, 1987.

2. Massie, P., *Operating Systems Theory and Practice.* New York: Macmillan, 1986.

3. Peterson, J., and A. Silberschatz, *Operating System Concepts.* Reading, MA: Addison-Wesley, 1985.

4. Finkel, R.A., *An Operating Systems Vade Mecum.* Englewood Cliffs, NJ: Prentice-Hall, 1986.

5. Comer, D., *Operating System Design—The XINU Approach.* Englewood Cliffs, NJ: Prentice-Hall, 1984.

6. Kaisler, S.H., *The Design of Operating Systems for Small Computer Systems.* Wiley, 1983.

7. Bach, M.J., *The Design of the UNIX Operating System.* Englewood Cliffs, NJ: Prentice Hall, 1986.

8. Ritchie, D.M., and K. Thompson, "The UNIX time-sharing system," *Bell System Technical Journal*, 57, 1978.

9. Bourne, S.R., *The UNIX System.* Reading, MA: Addison-Wesley, 1983.

Information Representation

This appendix presents different ways of representing information in binary form, suitable for use in microprocessor systems. The material is divided into three parts. Part I deals with the basic ideas for the representation of numerical values and textual information as binary patterns. Part II describes the ways in which signed numbers are handled in digital computers. A study of these techniques is essential for an understanding of the details of addition and subtraction operations, as they take place in a microprocessor. The remainder of the appendix, Part III, introduces the floating-point representation.

PART I: BASIC IDEAS

In this part we consider the representation of numerical values, including a discussion of the four basic arithmetic operations, in bases other than 10. The ways in which letters of the alphabet and symbols such as + or = are stored in a computer are also presented.

A.1 NUMBER REPRESENTATION

While there exist many schemes for representing numbers, one of the simplest and most frequently used is the *positional notation*. This is the familiar scheme that we learn in the early grades of mathematics. It is also the scheme used to represent numbers in computers.

A.1.1 Positional Notation

A number consists of several digits, written side by side. An n-digit number is written as: $a_{n-1}a_{n-2} \cdots a_1 a_0$. Each digit can take any of r possible values, and is represented by a unique symbol reflecting its value. Hence, r distinct symbols are needed. Decimal numbers, for example, use the ten Arabic numerals $0, 1, 2, \cdots, 9$. Each digit within a number is given a different weight, depending upon its position relative to the other digits. The most commonly used scheme for assigning weights to different digits uses exponential weights. In this case, digits $a_0, a_1, a_2, \cdots$ are given the weights $r^0, r^1, r^2, \cdots$, where r is called the base, or radix, of the system. In the decimal number representation, r is equal to 10.

Consider the decimal number 7293. The four digits in this number have different weights. The rightmost digit, 3, is the *least-significant digit,* which has a weight of 1 $(=10^0)$. Thus, it represents the value 3×1. The next digit has a weight of 10, and hence it represents the value 9×10, and so on. The leftmost digit, 7, is the *most-significant digit.* It has a weight of 10^3. Thus, the value of the number 7293 is given by

$$7293 = 7 \times 10^3 + 2 \times 10^2 + 9 \times 10^1 + 3 \times 10^0$$

Fractional values are represented in decimal notation by a straightforward extension of this idea. For example, the number 36.75 consists of an integer part, 36, and a fractional part, 75, separated by the decimal point. Its value is given by

$$36.75 = 3 \times 10^1 + 6 \times 10^0 + 7 \times 10^{-1} + 5 \times 10^{-2}$$

There is nothing special about the use of 10 as a base for number representation. Any reasonable value can be used as the base r. For example, base 60 was used by the ancient Greeks, as reflected today in the way we represent time. This relatively large base was found convenient, because it resulted in numbers that are easily divided by 2, 3, 4, 5, and 6. In this book, we encounter examples of the use of hexadecimal numbers (base 16), octal numbers (base 8), and most importantly, binary numbers (base 2). In the case of binary numbers, individual digits have only two possible values, 0 or 1. A binary digit is commonly called a *bit*.

The importance of binary number representation stems from the fact that it requires only two distinct symbols. This is well suited for use in computers, because electronic circuits can be designed such that they assume one of two possible states. For example, the circuit may produce a low output voltage in one state and a high output voltage in the other. These two output voltages can be regarded as representing the two possible values of a binary digit.

Consider the general case of a base-r number system. An arbitrary number N is represented in the form

$$N = a_{n-1}a_{n-2} \ \cdots \ a_2 a_1 a_0 . a_{-1} a_{-2} \ \cdots \ a_{-m}$$

The integer and fractional parts of this number are separated by the *radix point*. Each of the digits a_i ($i = n-1, \cdots 1, 0, -1, \cdots , -m$) can take any of the r values 0 to $r-1$. The value N is equal to the sum of the values of all $n+m$ digits, each multiplied by the weight associated with its position relative to the radix point. Therefore

$$N = a_{n-1}r^{n-1} + \ \cdots \ + a_2 r^2 + a_1 r + a_0$$

$$+ a_{-1}r^{-1} + a_{-2}r^{-2} + \ \cdots \ + a_{-m}r^{-m}$$

$$\text{(A.1)}$$

$$= \sum_{i=-m}^{n-1} a_i r^i$$

The bases 2, 8, 10, and 16 are commonly used in the computer world. Since a given set of digits represents different values for different bases, it is important to be able to identify the base of a given number unmistakably. Whenever there is room for ambiguity, the base may be given explicitly in the form of a subscript following the digits representing the number. Thus, the number 561_8 is in base 8. Its decimal representation may be derived by substituting in Equation A.1 and evaluating the result using the familiar rules of decimal arithmetic. Hence

$$561_8 = 5 \times 8^2 + 6 \times 8^1 + 1 \times 8^0$$

$$= 369_{10}$$

Similarly, the number 0100110_2 is expressed in binary notation, i.e., in base 2. Its decimal equivalent can be obtained as follows

$$0100110_2 = 0 \times 2^6 + 1 \times 2^5 + 0 \times 2^4 + 0 \times 2^3 + 1 \times 2^2 + 1 \times 2^1 + 0 \times 2^0$$

$$= 38_{10}$$

Numbers having a fractional part can also be converted to a decimal representation in the same way. For example,

$$62.4_8 = 6 \times 8^1 + 2 \times 8^0 + 4 \times 8^{-1}$$

$$= 50.5_{10}$$

and

$$1100.101_2 = 1 \times 2^3 + 1 \times 2^2 + 0 \times 2^1 + 0 \times 2^0 + 1 \times 2^{-1} + 0 \times 2^{-2} + 1 \times 2^{-3}$$

$$= 12.625_{10}$$

A.1.2 Horner's Rule

Conversion of a number represented in base r into its decimal representation involves the evaluation of the polynomial in Equation A.1. A convenient method for performing this computation is known as *Horner's Rule*. Let us consider the integer and fractional parts of the number separately. The integer part is given by

$$a_{n-1}r^{n-1} + a_{n-2}r^{n-2} + a_{n-3}r^{n-3} + \cdots + a_1 r + a_0$$

This can be rewritten in the form

$$((\cdots ((a_{n-1}r + a_{n-2})r + a_{n-3})r + \cdots a_1)r + a_0)$$

That is, the required computation can be performed by starting with the most significant digit of the number and proceeding as follows: multiply r times a_{n-1} and add a_{n-2}, multiply the result by r and add a_{n-3}, and continue until the least significant digit, a_0, is reached. The procedure is illustrated in Figure A.1, using the

Octal:

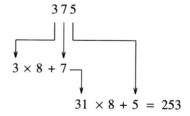

Binary:

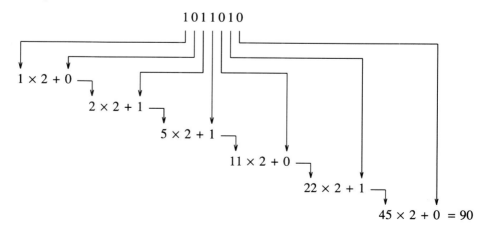

Figure A.1 Horner's Rule.

numbers 375_8 and 1011010_2. An example of a computer program to perform this computation is given in Figure A.2.

The fractional part of a number can be evaluated in a similar manner, using successive division instead of multiplication. The order of computation may be expressed as follows

$$a_{-1}r^{-1} + a_{-2}r^{-2} \cdots + a_{-(m-1)}r^{-(m-1)} + a_{-m}r^{-m}$$

$$= ((\cdots ((a_{-m}/r + a_{-(m-1)})/r + a_{-(m-2)})/r + \cdots + a_{-2})/r + a_{-1})/r$$

That is, starting at the least-significant digit, divide a_{-m} by r and add $a_{-(m-1)}$. Then, divide the result by r and add $a_{-(m-2)}$, and continue until the radix point is reached. As an illustration of this procedure, two examples are given in Figure A.3 using the numbers 0.346_8 and 0.01011_2.

A.1.3 Hexadecimal Numbers

The ten numerals $0, 1, \ldots, 9$ are sufficient to provide distinct symbols for representing numbers in any base less than or equal to 10. For a higher base, more symbols are needed. As mentioned earlier, base 16 numbers are commonly used in computer work. In this base, it is customary to use the letters A, B, C, D, E, and F to represent digits with an equivalent decimal value of 10, 11, 12, 13, 14, and 15, respectively.

A few examples of hexadecimal representation are given below, where all numbers to the right of the "=" sign are given in base 10

$$3B6_{16} = 3 \times 16^2 + 11 \times 16 + 6$$
$$= (3 \times 16 + 11) \times 16 + 6$$
$$= 950$$

$$F35C_{16} = ((15 \times 16 + 3) \times 16 + 5) \times 16 + 12$$
$$= 62{,}300$$

$$E9.3B_{16} = 14 \times 16 + 9 + (11/16 + 3)/16$$
$$= 233.23046875$$

```
SUM = 0
FOR I = N − 1 to 0 by −1
  SUM = SUM*R + A(i)
END FOR
```

Figure A.2 A program for polynomial evaluation using Horner's Rule.

We should note an important relationship between hexadecimal and binary representations. A single hexadecimal digit can represent one of 16 possible values. Exactly the same range of values is spanned by four binary bits. For example, the hexadecimal digits 0, 3, A, and F correspond to the binary numbers 0000, 0011, 1010, and 1111, respectively. This one-to-one correspondence makes it very easy to convert a hexadecimal number into its binary equivalent. We simply replace each hexadecimal digit by its 4-bit representation, as shown in the following examples:

$$A3_{16} = 1010\ 0011_2$$

$$F75B_{16} = 1111\ 0111\ 0101\ 1011_2$$

$$31.D1_{16} = 0011\ 0001.1101\ 0001_2$$

Blanks have been inserted in the binary numbers only to improve readability.

Octal:

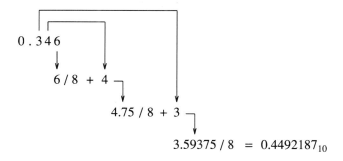

Binary

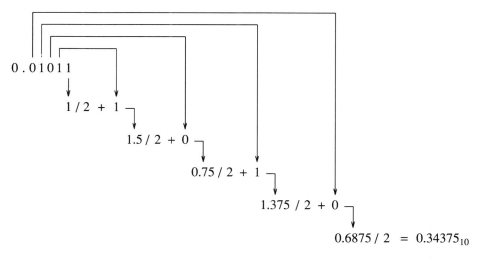

Figure A.3 Conversion of a fraction from base r to base 10 using Horner's rule.

By reversing the procedure described above, we can convert a binary number into its hexadecimal equivalent. The binary digits on each side of the binary point are considered separately. After appending 0s if necessary, they are organized in groups of 4 bits each. Then, each group is replaced by its hexadecimal equivalent. The conversion procedure is illustrated in Figure A.4. Because of the ease of conversion between hexadecimal and binary representations, the hexadecimal format is often used as a short form for expressing binary numbers. This is the main reason for the popularity of hexadecimal representation.

A.1.4 Conversion to Base-*r* Representation

The decimal representation for a number given in base r is easily obtained using Equation A.1. The summation needed in this equation may be evaluated directly, or techniques such as Horner's Rule may be employed to simplify the numerical computation. We will now consider the inverse operation. Given a decimal number, how can one obtain its representation in base r?

Consider a decimal number $A = a_{n-1} \cdots a_1 a_0$. Let its representation in base r be $b_{k-1} \cdots b_1 b_0$. The procedure to obtain the values b_0, b_1, etc. is based on the

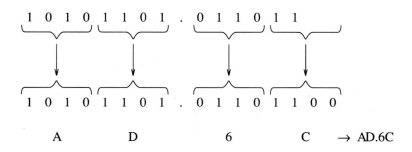

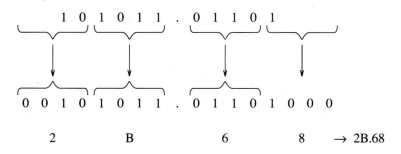

Figure A.4 Binary to hexadecimal conversion examples.

simple observation that dividing the number $b_{k-1} \cdots b_1 b_0$ by r results in moving the radix point one digit position to the left. That is,

$$b_{k-1} \cdots b_1 b_0 / r = b_{k-1} \cdots b_1 . b_0$$

For instance, in decimal notation, $725/10 = 72.5$. Similarly, in hexadecimal notation, $3AB/10 = 3A.B$ (recall that the symbol 10 in hexadecimal notation represents the value 16_{10}). Hence, the value b_0 may be obtained simply by dividing the number A by r and taking the fractional part of the result. The integer part may then be again divided by r to obtain b_1, and so on.

The conversion procedure is best understood by examining the defining equation for base-r representation, namely,

$$b_{k-1} r^{k-1} + \cdots + b_2 r^2 + b_1 r + b_0 = A \tag{A.2}$$

By dividing both sides of Equation A.2 by r, we obtain

$$b_{k-1} r^{k-2} + \cdots + b_2 r + b_1 + \frac{b_0}{r} = Q_0 + \frac{R_0}{r} \tag{A.3}$$

where Q_0 and R_0 are the quotient (integer part) and the remainder (fractional part) of the division operation. If Equation A.3 is to be satisfied, the fractional part on the left-hand side must be equal to the fractional part on the right-hand side. Hence

$$\frac{b_0}{r} = \frac{R_0}{r}$$

which yields

$$b_0 = R_0$$

Similarly, for the integer part, we have

$$Q_0 = b_{k-1} r^{k-2} + \cdots + b_2 r + b_1 \tag{A.4}$$

Having obtained the value of b_0, we may now proceed to evaluate b_1 by dividing both sides of Equation A.4 by r. Let the quotient and remainder in this case be Q_1 and R_1, respectively. That is,

$$Q_0 = Q_1 + \frac{R_1}{r}$$

Using the same argument as before, we obtain

$$b_1 = R_1$$

and

$$Q_1 = b_{k-1} r^{k-3} + \cdots + b_3 r + b_2$$

The procedure may be repeated until all the digits in the required base-r representation have been evaluated.

Conversion of decimal numbers to base-r is illustrated by two examples in Figure A.5. In part (a), the hexadecimal equivalent of the decimal number 2476 is shown to be CA9. In part (b) of the figure, the decimal number 49 is converted into its binary representation, 110001.

Fractional parts of a number can be converted to base r using a similar procedure. Instead of dividing by r, the process involves multiplication by r. Assume that it is desired to convert the fractional part F ($F < 1$) of a decimal number to its base-r representation. Let $F = 0.a_{-1}a_{-2} \ldots a_{-m}$ in base 10 and $0.b_{-1}b_{-2} \ldots b_{-m}$ in base r. The digits $b_{-1}, b_{-2}, \ldots, b_{-m}$ satisfy the equation

Decimal to hexadecimal

$$\frac{2476}{16} = 154 + \frac{12}{16} \qquad \rightarrow \quad a_0 = C$$

$$\frac{154}{16} = 9 + \frac{10}{16} \qquad \rightarrow \quad a_1 = A$$

$$\frac{9}{16} = 0 + \frac{9}{16} \qquad \rightarrow \quad a_2 = 9$$

Decimal to binary

$$\frac{49}{2} = 24 + \frac{1}{2} \qquad \rightarrow \quad a_0 = 1$$

$$\frac{24}{2} = 12 + \frac{0}{2} \qquad \rightarrow \quad a_1 = 0$$

$$\frac{12}{2} = 6 + \frac{0}{2} \qquad \rightarrow \quad a_2 = 0$$

$$\frac{6}{2} = 3 + \frac{0}{2} \qquad \rightarrow \quad a_3 = 0$$

$$\frac{3}{2} = 1 + \frac{1}{2} \qquad \rightarrow \quad a_4 = 1$$

$$\frac{1}{2} = 0 + \frac{1}{2} \qquad \rightarrow \quad a_5 = 1$$

Figure A.5 Decimal-to-hexadecimal and decimal-to-binary conversion.

$$\frac{b_{-1}}{r} + \frac{b_{-2}}{r^2} + \cdots + \frac{b_{-m}}{r^m} = F \tag{A.5}$$

Multiplying both sides of Equation A.5 by r, we obtain

$$b_{-1} + \frac{b_{-2}}{r} + \cdots + \frac{b_{-m}}{r^{m-1}} = rF \tag{A.6}$$

As before, Equation A.6 can be replaced by two equations for the integer and fractional parts of the numbers involved. Let the product rF have an integer component I_1 ($I_1 \geq 1$) and a fractional component F_1 ($F_1 < 1$). In this case

$$b_{-1} = I_1$$

and

$$\frac{b_{-2}}{r} + \cdots + \frac{b_{-m}}{r^{m-1}} = F_1 \tag{A.7}$$

This means that the most-significant digit of the representation in base r of a given fraction F is the integer part of the product rF. The process can be repeated by multiplying both sides of Equation A.7 by r, to obtain the values of a_{-2}, a_{-3}, etc.

As an example, the decimal fraction 0.8125 may be converted into a binary representation as follows:

$$
\begin{aligned}
0.8125 \times 2 &= 1.625 &&\rightarrow & a_{-1} &= 1 \\
0.625 \times 2 &= 1.25 &&\rightarrow & a_{-2} &= 1 \\
0.25 \times 2 &= 0.5 &&\rightarrow & a_{-3} &= 0 \\
0.5 \times 2 &= 1.0 &&\rightarrow & a_{-4} &= 1
\end{aligned}
$$

Hence, the desired representation is 0.1101.

In general, computation should be continued until the fractional part becomes 0. However, the value 0 will not always be reached, because the original number may not have an exact representation in the new base. The reader may attempt, for example, the conversion of the value 0.2_{10} into a binary number. The value of the fraction will start to repeat after four multiplication steps. In such a case, the conversion process should be terminated whenever sufficient precision has been attained. Exactly the same situation is encountered when representing the fraction 1/3 in decimal notation as $0.333 \cdots$.

A.2 ARITHMETIC OPERATIONS IN BASE r

Since we can represent numbers in an arbitrary base r, we should be able to perform the basic arithmetic operations of addition, subtraction, multiplication, and division directly in that base. The rules of arithmetic in base r are exactly the same as the familiar rules of arithmetic in base 10, except that the value r replaces the value 10 wherever it occurs. We will illustrate this assertion by a few examples.

Consider first the addition of two numbers in base 10. The steps involved in adding numbers 28 and 39 are illustrated in Figure A.6(a). The sum, 17, of the two least-significant digits exceeds the maximum value that can be represented by a single digit. It is reduced by the value of the base, 10, yielding 7 as the least-significant digit of the result, and a carry of one is added to the next higher digit position.

Addition in base 16 is illustrated in Figure A.6(b), using numbers 79 and 4C. The steps involved are exactly the same as in part (a) of the figure, except that a carry is generated whenever the sum of two digits exceeds 15. A few more examples are given in Figure A.7 illustrating addition in bases 16, 8, and 2.

Subtraction operations in base r follow exactly the same rules as in decimal arithmetic. A 1 borrowed from a higher-order digit position has a value of r. Thus, in base 16, 7A − 32 = 48, and B2 − 7C = 36. Several other subtraction examples in different bases are given in Figure A.8.

Multiplication and division follow in a straightforward manner, using the familiar rules. Some examples of multiplication in different number bases are worked

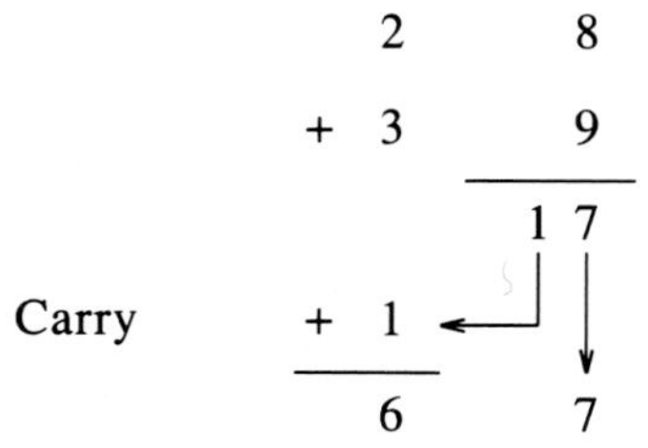

$$
\begin{array}{ccc}
 & 2 & 8 \\
+ & 3 & 9 \\
\hline
 & & 1\ 7 \\
\text{Carry} \quad + \ 1 & & \\
\hline
 & 6 & 7 \\
\end{array}
$$

(a) Decimal numbers

$$
\begin{array}{ccc}
 & 7 & 9 \\
+ & 4 & C \\
\hline
 & & 1\ 5 \\
\text{Carry} \quad + \ 1 & & \\
\hline
 & C & 5 \\
\end{array}
$$

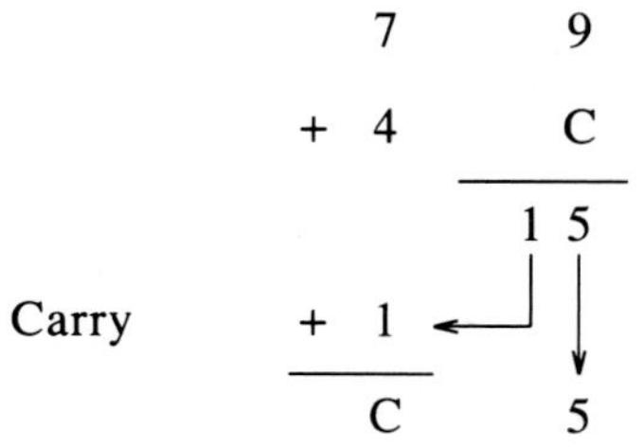

(b) Hexadecimal numbers

Figure A.6 Addition process.

Hexadecimal:

$$\begin{array}{r} 3\,D\,A\,9 \\ +\,B\,1\,2\,6 \\ \hline E\,E\,C\,F \end{array} \qquad \begin{array}{r} 5\,D\,9 \\ +\,F\,1\,A \\ \hline 1\,4\,F\,3 \end{array} \qquad \begin{array}{r} A\,8\,C\,7 \\ +\,1\,9\,D\,E \\ \hline C\,2\,A\,5 \end{array}$$

Octal:

$$\begin{array}{r} 1\,0\,4 \\ +\,5\,6\,3 \\ \hline 6\,6\,7 \end{array} \qquad \begin{array}{r} 4\,3\,7\,6 \\ +\,7\,1\,2\,5 \\ \hline 1\,3\,5\,2\,3 \end{array} \qquad \begin{array}{r} 1\,0\,3\,6 \\ +\,3\,4\,5\,2 \\ \hline 4\,5\,1\,0 \end{array}$$

Binary:

$$\begin{array}{r} 1\,0\,0\,0 \\ +\,0\,1\,1\,0 \\ \hline 1\,1\,1\,0 \end{array} \qquad \begin{array}{r} 0\,1\,1\,0\,1 \\ +\,1\,1\,1\,0\,1 \\ \hline 1\,0\,1\,0\,1\,0 \end{array} \qquad \begin{array}{r} 1\,0\,1\,0 \\ +\,0\,1\,1\,1 \\ \hline 1\,0\,0\,0\,1 \end{array}$$

Figure A.7 Examples of addition.

out in detail in Figure A.9. The rule for manual division is based on a trial and error approach. Individual digits of the quotient are obtained through a sequence of multiplication and subtraction operations.

A.3 BINARY CODED DECIMAL REPRESENTATION

The positional binary notation described in Section A.1.1 is normally used for representing numbers inside a computer. Thus, whenever a number is to be read by a computer from an input device or printed on an output device it must be converted from or to the more familiar decimal notation. The techniques needed for such conversions were discussed in Section A.1.4.

Often it is desirable to store numbers inside the computer in a format that maintains the identity of individual digits in the decimal representation. A number representation scheme that satisfies this requirement is the *binary coded decimal* (BCD) notation. The ten decimal digits, 0 to 9, are represented as 4-bit binary numbers, using the codes 0000, 0001, ... , 1001. Then, a decimal number is represented by replacing each of its digits with the equivalent 4-bit pattern. For example, the number 394 is represented as 0011 1001 0100.

Hexadecimal:

$$
\begin{array}{r}
F\,5\,C \\
-\,3\,B\,D \\
\hline
B\,9\,F
\end{array}
\qquad
\begin{array}{r}
3\,8\,D \\
-A\,5\,2 \\
\hline
-6\,C\,5
\end{array}
$$

Octal:

$$
\begin{array}{r}
6\,3\,1 \\
-2\,4\,5 \\
\hline
3\,6\,4
\end{array}
\qquad
\begin{array}{r}
2\,4\,3 \\
-4\,3\,1 \\
\hline
-1\,6\,6
\end{array}
$$

Binary:

$$
\begin{array}{r}
1\,0\,1\,1\,0 \\
-0\,1\,0\,1\,1 \\
\hline
0\,1\,0\,1\,1
\end{array}
\qquad
\begin{array}{r}
1\,0\,1\,1\,1 \\
-1\,1\,0\,0\,1 \\
\hline
-0\,0\,0\,1\,0
\end{array}
$$

Figure A.8 Examples of subtraction.

The BCD representation is simple and straightforward. It is not as efficient as the binary representation, because only ten out of the available 16 binary patterns of 4 bits are utilized. The patterns 1010 to 1111 are wasted. Thus, a given number of bits can represent fewer values in BCD than in the binary positional notation. For example, in BCD, 8 bits cover the range 0 to 99, while they can be used to represent values in the range 0 to 255 if the binary notation is used. Also, simpler circuitry is needed to perform various arithmetic operations on binary numbers than on BCD numbers. However, a capability to deal with BCD numbers is included in most modern computer systems, because there exist many commercial applications, particularly those involving accounting, where it is convenient to handle decimal data directly.

A.4 CHARACTER REPRESENTATION

In addition to storing numbers inside a computer, there is a need for storing textual information. Handling of such information requires means for representing letters of the alphabet, numerical digits, punctuation marks, mathematical symbols, etc. The term *character* is used to refer to an individual item, such as a letter of the alphabet, a digit, an exclamation mark, or an equal sign.

Hexadecimal:

```
            4 C 5                                    7 3
    ×       B 3 A                            ×       2 A
          -------                                  -----
          2 F B 2                                  4 7 E
          E 4 F                                    E 6
      3 4 7 7                                     -------
      ---------                                  1 2 D E
      3 5 8 B A 2
```

Octal:

```
              7 2                                  6 3 7
        ×     3 6                            ×       2 5
            -----                                  -------
            5 3 4                                  4 0 3 3
          2 5 6                                  1 4 7 6
          -------                                -----------
          3 3 1 4                                2 1 0 1 3
```

Binary:

```
          1 0 1 1                              1 1 0 1 0
      ×     1 0 1                          ×       1 1 0
          -------                              -----------
          1 0 1 1                              0 0 0 0 0
        0 0 0 0                                1 1 0 1 0
      1 0 1 1                                1 1 0 1 0
      -----------                            -------------
      1 1 0 1 1 1                            1 0 0 1 1 1 0 0
```

Figure A.9 Examples of multiplication.

As in the case of numbers, characters are represented using patterns of bits. A *character code* is a tabular assignment of binary patterns, usually of fixed length, to individual characters. For example, the 7-bit pattern 1000001 may be chosen to represent the letter "A." Thus, when the key for letter A is depressed on the keyboard of a terminal, this binary pattern is sent to the computer. Similarly, if this pattern is sent to a printer, the letter A is printed.

Since computer equipment is made by a variety of manufacturers, it is necessary to have a standard character code that can be used by all. One commonly used code is known as ASCII (American Standard Code for Information Interchange), shown in Table A.1. Each character is represented by 7 bits, providing sufficient patterns for 128 characters. For example, the letter A, the digit 6, and the equal sign are assigned the codes 1000001, 0110110, and 0111101, respectively. In most computers, characters are represented as 8-bit quantities, known as *bytes*. An 8-bit representation is derived from the ASCII code simply by adding a 0 on the left side. Thus, the three characters mentioned above are represented in a computer as 01000001, 00110110, and 00111101, or in hexadecimal notation, 41, 36, and 3D, respectively.

At this point, it should be emphasized that the codes shown in Table A.1 are arbitrary binary patterns chosen to represent various symbols. Their numerical value when interpreted as binary numbers has no significance. Nevertheless, the choice of assignment is made to facilitate certain processing tasks. For example, the codes for the letters A, B, C, $\cdots$ have the hexadecimal values 41, 42, 43, etc. The corresponding lower case letters have the values 61, 62, 63, etc. This simplifies the task of sorting words in alphabetical order, because, when compared as numbers, the codes follow an ascending sequence in the same order as the letters of the alphabet. Note also that an upper-case letter can be changed to lower case by setting bit b_5 to 1.

Several codes in Table A.1 represent characters that have no graphical symbol. These characters are used during various input and output operations in computers and during transmission of data from one computer to another. For example, Carriage Return, denoted as CR, corresponds to the Return key on a terminal keyboard. Form Feed, FF, is a character used to cause a printer to eject a page of output and start printing on a new one, and so on.

Some characters, such as STX and ETX, are not normally found on the keyboard of a computer terminal. Their primary use is in data communication applications. So, STX, which stands for Start of Text, indicates that the text of a message is about to begin, and ETX, End of Text, marks the end of the message text. Such characters may be generated on the keyboard of a terminal by means of a special key called CTRL (Control). When this key is depressed at the same time as any other key on the keyboard, it forces bit b_6 to 0. Thus, depressing the CTRL and B keys simultaneously causes the keyboard to generate the code 00000010, which represents STX. The codes for other nonprinting characters may be generated in a similar manner.

The importance of having a single, universally accepted character code standard can be readily appreciated. Unfortunately there are several such standards. Another widely used character code is known as EBCDIC (Extended Binary Coded Decimal Information Code), which is used on many IBM and IBM-compatible computers and terminals.

PART II: SIGNED NUMBERS

While working with a microprocessor, one often requires an understanding of the details of the way in which positive and negative quantities are handled. Such an understanding is particularly important in the process of program debugging, where it is sometimes necessary to examine binary patterns stored in the microprocessor's memory. The following two sections introduce the concepts and techniques needed.

Table A.1   THE ASCII CHARACTER CODE

Bit positions 3210 *	Bit positions 654 *							
	000	001	010	011	100	101	110	111
0000	NUL	DLE	SPACE	0	@	P	'	p
0001	SOH	DC1	!	1	A	Q	a	q
0010	STX	DC2	"	2	B	R	b	r
0011	ETX	DC3	#	3	C	S	c	s
0100	EOT	DC4	$	4	D	T	d	t
0101	ENQ	NAK	%	5	E	U	e	u
0110	ACK	SYN	&	6	F	V	f	v
0111	BEL	ETB	'	7	G	W	g	w
1000	BS	CAN	(	8	H	X	h	x
1001	HT	EM	)	9	I	Y	i	y
1010	LF	SUB	*	:	J	Z	j	z
1011	VT	ESC	+	;	K	[	k	{
1100	FF	FS	,	<	L	\	l		
1101	CR	GS	–	=	M	]	m	}
1110	SO	RS	.	>	N	^	n	~
1111	SI	US	/	?	O	_	o	DEL

NUL	NUL/IDLE	SI	Shift in
SOH	Start of header	DLE	Data link escape
STX	Start of text	DC1–DC4	Device control
ETX	End of text	NAK	Negative acknowledgment
EOT	End of transmission	SYN	Synchronous idle
ENQ	Enquiry	ETB	End of transmitted block
ACK	Acknowledgement	CAN	Cancel (error in data)
BEL	Audible signal	EM	End of medium
BS	Back space	SUB	Special sequence
HT	Horizontal tab	ESC	Escape
LF	Line feed	FS	File separator
VT	Vertical tab	GS	Group separator
FF	Form feed	RS	Record separator
CR	carriage return	US	Unit separator
SO	Shift out	DEL	Delete/idle

* Bit positions:

6	5	4	3	2	1	0

A.5 REPRESENTATION OF NEGATIVE NUMBERS

Many computational tasks involve the use of negative numbers. We write

$$23 - 57 = -34$$

The minus symbol on the left side of this equation is an arithmetic operator representing the subtraction operation. The same symbol on the right side represents a negative value. This is called the *sign-and-magnitude* notation. The term "sign and magnitude" refers to the fact that the representation of the number consists of two parts: the sign, which indicates whether the number is positive or negative, and the magnitude, which is the value 34 in the above example. The magnitude of a number is also called its *absolute value*. For a positive number the magnitude of the number should be preceded by a plus sign. However, by convention, the plus sign may be omitted.

When representing signed numbers inside a digital computer, we need some means for representing the sign of the number. Since the only symbols available are 0s and 1s, we can choose one of these symbols, e.g., 0, to represent the plus sign, and the other, 1, to represent the minus sign. In this way, the sign-and-magnitude notation may be used in a computer. For example, 3 bits may be used to represent the magnitude of the number, and one bit to represent the sign. This makes it possible to represent the values −7 through +7. If the sign bit is placed to the left of the magnitude bits, we obtain the representations shown in Figure A.10. Note that different bit patterns represent the value "zero," one for +0 and the other for −0. Of course, both patterns represent the same thing.

A few comments about Figure A.10 are in order. The bit pattern 1101 represents the value −5. The same pattern interpreted as an unsigned binary number corresponds to the value 13. In other words, there is no inherent meaning to a particular bit pattern inside a computer. It may be interpreted in a variety of ways, according to the needs of and the conventions agreed upon by the users. Unsigned numbers and sign-and-magnitude notation are but two ways of interpreting a given bit pattern. Some other possible meanings will be introduced later in this appendix.

	Sign	Magnitude					Sign	Magnitude		
− 7	1	1	1	1		+ 0	0	0	0	0
− 6	1	1	1	0		+ 1	0	0	0	1
− 5	1	1	0	1		+ 2	0	0	1	0
− 4	1	1	0	0		+ 3	0	0	1	1
− 3	1	0	1	1		+ 4	0	1	0	0
− 2	1	0	1	0		+ 5	0	1	0	1
− 1	1	0	0	1		+ 6	0	1	1	0
− 0	1	0	0	0		+ 7	0	1	1	1

Figure A.10 Number representation in sign-and-magnitude notation.

Let us now consider the problem of performing arithmetic operations on numbers represented in the sign-and-magnitude notation. Obviously, the rules for unsigned numbers, as discussed in Section A.2, cannot be used without modification. For example, if we were to add 1101 (-5_{10}) to 0010 ($+2_{10}$) , we would obtain 1111 (-7_{10}), which is clearly incorrect. In order to obtain the correct answer, the sign and magnitude parts of the numbers should be separated. Since the two numbers have different signs, addition must be replaced with subtraction. Also, the number having the higher magnitude must be determined, so that the numbers may be subtracted in the correct order. Thus, the subtraction operation may be performed as

$$101 - 010 = 011$$

Finally, the result is given the sign of the number having the higher magnitude, to yield the correct answer of 1011 (-3_{10}).

The awkward sequence of steps described above is an inherent aspect of the sign-and-magnitude notation. The comparisons and decisions involved are very easily done by humans, who perform them almost subconsciously. However, if the same approach is to be used by a machine, each step must be precisely defined, then implemented either in the hardware or in the software of the computer. Because of the complexity of the sign-and-magnitude approach, an alternative scheme known as *two's complement* representation has been developed for handling signed numbers. It will be discussed in the next section.

Before concluding the discussion of the sign-and-magnitude scheme, it should be pointed out that, despite its complexity, this scheme is used for representing numbers inside a computer in many circumstances. Perhaps its most common application is in the representation of floating-point numbers, which will be discussed in Section A.8.

A.6 TWO'S COMPLEMENT ARITHMETIC

The main motivation for developing 2's complement arithmetic is to enable addition and subtraction operations for signed numbers to be carried out in a straightforward manner. Before describing the details of this scheme, we will introduce a few underlying concepts using the familiar decimal numbers.

A.6.1 Addition and Subtraction

It is important to keep in mind that in most practical computer systems, a numerical value is represented using a fixed number of digits. Let us assume that a given computer uses two decimal digits for representing a number. This provides a total of 100 distinct symbols, from 00 to 99. If only unsigned numbers were to be used in this computer, we could simply use each of these symbols to represent the corresponding value, N, as calculated according to the rules of the positional notation presented in Section A.1. For example, the symbol 87 would represent the value

$$N = 8 \times 10^1 + 7 \times 10^0$$

We should emphasize that the assignment of values to symbols in this manner, though simple and convenient, is totally arbitrary.

In order to represent negative numbers in the above computer, some of the available 100 symbols must be used to represent negative values. Consider the following assignment. Each of the first 50 symbols, 00 to 49, will be used to represent the same positive value, N, as before. The symbols 50 to 99 will be assigned negative values according to the rule

$$\text{Value} = -(100 - N) \qquad\qquad \textbf{(A.8)}$$

Thus, the symbol 99 represents (−1), 98 represents (−2), and so on. This assignment is illustrated in Figure A.11. In order to avoid confusion in this discussion, numerical values are enclosed between brackets, while the symbols representing them are not.

Let us examine the properties of the choice of assignments described above. For positive numbers, addition operations can be carried out in the usual manner. However, the result of addition of two positive numbers cannot exceed the available range for positive numbers, which is 0 to 49, inclusive. If it does, *arithmetic overflow* will be said to have occurred, and the result is incorrect. For example, the sum of (+5) and (+30) is (+35), which is within the allowed range. On the other hand, (+32) + (+46) yields (+78), which cannot be represented. Hence, this addition operation results in arithmetic overflow.

Value	Representation
0	0
+1	1
+2	2
+3	3
⋮	⋮
+48	48
+49	49
− 50	50
− 49	51
⋮	⋮
− 4	96
− 3	97
− 2	98
− 1	99

Figure A.11 A possible choice for the representation of negative numbers.

If the two values to be added have opposite signs, the operation is equally easy. We add the symbols that represent them, without regard to the sign. For example, consider the addition of the two values (–20) and (+12). Using the corresponding representations from Figure A.11, we obtain

$$80 + 12 = 92$$

The result, 92, represents the value (–8), which is the correct answer. This is not surprising, because the representation for the value (–20) is 100–20. Thus

$$(100 - 20) + 12 = 100 - (20 - 12)$$
$$= 100 - 8$$
$$= 92$$

A slightly different situation arises when a positive number is added to a negative number of a smaller magnitude. The addition of (+8) and (–5) produces the following, again using the corresponding symbols from Figure A.11:

$$(+8) + (-5) \quad \rightarrow \quad 8 + 95 = 103$$

The result is outside the available range of symbols. However, it does not represent an overflow situation, because the correct sum, (+3), is within the available range for positive numbers. It can be easily seen that the above operation produces the correct result, with the value 100 added to it. Hence, the correct result can be obtained if we adopt a rule that the computations will be carried out modulo 100. That is, whenever the result of an addition operation is larger than 100, only the amount by which the result exceeds 100 is retained. This means that the carry from the most-significant digit position should be ignored.

Perhaps a more illustrative way to view modulo 100 addition is to organize the numbers 0 to 99 on a circle, as shown in Figure A.12. The operation 3 + 2 may be performed by locating the number 3 on the circle, then moving clockwise two positions to get the sum, 5. Using a similar procedure, the operation 99 + 2 yields a result of 1. When the addition operation is performed in the normal manner, the result is 101. Thus, the correct result is obtained by ignoring the carry from the most-significant digit position, which is the carry from the second digit position in this case.

When negative numbers are represented as in Figure A.11, subtraction operations can also be carried out in a straightforward manner. For example, to subtract (+5) from (+3), we locate the number 3 on the circle in Figure A.12 and count 5 positions in the counterclockwise direction. The result is 98, which is the representation for the correct answer, (–2). We should expect to get the same result when adding (–5) to (+3). Taking the corresponding symbols from Figure A.11, we obtain

$$3 + 95 = 98$$

which is the same result.

To summarize, the number representation defined in Figure A.11, combined with the use of arithmetic modulo 100 leads to a scheme in which the operations of addition and subtraction can be performed without regard to the signs of the numbers

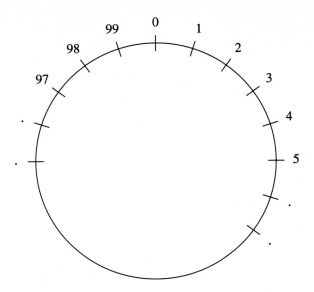

Figure A.12   Illustration of arithmetic modulo 100.

involved. This is unlike the sign-and-magnitude representation, in which addition may have to be replaced by subtraction, or vice versa, depending upon the signs of the two numbers. Also, in the representation of Figure A.11 one need not establish which number has the greater magnitude before performing a subtraction. The result obtained is always correct, both in sign and in magnitude.

A good understanding of the number representation scheme described above is essential as a basis for studying the way in which arithmetic is performed in computers. The reader is encouraged to try several examples of addition and subtraction to become familiar with this technique and its characteristics, before proceeding to study the binary case described below.

A.6.2 The Binary Case

In the subsection above, we used two-digit numbers in base 10. The ideas presented are applicable to any base. In general, if numbers are represented by n digits in base r, then r^n distinct symbols are available. Hence, addition and subtraction can be performed modulo r^n. A number N is negated by subtracting it from r^n. The difference, $r^n - N$, is the *complement* of N with respect to r^n. It is called the *r's complement* of N. Let us now apply the above ideas to binary numbers. In this case, numbers are said to be represented in *2's complement* notation. Assume that numbers are represented using 8 bits, providing 2^8 or 256 distinct symbols. Following the same approach as in Figure A.11, we arrive at the assignments shown in Figure A.13. The binary numbers 00000000 to 01111111 represent positive values, and the remaining numbers, 10000000 to 11111111 represent negative values. The

Binary Pattern	Assigned Value	
0 0 0 0 0 0 0 0	0	
0 0 0 0 0 0 0 1	1	
.	.	Positive
.	.	Values
.	.	
0 1 1 1 1 1 1 0	126	
0 1 1 1 1 1 1 1	127	
1 0 0 0 0 0 0 0	−128	
1 0 0 0 0 0 0 1	−127	
.	.	Negative
.	.	Values
.	.	
1 1 1 1 1 1 1 0	−2	
1 1 1 1 1 1 1 1	−1	

Figure A.13 Two's complement representation using eight bits.

value of any number in the positive range is as derived using the usual rules of positional notation. The magnitude of a number in the negative range can be computed by applying the rules of positional notation, then subtracting the result from 256. Note that the sign of a number is easily identified by the fact that in the positive range, the most-significant bit is always equal to 0, while in the negative range it is equal to 1. This is a very convenient side effect of the particular choice of assignment used in Figure A.13.

Let us examine a few examples. Consider the binary patterns $A = 00011011$ and $B = 11010010$. When interpreted according to the rules of 2's complement representation, A is a positive number, because its most-significant bit is equal to 0. On the other hand, B is a negative number. The value of A is equal to the weighted sum of its digits. Hence

$$A = 1 \times 2^4 + 1 \times 2^3 + 1 \times 2^1 + 1 \times 2^0$$

$$= 27$$

In order to compute the value represented by B, we first obtain the weighted sum of its digits, i.e.,

$$1 \times 2^7 + 1 \times 2^6 + 1 \times 2^4 + 1 \times 2^1 = 210$$

Therefore, the value of B is given by Equation A.8, after replacing the modulus 100 with 256, as follows:

$$B = -(256 - 210)$$

$$= -46$$

Similarly, the value (+37) is represented by 00100101. In order to find the representation of (−37) we first subtract 37 from 256, which yields 219. After conversion to binary we obtain 11011011.

Figure A.14 shows a few examples of addition and subtraction using 8-bit 2's complement numbers. The binary representation and the signed decimal value are given for each number. When studying these examples, it is important to remember that a 1 to the left of the most-significant digit position has the value 100000000, which is 256. Hence, in order to carry out computations modulo 256, any carry or borrow that may result from operations at the most-significant digit position should simply be ignored. Ignoring the carry reduces the result by 256, while ignoring the borrow increases it by 256. In either case, the result is correct when interpreted as a

	2's Complement Representation	Equivalent Decimal Value	
(a)	0 1 0 0 0 1 1 0	(+70)	
	+ 0 0 0 1 0 1 0 1	+ (+21)	
	0 1 0 1 1 0 1 1	(+91)	
(b)	1 1 1 0 0 1 0 0	(−28)	
	+ 1 0 1 0 1 1 0 1	+ (−83)	
	1←1 0 0 1 0 0 0 1	(−11)	Ignore carry
(c)	0 0 1 1 1 1 0 1	(+61)	
	+ 1 0 0 1 0 1 1 1	+ (−105)	
	1 1 0 1 0 1 0 0	(−44)	
(d)	0 1 1 1 1 0 0 1	(+121)	
	+ 1 1 0 0 0 1 1 1	+ (−57)	
	1←0 1 0 0 0 0 0 0	(+64)	Ignore carry
(e)	0 0 1 0 0 0 1 1	(+35)	
	− 0 1 0 0 1 0 0 0	− (+72)	
	1←1 1 0 1 1 0 1 1	(−37)	Ignore borrow

Figure A.14 Examples of 2's complement addition and subtraction.

modulo 256 number. This is exactly what happens when arithmetic computations are performed in an 8-bit computer.

The reader should examine the operations in Figure A.14 in detail and verify their correctness. All examples in this figure yield a result within the range of values that can be represented in 8 bits. Two examples that result in arithmetic overflow are given in Figure A.15. The addition of (+102) to (+57) yields (−97), which is obviously incorrect. The correct sum is 159, which is outside the available range for positive numbers. Similarly, the operation (−63) + (−97) yields (+96) instead of (−160). Arithmetic overflow is easily detected, because it is always accompanied by sign reversal. That is, the addition of two numbers with like signs yields a result with an opposite sign. When the two numbers being added have opposite signs, overflow can never occur, because the result will always be within the available range.

We have used 8-bit numbers as examples in our discussion. The techniques are equally applicable to numbers having any number of bits. The positive range consists of all numbers whose most-significant bit is equal to 0, and the negative range consists of all numbers whose most-significant bit is equal to 1. Thus, if only 4 bits are available, we can represent the positive values 0 (0000) to +7 (0111) and −1 (1111) to −8 (1000). In a 16-bit representation, the positive range is 0 to +32,767 and the negative range is −1 to −32,768.

A.6.3 Negation

Numerical computations involve both addition and subtraction operations. Hence, in general, it appears that a computer must incorporate an adder circuit and a subtractor circuit. We will show that when the 2's complement notation is used, a slightly augmented adder circuit can be used for both operations. The operation $A - B$ is performed by first negating the subtrahend, B, then adding it to the minuend, A. Thus, what is needed now is a simple way to negate numbers.

	2's Complement Representation	Equivalent Decimal Value	
(a)	0 1 1 0 0 1 1 0	(+102)	
	+ 0 0 1 1 1 0 0 1	+(+57)	
	1 0 0 1 1 1 0 0	(−97)	Overflow
(b)	1 1 0 0 0 0 0 1	(−63)	
	+ 1 0 0 1 1 1 1 1	+(−97)	
	1←0 1 1 0 0 0 0 0	(+96)	Overflow

Figure A.15 Two's complement addition and subtraction examples resulting in arithmetic overflow.

Negation of a number N is the operation of finding a number N', such that $N + N' = 0$. From the definition of the 2's complement representation, N' is obtained by subtracting N from 2^n, where n is the number of bits used in the representation. For example, when 8-bit numbers are used, negation of the value (+9) yields $256 - 9 = 247$ ($=11110111_2$), which is the representation for (–9). If the number 247 is negated, we obtain $256 - 247 = 9$, which represents the original value.

We will now show that evaluation of the expression $256 - N$ for any number N does not in fact require a subtraction operation to be carried out. The technique used is based on the observation that

$$256 - N = 255 + 1 - N$$
$$= (255 - N) + 1$$

That is, the result of subtraction from 256 may be obtained by subtracting N from 255, then adding 1. The reason for replacing the single subtraction operation with two operations is that $255 - N$ is very easy to evaluate. The binary representation for 255 is 11111111. Since it has 1s in all bit positions, the subtraction operation $255 - N$ does not involve any borrows. Let N have the binary representation $a_7 a_6 \cdots a_0$. Then, $255 - N$ has the representation

$$(1 - a_7)(1 - a_6) \cdots (1 - a_0)$$

That is, bit i of the result is obtained by subtracting a_i from 1; if a_i is equal to 0, the corresponding bit of the result is equal to 1, and if a_i is 1 the result is 0. The value $1 - a_i$ is the complement of a_i with respect to 1. Hence, it is called the *1's complement,* or simply the complement of a_i, and the resulting number, which is equal to $255 - N$, is called the 1's complement of N.

The procedure described above applies to 2's complement representations of any length. A number is negated by obtaining its 1's complement, then adding 1 to the result. No subtraction operation is needed in the process. This procedure is illustrated in Figure A.16. Negation of (+56) yields 11001000, which is the correct 2's complement representation for (–56). Also, negation of a negative number, such as (–25) or (–6), yields the correct positive value, as shown.

On the basis of this discussion, subtraction of 2's complement numbers can be performed by adding the 1's complement of the subtrahend plus 1 to the minuend. Thus, the subtraction operation in example of Figure A.14(e) may be done in a computer as indicated in Figure A.17(a). The 1's complement of the subtrahend is 10110111. This value plus 1 represents (–72), which must be added to the minuend to obtain the correct result, (–37). In this example, no carry results from the addition operation at the most-significant bit position. Figure A.17(b) shows a case where a carry results at this position. According to the rules of 2's complement arithmetic, this carry is simply ignored, leaving the correct result, (87).

A useful shortcut may be observed at this point. Consider again the examples in Figure A.16. In each case, compare the bits of the original number, N, with those of its negative, $-N$. Starting from the least-significant bit, zeros up to and including the right-most 1 in N appear unchanged in $-N$. All the remaining bits are complemented. This observation provides a convenient method for negating numbers in

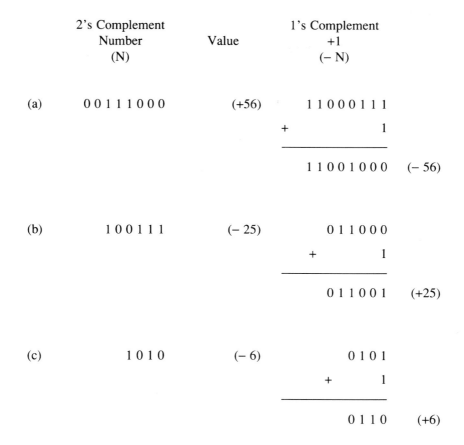

Figure A.16 Negation using 1's complement.

manual computations. Its validity may be illustrated by examining the way in which the carry propagates during the addition operation used in the derivation of the 2's complement in Figure A.16. For example, the three least significant 0s in part (a) become 1's when complemented. They change back to 0s after the addition. Meanwhile, the propagating carry causes a 1 to appear in the fourth bit position. Since no carry propagates to the left of this bit, the remaining bits retain their complemented value.

When negating numbers, one special case should be noted. The 4-bit number 1000 represents the value (−8). Negation of this number requires adding 1 to its 1's complement, 0111. The addition operation results in arithmetic overflow, because the value (+8) is outside the range of positive numbers that can be represented in 4 bits. A similar exception occurs for numbers having any number of bits. The negative number having the largest magnitude does not have a positive counterpart.

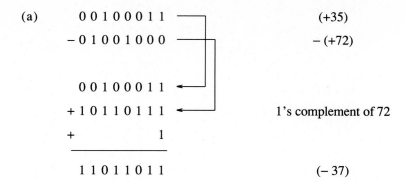

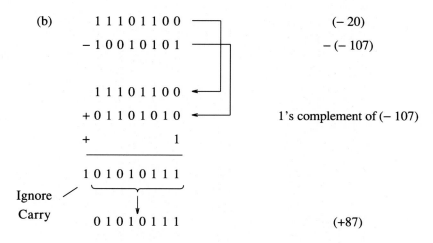

Figure A.17 Examples of the way subtraction is usually implemented in a computer.

A.6.4 Sign Extension

When dealing with unsigned numbers, it is customary to omit any 0s appearing on the left-hand side, because they do not change the value of the number. Both 001001 and 1001, interpreted as unsigned numbers, represent the value (+9). This is not the case when dealing with signed numbers. In 2's complement representation, the number 001001 is a 6-bit representation of the value (+9), while 1001 is a 4-bit representation of the value (−7). Thus, we cannot change the number of bits in which a given value is represented by simply adding or dropping zeros.

Let us return to the definition of 2's complement representation. Numbers in the positive range are evaluated by computing the weighted sum of individual bits. Hence, zeros may be added or dropped on the left-hand side of a positive number without changing its value, provided that the most-significant bit remains equal to 0. Both 00010110 and 010110 represent the value (+22).

The situation with negative numbers is slightly different. The value (−6) is represented in 4 bits by 1010, in 6 bits by 111010, and in 8 bits by 11111010. Thus, adding zeros to the left of a positive number or adding ones to a negative number does not change its value. This simple rule can be easily understood by recalling the fact that when negating a number, the most-significant bit is always complemented. For example, the value (+6) is represented in 4 bits by 0110, which when negated yields 1010. If extended to 6 bits, negation of 000110 yields 111010, and so on.

This discussion may be summarized in the form of a simple rule. The number of bits representing a given value may be increased by duplicating the most-significant bit as many times as necessary. Since the most-significant bit indicates the sign of the number, this process is called *sign extension.*

A.6.5 Multiplication and Division

We have discussed the operations of addition and subtraction of 2's complement numbers at length. We have also pointed out that it is possible to perform subtraction as a combination of negation and addition.

Conceptually, an easy way to multiply or divide 2's complement numbers is to convert them first to sign-and-magnitude representation, perform the required operation using the familiar rules, then convert the result back to 2's complement representation. Multiplication and division may also be carried out directly on 2's complement numbers. However, the techniques available are considerably more involved than the simple approach described above. Details of such techniques can be found in references [1,2].

PART III: FLOATING-POINT REPRESENTATION

In the remainder of this appendix, we introduce a method for representing floating-point numbers, which is widely used in engineering and scientific computation.

A.7 FLOATING-POINT NUMBERS

When writing or printing noninteger numbers, such as 31.739_{10} or 1101.011_2, the position of the radix point is given explicitly. However, computers have no special symbols with which to indicate where the radix point is. Instead, the position of the radix point is implied, by using certain conventions. For example, we may adopt the convention that an 8-bit number consists of a 5-bit integer part and a 3-bit fractional part. Thus, the binary pattern 01101011 represents the number 01101.011_2. With such a convention the radix point is always at the same place, and the scheme is called *fixed point* representation.

The range of values that is spanned by an n-bit fixed point number is 0 to 2^n. This is too small a range for many applications. In scientific and engineering work, values such as 3×10^{15}, 5×10^{-23} or, in general, $M \times 10^E$ are used frequently. In this form, M is called the *mantissa* and E the *exponent*. Such numbers can span a very wide range of values, determined by the exponent. Meanwhile, the precision with which a given value is expressed can be controlled independently by varying the number of digits of the mantissa. For example, the speed of light is approximately equal to 3×10^8 m/s. A more precise value for it is 2.997925×10^8 m/s.

The same approach is used in a computer, when a wide range of values is needed. A given value is represented using two separate signed numbers, one for the mantissa and one for the exponent. In decimal representation, increasing the exponent by 1 is equivalent to multiplying the number by 10, that is, moving the decimal point one digit place to the right. In binary representation, moving the binary point to the right increases the value of the number by a factor of 2. Using a base of 2, a binary floating-point number that has a mantissa M and an exponent E, represents the value

$$M \times 2^E$$

where M and E may be either positive or negative numbers. The exponent, E, is restricted to being an integer, and the mantissa is, in general, a fixed-point number.

While we will restrict our discussion to a base of 2, we should note that other powers of 2 can be used as the base that is raised to the power E. For example, if the base is 16, then the number would be interpreted as

$$M \times 16^E$$

In this case, increasing the value of E by 1 is equivalent to moving the radix point four bit positions to the right.

A.7.1 Basic Format

The number of bits used to represent M determines the precision with which a floating-point number is represented. The range covered by the number is determined by the number of bits used for the exponent E. Any leading 0s in the mantissa leave fewer bits to represent the desired value. In order to maximize the precision with which a given value is represented, it is customary to use all bits of the mantissa as significant digits, with no leading zeros, then adjust the exponent to obtain the desired magnitude. This process is called *normalization*. For example, using a four-digit mantissa for decimal numbers, with the radix point to the left of the most-significant digit, the value $1/3 \times 10^{12}$ may be represented as 0.0033×10^{14} or it may be normalized as 0.3333×10^{12}. By eliminating the leading zeros we are able to provide a more precise representation for the desired value.

There are several formats for representing floating point numbers. One arrangement has been recommended by the Institute of Electrical and Electronics Engineers (IEEE) Standards Committee [3]. There are two formats defined by this standard—a 32-bit format, called single-precision, and a 64-bit format, called double precision. We will describe the single precision format as an example of floating-point representations.

Of the 32 available bits, 24 are allocated for the mantissa and 8 for the exponent. The overall format is shown in Figure A.18. The mantissa is represented in the sign-and-magnitude format described in Section A.5. In order to maximize precision, the mantissa is normalized, which means that its most-significant bit is always equal to 1. Because the value of this bit is known, it is not actually stored. Instead, the mantissa is assumed to be of the form $1.xxx\cdots$, and only the fraction part, $xxx\cdots$, is stored. Of the 24 bits allocated for the mantissa, one bit is used for the sign and the remaining 23 bits for the fractional part. This means that the smallest magnitude for the mantissa is $1.00\cdots0$ and the largest magnitude is $1.11\cdots1$; i.e.,

$$1.0 \le |M| < 2.0$$

The exponent is a signed integer number represented in a variation of the 2's complement scheme of Figure A.13, known as excess 127. In the discussion in Section A.6, it was emphasized that the assignment of values to symbols, or binary patterns, is a matter of interpretation. Any convenient assignment may be used. The choice made for the IEEE floating-point standard is shown in Figure A.19. The assigned values in this figure differ from those in Figure A.13 by 127, modulo 256. An exponent value of 0 is represented by the binary number representing the value 127, hence the name of the representation.

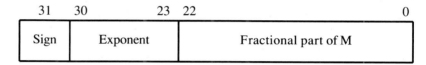

31	30	23	22	0
Sign	Exponent		Fractional part of M	

Figure A.18 Single-precision floating-point number representation in IEEE standard.

Binary Pattern	Exponent Value
00000000	Reserved
00000001	(− 126)
⋮	⋮
01111111	(0)
10000000	(+1)
10000001	(+2)
⋮	⋮
11111110	(+127)
11111111	Reserved

Figure A.19 Excess 127 representation.

The two numbers 0 and 255 in the exponent field are reserved for representing special values. Because of the use of normalization, the mantissa can never be equal to 0. Instead, the entry 00000000 in the exponent field denotes a floating-point number that is equal to 0. At the other end of the spectrum, an exponent entry of 11111111 is used to denote infinity. The ability to represent these special values is very useful in reducing errors when extensive numerical computations are carried out.

The IEEE standard also specifies a 64-bit format in the double-precision representation. In this case, the exponent field is expanded to 11 bits, to enable a wider range of values to be represented (from 2^{-1022} to 2^{+1023}, i.e., from 10^{-308} to 10^{+308}). This leaves 52 bits for the fractional part of the mantissa, thus providing a substantial increase in the precision with which a given value is represented.

A.7.2 Arithmetic Operations

Consider two floating-point numbers $A = M_A \times 2^{E_A}$ and $B = M_B \times 2^{E_B}$. We will discuss briefly the four basic operations of addition, subtraction, multiplication, and division for these numbers. In performing any of these operations, the "hidden bit," which is the 1 to the left of the binary point of the mantissa, must be taken into consideration.

Multiplication and division are easily performed, according to the relationships

$$A \times B = M_A \times M_B \times 2^{E_A + E_B}$$

and

$$\frac{A}{B} = \frac{M_A}{M_B} \times 2^{E_A - E_B}$$

For multiplication, the mantissas of the two numbers are multiplied and the exponents added, while for division, the mantissas are divided and the exponents subtracted. The sign of the result is positive if the two numbers have the same signs and negative if they are different.

Addition and subtraction of floating-point numbers are slightly more involved. Because sign-and-magnitude notation is used, we must inspect the signs of the two numbers to determine which operation to perform. In order to add two numbers of the form $M \times 2^E$, we must first adjust the values of M and E such that the two numbers have the same exponent. Thus, if $E_A < E_B$, we can adjust the representation of A as follows

$$A = M_A \times 2^{E_A} = M_A \times 2^{-(E_B - E_A)} \times 2^{E_B}$$

$$= M'_A \times 2^{E_B}$$

The new mantissa, M'_A, has a number of leading zeros equal to the difference between the two exponents. Thus, in effect, we have shifted the number A to the right relative to B in order to perform the addition correctly. This process is called *alignment*. Addition may now be performed as follows:

$$A + B = M'_A \times 2^{E_B} + M_B \times 2^{E_B}$$

$$= (M'_A + M_B) \times 2^{E_B}$$

The mantissa of the result is equal to the sum of the mantissa of B and the aligned mantissa of A. The exponent of the result is that of the larger number, E_B in this case.

Subtraction follows the same procedure, with the additional step that if $E_A = E_B$, we must identify the larger mantissa and perform subtraction in the correct order.

The result of any of the above operations will not necessarily be in a normalized form; i.e., the mantissa may be outside the allowable range, which is $1.0 \leq |M| < 2.0$ in the case of the IEEE standard. Hence, as a final step, the result must be examined and normalized if necessary. This involves shifting the mantissa either to the left or to the right and adjusting the exponent accordingly.

When performing operations on floating-point numbers several types of errors may occur, for which there is no equivalent in integer arithmetic. Consider, for example, the operation of adding two numbers having the mantissas $M_1 = 1.0110$ and $M_2 = 1.1101$, and the exponents $E_1 = E_2 = 0011$. (For illustration purposes, we have assumed that the exponents and the fractional parts of the mantissas are each 4 bits long.) Because the two numbers have the same exponent, no alignment is needed, and the two mantissas are added directly as follows:

$$1.0110 + 1.1101 = 11.0011$$

The result may now be normalized by shifting it one place to the right, and increasing the exponent by 1. Since there are only 4 bits available to store the fractional part, its least-significant bit will be lost, leaving the result $M = 1.1001$ and $E = 0100$. Obviously, the value of the mantissa is in error by 0.00001. Since M was obtained by truncating the result of addition to the desired length, this error is called *truncation error*.

When a large number of operations are performed, the errors introduced by truncation accumulate and may lead to significant errors in the final result. Numerical errors of this kind must be handled with extreme care, both in the design of a computer and in the preparation of the software that deals with floating-point numbers. Techniques such as rounding could be used to reduce the accumulation of errors. A detailed discussion of such techniques is beyond the scope of this text, but it can be found in many existing books [1,2].

A.8 CONCLUDING REMARKS

An understanding of the ways in which different kinds of information are represented in modern computers is essential for the study of these machines. In this appendix, we have presented the most commonly used representations for numeric and textual information.

A.9 PROBLEMS

1. Find the decimal equivalent of the following numbers:

$$3A7_{16}, \quad B8C_{16}, \quad 1001011_2, \quad 1100101_2, \quad 1735_8, \quad 2416_8$$

2. Use Horner's rule to convert the following numbers to decimal representation:

$$27D1_{16}, \quad 1E3F_{16}, \quad 101110101_2, \quad 010110111_2, \quad 234617_8, \quad 4215_8$$

3. Convert the following decimal numbers to the representation indicated:
   (a) 356,   4671,   57319   to hexadecimal
   (b) 203,   94,   175   to binary

4. Use Horner's rule to find the decimal equivalent of the following numbers:

$$37.516_8, \quad 6.1523_8, \quad 101.01101_2, \quad 0.101101_8$$

5. Perform the following operations directly in the number base indicated (i.e., without converting to decimal notation):
   (a) $(3AB2 + 2CF)_{16}$ (b) $(3256 + 714)_8$
   (c) $(101101 + 110111010)_2$ (d) $(10111 - 1010)_2$
   (e) $(110101 - 101101)_2$ (f) $(11011 - 1011101)_2$
   (g) $(1011 \times 11010)_2$ (h) $(11011 \times 100011)_2$

6. Evaluate the following expressions in arithmetic modulo 100:

$$73 + 57, \quad 36 \times 41, \quad 70 - 23, \quad 25 - 81$$

7. Give a 6-bit 2's complement representation for the following values:

$$27, \quad 15, \quad -30, \quad -7$$

8. Let A = 17, B = −23, C = 9. Find a 6-bit 2's complement representation for each number, then evaluate the following expressions:
   (a) $A + B$ (b) $A - C$
   (c) $C - A$ (d) $A - B$

   Verify the results by converting back to decimal notation. Repeat using 7 bits and comment on any differences.

9. The following operations are performed in 2's complement arithmetic. Which operations result in arithmetic overflow?
   (a) $101101 + 011001$ (b) $0110111 + 0101101$
   (c) $111011 + 100110$ (d) $1101001 + 1110001$

10. Consider a 10-bit floating-point number, represented in a format similar to Figure A.18, with a 4-bit exponent and a 6-bit mantissa. The exponent uses an excess-7 representation. What are the values of the following numbers:

0010110111,   1101101010,   010011011

11. Two floating-point numbers, A and B, have the format described in Problem A.10. Calculate $A + B$ and $A - B$, assuming that:

$$A = 0110100101, \quad B = 1001100110$$

A.10 REFERENCES

1. Hamacher, V.C., Z.G. Vranesic, and S.G. Zaky, *Computer Organization*. 2nd ed., New York: McGraw-Hill, 1984.

2. Cavanagh, J.J.F., *Digital Computer Arithmetic: Design and Implementation*. New York: McGraw-Hill, 1984.

3. "A proposed standard for floating-point arithmetic," *IEEE Computer*, vol. 14, no. 3, Mar. 1981, pp. 51-62.

INDEX

Date Due

DEC 17 2004			
OCT 04 2007			
DEC 03 2007			